Inna Shingareva
Carlos Lizárraga-Celaya

Maple and Mathematica

A Problem Solving Approach
for Mathematics

Second Edition

SpringerWien NewYork

Dr. Inna Shingareva
Department of Mathematics, University of Sonora, Sonora, Mexico

Dr. Carlos Lizárraga-Celaya
Department of Physics, University of Sonora, Sonora, Mexico

Additional material to this book can be downloaded from http://extras.springer.com

© 2007, 2009 Springer-Verlag/Wien

SpringerWienNewYork is part of
Springer Science + Business Media
springer.at

Typesetting: Camera ready by the authors

Printed on acid-free and chlorine-free bleached paper
SPIN: 12689057

Library of Congress Control Number: 2009933349

ISBN 978-3-211-73264-9 1st edn. SpringerWienNewYork
ISBN 978-3-211-99431-3 2nd edn. SpringerWienNewYork

To our parents,
with inifinite admiration, love, and gratitute.

Preface

In the history of mathematics there are many situations in which calculations were performed incorrectly for important practical applications. Let us look at some examples, the history of computing the number π began in Egypt and Babylon about 2000 years BC, since then many mathematicians have calculated π (e.g., Archimedes, Ptolemy, Viète, etc.). The first formula for computing decimal digits of π was discovered by J. Machin (in 1706), who was the first to correctly compute 100 digits of π. Then many people used his method, e.g., W. Shanks calculated π with 707 digits (within 15 years), although due to mistakes only the first 527 were correct. For the next examples, we can mention the history of computing the fine-structure constant α (that was first discovered by A. Sommerfeld), and the mathematical tables, exact solutions, and formulas, published in many mathematical textbooks, were not verified rigorously [25]. These errors could have a large effect on results obtained by engineers.

But sometimes, the solution of such problems required such technology that was not available at that time. In modern mathematics there exist computers that can perform various mathematical operations for which humans are incapable. Therefore the computers can be used to verify the results obtained by humans, to discovery new results, to improve the results that a human can obtain without any technology. With respect to our example of computing π, we can mention that recently (in 2002) Y. Kanada, Y. Ushiro, H. Kuroda, and M. Kudoh, calculated π over 1.241 trillion digits (explicitly, $1,241,100,000,000$).

In the middle of the 20th century, mathematicians began constructing sufficient sophisticated programs for performing *symbolic mathematics*. In general, we can describe symbolic mathematics as an *analytical problem-solving approach* for various mathematical problems (e.g., arithmetical, algebraical, geometrical, etc.) in which the unknown quantities are represented by abstract symbols and it is required systematic manipulation of symbolic representation of the problem. So, symbolic mathematics performs mathematics at a higher level as compared to the numerical approach that provides numerical results. For example, if we solve analytically (exactly or approximately) a complicated equation or a system of equations and obtain an analytical solution, then we can analyze this solution and create any number of numerical solutions.

The main idea is that in modern times there exist *computer algebra systems* that can perform mathematics at the analytical level. But on the other hand, one of the advantages of computer-algebra systems for scientific computing is that the symbolic computation can be easily combined with the traditional numerical computation and with arbitrary-precision numerical computation.

Also it is well known that computer algebra systems have revolutionized teaching and the learning processes in mathematics, science, and engineering, allowing students to computationally investigate complicated problems to find exact or approximate analytic solutions, numeric solutions, and illustrative two- and three-dimensional graphics. Computer algebra systems can be applied, for example, for symbolic integration and differentiation, formal and algebraic substitution, simplification of expressions, etc.

The computer algebra systems are computational interactive programs that facilitate symbolic mathematics and can handle other type of problems. The first popular systems were Reduce, Derive, and Macsyma, which are still commercially available. Macsyma was one of the first and most mature systems. It was developed at the Massachusetts Institute of Technology (MIT), but practically its evolution has stopped since the summer of 1999. A free software version of Macsyma, Maxima, is actively being maintained.

Since the 1960s there has existed individual packages for solving specific analytic, numerical, graphical and other problems. The need to solve all those problems with the aid of a single system, has led to the idea of construction of a modern general purpose computer algebra system. The first two papers describing analytic calculations realized on a computer were published in 1953 [9]. In the early 1970s, *systems of analytic computations (SAC)*, or *computer algebra systems (CAS)*, began to appear.

To the present day, more than a hundred computer algebra systems have been developed [9], [23]. Among these we can find Axiom, Derive,

GAP, Macaulay 2, Magma, Maple, Mathematica, Maxima, MuMATH, MuPAD, NTL, Reduce, SINGULAR, etc. All these systems can be subdivided into specialized and general-purpose computer algebra systems ([9], [23], [3]).

In this book we consider the two general purpose computer algebra systems, *Maple* and *Mathematica* which appeared around 1988; today both being the most popular, powerful, and reliable systems that are used worldwide by students, mathematicians, scientists, and engineers. It is not, however, our intention to prove or assert that either *Maple* or *Mathematica*, is better than the other.

Following our experience in working with these two computer algebra systems, we have structured this book in presenting and work in parallel in both systems. We note that *Mathematica*'s users can learn *Maple* quickly by finding the *Maple* equivalent to *Mathematica* functions, and vice versa. Additionally, in many research problems it is often required to make independent work and compare the results obtained using the two computer algebra systems, *Maple* and *Mathematica*.

Let us briefly describe these two computer algebra systems.

The computer algebra system *Maple* (see [11]) was developed initially at the University of Waterloo in the 1980s as a research project. At the beginning, the software development group discussed the idea of how to develop a system like Macsyma. Now *Maple* has a rich set of functions that makes it comparable to Macsyma in symbolic power [23] and has a leading position in education, research, and industry. It incorporates the best features of the other systems, it is written in C language and has been ported to the major operating systems.

The *Maple* system evolution has been more community oriented, that is, it allows community participation in the development and incorporation of new tools. Today, modern *Maple* offers three main systems: *The Maple Application Center, Maple T.A.,* and *MapleNet.*

The Maple Application Center offers free resources, for example, research problems, simple classroom notes, entire courses in calculus, differential equations, classical mechanics, linear algebra, etc. *Maple T.A.,* provided for Web-based testing and assessment, includes access control, interactive grade book, algorithmic question generation, and intelligent assessment of mathematical responses. *MapleNet* is a full environment for online e-learning, providing an interactive Web-based environment for education.

In contrast, *Mathematica* evolution has depended upon one person, Stephen Wolfram. Wolfram in 1979 began the construction of the modern computer algebra system for technical computing, that was essentially the zero-th version of *Mathematica*. The computer algebra system *Mathematica* (see [54]) is also written in C language. Despite the fact that this is a newer system, it already has about the same capabilities of symbolic computations as Macsyma. Unlike other computer algebra systems, *Mathematica* has powerful means of computer graphics including 2D and 3D plots of functions, geometrical objects, contour, density plots, intersecting surfaces, illumination models, animation, dynamic interactivity, and more.

In *Mathematica* language, there exists a concept of basic primitive. For example, the expression (the basic computational primitive) can be used for representing mathematical formulas, lists, graphics; or the list (the basic data type primitive) can be used for representing as well many objects (vectors, matrices, tensors, etc.).

At first, *Mathematica* was used mainly in the physical sciences, engineering and mathematics. But over the years, it has become important in a wide range of fields, for instance, physical, biological, and social sciences, and other fields outside the sciences.

With the symbolic, numerical, and graphical calculus capabilities, *Maple* and *Mathematica* have become as a pair, one of the most powerful and complementary tools for students, professors, scientists, and engineers. Moreover, based on our experience of working with both *Maple* and *Mathematica*, these systems are optimal from the point of view of requirements of memory resources and have outstanding capabilities for graphic display ([1], [13], [27], [14], [15], [52]).

In the book we will consider the known mathematical concepts or which can be consulted in any mathematics handbook (e.g., [2], [8], [19], [26], [34], [45], [46], [49], [57]). Also we recall some important concepts and reformulate, interpret, and introduce the new concepts that belong to the broad area of symbolic mathematics.

The core of this book, is a large number of problems and their solutions that have been obtained simultaneously with the help of *Maple* and *Mathematica*. The structure of the book consists of the two parts.

In Part I, the foundations of the computer algebra systems *Maple* and *Mathematica* are discussed, starting from the introduction, basic concepts, elements of language (Chapters 1 and 2).

In Part II, the solutions of mathematical problems are described with the aid of the two computer algebra systems *Maple* and *Mathematica* and are presented in parallel, as in a dictionary. We consider a variety areas of mathematics that appeal to students such as abstract and linear algebra, geometry, calculus and analysis, complex functions, special functions, integral and discrete transforms, mathematical equations, and numerical methods (Chapters 3–12).

It should be noted that the results of the problems are not explicitly shown in the book (formulas, graphs, etc.). In music, musicians must be able to read musical notes and have developed the skill to follow the music from their studies. This skill allows them to read new musical notes and be capable of hearing most or all of the sounds (melodies, harmonies, etc.) in their head without having to play the music piece. By analogy, in Mathematics, we believe a scientist, engineer or mathematician must be able to read and understand mathematical codes (e.g., *Maple*, *Mathematica*) in their head without having to execute the problem. As an example, let us present one music piece (or the music code), "Valse Lento" and the mathematical code (with *Maple* and *Mathematica*) of some simple problem of constructing the stability diagram of the Mathieu differential equation $x'' + [a - 2q\cos(2t)]x = 0$ in the (a, q)-plane. In both cases, the reader can follow, read and understand in their head the music and mathematics.

To Carlos Lizárraga-Celaya

Lento   *poco rit.*   *I. Shingareva, Op. 1, No. 9*

Maple:

```
with(plots): S1:=[MathieuA(i,q) $ i=0..5];
S2:=[MathieuB(i,q) $ i=1..6];
G1:=plot(S1,q=0..35,-20..45,color=red):
G2:=plot(S2,q=0..35,-20..45,color=blue):
display({G1,G2},thickness=3);
```

Mathematica:

```
s1=Table[MathieuCharacteristicA[i,q],{i,0,5}]
s2=Table[MathieuCharacteristicB[i,q],{i,1,6}]
g1=Plot[Evaluate[s1],{q,0,35},PlotStyle->Red];
g2=Plot[Evaluate[s2],{q,0,35},PlotStyle->Blue];
Show[{g1,g2},PlotRange->{{0,35},{-20,45}},Frame->True,
 PlotLabel->"Stability Regions",Axes->False,AspectRatio->1]
```

Following an approach based on solutions of problems rather than technical details, we believe, is the ideal way for introducing students to mathematical and computer algebra methods. The computer algebra systems, *Maple* and *Mathematica*, can and must be used also to help students to learn abstract mathematical ideas. We think that the active role of students in constructing mathematical methods is essential for learning the concepts of modern mathematics. When students construct some symbolic functions, they will learn mathematical concepts by means of the action, since they must describe them exactly to obtain the correct result in a computer.

For example, we know from our experience, students (that write their own symbolic functions for various courses) comment that the best way to understand exactly some method is to write symbolic functions. In addition, the widespread use of computer algebra systems with high-level languages (such as *Maple*, *Mathematica*, etc.) can facilitate much more the process of practical realization (for comparison, we can recall the traditional languages as Fortran).

Maple and *Mathematica* and another computer algebra systems are now so widespread that and they can be considered as reference materials or handbooks for elementary and non elementary results for new generations of students. A reference book and computer algebra systems show formulas without derivation, the solutions to equations are presented without explanation. But, if it necessary, mathematicians, students, and other readers can write their explanations and demonstrate solution methods.

Moreover, when we wrote this book, the idea was not to depend on a specific version of *Maple* or *Mathematica*, our idea is to show how to solve any problem in *Maple* and *Mathematica* for any sufficiently recent version. But we would like to note that the dominant versions for *Maple* and *Mathematica* in the First Edition of the book were, respectively, 11 and 5, and in the Second Edition, are 12 and 6 and with additional enhancements in 7. Moreover, in the Second Edition, a CD-ROM with the solutions (segments of *Maple* and *Mathematica* code) have been included in the book. Also we have added some other supplementary material (in every chapter) and included new Chapters 5 and 12, some new functions, problems, etc.

The book consists of core material for implementing the computer algebra systems *Maple* and *Mathematica* into different undergraduate mathematical courses concerning the areas mentioned above. Also, we believe that the book can be useful for graduate students, professors, and scientists and engineers. "Finally, this book is ideal for scientists who want to corroborate their *Maple* and *Mathematica* work with the independent verification provided by another CAS," J. Carter wrote in SIAM Review[10]. Any suggestions and comments related to this book are most appreciated. Please send your e-mail to `inna@fisica.uson.mx` or `carlos@fisica.uson.mx`.

We would like to express our gratitude to the Mexican Department of Public Education (SEP) and National Council for Science and Technology (CONACYT), for supporting this work under grant no. 55463. Also we would like to express our sincere gratitude to Profs. Andrei D. Polyanin and Alexander V. Manzhirov for their helpful ideas and commentaries. Finally, we wish to express our special thanks to Mr. Stephen Soehnlen and Mag. Wolfgang Dollhäubl from Springer Vienna for their invaluable and continuous support.

May 2009   Inna Shingareva
                                                                       Carlos Lizárraga-Celaya

Contents

Part I

Foundations of *Maple* and *Mathematica*

Chapter 1

Maple

1.1 Introduction

1.1.1 History

We begin with a brief history of *Maple*, from a research project at a university to a leading position in education, research, and industry.

The first concept of *Maple* and initial versions were developed by the Symbolic Computation Group at the University of Waterloo in the early 1980s.

In 1988, the new Canadian company Waterloo Maple Inc., was created to commercialize the software. While the development of *Maple* was done mainly in research labs at Waterloo University and at the University of Western Ontario, with important contribution from worldwide research groups in other universities.

In 1990, the first graphical user interface was introduced for Windows in version V.

In 2003, a Java "standard" user interface was introduced in version 9.

In 2005, Maple version 10 comes with a "document mode" as part of the user interface.

In 2007, Maple version 11 is introduced, it comes with an improved smart document environment to facilitate the user-interface learning curve. This tool integrates in an optimal way, all different sources and types of related information to the problem that the user is solving in that moment. This version includes more mathematical tools for modeling and problem analysis.

I.K. Shingareva, C. Lizárraga-Celaya, *Maple and Mathematica*, 2nd ed.,
DOI 10.1007/978-3-211-99432-0_1, © Springer-Verlag Vienna 2009

In 2008, Maple version 12 comes with additional user-interface enhancements (e.g., new predefined document templates, running headers
and footers to printed documents, code edit regions and start-up
code region, etc.), plotting enhancements, and new forms of con-
nectivity with MATLAB, databases, CAD systems. This newer
version improves and augments computational and mathematical
tools for problem-solving process.

1.1.2 Basic Features

Fast symbolic, numerical computation, and interactive visualization,

Easy to use, help can be found within the program or on the Internet,

Extensibility, the system can easily incorporate new user defined capabilities to compute specific purpose applications,

Accessible to large numbers of students and researchers,

Available for almost all operating systems (MS Windows, Linux. Unix,
Mac OS),

Powerful programming language, intuitive syntax, easy debugging,

Extensive library of mathematical functions and specialized packages,

Two forms of interactive interfaces: a command-line and a graphic environment,

Free resources, collaborative character of development, *Maple* Application Center, Teacher Resource Center, Student Help Center, *Maple*
Community,

Understandable, open-source software development path.

1.1.3 Design

Maple consists of three parts: the interface, the kernel (basic computational engine), and the library. *The interface* and *the kernel* form
a smaller part of the system, which has been written in the C
programming language; they are loaded when a *Maple* session is
started.

The interface handles the input of mathematical expressions, output display, plotting of functions, and support of other user communication with the system. The interface medium is *the Maple worksheet.*

The kernel interprets the user input and carries out the basic algebraic operations, and deals with storage management.

The library consists of two parts: the basic library and a collection of packages. The basic library includes many functions in which resides most of the common mathematical knowledge of *Maple* and that has been coded in the *Maple* language.

1.2   Basic Concepts

1.2.1 First Steps

We type the *Maple* function to the right of the prompt symbol >, and at the end of the function we place a semicolon[1], and then press Enter (or Shift+Enter to continue the function onto the next line). *Maple* evaluates the function, displays the result, and inserts a new prompt.

```
>?introduction
> evalf(gamma,40);
> solve(11*x^3-9*x+17=0, x);
> sort(expand((y+1)^(10)));
> plot({4*sin(2*x),cos(2*x)^2},x=0..2*Pi);
> plot3d(cos(x^2+y^2),x=0..Pi,y=0..Pi);
```

Above (typing line by line), the first line gives you introductory information about *Maple*, the second line returns a 40-digit approximation of Euler's constant γ, the third line solves the equation $11x^3 - 9x + 17 = 0$ for x, the fourth line expresses $(y + 1)^{10}$ in a polynomial form, the fifth line plots the functions $4\sin(2x)$ and $\cos^2(2x)$ on the interval $[0, 2\pi]$, and the last line plots the function $\cos(x^2 + y^2)$ on the rectangle $[0, \pi] \times [0, \pi]$.

1.2.2 Help System

Maple contains a complete online help system. You can use it to find information about a specific topic and to explain the available functions.

---

[1]In earlier versions of *Maple* and in *Classic Worksheet Maple*, we have to end a function with a colon or semicolon. In the book we follow this tradition in every example and problem.

You can use the various forms:

by typing `?function` or `help(function);` or `?help`

by typing `usage(function);` or `example(function);`

by using the `Help` menu, and especially useful (`Full Text Search`),

by highlighting a function and then pressing `Ctrl-F1` or `F1` or `F2` (for $ver \geq 9$),

by pressing `Ctrl-F2`, you can read the Quick Reference summary of Maple interface help.

1.2.3 Worksheets and Interface

Maple worksheets are files that keep track of a working process and organize it as a collection of expandable groups (see `?worksheet`, `?shortcut`), e.g.,

`Insert` $\rightarrow$ `Section`, inserting a new group at the current level,

`Insert` $\rightarrow$ `Subsection`, inserting a subgroup,

`Ctrl-k`, inserting a new input group above the current position[2],

`Ctrl-j`, inserting a new input group below the current position,

`Ctrl-t` (or `F5`), converting the current group into a text group,

`Ctrl-r, Ctrl-g` (`Ctrl-R, Ctrl-G`), inserting `Standard Math` typesetting,

`Format` $\rightarrow$ `Convert to` $\rightarrow$ `StandardMath`, converting a selected formula into standard mathematical notation,

`Ctrl-Space`, completing function.

Palettes can be used for building or editing mathematical expressions without the need of remembering the *Maple* syntax, e.g.

`View` $\rightarrow$ `Palettes` $\rightarrow$ `ShowAllPalettes`

The Maplet User Interface (for $ver \geq 8$) consists of *Maplet applications* that are collections of windows, dialogs, actions (see `?Maplets`).

---

[2]The lowercase letters correspond to the *Classic Worksheet Maple*.

1.2.4 Packages

In addition to the standard library functions, a number of specialized functions are available in various packages (subpackages), in more detail, see ?index[package]. A package (subpackage) function can be loaded in the form (see ?with).

```
with(package); func(args);      package[subpackage][func](args);
with(package[subpackage]); func(args);      package[func](args);
```

Problem 1.1 Define a square matrix $(n = 3)$ of random entries and compute the determinant.

```
with(LinearAlgebra): n:=3; A1:=RandomMatrix(n,n,generator=1..9,
 outputoptions=[shape=triangular[upper]]); B1:=Determinant(A1);
restart: n:=3; A2:=LinearAlgebra[RandomMatrix](n,n,
 generator=1..9,outputoptions=[shape=triangular[upper]]);
B2:=LinearAlgebra[Determinant](A2);
A3:=Student[LinearAlgebra][RandomMatrix](n,n,generator=1..9,
 outputoptions=[shape=triangular[upper]]);
B3:=Student[LinearAlgebra][Determinant](A3);
```

□

1.2.5 Numerical Evaluation

One of the advantages of computer algebra systems is the ability to compute exactly or with user-specified precision (or arbitrary precision) (see Sect. 4.2).

Maple gives an exact answer to arithmetic expressions with integers and reduces fractions. When the result is an irrational number, the output is returned in unevaluated form.

```
135+638; -13*77; 467/31; (3+8*4+9)/2; (-5)^(22); -5^(1/3);
```

Most computers represent both integer and floating point numbers internally using the binary number system. *Maple* represents the numbers in the decimal number system using the user-specified precision.

Numerical approximations:

> `evalf(expr)`, numerical approximation of `expr` (to 10 significant digits),
>
> `Digits:=value`, global changing a user-specified precision, with the environment variable `Digits`, where `value` ∈ $\mathbb{Z}$ (see `?Digits`, `?environment`),
>
> `evalf(expr,n)`, local changing a user-specified precision, with the function `evalf(expr,n)`,
>
> `evalhf(expr)`, numerical approximation of `expr` using a binary hardware floating-point system.
>
> `UseHardwareFloats:=value`, performing numerical approximations using the hardware or software floating-point systems, with the environment variable `UseHardwareFloats`, where `value` can be `true`, `false`, or `deduced` (see `?UseHardwareFloats`, `?environment`). The predefined value is `deduced`, i.e. if `Digits` ≤ `evalhf(Digits)` then hardware float computations are performed (see Sect. 4.2).

```
evalf((-5)^(1/3)); evalf(sqrt(122)); evalf((1+sqrt(5))/2);
evalf(exp(1),50); Digits:=50;         evalf(exp(1));    0.9;
evalf(0.9);        evalhf(0.9);   restart: evalhf(Digits);
UseHardwareFloats:=true;                    evalf(Pi); Digits;
```

Note. In applications (e.g., simple computations and plotting a function), the hardware floating-point computation environment is usually much faster.

1.3 *Maple* Language

Maple language is a high-level programming language, well-structured, comprehensible. It supports a large collection of *data structures* or *Maple objects* (functions, sequences, sets, lists, arrays, tables, matrices, vectors, etc.) and operations among these objects (type-testing, selection, composition, etc.).

The *Maple* procedures in the library are available in readable form. The library can be complemented with locally user developed programs and packages.

1.3.1 Basic Principles

Arithmetic operators: + - * / ^ mod.

Logic operators: and, or, xor, implies, not.

Relation operators: <, <=, >, >=, =, <>.

A variable name, var, is a combination of letters, digits, or the underline
symbol (_), beginning with a letter, e.g., a12_new.

Abbreviations for the longer *Maple* functions or any expressions: alias,
e.g., alias(H=Heaviside); diff(H(t),t); to remove this abbreviation, alias(H=H);

Maple is case sensitive, there is a difference between lowercase and uppercase letters, e.g., evalf(Pi) and evalf(pi).

Various reserved keywords, symbols, names, and functions, these words
cannot be used as variable names, e.g., operator keywords, additional language keywords, global names that start with (_) (see
?reserved, ?ininames, ?inifncs, ?names).

The assignment/unassignment operators: a variable can be "free" (with
no assigned value) or can be assigned any value (symbolic, numeric)
by the assignment operators a:=b or assign(a=b). To unassign
(clear) an assigned variable (see ?:= and ?'):

x:='x', evaln(x), or unassign('x').

The difference between the operators (:=) and (=). The operator var:=expr
is used to assign expr to the variable var, and the operator A=B — to
indicate equality (not assignment) between the left- and the right-hand
sides (see ?rhs), e.g.,

Equation:=A=B; Equation; rhs(Equation); lhs(Equation);

The range operator (..), an expression of type range expr1..expr2, e.g.,
a[i] $ i=1..9; plot(sin(x),x=-Pi..Pi);

Statements, stats, are input instructions from the keyboard that are executed
by *Maple* (e.g., break, by, do, end, for, function, if, proc, restart,
return, save, while, not).

The new worksheet (or the new problem) it is best to begin with the statement
restart for cleaning *Maple*'s memory. All examples and problems in the
book assume that they begin with restart.

```
restart: with(LinearAlgebra): d:=10;
for i from 1 to 7 do
 n:=2^i; A:=RandomMatrix(n,n,generator=-d..d):
 t:=time(ReducedRowEchelonForm(A)); print(i,n,t); od:
```

The statement separators semicolon (;) and colon (:). The result of a statement
     followed with a semicolon (;) will be displayed, and it will not be displayed
     if it is followed by a colon (:), e.g.,

```
plot(sin(x),x=0..Pi); plot(sin(x),x=0..Pi):
```

An expression, `expr`, is a valid statement, and is formed as a combination of
     constants, variables, operators and functions.

Data types, every expression is represented as a tree structure in which each
     node (and leaf) has a particular data type. For the analysis of any node
     and branch, the functions `type`, `whattype`, `nops`, `op` can be used.

A boolean expression, `bexpr`, is formed with the *logical operators* and the re-
     lation operators.

An equation, `eq`, is represented using the binary operator (=), and has two
     operands, the left-hand side, `lhs`, and the right-hand side, `rhs`.

Inequalities, `ineq`, are represented using the relation operators and have two
     operands, the left-hand side, `lhs`, and the right-hand side, `rhs`.

A string, `str`, is a sequence of characters having no value other than itself,
     cannot be assigned to, and will always evaluate to itself. For instance,
     `x:="string";` and `sqrt(x);` is an invalid function. Names and strings
     can be used with the `convert` and `printf` functions.

Incorrect response. If you get no response or an incorrect response you may
     have entered or executed the function incorrectly. Do correct the function
     or interrupt the computation (the stop button in the Tool Bar menu).

Maple is sensitive to types of brackets and quotes.

Types of brackets:

     Parentheses (`expr`), for grouping expressions, `(x+9)*3`, for delimiting
     the arguments of functions, `sin(x)`.

     Square brackets [`expr`], for constructing lists, `[a,b,c]`, vectors, matri-
     ces, arrays.

     Curly brackets {`expr`}, for constructing sets, `{a,b,c}`.

Types of quotes:

Forward-quotes `'expr'`, to delay evaluation of expression,
`'x+9+1'`, to clear variables, `x:='x'`;

Back-quotes `` `expr` ``, to form a symbol or a name,
`` `the name:=7` ``; `k:=5`; `print(`the value of k is`,k)`;

Double quotes `"expr"`, to create strings, and a single double quote `"`, to delimit strings.

Previous results (during a session) can be referred with symbols `%` (the last result), `%%` (the next-to-last result), `%%...%`, k times, (the k-th previous result), `a+b`; `% ^ 2`; `%% ^ 2`;

Comments can be included with the sharp sign `#` and all characters following it up to the end of a line. Also the text can be inserted with `Insert` → `Text`.

Maple source code can be viewed for most of the functions, general and specialized (package functions).

```
interface(verboseproc=2);
print(factor); print(`plots/arrow`);
```

1.3.2 Constants

Types of numbers: integer, rational, real, complex, root, e.g.,

```
-55, 5/6, 3.4, -2.3e4, Float(23,-45), 3-4*I, Complex(2/3,3);
RootOf(_Z^3-2,index=1);
```

Predefined constants: symbols for definitions of commonly used mathematical constants, `true`, `false`, `gamma`, `Pi`, `I`, `infinity`, `Catalan`, `FAIL`, `exp(1)` (see `?ininames`, `?constants`).

An angle symbolically has dimension 1. *Maple* knows many units of angle (see `?Units/angle`), e.g., `convert(30*degrees,radians)`;

```
convert(30,units,degrees,radians);
```

The packages `ScientificConstants` and `Units` (for *ver* ≥ 7) provide valuable tools for scientists and engineers in Physics and Chemistry (see `?ScientificConstants`, `?Units`).

1.3.3 Functions

Functions or function expressions have the form f(x) or expr(args) and
   represent a function call, or application of a function (or procedure)
   to arguments (args).

Active functions (beginning with a lowercase letter) are used for computing, e.g., diff, int, limit.

Inert functions (beginning with a capital letter) are used for showing
   steps in the problem-solving process; e.g., Diff, Int, Limit.

Two classes of functions: the library functions (predefined functions)
   and user-defined functions.

Predefined functions: most of the well known functions are predefined
   by *Maple* and they are known to some *Maple* functions (diff, evalc,
   evalf, expand, series, simplify). In addition, numerous special
   functions are defined (see ?FunctionAdvisor). We will discuss some
   of the more commonly used functions.

Elementary trascendental functions: the exponential function, the natural logarithm, the general logarithm, the common logarithm, the
   trigonometric and hyperbolic functions and their inverses.

exp	ln	log	log[b]	log10	ilog2
sin	cos	tan	cot	sec	csc
sinh	cosh	tanh	coth	sech	csch
arcsin	arccos	arctan	arccot	arcsec	arccsc
arcsinh	arccosh	arctanh	arccoth	arcsech	arccsch

```
exp(x^2); evalf(ln(exp(1)^3)); evalf(tan(3*Pi/4));
evalf(arccos(1/2));  evalf(tanh(1));
```

Other useful functions (see ?inifcn)

```
round(x);        floor(x);        ceil(x); trunc(x); frac(x);
max(x1,x2,...);  min(x1,x2,...); abs(x); sign(expr); quo(x);
rem(x); gcd(x);  lcm(x);        ithprime(n); time(expr);     n!;
factorial(n); doublefactorial(n); with(combinat):fibonacci(n);
type(expr,type); print(expr1,expr2,...);    lprint(expr);
```

max, min, finding the maximum/minimum of numbers,

round, floor, ceil, trunc, frac, converting real numbers to the nearest
   integers, and determining the fractional part of a number,

abs, the absolute value function,

sign, signum, the sign of a number or a polynomial and the sign function for
   real and complex expressions,

n!, factorial, doublefactorial, computing factorial and double factorial,

ithprime, nextprime, prevprime, isprime, determining the i-th prime number, the next largest and previous smallest prime numbers, and the pri-
   mality test,

fibonacci (of the combinat package), generating the Fibonacci numbers or
   polynomials,

quo, rem, gcd, lcm, determining the quotient and the remainder of polynomials
   (in the division algorithm), the greatest common divisor and the least
   common multiple of polynomials,

time, the total CPU time (in seconds) for the *Maple* session,

type (see ?type), testing whether an expression belongs to a given type (or
   class), the results is the value true o false,

print, lprint, printf, printing of expressions, the function print can be
   used in loops, strings, and other outputs for displaying the results.

Functions for generating random numbers and variables,

```
randomize():                 k:=rand(a..b): k();
with(Statistics):            X:=RandomVariable(Normal(a,b));
with(RandomTools):           Generate(float(range=(a..b),digits=n));
AddFlavor(x=rand(a..b);      SetState(state=value);
with(LinearAlgebra):         M:=RandomMatrix(n,n,generator=a..b,
                             outputoptions=[shape=triangular[upper]]);
```

```
iquo(25,3); irem(25,3); type(x^3*y+sqrt(y)*y,'polynom');
RandomTools[Generate](float(range=(2..9),digits=10));
with(combinat): fibonacci(10000); time(%);
```

User-defined functions: the functional operator (see ?->), e.g., functions of one or many variables $f(x)$ =expr, $f(x_1,\ldots,x_n)$ =expr, the vector functions of one or many variables, $f(x) = (x_1,\ldots,x_n)$, $f(x_1,\ldots,x_n) = \langle f_1(x_1,\ldots,x_n),\ldots,f_n(x_1,\ldots,x_n)\rangle$ (for more details see Sect. 4.8).

```
f:=x->expr;   f:=(x1,...,xn)->expr;      f:=t-><x1(t),...,xn(t)>;
              f:=(x1,...,xn)-><f1(x1,...,xn),...,fm(x1,...,xn)>;
```

Alternative definitions of functions: unapply converts an expression to a function, and a procedure is defined with proc.

```
f:=proc(x) option operator,arrow; expr end proc;
f:=proc(x) expr end;                f:=proc(x1,...,xn) expr end;
f:=proc(t) <x1(t),...,xn(t)> end;           f:=unapply(expr,x);
```

Evaluation at $x = a$, $\{x = a, y = b\}$, etc.

```
f(a);         subs(x=a, f(x));                eval(f(x),x=a);
f(a,b);       subs(x=a,y=b,f(x,y));    eval(f(x,y),{x=a,y=b});
```

Composition operator @, e.g., the composition function $(f\circ f\circ \cdots \circ f)(x)$ (n times) or $(f_1\circ f_2\circ \cdots \circ f_n)(x)$.

```
(f @@ n)(x);                         (f1 @ f2 @ f3 @ ... @ fn)(x);
```

Problem 1.2 Define the function $f(x,y) = 1 - \sin(x^2 + y^2)$ and evaluate $f(1,2)$, $f(0,a)$, $f(a^2 - b^2, b^2 - a^2)$.

```
f:=(x,y)->1-sin(x^2+y^2);
evalf(f(1,2));  f(0,a);  simplify(f(a^2-b^2,b^2-a^2));
```
□

Problem 1.3 Define the vector function $h(x,y) = \langle\cos(x - y), \sin(x - y)\rangle$ and calculate $h(1,2)$, $h(\pi, -\pi)$, and $h(\cos(a^2), \cos(1 - a^2))$.

```
h:=(x,y)-><cos(x-y),sin(x-y)>; evalf(h(1,2)); h(Pi,-Pi);
combine(h(cos(a^2),cos(1-a^2)),trig);
```
□

Problem 1.4 Graph the real roots of the equation $x^3 + (a-3)^3 x^2 - a^2 x + a^3 = 0$ for $a \in [0, 1]$.

```
Sol:=[solve(x^3+(a-3)^3*x^2-a^2*x+a^3=0,x)];
for i from 1 to 3 do
    R||i:=unapply(Sol[i],a):
    print(plot(R||i(a),a=0..1,numpoints=500)); od:
```

□

Problem 1.5 For the functions $f(x) = x^2$ and $h(x) = x + \sin x$ calculate the composition function $(f \circ h \circ f)(x)$ and $(f \circ f \circ f \circ f)(x)$.

```
f:=x->x^2;   h:=x->x+sin(x);
f(h(f(x)));  (f@h@f)(x);   f(f(f(f(x))));   (f@@4)(x);
```

□

Piecewise functions can be defined with `piecewise` or as a procedure
    with the control statement `if`.

```
f:=x->piecewise(cond1,expr1,expr2);
g:=x->piecewise(cond1,expr1,cond2,expr2,expr3);
f:=proc(x) if cond1 then expr1 else expr2 end if; end proc;
g:=proc(x) if cond1 then expr1 elif cond2 then expr2
                    else expr3 end if; end proc;
```

Problem 1.6 Define and graph the function $g(x) = \begin{cases} 0, & |x| > 1 \\ 1-x, & 0 \le x \le 1 \\ 1+x, & -1 \le x \le 0 \end{cases}$

in $[-3, 3]$.

```
g:=x->piecewise(x>=-1 and x<=0,1+x,x>=0 and x<=1,1-x,0);
plot(g(x),x=-3..3,thickness=3,color=blue);
g:=proc(x) if x>=-1 and x<=0 then 1+x
    elif x>=0 and x<=1 then 1-x else 0 end if: end proc;
plot(g,-3..3,thickness=3,color=blue);
```

□

Problem 1.7 Graph the periodic extension of $f(x) = \begin{cases} x, & 0 \leq x < 1 \\ 1, & 1 \leq x < 2 \\ 3-x, & 2 \leq x < 3 \end{cases}$

in $[0, 20]$.

```
f:=proc(x) if x>=0 and x<1 then x elif x>=1 and x<2 then 1
 elif x>=2 and x<3 then 3-x end if: end proc:
g:=proc(x) local a,b; a:=trunc(x) mod 3; b:=frac(x);
 f(a+b); end proc: plot(f,0..3,thickness=3,color=blue);
plot(g,0..20,thickness=3,color=blue);
```

$\square$

1.3.4 Procedures and Modules

In *Maple* language there are two forms of modularity: *procedures* and *modules*.

A procedure (see ?procedure) is a block of statements which one needs to use repeatedly. A procedure can be used to define a function (if the function is too complicated to write by using the arrow operator), to create a matrix, a graph, a logical value, etc.

```
proc(args) local v1; global v2; options ops; stats; end proc;
proc(args) local v1; global v2; options ops; stats; end;
```

Here args is a sequence of arguments, v1 and v2 are the names of local and global variables, ops are special options (see ?options), and stats are statements that are realized inside the procedure.

Problem 1.8 Define a procedure maximum that calculate the maximum of two arguments x, y, and then calculate the maximum of $(34/9, 674/689)$.

```
Maximum:=proc(x, y) if x>y then x else y end if end proc;
Maximum(34/9,674/689);
```

$\square$

Problem 1.9 In Lagrangian mechanics, N-degree of freedom holonomic systems are described by the Lagrange equations [50]:

$$\frac{d}{dt}\frac{\partial L}{\partial \dot{q}_i} - \frac{\partial L}{\partial q_i} = Q_i, \quad i = 1, \ldots, N,$$

where $L = T - P$ is the Lagrangian of the system, T and P are kinetic and potential energy, respectively, q_i and Q_i are, respectively, the generalized coordinates and forces (not arising from a potential), and

$$\frac{d}{dt}\frac{\partial L}{\partial \dot{q}_i} = \sum_{j=1}^{N}\left(\frac{\partial}{\partial q_j}\frac{\partial L}{\partial \dot{q}_i}\dot{q}_j + \frac{\partial}{\partial \dot{q}_j}\frac{\partial L}{\partial \dot{q}_i}\ddot{q}_j\right).$$

Let the generalized forces, $Q_i(q_1, \ldots, q_n; \dot{q}_1, \ldots, \dot{q}_n)$, the kinetic and potential energy, $T(q_1, \ldots, q_n; \dot{q}_1, \ldots, \dot{q}_n)$ and $P(q_1, \ldots, q_n)$, be given.

Construct a procedure for deriving the Lagrange equations. As an example, consider the motion of a double pendulum of mass m and length L in the vertical plane due to gravity. For the pendulum ($N = 2$), the generalized coordinates are angles, A and B, the generalized non potential forces are equal to zero.

```
Eq_Lagrange:=proc(q,L,Q)
 local i,j,N,dq,d2q,Lq,Ldq,dLdq,Eq;  N:=nops(q);
 for i from 1 to N do dq[i]:=cat(q[i],`'`); end do;
 for i from 1 to N do d2q[i]:=cat(q[i],`''`); end do;
 for i from 1 to N do
 Lq:=diff(L,q[i]); Ldq:=diff(L,dq[i]);
 dLdq:=add(diff(Ldq,q[j])*dq[j]+diff(Ldq,dq[j])*d2q[j],j=1..N);
 Eq[i]:=dLdq-Lq-Q[i]; end do;
 RETURN(normal(convert(Eq,list))); end proc;
T:=m*x^2*`A'`^2+m*x^2*`A'`*`B'`*cos(A-B)+1/2*m*x^2*`B'`^2;
P:=-m*g*x*cos(A)-m*g*x*(cos(A)+cos(B));
q:=[A,B]; g:=omega^2*x; Q:=[0,0];
Eq_L:=Eq_Lagrange(q,T-P,Q); Eq_L1:=factor(Eq_L);
```

□

Recursive procedures are the procedures that call themselves until a condition is satisfied (see ?remember).

Problem 1.10 Calculate Fibonacci numbers and the computation time for $F(n) = F(n-1) + F(n-2)$, $F(0) = 0$, $F(1) = 1$.

```
Fib:=proc(n::integer) option remember;
 if n<=1 then n else Fib(n-1)+Fib(n-2) ond if; ond proc;
NF:=NULL: for i from 10 to 40 do NF:=NF,Fib(i); end do:
NF:=[NF]; ti:=time(): Fib(3000); tt:=time()-ti;
op(4,eval(Fib)); forget(Fib);
```

If we include the `remember` option (see `?remember`), *Maple* will construct a remember table in which all the values computed are stored and *Maple* will not recompute them again. The remember table is accessible as `op(4,eval(Fib))` and can be manipulated as a table object. This way, we can improve the computational efficiency, save more time, but use more memory. The functions `forget` or `restart` can be used to clear the remember table of the procedure. □

A module (see `?module`) is a generalization of the procedure concept. Since the procedure groups a sequence of statements into a single statement (block of statements), the module groups related functions and data.

```
module() export v1; local v2; global v3;
                            option ops; stats; end module;
```

Here `v1`, `v2`, and `v3` are the names of export, local, and global variables, respectively, `ops` are special options (see `?module[options]`), and `stats` are statements that are realized inside the module.

Problem 1.11 Define a module `maxmin` that calculates the maximum and the minimum of two arguments x, y, and then calculate the maximum and minimum of $(34/9, 674/689)$.

```
MaxMin:=module() local x, y; export maximum, minimum;
 maximum:=(x,y)->if x>y then x else y end if;
 minimum:=(x,y)->if x<y then x else y end if; end module;
MaxMin:-maximum(34/9,674/689); MaxMin:-minimum(34/9,674/689);
```
□

1.3.5 Control Structures

In *Maple* language there are essentially *two control structures*: the selection structure `if` and the repetition structure `for`,

```
if cond1 then expr1 else expr2  end if;
if cond1 then expr1 elif cond2 then expr2 else expr3 end if;
for i from i1 by step to i2 do stats end do;
for i from i1 by step to i2 while cond1 do stats end do;
for i in expr1 do stats end do;  for i in expr1 do stats od;
for i in expr1 while expr2 do stats end do;
```

where cond1 and cond2 are conditions, expr1, expr2 are expressions, stats are statements, i, i1, i2 are, respectively, the loop variable, the initial and the last values of i. These operators can be nested. The operators break, next, while inside the loops are used for breaking out of a loop, to proceed directly to the next iteration, or for imposing an additional condition. The operators end if and fi, end do and od are equivalent.

Problem 1.12 Define the function *double factorial* for any integer n,

$$n!! = \begin{cases} n(n-2)(n-4) \ \dots \ (4)(2), & n = 2i, i = 1, 2, \dots, \\ n(n-1)(n-3) \ \dots \ (3)(1), & n = 2i+1, i = 0, 1, \dots \end{cases}$$

```
N1:=20; N2:=41;
FD:=proc(N) local P, i1, i; P:=1;
 if modp(n,2)=0 then i1:=2 else i1:=1 end if:
 for i from i1 by 2 to N do P:=P*i; end do: end proc:
printf(" %7.0f!! = %20.0f", N1, FD(N1));
printf(" %7.0f!! = %20.0f", N2, FD(N2));
```

□

Problem 1.13 Calculate the values x_i, where $x_i = (x_{i-1} + 1/x_{i-1})$, $x_0 = 1$, until $|x_i - x_{i-1}| \leq \varepsilon$ ($i = 1, 2, \dots, n$, $n = 10$, $\varepsilon = 10^{-3}$).

```
n:=10; x:=1; epsilon:=10^(-3);
for i from 1 to n do xp:=x: x:=evalf((x+1/x));
 if abs(x-xp)<=epsilon then break else
  printf(" x= %20.8f \n", x) end if: end do:
```

□

Problem 1.14 Find an integer $N(x)$, $x \in [a, b]$ such that $\sum_{i=1}^{N(x)} i^{-x} \geq c$, where $a = 1$, $b = 2$, and $c = 2$.

```
a:=1; b:=2; c:=2;
for x from a to b by 0.01 do S:=0:
 for i from 1 to 100 while S<c do
  S:=S+evalf(i^(-x)); end do:
 printf("x= %7.4f   N(x)= %7d \n", x, i); end do:
```

□

1.3.6 Objects and Operations

Maple objects, sequences, lists, sets, tables, arrays, vectors, matrices, are used for representing more complicated data.

```
a:=val1; b:=val2; n:=val1; m:=val2; Sequence1:=expr1,...,exprn;
Sequence2:=seq(f(i),i=a..b);          Sequence3:='f(i)'$'i'=a..b;
List1:=[Sequence1];   Range1:=a..n;        Set1:={Sequence1};
Table1:=table([expr1=A1,...,exprN=AN]); Arr1:=Array(a..n,a,,m);
Arr2:=Array();                        Arr3:=Array(1..n,fill=1);
Vec1:=Vector(<a1,...,an>);       Vec2:=Vector(1..n,[a1,...,an]);
Matrix(n,m,[a11,a12,...,anm]]);         Matrix(n,m,symbol=a);
Matrix(<<a11,a21,a31>|<a12,a22,a32>>);     Matrix(n,m,fill=1);
Matrix([[a11,a12],[a21,a22],[a31,a32]]);        f:=(x,y)->expr;
Matrix(n,m,(i,j)->f(i,j));   Matrix([[a11,...],..,[an1,...]]);
```

Sequences, lists, sets are groups of expressions. *Maple* preserves the order and repetition in sequences and lists and does not preserves it in sets. The order in sets can change during a *Maple* session.

A table is a group of expressions represented in tabular form. Each entry has an index (an integer or any arbitrary expression) and a value (see ?table).

An array is a table with integer range of indices (see ?Array). In *Maple* arrays can be of any dimension (depending of computer memory).

A vector is a one-dimensional array with positive range integer of indices (see ?vector, ?Vector).

A matrix is a two-dimensional array with positive range integer of indices (see ?matrix, ?Matrix).

```
a:=5; b:=7; Range1:=a..b; S1:=x,y,z,a,b,c; LS1:=[S1]; SS1:={S1};
L1:=[sin(x),cos(x),sin(2*x),cos(2*x)]; Set1:={x,y,z};
A1:=Array(-1..3); A11:=array(-1..3); A2:=Array(1..4,[1,2,3,4]);
```

Basic operations with data structures:

Create the empty structures, NULL, :=, [ ].

```
Seq1:=NULL; List1:=NULL; List2:=NULL; List3:=[]; Set1:={};
Set2:=NULL; Tab1:=table(); Array1:=Array(-10..10);
Vec1:=Vector(10); Matrix1:=Matrix(10,10);
```

Concatenate structures, ||, op, [ ], cat.

```
k:=1; Seq1:=seq(i,i=1..9); List1:=[Seq1]; a||Seq1;
"Seq1"||(1..9); k||(1..9); cat(a,Seq1);
op([List1,List1]); cat(Seq1,Seq1);
```

Extract an i-th element from a structure, [ ], op, select,has.

```
Seq1:=seq(i^2,i=1..9); List1:=[Seq1]; Set1:={Seq1};
Array1:=Array(1..9,1..9,[List1,0,List1,0,List1,List1]);
i:=5; j:=2; List1[i]; op(i,List1); Array1[i,j]; op(i,Set1);
element:=9; select(has,Set1,element);
```

Determine the number of elements in a structure, nops.

```
List1:=[x,y,z]; Set1:={op(List1)}; nops(List1); nops(Set1);
```

Create a substructure, op, [ ].

```
Seq1:=x||(1..9); List1:=[Seq1]; Set1:={Seq1}; n:=nops(List1);
List2:=[Seq1,Seq1]; n1:=2; n2:=5; List3:=[op(n1..n2,List1)];
List4:=List1[n1..n2]; Seq2:=op([(n1..n2)],List2);
Set2:={op(n1..n2,Set1)}; Set3:=Set1[n1..n2];
Seq3:=op([(n1..n2)],Set1);
```

Here $n_1 \le n_2 \le n$ and n is a number of elements of List1.

Replace the i-th element of a structure, :=, [ ], subsop, subs, evalm.

```
Seq1:=x[i] $ i=1..9; List1:=[Seq1]; i:=5; j:=2; val:=20;
List1[i]:=val; evaln(List1)=List1; List2:=subsop(i=val+1,List1);
A:=Matrix(1..i,1..i,symbol=s); A1:=subs(s[i,j]=cos(a+b),A);
```

Insert an element or some elements into a structure, [ ], op.

```
n1:=2; n2:=5; Seq1:=x[i] $ i=1..9; List1:=[Seq1];
List2:=[op(List1),A1]; List3:=[A1,op(List1)];
List4:=[op(n1..n2,List1),A1,A2,A3,A4,op(List2)];
```

Create a structure according to a formula or with some special properties, e.g., zero, identity, sparse, symmetric, diagonal, etc. (in
more detail, see Chapter 5).

```
f:=x->cos(x); n:=0; m:=3; List1:=[seq(f(i*t),i=n..m)];
Set1:=map(x->x^2,{x,y,z}); List2:=[op(Set1)];
Matrix1:=Matrix(m,m,(i,j)->i+j); Vector1:=Vector(m,i->i^2);
with(LinearAlgebra): ZeroMatrix(m,m); IdentityMatrix(m,m);
```

Operations with sets, matrices: the union and intersection of sets, removing elements from sets, the sum, difference, multiplication, division, scalar multiplication of matrices, union, intersect, minus,
remove,has, evalm, &*.

```
Set1:=map(x->x^2,{x,y,z}); Set2:=map(x->cos(x),{x,y,t});
Set3:=Set1 union Set2; Set3:=Set1 intersect Set2;
Set3:=Set1 minus Set2; Set2:=remove(has,Set1,A1);
M1:=Matrix(3,3,symbol=a); M2:=Matrix(3,3,symbol=b);
Mat3:=M1.M2; Mat4:=evalm(M1&*M2);
```

Apply a function to each element of a structure, map, apply, applyop.

```
Set1:={a||(1..9)}; func:=x->cos(x);
Set2:=map(x->x^2,Set1); Set2:=map(func,Set1);
apply(f,op(Set1)); applyop(func,{2,3},Set1);
```

Problem 1.15 A sequence of numbers $\{x_i\}$ $(i = 0, \ldots, N)$ is defined by
$x_{i+1} = a x_i (1 + x_i)$ $(0 < a < 10)$, where a is a given parameter. Define
a list of coordinates $[i, x_i]$ such that $a = 3$, $x_0 = 0.1$, $N = 100$. Plot the
graph of the sequence $\{x_i\}$.

```
a:=3; x[0]:=0.1; N:=100;
for i from 1 to N do x[i]:=a*x[i-1]*(1-x[i-1]): od:
Seq1:=seq([i,x[i]],i=2..N):
plot([Seq1],style=point,symbol=circle,symbolsize=25);
```
                                                                                                    □

Problem 1.16 Observe the function behavior $y(x) = \cos(6(x - a \sin x))$,
$x \in [-\pi, \pi]$, $a \in [\frac{1}{2}, \frac{3}{2}]$.

```
with(plots): y:=x->cos(6*(x-a*sin(x))); G:=NULL; N:=20;
for i from 0 to N do
  a:=1/2+i/N; G:=G,plot(y(x),x=-Pi..Pi); od:
G:=[G]: display(G,insequence=true);
```
                                                                                                    □

Chapter 2

Mathematica

2.1 Introduction

2.1.1 History

In 1979–1981, Stephen Wolfram constructed SMP (*Symbolic Manipulation Program*), the first modern computer algebra system (SMP was essentially Version Zero of *Mathematica*).

In 1986–1988, Stephen Wolfram developed the first version of *Mathematica*. The concept of *Mathematica* was a single system that could handle many specific problems (e.g., symbolic, numerical, algebraic, graphical). In 1987, Wolfram founded a company, *Wolfram Research*, which continues to extend *Mathematica*.

In 1991, the second version of *Mathematica* appears with more built-in functions, MathLink protocol for interprocess and network communication, sound support, notebook front end.

In 1996, the third version introduced interactive mathematical typesetting system, exporting HTML, hyperlinks, and many other functions.

In 1999, the fourth version of *Mathematica* appears with important enhancements in speed and efficiency in numerical calculation, publishing documents in a variety of formats, and enhancements to many built-in functions.

In 2003, in the fifth version of *Mathematica* the core coding was improved and the horizons of *Mathematica* are more extended (e.g., in numerical linear algebra, in numerical solutions for differential

I.K. Shingareva, C. Lizárraga-Celaya, *Maple and Mathematica*, 2nd ed.,
DOI 10.1007/978-3-211-99432-0_2, © Springer-Verlag Vienna 2009

equations, in supporting an extended range of import and export graphic and file structures, in solving equations and inequalities symbolically over different domains).

In 2007, version 6 is introduced. In this upgrade, *Mathematica* has been substantially redesigned in its internal architecture for better functionality. It increases interactivity, more adaptive visualization, ease of data integration, symbolic interface construction among others new features.

In 2008, version 7 comes with a lot of achievements and several new areas of Mathematica. Among them we would like to mention integrated image processing, parallel high-performance computing, discrete calculus, new computable data sources, new algorithms for visualization and graphics and for solving transcendental, differential, and difference equations.

2.1.2 Basic Features

Symbolic, numerical, acoustic, graphical, parallel computations,

Static and Dynamic computations,

Extensibility and elegance,

Available for MS Windows, Linux, UNIX, Mac OS operating systems,

Powerful and logical language,

Extensive library of mathematical functions and specialized packages,

An interactive front end with notebook interface,

Interactive mathematical typesetting system,

Free resources, The *Mathematica* Learning Center, Wolfram Demonstrations Project, Wolfram Information Center.

2.1.3 Design

Mathematica consists of two basic parts: *the kernel*, computational engine and the interface, *front end*. These two parts are separate,
but communicate with each other via the *MathLink* protocol.

The kernel interprets the user input and performs all computations.
The kernel assigns the labels In[number] to the input expression
and Out[number] to the output. These labels can be used for keep-
ing the computation order. In the book, we will not include these
labels in the examples. The result of kernels work can be viewed
with the function InputForm.

```
Plot3D[Cos[x y],x,-Pi,Pi,y,-Pi,Pi]//InputForm
```

The interface between the user and the kernel is called *front end* and is
used to display the input and the output generated by the kernel.
The medium of the front end is the *Mathematica notebook*.

2.1.4 Changes for New Versions

There are significant changes to numerous *Mathematica* functions incorporated to the new versions of the system. Let us describe in general
differences for $ver < 6$ and $ver \geq 6$ (in the next Chapters we note some of
them for particular cases). A complete list of all changes can be found
in the *Documentation Center* and on the *Wolfram Website*. Moreover,
if you apply some functions (for $ver < 6$), *Mathematica* explains the
corresponding changes for the current version.

Most notable changes (for $ver \geq 6$):

Changes in menus, dialogue boxes, and palettes,

Numerous changes in graphics functions,

The semicolon symbol (;) for complete suppressing graphics output,

Numerous functions (for $ver < 6$) located in packages have been incorporated to the *Mathematica* kernel. So many packages going obsolete.

Numerous functions (for $ver < 6$) are now located in different packages.

Numerous functions (for $ver < 6$) are available on the *Wolfram Website*.

Numerous functions (for *ver* < 6) are located in the "Legacy" packages.

Numerous functions (for *ver* < 6) have been eliminated and incorporated to other functions.

Significant changes and enhancements in animation functions.

2.2   Basic Concepts

2.2.1 First Steps

We type *Mathematica* command and press the `RightEnter` key or `Shift+ Enter` (or `Enter` to continue the command on the next line). *Mathematica* evaluates the command, displays the result, and inserts a horizontal line (for the next input).

```
?Arc*
N[EulerGamma,40]
Solve[11*x^3-9*x+17==0,x]
Expand[(y+1)^10]
Plot[{4*Sin[2*x],Cos[2*x]^2},{x,0,2*Pi}]
Plot3D[Cos[x^2+y^2],{x,0,Pi},{y,0,Pi}]
```

The first line gives you information about *Mathematica*'s functions beginning with `Arc`, the second line returns a 40-digit approximation of Euler's constant γ, the third line solves the equation $11x^3 - 9x + 17 = 0$ for x, the fourth line expresses $(y+1)^{10}$ in a polynomial form, the fifth line plots the functions $4\sin(2x)$ and $\cos^2(2x)$ on the interval $[0, 2\pi]$, and the last line plots the function $\cos(x^2 + y^2)$ on the rectangle $[0, \pi] \times [0, \pi]$.

2.2.2 Help System

Mathematica contains many sources of online help:

> *Wolfram Documentation Center, Wolfram Demonstrations Project* (for *ver* ≥ 6), *Mathematica Virtual Book* (for *ver* ≥ 7),

> `Help` menu, to mark a function and to press F1,

> to type `?func` (information about a function), `??func` (for more extensive information), `Options[func]` (for information about options),

> to use the symbols (`?`,`*`), e.g., `?Inv*`, `?*Plot`, `?*our*`.

2.2.3 Notebook and Front End

Mathematica notebooks are electronic documents that may contain *Mathematica* output, text, graphics (see ?Notebook). You can work simultaneously with many notebooks.

Cells: a *Mathematica* notebook consists of a list of cells. Cells are indicated along the right edge of the notebook by blue brackets. Cells can contain subcells, and so on. The kernel evaluates a notebook cell by cell.

Operations with cells: a horizontal line across the screen is the beginning of the new cell, to open (close) cells or to change their format click (double-click) on the cell brackets, e.g., for changing the background cell color, click on a cell bracket, then

Format → BackgroundColor → Yellow.

Different types of cells: input cells (for evaluation), text cells (for comments), Title, Subtitle, Section, Subsection, etc., can be found in the menu Format → Style.

Palettes can be used for building or editing mathematical expressions, texts, graphics and allows one to access by clicking the mouse to the most common mathematical symbols, e.g.,

Palettes → BasicMathInput.

Useful features of the Front End

Completing existing symbol names, Ctrl-k,

Reminding the syntax of a function, Ctrl-Shift-k,

Typing shortcut keys and abbreviations, Esc abbreviation Esc,

Editing graphics boxes with the mouse.

In *Mathematica* (for *ver* ≥ 6), the *Mathematica language* and the *Front End* are one and the same, this allows us to access programmatically to the whole Front End, and apply its components in any program.

2.2.4 Packages

In Mathematica, there exist many specialized functions and modules which are not loaded initially. They must be loaded separately from files in the *Mathematica* directory. These files are of the form `filename.m`.

The full name of a package consists of a `context` and a `short` name, and it is written as `context`short`.

Operations with packages:

   `<<context`, to load a package corresponding to a context,

   `<<c1`c2``, to load a package corresponding to a sequence of contexts,

   `$Packages`, to display a list of loaded packages,

   `$Context`, to display the current context,

   `$ContextPath`, to get a list of contexts,

   `Names["context`*"]`, to get a list of the functions in a package.

Problem 2.1 Show various maps of North America.

```
<<WorldPlot`
WorldPlot[{NorthAmerica,RandomColors}]
WorldPlot[NorthAmerica,WorldBackground->Hue[0.85],
 WorldGrid->None,WorldFrame->{Purple,Thickness[.012]},
 WorldProjection->LambertAzimuthal]
```

$\square$

Problem 2.2 Print the names of all packages and the names of all functions located in the `FiniteFields` package.

```
SetDirectory[ToFileName[{$InstallationDirectory,
 "AddOns","Packages"}]]; FileNames[]
 SetDirectory[ToFileName[{$InstallationDirectory,
"AddOns","LegacyPackages"}]]; FileNames[]
SetDirectory[ToFileName[{$InstallationDirectory,
 "AddOns","ExtraPackages"}]]; FileNames[]
fNames[p_String]:=(Needs[StringJoin[p,"`"]];
 Names[StringJoin[p,"`*"]]); fNames["FiniteFields"]
```

$\square$

2.2.5 Numerical Evaluation

Mathematica gives exact answers to arithmetic expressions with integers and reduces fractions. When the result is an irrational number, the output is returned in unevaluated form.

```
{135+638,-13*77,467/31,(3+8*4+9)/2,(-5)^(22),-5^(1/3)}
```

Mathematica represents the numbers in the decimal number system using a user-specified precision.

Numerical approximations:

N[expr], expr//N, numerical approximation of expr (to 6 significant digits),

N[expr,n], NumberForm[expr,n], numerical approximation of the expression to n significant digits,

EngineeringForm[expr,n], ScientificForm[expr,n], engineering and scientific notations of numerical approximation of expr to n significant digits.

```
{Sqrt[17],Pi+2*Pi*I,1/3+1/5+1/7,N[E,30],N[Pi],Pi//N}
{x=N[Sin[Exp[Pi]]], ScientificForm[x,10],
 EngineeringForm[x,10], NumberForm[x,10]}
```

Compile[{x1,...},fb], Compile[{{x1,t1},{x2,t2}...},fb],

machine code of a function for improving the speed in numerical calculations (fb is the function body, and t1,t2,... are the types of variables, e.g., _Real, _Complex, _Integer).

```
n=0.5; f1[x_]:=(x^n*Exp[-x^n])^(1/n);
f2=Compile[{x},Evaluate[(x^n*Exp[-x^n])^(1/n)]];
Timing[Table[f1[x],{x,0.,50.,0.01}]]
Timing[Table[f2[x],{x,0.,50.,0.01}]]
```

2.3 *Mathematica* Language

Mathematica language is a very powerful programming language based on systems of transformation rules, functional, procedural, and object-oriented programming techniques. This distinguishes it from traditional programming languages. It supports a large collection of *data structures* or *Mathematica objects* (functions, sequences, sets, lists, arrays, tables, matrices, vectors, etc.) and operations on these objects (type-testing, selection, composition, etc.). The library can be enlarged with custom programs and packages.

2.3.1 Basic Principles

Symbol in *Mathematica*, symb, refers to a symbol with the specified name, e.g., expressions, functions, objects, optional values, results, argument names.

A name of symbol, name, is a combination of letters, digits, or certain special characters, not beginning with a digit, e.g., a12new. Once defined, a symbol retains its value until it is changed, cleared, or removed.

Expression, expr, is a symbol that represents an ordinary *Mathematica* expression in readable form. The head of expr can be obtained with Head[expr]. The structure and various forms of expr can be analyzed with TreeForm, FullForm[expr], InputForm[expr], e.g.,

```
l1={5, 1/2, 9.1, 2+3*I, x, {A,B}, a+b, a*b}
{Head /@ l1,FullForm[l1],InputForm[l1],TreeForm[l1],
 TraditionalForm[l1]}
```

Mathematica is case sensitive, there is a difference between lowercase and uppercase letters, e.g., Sin[Pi] and sin[Pi] are different. All *Mathematica* functions begin with a capital letter. Some functions (e.g., PlotPoints) use more than one capital. To avoid conflicts, it is best to begin with a lower-case letter for all user-defined symbols.

The result of each calculation is displayed, but it can be suppressed by using a semicolon (;), e.g., Plot[Sin[x],x,0,2*Pi]; a=9; b=3; c=a*b

Basic arithmetic operators and the corresponding functions:

+	-	-	*	/	^
Plus	Subtract	Minus	Times	Divide	Power

A missing symbol (*), 2a, or a space, a b, also imply multiplication. So it is best to indicate explicitly all arithmetic operations.
Times[2,3,4,5], Power[2,3].

Additional arithmetic operators and their equivalent functions:

x++	x--	++x	--x
Increment	Decrement	PreIncrement	PreDecrement
x+=y	x-=y	x*=y	x/=y
AddTo	SubtractFrom	TimesBy	DivideBy

In these operations x must have a numeric value, e.g., x=5; x++;

Logic and relation functions and their equivalent operators:

And[x,y]	Or[x,y] Xor[x,y]	Not[x,y]	Implies[x,y]
x&&y	x||y	!x	x=>y
Equal[x,y]	Unequal[x,y]	Less[x,y]	Greater[x,y]
x==y	x!=y	x<y	x>y
LessEqual[x,y] x<=y		GreaterEqual[x,y] x>=y	

Logic expressions can be compared using LogicalExpand.

Patterns: Mathematica language is based on pattern matching.

A pattern is an expression which contains an underscore character (_).
   The pattern can stand for any expression. Patterns can be constructed from the templates, e.g.,

x_	x_/;cond	pattern?test	x_:IniValue
x_^n_	x_^n_	{x_, y_}	f[x_] f_[x_]

Problem 2.3 Define the function f with any argument named x, define
an expression satisfying a given condition.

```
f[x_]:=Abs[x]; f[t]
ex=t+Log[a]+Log[b];{ex/.Log[b_]->Sin[b],ex/.Log[b]->Sin[b]}
```
                                                                            □

Basic transformation rules: ->,   :>,   =,   :=,   ^:=,   ^-

The rule lhs->rhs transforms lhs to rhs. *Mathematica* regards the left-
   hand side as a *pattern*.

The rule `lhs:>rhs` transforms `lhs` to `rhs`, evaluating `rhs` only after the rule is actually used.

```
n=3; a1=(a+b)^n; {a1/.Power[x_,n]->Power[Expand[x],n],
                  a1/.Power[x_,n]:>Power[Expand[x],n]}
```

The assignment `lhs=rhs` (or `Set`) specifies that the rule `lhs->rhs` should be used whenever it applies.

The assignment `lhs:=rhs` (or `SetDelayed`) specifies that `lhs:>rhs` should be used whenever it applies, i.e., `lhs:=rhs` does not evaluate `rhs` immediately but leaves it unevaluated until the rule is actually called.

```
a1=TrigReduce[Sin[x]^2]; a2:=TrigReduce[Sin[x]^2]; {a1,a2}
x=a+b; {a1,a2}
```

Note. In many cases both assignments produce identical results, but the operator (`:=`) must be used, e.g., for defining piecewise and recursive functions (see Sect. 4.8).

The rule `lhs^:=rhs` assigns `rhs` to be the delayed value of `lhs`, and associates the assignment with symbols that occur at level one in `lhs`, e.g.,

```
D[int[f_,1_List],var_]^:=int[D[f,var],1];D[int[f[t],{t,0,n}],t]
Unprotect[D]; D[int[f_,1_List],var_]:=int[D[f,var],1];
Protect[D]; D[int[f[t],{t,0,n}],t]
```

The rule `lhs^=rhs` assigns `rhs` to be the value of `lhs`, and associates the assignment with symbols that occur at level one in `lhs`, e.g.,

```
Unprotect[{Cos,Sin}]; Cos[k_*Pi]:=(-1)^k/;IntegerQ[k];
Sin[k_*Pi]:=0/;IntegerQ[k]; Protect[{Cos, Sin}];
IntegerQ[n]^=True;            {Cos[n*Pi], Sin[n*Pi]}
```

Transformation rules are useful for making substitutions without making the definitions permanent and are applied to an expression using the operator `/.` (`ReplaceAll`) or `//.` (`ReplaceRepeated`).

```
sol=Solve[x^2+4*x-10==0,x]; x+1/.sol
```

Unassignment of definitions:

> Clear[symb] clears the symbol's definition and values, but does not clear its attributes, messages, or defaults,

> ClearAll[symb] clears all definitions, values, attributes, messages, defaults,

> Remove[symb] removes symbol completely,

> symb=. clears a symbol's definition, e.g., a=5; a=.; ?a.

To clear all global symbols defined in a *Mathematica* session can be produced by several ways, e.g.,

> Clear["Global`*"]; ClearAll["Global`*"]; Remove["`*"];

To recall a symbol's definition: ?symb, e.g., a=9; ?a

To recall a list of all global symbols that have been defined (during a session): ?`*.

Initialization: in general, it is useful to start working with the following initialization:

ClearAll["Global`*"];                          Remove["Global`*"];

The difference between the operators (=) *and* (==): the operator lhs=rhs is used to assign rhs to lhs, and the equality operator lhs==rhs indicates equality (not assignment) between lhs and rhs, e.g.,

{eq=A==B,eq,eq[[2]],eq[[1]]}

An expression, expr, is a symbol that represents an ordinary *Mathematica* expression and there is a wide set of functions to work with it.

Data types: every expression is represented as a tree structure in which each node (and leaf) has a particular data type. For the analysis of any node and branch can be used a variety number of functions, e.g., Length, Part, a group of functions ending in the letter Q (DigitQ, IntegerQ, etc.). For example, the function SameQ or its equivalent lhs===rhs yields True if lhs is identical to rhs, and yields False otherwise.

A boolean expression, `bexpr`, is formed with the *logical operators* and the relation operators, e.g.,

`LogicalExpand[!(p&&q)===LogicalExpand[!p||!q]]`

An equation, `eq`, is represented using the binary operator `==`, and has two operands, the left-hand side `lhs` and the right-hand side `rhs`.

Inequalities, `ineq`, are represented using the relational operators and have two operands, the left-hand side `lhs` and the right-hand side `rhs`.

Strings: a string, `str`, is a sequence of characters having no value other than itself and can be used as labels for graphs, tables, and other displays. The strings are enclosed within double-quotes, e.g., `"abc"`.

Some useful string manipulation functions:

`StringLength[str]`, a number of characters in `str`,

`StringJoin[str1,...]` or `str1<>...`, concatenation of strings,

`StringReverse[str]` reverses the characters in `str`,

`StringDrop[str,n,m]` eliminates characters in `str` from n to m,

`StringTake[str,n,m]` takes characters in `str` from n to m,

`StringInsert[str1,str2,n1,n2,...]` inserts `str2` at each of the positions n_1, n_2, ..., of `str1`,

`StringReplace[str,str1->nstr1,...]` replaces `str1` by `nstr1`,

`StringPosition[str,substr]`, a list of the start and end positions of substring.

Types of brackets:

Parentheses (`expr`): grouping, `(x+9)*3`,

Square brackets [`expr`]: function arguments, `Sin[x]`,

Curly brackets {`expr`}: lists, `{a,b,c}`.

Types of quotes:

Back-quotes `` `expr` ``:

     `fullname=context`short`, `` ` `` is a context mark,

     `` `x` `` is a format string character, ``StringForm["`1`,`2`",a,b]``,

     `` ` `` is a number mark, `12.8``, machine precision approximate number,

     `` ` `` is a precision mark, ``12.8`10``, arbitrary precision number with precision 10,

     `` `` `` is an accuracy mark, ``12.8``15``, arbitrary precision number with accuracy 15.

Double-quotes `"expr"`: to create strings.

Previous results (during a session) can be used with symbols `%` (the last result), `%%` (the next-to-last result), and so on.

Comments can be included within the characters (`*comments*`).

Function application: `expr//func` is equivalent to `fun[expr]`.

Incorrect response: if some functions take an "infinite" computation time, you may have entered or executed the command incorrectly. To terminate a computation, you can use:

`Evaluation` → `Quit Kernel` → `Local`.

2.3.2 Constants

Types of numbers: integer, rational, real, complex, root, e.g.,

`{-5,5/6,-2.3^-4,ScientificForm[-2.3^-4],3-4*I,Root[#^2+#+1&,2]}`

Mathematical constants: symbols for definitions of selected mathematical constants, e.g., `Catalan`, `Degree`, `E`, `EulerGamma`, `I`, `Pi`, `Infinity`, `GoldenRatio`, e.g., `{60Degree//N, N[E,30]}`.

Scientific constants: valuable tools for scientists and engineers in Physics and Chemistry can be applied with the packages `Units`` `` and `PhysicalConstants`` ``.

Note. Many functions from the `Miscellaneous`ChemicalElements` `` package (for *ver* < 6) have been incorporated to the new `ElementData` function.

2.3.3 Functions

Two classes of functions: pure functions and functions defined in terms of a variable (*predefined* and *user-defined* functions).

Pure functions are defined without a reference to any specific variable. The arguments are labeled #1,#2,..., and an ampersand & is used at the end of definition.

```
f:=Sin[#1]&; g:=Sin[#1^2+#2^2]&;
{f[x],f[Pi],g[x,y],g[Pi,Pi]}
```

Predefined functions. Most of the mathematical functions are predefined.

Special functions. Mathematica includes all the common special functions of mathematical physics. We will discuss some of the more commonly used functions.

The names of mathematical functions are complete English words or the traditional abbreviations (for a few very common functions), e.g., Conjugate, Mod. Person's name mathematical functions have names of the form PersonSymbol, for example, the Legendre polynomials $P_n(x)$, LegendreP[n,x].

Elementary trascendental functions:

Exp[x], the exponential function,

Log[x], Log[b,x], the natural logarithm and the general logarithm,

Sin, Cos, Tan, Cot, Sec, Csc, the trigonometric functions,

Sinh, Cosh, Tanh, Coth, Sech, Csch, the hyperbolic functions,

ArcSin, ArcSinh, ArcCos, ArcCosh, ... (see ?Arc*), the inverse trigonometric and hyperbolic functions.

{Sin[30 Degree], Sin[Pi/3], Exp[5]//N}

Other useful functions:

```
Max[x]  Min[x]     Round[x] Floor[x]    Ceiling[x] IntegerPart[x]
FractionalPart[x] Abs[x]   Sign[x]     n!   Factorial[x]     n!!
Factorial2[x]     Prime[n] Fibonacci[n]    Quotient[x]   Mod[x]
GCD[x] LCM[x] Timing[expr] RandomInteger[] RandomReal[]Random[]
RandomReal[{xmin,xmax}]     RandomComplex[]        IntegerQ[expr]
Print[expr1,...] PolynomialQ[expr] NumericQ[expr] VectorQ[expr]
```

Abs[x], Sign[x], the absolute value and the sign functions,

n!, Prime[n], Fibonacci[n], the factorial, the *n*-th prime, and the Fibonacci functions,

Round, Floor, Ceiling, IntegerPart, FractionalPart, converting real numbers to nearby integers, and the fractional part or real numbers,

Quotient, Mod, GCD, LCM, the quotient and the remainder (in the Division Algorithm), the greatest common divisor and the least common multiple,

Random[], Random[type,range], the random functions,

Timing, the kernel computation time of the expression,

IntegerQ, PolynomialQ,... (see ?*Q), a group of functions ending in the letter Q can be used for testing for certain conditions and return a value True or False,

Print, this function can be used in loops, strings, and other displays.

```
{Quotient[25,3], Mod[25,3], Random[Real,{2,9},10]}
{Fibonacci[10000]//Timing, PolynomialQ[x^3*y+Sqrt[y]*y,y]}
```

User-defined functions are defined using the pattern x_, e.g., the functions of one or n variables $f(x)$=expr, $f(x_1,\ldots,x_n)$=expr, the vector functions of one or n variables (for more details see Sect. 4.8).

```
f[x_]:=expr;     f=Function[x,expr];     f[x1_,...,xn_]:=expr;
f[t_]:={x1[t],...,xn[t]};     f=Function[t,{x1[t],...,xn[t]}];
     f[x1_,...,xn_]:={f1[x1_,...,xn_],...fn[x1_,...,xn_]};
```

Evaluation of a function or an expression without assigning a value can be performed using the replacement operator /.,

```
f[a]          f[a,b]          expr /. x->a          expr /.{x->a,x->b}
```

Composition functions are defined with the operation `Composition` (the arguments of this operation are pure functions f1, f2,...) and using the functions `Nest`, `NestList`, e.g., the composition function $(f_1 \circ f_2 \circ \cdots \circ f_n)(x)$ or $(f \circ f \circ \cdots \circ f)(x)$ (n times).

```
Composition[f1,f2,...]        Composition[f,...,f]        f1@f2@...
Nest[f,expr,n]                NestList[f,expr,n]          f@f@f...
```

Problem 2.4 Define the function $f(x,y) = 1 - \sin(x^2 + y^2)$ and evaluate $f(1,2)$, $f(0,a)$, $f(a,b)$.

```
f[x_,y_]:=1-Sin[x^2+y^2];
f1=Function[{x,y},1-Sin[x^2+y^2]]; {N[f[1,2]],f[0,a],
 Simplify[f[a,b]], N[f1[1,2]], f1[0,a], Simplify[f1[a,b]]}
```
$\square$

Problem 2.5 Define the vector function $h(x,y) = \langle\cos(x-y), \sin(x-y)\rangle$ and calculate $h(1,2)$, $h(\pi, -\pi)$, and $h\big(\cos(a^2), \cos(1-a^2)\big)$.

```
h[x_,y_]:={Cos[x-y],Sin[x-y]};
{N[h[1,2]],h[Pi,-Pi],h[Cos[a^2],Cos[1-a^2]]//FullSimplify}
```
$\square$

Problem 2.6 Graph the real roots of the equation $x^3 + (a-3)^3 x^2 - a^2 x + a^3 = 0$ for $a \in [0,1]$.

```
SetOptions[Plot,ImageSize->300,PlotStyle->
 {Hue[0.9],Thickness[0.01]},PlotPoints->100,PlotRange->All];
i=1; nD=10; sol=Solve[x^3+(a-3)^3*x^2-a^2*x+a^3==0,x];
r[i_,b_]:=N[sol[[i,1,2]]]/.{a->b},nD];
g=Table[Plot[r[i,b],{b,0,1}],{i,1,3}]; GraphicsRow[g]
```
$\square$

Problem 2.7 For the functions $f(x) = x^2$ and $h(x) = x + \sin x$ calculate the composition functions $(f \circ h \circ f)(x)$ and $(f \circ f \circ f \circ f)(x)$.

```
f[x_]:=x^2; fF:=#1^2&; hF:=#1+Sin[#1]&;
{fF@hF@fF[x],fF@fF@fF@fF[x],Nest[f,x,4],NestList[f,x,4]}
```
◻

Piecewise functions can be defined using the conditional operator (/;) or the functions Piecewise and UnitStep.

```
f[x_]:=expr/;cond                f[x_]:=UnitStep[x-a]*UnitStep[b-x];
f[x_]:=Piecewise[{{cond1,val1},...,{condn,valn}}];              f[x]
```

Problem 2.8 Define and graph the function $f(x) = \begin{cases} 0, & |x| > 1 \\ 1-x, & 0 \leq x \leq 1 \\ 1+x, & -1 \leq x \leq 0 \end{cases}$

in [-3, 3].

```
f[x_]:=0/;Abs[x]>1; f[x_]:=1-x/;0<=x<=1;
f[x_]:=1+x/;-1<=x<=0; Plot[f[x],{x,-3,3},
 PlotStyle->{Hue[0.85],Thickness[0.01]}]
```
◻

Problem 2.9 Graph the periodic extension of $f(x) = \begin{cases} x, & 0 \leq x < 1 \\ 1, & 1 \leq x < 2 \\ 3-x, & 2 \leq x < 3 \end{cases}$

in [0, 20].

```
f[x_]:=x/;0<=x<1; f[x_]:=1/;1<=x<2;
f[x_]:=3-x/;2<=x<=3; f[x_]:=f[x-3]/;x>3; Plot[f[x],{x,0,20}]
```
◻

Recursive functions are the functions that are defined in terms of themselves.

Problem 2.10 Calculate Fibonacci numbers, $F(n) = F(n-1) + F(n-2)$, $F(0) = 0$, $F(1) = 1$, using the memory feature and the analytic function. Compare the computation time in both cases.

```
$RecursionLimit=Infinity;
Fib[0]=0; Fib[1]=1; Fib[n_]:=Fib[n]=Fib[n-1]+Fib[n-2];
Table[Fib[i],{i,10,40}]
F[n_]:=(((1+Sqrt[5])/2)^n-((1-Sqrt[5])/2)^n)/Sqrt[5];
Fib[3000]//Timing
Expand[F[3000]]//Timing
```
◻

2.3.4 Modules

A module is a local object that consists of several functions which one
needs to use repeatedly (see ?Module). A module can be used to
define a function (if the function is too complicated to write by
using the notation f[x_]:=expr), to create a matrix, a graph, a
logical value, etc.

Block is similar to Module, the main difference between them is that
Block treats *the values* assigned to symbols as local, but *the names*
as global, whereas Module treats *the names* of local variables as
local.

With is similar to Module, the principal difference between them is that
With uses *local constants* that are evaluated only once, but Module
uses *local variables* whose values may change many times.

```
Module[{var1,...},body];        Module[{var1=val1,...},body];
Block[{var1,...},expr];          Block[{var1=val1,...},expr];
With[{var1=val1,var2=val2,...},expr];
```

Here var1,... are local variables, val1,... are initial values of local
variables, body is the body of the module (as a sequence of statements
separated by semicolons). The final result of the module is the result
of the last statement (without a semicolon). Also Return[expr] can be
used to return an expression.

Problem 2.11 Define a module and module function, mF, that calculate
the maximum of x and y, and then find the maximum of the values
(34/9, 674/689).

```
Module[{x=34/9,y=674/689},If[x>y,x,y]]
mF[x_,y_]:=Module[{m},If[x>y,m=x,m=y]]; mF[34/9,674/689]
```
                                                                        □

Problem 2.12 Compare the following calculations using With, Module,
and Block.

```
f1[x_]:=With[{a=2,b=1},Tan[a*x]-b];
f2[x_]:=Module[{i},Sum[x^(-2*i),{i,1,9}]];
f3[x_]:=Block[{i},Sum[x^(-2*i),{i,1,9}]];
{f1[x], {a,b}, f2[4], f2[x], f2[i], f3[4], f3[x], f3[i]}
```
                                                                        □

Problem 2.13 Consider a 2-degree of freedom holonomic system described by the Lagrange equations, namely, the motion of a double pendulum of mass m and length L in the vertical plane due to gravity (see Sect. 1.3.4). Construct a module for deriving the Lagrange equations.

```
eqLagrange[q_,L_,Q_]:=Module[
 {Lq,Ldq,Ldq1,dLdq,dLdq1,eq},
 L1=L/.{dq[[1]]->"A'",dq[[2]]->"B'"}; n=Length[q];
 Lq=Map[D[L1,#1]&,q]/.{"A'"->dq[[1]],"B'"->dq[[2]]};
 Ldq=Map[D[L,#1]&,dq]; Ldq1=Ldq/.{dq[[1]]->"A'",dq[[2]]->"B'"};
 dLdq=Sum[Map[D[#1,q[[j]]]&,Ldq1]*dq[[j]]+Map[
 D[#1,dq[[j]]]&,Ldq]*d2q[[j]],{j,1,n}]; dLdq1=dLdq/.
 {"A'"->dq[[1]],"B'"->dq[[2]]}; eq=Factor[dLdq1-Lq-Q]];
 q={A,B}; Q={0,0}; dq=Map[#1'&,q]; d2q=Map[#1''&,q];
 T=m*x^2*dq[[1]]^2+m*x^2*dq[[1]]*dq[[2]]*Cos[q[[1]]
 -q[[2]]]+1/2*m*x^2*dq[[2]]^2;
 P=-m*g*x*Cos[q[[1]]]-m*g*x*(Cos[q[[1]]]+Cos[q[[2]]]);
 g=omega^2*x; L=T-P; eqLagrange[q,L,Q]
```
                                                                            □

2.3.5 Control Structures

In Mathematica language there are the following *two control structures*: the selection structures If, Which, Switch and the repetition structures Do, While, For,

If[cond,exprTrue]	If[cond,exprTrue,exprFalse]
	If[cond,exprTrue,exprFalse,exprNeither]
Which[cond1,expr1,...]	Switch[expr,patt1,val1,patt2,val2,...]
Do[expr,{i,i1,i2,iStep}]	Do[expr,{i,i1,i2,iS},{j,j1,j2,jS},...]
While[cond,expr]	For[i=i1,cond,iStep,expr]

where exprTrue, exprFalse, exprNeither are expressions that execute, respectively, if the condition cond is True, False, and is neither True or False; i,i1,i2 (j,j1,j2) are the loop variable and the initial and the last values of i (j).

The result of Which is the expression expr1,expr2,... corresponding to the first true condition cond1,cond2,....

In Switch, the expression expr is compared with patterns patt1,... until a match is found and the corresponding value val1,... is the result. These operators can be nested.

The operators Break[], Continue[], Goto[name] inside the loops are used
    for breaking out of a loop, to proceed directly to the next iteration,
    or for transferring the control to the point Label[name].

Problem 2.14 Compare the following calculations using Which and
Switch.

```
f1[x_]:=Which[x>0,1,x==0,1/2,x<0,-1]; f1[0]
Plot[f1[x],{x,-2,2},PlotStyle->Thickness[0.01]]
f2[x_]:=Switch[x,_Integer,x,_List,Join[x,{c}],_Real,N[x]];
{f2[4], f2[Sum[2.3+0.01*i,{i,1,10}]], f2[{a,b}]}
```
                                                                          □

Problem 2.15 Find all numbers from 1 to 30 which are not multiples
of 2 or 5.

```
Do[If[Mod[i,2]==0||Mod[i,5]==0,,Print[i]],{i,1,30}]
```
                                                                          □

Problem 2.16 Calculate 20! using Do, While, For loops.

```
fact=1; n=20; Do[fact=fact*i, {i,1,n}]; fact
fact=1; i=20; While[i>0, fact=fact*i; i--]; fact
For[fact=1; i=1, i<=20, i++, fact=fact*i]; fact
```
                                                                          □

Problem 2.17 Define the function *double factorial* for any integer n,

$$n!! = \begin{cases} n(n-2)(n-4) \ \dots \ (4)(2), & n = 2i, i = 1, 2, \dots, \\ n(n-1)(n-3) \ \dots \ (3)(1), & n = 2i+1, i = 0, 1, \dots \end{cases}$$

```
doubleFactorial[n_]:=Module[{p=1,i1},
  If[Mod[n,2]==0,i1=2,i1=1]; Do[p=p*i,{i,i1,n,2}];p];
n1=20; n2=41; {doubleFactorial[n1],doubleFactorial[n2],
  Factorial2[n1],Factorial2[n2]}
Print["20!!=",doubleFactorial[n1]];
Print["41!!=",doubleFactorial[n2]];
```
                                                                          □

Problem 2.18 Calculate the values x_i, where $x_i = (x_{i-1}+1/x_{i-1})$, $x_0=1$,
until $|x_i - x_{i-1}| \le \varepsilon$ ($i = 1, 2, \dots, n$, $n = 10$, $\varepsilon = 10^{-3}$).

```
nD=10; n=10; x=1; xp=1; epsilon=10^(-3); i=1;
While[Abs[x-xp]<epsilon||i<=n, xp=x; x=N[x+1/x,nD];
    Print["x=",x]; i++]
```
                                                                          □

Problem 2.19 Find an integer $N(x)$, $x \in [a,b]$ such that $\displaystyle\sum_{i=1}^{N(x)} i^{-x} \geq c$, where $a = 1$, $b = 2$, and $c = 2$.

```
nD=10; a=1; b=2; c=2; Do[s=0; i=1; x=k;
 While[s<c && i<=100, s=s+N[i^(-x),nD]; i++];
 Print["s=",s,"   ","x=",x,"   ","N(x)=",i],{k,a,b,N[1/100,nD]}]
```

□

2.3.6 Objects and Operations

Lists are the fundamental objects in *Mathematica*. The other objects (e.g., sets, matrices, tables, vectors, arrays, tensors, objects containing data of mixed type) are represented as lists. A list is an ordered set of objects separated by commas and enclosed in curly braces, {elements}, or defined with the function List[elements].

Nested lists are lists that contain other lists. There are many functions which manipulate lists and here we review some of the most basic.

Basic manipulation functions:

Create empty lists: { }, List[].

Listable function. Many *Mathematica* functions performed on a list will be performed on each element of the list.

```
{list1={1,2,3,4,5}, Sqrt[list1]}
```

The basic standard operations should be applied to lists with the same number of elements.

```
{list1={1,3,5,7}, list2={2,4,6,8}, list1/list2}
```

Create lists and nested lists:

```
Range[n,m,step], lists of numbers,
Table[expr,{i,n,m,step}], lists according to a formula,
Array[f,n,nIni], list elements are functions f[n],
Table[expr,{i,n,m},{j,k,l},...], nested lists according to a formula,
Array[f,{n,m},{nIni,mIni},...], nested lists with elements f[i,j],
Characters[str], CharacterRange[ch1,ch2], lists of characters.
```

```
f[x_]:=x^3-x^2-x-1; g[x_,y_]:=x^2+y^2;
{Range[0,Pi,Pi/3],Table[Sqrt[i],{i,1,20,4}],Array[f,20,0],
 Table[Sqrt[i+j],{i,1,4},{j,1,7}],Array[g,{3,5},{0,0}]}
{Characters["Maple&Mathematica"],CharacterRange["c","l"]}
```

Determine the structure of lists:

```
Length[list], Depth[list], AtomQ[list], LeafCount[list],
StringLength[str], Dimensions[list], ArrayDepth[list].
```

```
vec={a,b,c}; mat={{a,b,c},{d,e,f}}; tens={{{a,b,c},{d,e,f}},
 {{g,h,i},{j,k,l}}}; l1={vec,mat,tens}; {Length/@l1,
Dimensions/@l1, TensorRank/@l1,ArrayDepth/@l1, Depth/@l1}
```

Extract an *i*-th element from a list or a nested list:

```
Part[list,i] or list[[i]],
First[list], Last[list],
```
the first and the last elements of `list`.

```
{list1=Range[1,20,3], list2=Table[(i+j)^2,{i,1,4},{j,1,7}]}
{Part[list1, 4], list1[[4]], list2[[2,4]]}
```

Create a substructure (a part of a list):

```
Rest[list],
```
a list without the first element of `list`,
```
Delete[list,n],
```
a list without the element in the *n*-th position of `list`,
```
Take[list,{n,m}],
```
a list with the elements taken within the range,
```
Drop[list,{n,m}],
```
a list with the elements deleted within the range,
```
Select[list,crit],
```
a list with the elements satisfying a criterion,
```
Cases[list,patt],
```
a list with the elements matching a pattern,
```
DeleteCases[list,patt],
```
removing elements matching a pattern.

```
f1[x_]:=Sqrt[x^2-x-1]; list1=Array[f1,10,0]
{Take[list1,{2,7}],Drop[list1,{2,7}],Select[{0,1,-2,Pi,a},#>0&]}
Cases[{1,-2,I,"a",b},x_?StringQ]
DeleteCases[{1,-2,I,"a",b},x_?StringQ]
```

Insert an element into a list:

```
Append[list,x],
```
insert an element *x* to the right of the last element of `list`,
```
Prepend[list,x],
```
insert an element *x* to the left of the last element of `list`,
```
Insert[list,x,n],
```
insert an element *x* in position *n*.

```
list1=Table[i^2,{i,1,20,2}]; {Append[list1,a],Insert[list1,a,5]}
```

Replace the *n*-th element of a list by *x*: `ReplacePart[list,x,n]`.

```
{list1=CharacterRange["0","9"], ReplacePart[list1,a,5]}
```

Rearrange lists:

> `Sort[list]` and `Reverse[list]`, sorting and reversing of lists,
> `RotateLeft[list,n]`, `RotateRight[list,n]`, cycling of lists.

```
list1=Table[Random[Integer,{1,20}],{i,1,10}]
{Sort[list1],RotateLeft[list1,2],RotateRight[list1,-2]}
```

Concatenation of lists:

> `Join[list1,list2,...]`, `StringJoin[str1,str2,...]`,
> `Flatten[{list1,list2},...},1]`.

```
{list1=Range[-10,10],list2=Table[i^2,{i,1,10}]}
{Join[list1,list2],Flatten[{list1,list2},1],StringJoin["A","B"]}
```

Manipulation with nested lists:

> `TreeForm[list]`, visualization of nested lists as a tree,
> `Depth[list]`, the number of levels in a nested list,
> `Level[list,levels]`, a list of all sublists of `list` on levels,
> `Level[list,levels,f]`, applies function `f` to the sequence of sublists,
> `Flatten[list]`, convert a nested list into a simple list,
> `Flatten[list,n]`, partial flattening to level *n*,
> `Partition[list,n]`, converts a simple list into sublists of length *n*.

```
tensor={{{a,b,c},{d,e,f}},{{g,h,i},{j,k,l}}}; TreeForm[tensor]
list1={{a1,a2,a3},{a4,a5,a6}}; list2=Flatten[list1];
{Depth[list1], Level[list1,2], Level[list1,1],
 Level[list1,{1},Subtract], Partition[list2, 2]}
```

Sets are represented as lists:

Union[list1,...], a list of the distinct elements of lists,
Intersection[list1,...], intersection of lists,
Subset[list], a list of all subsets of the elements in list.

Note. The package Combinatorica` includes various useful set functions (for more details see Sect. 4.1).

Vectors are represented as lists, vectors are simple lists. Vectors can be expressed as single columns with ColumnForm[list,horiz,vert].

d[i_]:=1/i; f[i_]:=1/i^2; {vec={a,b,c},
 Length[vec],Dimensions[vec],ArrayDepth[vec],VectorQ[vec]}
v1=N[Array[d,5]]; v2=N[Array[f,5]]; {v1,ColumnForm[v1,Right],
 v2,ColumnForm[v2,Right],a*v1+b*v2,v1.v2}

Tables, matrices, and tensors are represented as nested lists. There is no difference between the way they are stored: they can be generated using the functions MatrixForm[list], TableForm[list], or using the nested list functions (see above). Matrices and tables can also be conveniently generated using the *palette*,

Insert → Table/Matrix or Palettes → BasicMathInput.

A1={{a1,a2},{a3,a4}}; {MatrixForm[A1],TableForm[A1]}

Basic manipulation functions with matrices, tables, and tensors:

A matrix is a list of vectors. Matrices can be combined using the operations: addition (+), subtraction (-), scalar (*) and matrix multiplication (.), in more detail about matrix manipulations, see Chapter 5.

m1={{a1,a2},{a3,a4}}; m2={{b1,b2},{b3,b4}}; {MatrixForm[m1],
MatrixForm[m2], 4*m1-m2//MatrixForm, m1.m2//MatrixForm}

Some useful functions for tables:

TableForm[list,ops], generating tables with some properties,
PaddedForm[list,{n,m}], formatting numerical tables.

```
t1={{a1,2*a2,a3},{-a4,a5,-a6}}; TableForm[t1]
TableForm[t1,TableAlignments->Right, TableHeadings->
  {{"r1","r2"}, {"c1","c2","c3"}},TableSpacing->{3,3}]
t2=Table[{i,N[Sin[i]],N[Cos[i]]},{i,1,5}]; TableForm[t2]
f1:=PaddedForm[i,3]; f2:=PaddedForm[N[Sin[i]],{12,5}];
t3=Table[{f1,f2},{i,1,5}]; TableForm[t3]
```

A tensor is a list of matrices with the same dimensionality (in more detail about tensor manipulations, see Sect. 5.13).

Some useful functions for tensors:

Table, Array, creating tensors,
TreeForm, MatrixForm, visualizing tensors as a tree or a matrix,
Length, Dimensions, TensorRank, determining the tensor structure.

```
{Table[i1*i2,{i1,2},{i2,3}], Array[(#1*#2)&,{2,3}]}
tens={{{a,b,c},{d,e,f}},{{g,h,i},{j,k,l}}}
{TreeForm[tens], MatrixForm[tens], Length[tens],
Dimensions[tens],TensorRank[tens],TensorQ[tens]}
```

Apply a function to each element of an object:

Map[f,expr] or f/@expr, at the first level in expr,
Map[f,expr,levels], apply f to parts of expr,
Apply[f,expr] or f@@expr, replace the head of expr with the function f,
Thread[f[args],head], "threads" f over any objects with head
   that appear in the arguments, args, of the function f.

```
l1={{a,b},{c,d}}; l2={{e,f},{g,h}}; eq1=a==b;
{f1/@l1, Map[f2,l1,{2}], f3/@(x^3+x+2)}
(Apply[Plus,(Range[1,100])^2]//N)===(Sum[i^2,{i,1,100}]//N)
{Thread[#^n&@eq1,Equal],Thread[eq1^n,Equal],eq1^n}
{l1==l2,Thread[l1==l2]}
```

Problem 2.20 A sequence of numbers $\{x_i\}$ $(i = 0, \ldots, n)$ is defined by $x_{i+1} = ax_i(1 + x_i)$ $(0 < a < 10)$, where a is a given parameter. Define a list of coordinates $[i, x_i]$ such that $a = 3$, $x_0 = 0.1$, $n = 100$. Plot the graph of the sequence $\{x_i\}$.

```
n=100; l=Array[x,n,0]; a=3; l[[1]]=0.1;
Do[l[[i]]=a*l[[i-1]]*(1 l[[i 1]]),{i,2,n}]
l1=Table[{i,l[[i]]},{i,2,n}]
ListPlot[l1,PlotStyle->{PointSize[0.02],Hue[0.7]}]
```

□

Problem 2.21 Observe the function behavior $y(x) = \cos(6(x - a \sin x))$, $x \in [-\pi, \pi]$, $a \in [\frac{1}{2}, \frac{3}{2}]$ (in more detail about animation, see Sect. 3.6).

```
y[x_,a_]:=Cos[6*(x-a*Sin[x])]; n=20;
Animate[Plot[y[x,a],{x,-Pi,Pi},PlotRange->{-1,1},
  PlotStyle->Hue[0.7]],{a,1/2,3/2,1/n}]
```
                                                                          □

2.3.7 Dynamic Objects

In *Mathematica* (for $ver \geq 6$), the new kind of output, the *dynamic output* has been introduced allowing to create dynamic interfaces of different types. Numerous new functions for creating various dynamic interfaces have been developed. We mention the most important of them:

```
Dynamic[expr]    Slider[Dynamic[x]]        Slider[x,{x1,x2,xStep}]
Manipulate[expr,{x,x1,x2,xStep}]       TabView[{expr1,expr2,...}]
SlideView[{expr1,expr2,...}]       DynamicModule[{x=x0,...},expr]
Manipulator[x,{x1,x2}]       Animator[x,{x1,x2}]       Pane[expr]
```

```
{Slider[Dynamic[par]],Dynamic[Plot[Sin[1/var*par],{var,-1,1},
  PlotRange->{{-1,1},{-1,1}},ImageSize->500]]}
DynamicModule[{x=.1},{Slider[Dynamic[x]],
 Dynamic[Plot[Sin[1/var*x],{var,-1,1},PlotRange->
 {{-1,1},{-1,1}},ImageSize->500]]}]
Manipulate[Expand[(z+1)^n],{n,3,10,1}]
TabView[Table[Plot[Sin[1/x*par],{x,-1,1}],{par,-1,1,0.1}]]
SlideView[Table[Plot[Sin[1/x*par],{x,-1,1}],{par,-1,1,0.1}]]
```

Problem 2.22 Display a dynamic object without controls and the animation frame.

```
t=0; Row[{Pane[Animator[Dynamic[t],{-1,1}],{0,0}],
 Dynamic[Plot[Sin[1/x*t],{x,-1,1},ImageSize->300,
 PlotRange->{{-1,1},{-1,1}}]]}]
```
                                                                          □

Part II

Mathematics: *Maple* and *Mathematica*

Chapter 3

Graphics

3.1 Simple Graphs

Graphs of real values of `expr` or the functions $f(x)$, $f(x, y)$, $x \in [x_1, x_2]$, $(x, y) \in [x_1, x_2] \times [y_1, y_2]$.

Maple:

```
f:=expr;                 plot(f, x1..x2);
f:=x->expr;              plot(f(x),x= x1..x2, ops);
f:=(x,y)->expr;          plot3d(f(x,y),x=x1..x2,y=y1..y2,ops);
```

```
f1:=x->sin(x): f2:=sin(x); f3:=(x,y)->x^2-x-y^2-y-8;
plot(f1(x),x=-2*Pi..2*Pi); plot(f2,x=-2*Pi..2*Pi);
plot3d(f3(x,y),x=-6..6,y=-6..6,axes=boxed);
```

Mathematica:

```
Plot[f[x],{x,x1,x2}]          Plot3D[f[x,y],{x,x1,x2},{y,y1,y2}]
```

```
f[x_,y_]:=x^2-x-y^2-y-8; Plot[Sin[x],{x,-2*Pi,2*Pi}]
Plot3D[f[x,y],{x,-6,6},{y,-6,6},Boxed->True]
```

3.2 Various Options

In Maple, all the graphs can be drawn with various versions of `plot` and the package `plots`. The function `plot` has various forms (e.g., `logplot`, `odeplot`, `plot3d`) and various optional arguments which

I.K. Shingareva, C. Lizárraga-Celaya, *Maple and Mathematica*, 2nd ed.,
DOI 10.1007/978-3-211-99432-0_3, © Springer-Verlag Vienna 2009

define the final figure (see ?plot[options], ?plot3d[options]), e.g., light setting, legends, axis control, titles, gridlines, real-time rotation of 3D graphs, wide variety of coordinate systems, etc.

```
g1:=x->exp(-(x-3)^2*cos(Pi*x)^2);
plot(g1(x),x=0..6,tickmarks=[4,4],title=`Graph of g(x)`);
plot(sin(x)/x,x=-3*Pi..3*Pi,scaling=constrained);
plot(sin(x)/x,x=-3*Pi..3*Pi,style=point);
plot(tan(x),x=-2*Pi..2*Pi,y=-4..4,discont=true);
Points:=[[1,2],[2,3],[3,5],[4,7],[6,13],[7,17],[8,19]];
plot(Pints,style=point);
plot(Points,style=line);
g2:=(x,y)->x^2*sin(2*y)+y^2*sin(2*x);
plot3d(g2(x,y),x=0..Pi,y=0..Pi,grid=[20,20],style=patch);
```

In Mathematica, there are many options available for function graphics which can define the final picture (in more detail, see Options [Plot], Options[Plot3D]), e.g., light modeling, legends, axis control, titles, gridlines, etc. The general rule for defining options is:

```
Plot[f[x],{x,x1,x2},opName->value,...]
         Plot3D[f[x,y],{x,x1,x2},{y,y1,y2},opName->value,...]
```

Here opName is the option name.

Note. In *Mathematica* ($ver < 6$), a semicolon (;) was used to suppress the special line -Graphics- which follows the graphs. In *Mathematica* ($ver \geq 6$), the semicolon suppresses the graph output completely. The options DisplayFunction->Identity and DisplayFunction->$DisplayFunction (also used in $ver < 6$ for suppressing the graph output) are no longer needed.

Formula for color graphs: Mathematica makes it easy to compute the RGB formula for color graphs, using Insert $\rightarrow$ Color. In addition, the names of predefined colors are available in the new ColorData function. The list of all colors can be obtained by typing ColorData["Legacy","Names"], and the RGB formula of a particular color, e.g. Coral, by typing ColorData["Legacy","Coral"]. Additionally, all predefined color schemes can be inserted by using Palettes $\rightarrow$ ColorSchemes.

Some useful options for 2D graphs:

```
f[x_]:=Exp[-(x-3)^2*Cos[Pi*x]^2]; <<PlotLegends`
Plot[f[x],{x,-Pi,2*Pi},PlotRange->All,PlotLabel->"f[x]"]
Plot[f[x],{x,-Pi,2*Pi},AxesLabel->{"x,sm","y,sec"}]
Plot[f[x],{x,0,2*Pi},AspectRatio->Automatic]
Plot[f[x],{x,-Pi,2*Pi},AxesOrigin->{3,0},PlotRange->{{2,4},
 {0,1}},Frame->True,Axes->False,GridLines->{{2,2.5,3,Pi,
 3.5,4},Automatic},FrameTicks->{Automatic,{0.2,0.8}}]
f1[x_]:=Sin[x]; f2[x_]:=Cos[x]; f3[x_]:=Sin[x]-Cos[x];
Plot[{f1[x],f2[x],f3[x]},{x,-Pi,Pi},PlotStyle->
 {GrayLevel[0.5],Dashing[{0.02,0.03}],Thickness[0.02]},
 PlotLegend->{f1[x],f2[x],f3[x]},Filling->{1->{3},2->{3}},
 Ticks->{{0,Pi/4,Pi/2,3*Pi/4,Pi},Automatic}]
Plot[{f1[x],f2[x]},{x,-Pi,Pi},PlotStyle->{Red,Blue}]
Plot[{f1[x],f2[x],f3[x]},{x,-Pi,Pi},PlotStyle->
 {RGBColor[0.501961,1,0],RGBColor[1,0.501961,1],
 RGBColor[1,0.501961,0]}]
```

Options for 3D graphs: most 2D graph options are valid for Plot3D. Here we present some useful special options for 3D graphs:

```
f[x_,y_]:=Exp[-(x+y)]; Plot3D[f[x,y],{x,-Pi,2*Pi},
 {y,-Pi,2*Pi},PlotPoints->{25,40},Filling->Bottom]
Plot3D[f[x,y],{x,-Pi,2*Pi},{y,-Pi,2*Pi},
 Mesh->False,Boxed->False]
Plot3D[f[x,y],{x,-Pi,2*Pi},{y,-Pi,2*Pi},Mesh->8,
 MeshShading->{{Orange,Green},{Brown,Yellow}},
 PlotRange->All]
Plot3D[f[x,y],{x,-Pi,2*Pi},{y,-Pi,2*Pi},Mesh->8,
 MeshStyle->Gray]
Plot3D[f[x,y],{x,-Pi,2*Pi},{y,-Pi,2*Pi},BoxRatios->{1,2,1}]
Plot3D[f[x, y],{x,-Pi,2*Pi},{y,-Pi,2*Pi},FaceGrids->All,
 Axes->False]
Plot3D[f[x, y],{x,-Pi,2*Pi},{y,-Pi,2*Pi},
 PlotStyle->Glow[White],Lighting->"Neutral"]
Plot3D[f[x,y],{x,-Pi,2*Pi},{y,-Pi,2*Pi},
 PlotStyle->FaceForm[],Lighting->"Neutral"]
Plot3D[f[x,y],{x,-Pi,2*Pi},{y,-Pi,2*Pi},ViewPoint->{-1,2,1}]
```

In Mathematica, the 3D graphs can be examined from any viewpoint using real-time 3D manipulation (with the mouse) and various options such as `ViewPoint`, `ViewCenter`, `ViewVertical`, `ViewVector`, `ViewRange`, `ViewAngle` (the `3DViewPointSelector` has been eliminated in *Mathematica* 6).

The global options for 2D and 3D graphs:

Maple:

```
with(plots);       setoptions(NameOpt1=val1,...,NameOptn=valn);
                   setoptions3d(NameOpt1=val1,...,NameOptn=valn);
```

```
with(plots); setoptions(axes=boxed,title="graph of f(x)");
      setoptions3d(axes=normal,title="graph of g(x,y)");
f:=x->x^2*sin(x^2); g:=(x,y)->sin(x^2+y^2);
plot(f(x),x=-Pi..Pi); plot3d(g(x,y),x=-Pi..Pi,y=-Pi..Pi);
```

Mathematica:

```
SetOptions[symb,NameOpt1->val1,...,NameOptn->val1]
```

```
SetOptions[Plot,Frame->True,PlotLabel->"f[x]"]
SetOptions[Plot3D,Boxed->False,PlotLabel->"f[x,y]"]
Plot[x^2*Sin[x^2],{x,-Pi,Pi}]
Plot3D[Sin[x^2+y^2],{x,-Pi,Pi},{y,-Pi,Pi}]
```

3.3   Multiple Graphs

A list or a set of graphs in the same figure.

Maple:

```
L1:=[f1(x),...,fn(x)]:   L2:=[F1,...,Fn]:
S1:={f1(x),...,fn(x)}:   S2:={F1,...,Fn}:        plot(L1,x=a..b);
plot(S1,x=a..b);         plot(L2,x=a..b);        plot(S2,x=a..b);
```

Here `L1,L2` and `S1,S2` are the lists and sets of functions and expressions, respectively.

```
f:=x->sin(x)/x;  g:=x->cos(x)/x;
plot([f(x),g(x)],x=0..10*Pi,y=-1..2,
 linestyle=[SOLID,DOT],color=[red,blue]);
```

Mathematica:

```
Plot[{f1[x],f2[x],...,fn[x]},{x,x1,x2}]
GraphicsGrid[{{Plot[f11...],...,Plot[f1m...]},
        ...{Plot[fn1...],...,Plot[fnm...]}}]
                GraphicsRow[{Plot[f1...],...,Plot[fn...]}]
                GraphicsColumn[{Plot[f1...],...,Plot[fn...]}]
```

```
Plot[{Sin[x]/x,Cos[x]/x},{x,0,10*Pi},
 PlotRange->{{0,10*Pi},{-1,2}}]
GraphicsGrid[{{Plot[Cos[x],{x,-Pi,Pi}],Plot[Sin[x],
{x,-Pi,Pi}]}}]
GraphicsGrid[{{Plot[Cos[x],{x,-Pi,Pi}]},{Plot[Sin[x],
{x,-Pi,Pi}]}}]
GraphicsRow[{Plot[Cos[x],{x,-Pi,Pi}],Plot[Sin[x],{x,-Pi,Pi}]}]
GraphicsColumn[{Plot[Cos[x],{x,-Pi,Pi}],Plot[Sin[x],
{x,-Pi,Pi}]}]
```

Merging various saved graphic objects, an array of graphic objects.

Maple:

```
with(plots):              L1:=[G1,...,Gn]:        L2:=[H1,...,Hn]:
S1:={G1,...,Gn}:          S2:={H1,...,Hn}:   display(L1,x=a..b);
display(S1,x=a..b);  display3d(L2,x=a..b);  display(S2,x=a..b);
G:=array(1..n): G[i]:=plot(fun[i],x=a..b):          display(G);
```

Here L1,L2 and S1,S2 are the lists and sets of saved 2D and 3D graphs, respectively.

```
with(plots);  f:=x->abs(sin(x));  g:=x->-cos(x);
G1:=plot(f(x),x=-Pi..Pi): G2:=plot(g(x),x=-Pi..Pi):
display({G1,G2},title="f(x) and g(x)");
G:=array(1..2);  G[1]:=plot(sin(x),x=-Pi..Pi):
G[2]:=plot(cos(x),x=0..Pi): display(G);
```

Mathematica:

```
Show[{g1,...,gn}]
                    GraphicsGrid[{{g11,...,g1m},...{gn1,...,gnm}}]
GraphicsRow[{g1,...,gn}]                    GraphicsColumn[{g1,...,gn}]
ListPlot[{list1,list2...}]         ListLinePlot[{list1,list2...}]
```

```
f1[x_]:=Abs[Sin[x]]; f2[x_]:=-Cos[x];
g1=Plot[f1[x],{x,-Pi,Pi}]; g2=Plot[f2[x],{x,-2*Pi,2*Pi}];
Show[{g1,g2},Frame->True,PlotLabel->"f1 and f2",
 AspectRatio->1,PlotRange->All]
GraphicsGrid[{{g1,g2}},Frame->True,AspectRatio->1]
GraphicsGrid[{{g1},{g2}},Frame->True,Frame->All]
GraphicsRow[{g1,g2},Background->RGBColor[0.5,1.,1.]]
GraphicsColumn[{g1,g2},Alignment->Top]
list1=Table[{i,i^3},{i,-5,5}]; list2=Table[{i,i^2},{i,-5,5}];
g11=ListPlot[{list1,list2}]; g12=ListLinePlot[{list1,list2}];
GraphicsColumn[{g11,g12}]
g3=Plot3D[2*x^2-3*y^2,{x,-1,1},{y,-1,1}]; g4=Plot3D[3*x+y,
 {x,-1,1},{y,-1,1}]; Show[g3,g4,BoxRatios->{1,1,2}]
```

Problem 3.1 Graph the stability diagram of the Mathieu differential equation $x'' + [a - 2q\cos(2t)]x = 0$ in the (a, q)-plane (see Sect. 9.4).

Maple:

```
with(plots):
S1:=[MathieuA(i,q) $ i=0..5]; S2:=[MathieuB(i,q) $ i=1..6];
G1:=plot(S1,q=0..35,-20..45,color=red):
G2:=plot(S2,q=0..35,-20..45,color=blue):
display({G1,G2},thickness=3);
```

Mathematica:

```
s1=Table[MathieuCharacteristicA[i,q],{i,0,5}]
s2=Table[MathieuCharacteristicB[i,q],{i,1,6}]
g1=Plot[Evaluate[s1],{q,0,35},PlotStyle->Red];
g2=Plot[Evaluate[s2],{q,0,35},PlotStyle->Blue];
Show[{g1,g2},PlotRange->{{0,35},{-20,45}},Frame->True,
 PlotLabel->"Stability Regions",Axes->False,AspectRatio->1]
```

□

3.4   Text in Graphs

Drawing text strings on 2D and 3D graphs.

Maple:

```
textplot([[x1,y1,str1],..,[xn,yn,strn]],ops);
        textplot3d([[x1,y1,z1,str1],..,[xn,yn,zn,strn]],ops);
```

```
with(plots): f:=x->4*x^3+6*x^2-9*x+2;
G1:=plot([f(x),D(f)(x),(D@@2)(f)(x)],x=-3..3):
G2:=textplot([1.2,100,"f(x) and derivatives"],
    font=[HELVETICA,BOLD,13],color=plum): display([G1,G2]);
P1:=plot3d(exp(-(x^2+y^2)),x=-6..6,y=-6..6,grid=[25,25]):
P2:=plot3d(-5-4*sin(sqrt(x^2+y^2)),x=-6..6,y=-6..6):
P3:=textplot3d([1,1,2,"a"],font=[SYMBOL,25],color=blue):
display3d([P1,P2,P3],orientation=[34,79]);
```

Mathematica:

```
Graphics[Text["string",{xt,yt}],BaseStyle->val,ops]
        Graphics3D[Text["string",{xt,yt,zt}],FormatType->val]
        Text[Grid[{{"s11",...,"s1m"},...,{"sn1",...,"snm"}}]]
```

```
f:=4*x^3+6*x^2-9*x+2; fD1=D[f,x]; fD2=D[f,{x,2}];
g11=Plot[{f,fD1,fD2},{x,-3,3},PlotStyle->{Red,Blue,Green},
 BaseStyle->{FontFamily->"Helvetica",FontSlant->"Italic",
 FontWeight->"Bold",FontSize->15}];
g12=Graphics[Text["f[x] and derivatives",{0,50}]];
Show[{g11,g12},Frame->True,Axes->False,PlotRange->All]
g31=Plot3D[Exp[-(x^2+y^2)],{x,-6,6},{y,-6,6},BaseStyle->
 {FontFamily->"Symbol",FontSize->25},PlotRange->All];
g32=Plot3D[-5-4*Sin[Sqrt[x^2+y^2]],{x,-6,6},{y,-6,6},
 PlotRange->All]; g33=Graphics3D[Text["a",{0,9,0}]];
Show[{g31,g32,g33}]
g41=Graphics3D[Table[With[{p={i,j,k}/3},{RGBColor[p*3],
 Cuboid[p,p+.32]}],{i,3},{j,3},{k,3}],BaseStyle->
 {FontFamily->"Ariel",FontSize->25}]; g42=Graphics3D[
 Text["Rubik's Cube",{1,1,1}]]; Show[{g41,g42}]
g43=Graphics3D[Text[Grid[{{"X","O","X"},{"O","X","X"},
 {"O","O","X"}}],{1,1,1}]]; Show[{g41, g43}]
```

3.5   Special Graphs

Grid lines for 2D graphs

Maple:

```
plot(f(x),x=x1..x2,gridlines=true);
with(plots):         conformal(z,z=z1..z2,ops):
                     coordplot(coordsystem,[xrange,yrange],ops);
```

Mathematica:

```
Plot[f[x],{x,x1,x2},GridLines->{{xGL},{yGL}},GridLinesStyle->s]
```

Problem 3.2 Plot $f(x) = x\sin(1/x)$ and $g(x) = \dfrac{x^2 - x + 1}{x^2 + x - 1}$ together with the corresponding grid lines.

Maple:

```
with(plots): A:=4: f:=x->sin(1/x)*x;
plot(f(x),x=-Pi/2..Pi/2,color=blue,thickness=3,gridlines=true):
G1:=plot(f(x),x=-Pi/2..Pi/2,color=blue,thickness=3):
M1:=conformal(z,z=-A-I..A+I,grid=[20,10],color=grey):
display([G1,M1]); g:=x->(x^2-x+1)/(x^2+x-1); R:=-5..5;
G2:=plot(g(x),x=R,R,discont=true,thickness=3):
M2:=coordplot(cartesian,[R,R],view=[R,R],grid=[10,10],
  color=[grey,grey]): display([G2,M2],axes=boxed,
  scaling=constrained,xtickmarks=5,ytickmarks=5);
```

Mathematica:

```
f[x_]:=x*Sin[1/x]; g[x_]:=(x^2-x+1)/(x^2+x-1);
Plot[f[x],{x,-Pi/2,Pi/2},Frame->True,Axes->False,PlotStyle->
  {Blue,Thickness[0.01]},GridLines->{Automatic,Automatic}]
Plot[g[x],{x,-5,5},Frame->True,Axes->False,PlotStyle->
  {Blue,Thickness[0.007]},GridLines->{Automatic,Automatic}]
Plot[g[x],{x,-5,5},Frame->True,PlotStyle->{Blue,Thickness[0.01]},
  GridLines->{Automatic,Automatic},Exclusions->{x^2+x-1==0}]
```

□

Bounded regions for 2D graphs

Maple:

```
plot(f(x),x=a..b,filled=true);
with(plots):                    inequal(ineqs,x=a..b,y=c..d,ops);
```

More information about the function `inequal` see `?plot[options]`.

Mathematica:

```
Plot[f1[x],...,{x,x1,x2},Filling->val]
        RegionPlot[Ineq,{x,x1,x2},{y,y1,y2}]
        RegionPlot3D[Ineq,{x,x1,x2},{y,y1,y2},{z,z1,z2}]
```

Here `value` can take various forms, e.g. `Top, Bottom, Axis, {i}, {i->{j}}`, etc. Various bounded regions for 2D (and 3D) graphs can be constructed with this new functions `RegionPlot`, `RegionPlot3D`, and the new option `Filling`.

```
Plot[{Cos[x],Cos[2*x],Cos[3*x]},{x,-Pi/2, Pi/2}]
Plot[{Cos[x],Cos[2*x],Cos[3*x]},{x,-Pi/2,Pi/2},
 Filling->{1->{{2},GrayLevel[0.6]},{2->{{3},Red}}}]
```

Problem 3.3 Plot the region bounded by $f(x) = -(x+1)^2 + 10$ and the axis $x = 0$.

Maple:

```
with(plots): f:=x->-(x+1)^2+10: S:=[fsolve(f(x)=0,x)];
f1:=plot([f(x),0],x=-4.5..4.5,-2..10,thickness=3):
f2:=plot(f(x),x=S[1]..S[2],filled=true,color=grey):
display([f1,f2]);
```

Mathematica:

```
f[x_]:=-(x+1)^2+10; S=NSolve[f[x]==0,x]
Plot[{f[x],0},{x,S[[1,1,2]],S[[2,1,2]]},PlotStyle->
 {{Red,Thickness[0.01]},{Green,Thickness[0.01]}},
 Filling->{1->{0,GrayLevel[0.6]}}]
```

□

Problem 3.4 Plot the region that satisfies the inequality $2x - 2y > 1$.

Maple:

```
with(plots): Ineq:=x->2*x-2*y>1; A:=(color=blue);
B:=(color=grey); C:=(color=green,thickness=10);
inequal(Ineq(x),x=-2..2, y=-2..2,
    optionsfeasible=A,optionsexcluded=B,optionsopen=C);
```

Mathematica:

```
RegionPlot[2*x-2*y>1,{x,-2,2},{y,-2,2},BoundaryStyle->
 {Green,Thickness[0.02]},PlotStyle->{Blue},
 Background->GrayLevel[0.6]]
```

□

Logarithmic plots in the plane

Maple:

```
with(plots):                                    logplot(f,range,ops);
semilogplot(f,range,ops);                       loglogplot(f,range,ops);
```

```
with(plots); with(stats):
al:=stats[random, normald](20);
Points:=[seq([0.2*i,exp(0.1*i)+0.1*al[i]],i=1..20)];
G1:=logplot(Points, style=point, color=green):
G2:=logplot(x+sin(x),x=0.5..3,style=line,color=red):
display({G1,G2}); f:=x->x^5+exp(-x^5);
loglogplot(f(x),x=0.1..100,axes=boxed);
```

Mathematica:

```
LogPlot[f,{x,x1,x2}]                LogPlot[{f1,...,fn},{x,x1,x2}]
LogLogPlot[f,{x,x1,x2}]        LogLogPlot[{f1,...,fn},{x,x1,x2}]
LogLinearPlot[f,{x,x1,x2}]LogLinearPlot[{f1,...,fn},{x,x1,x2}]
ListLogPlot[{{x1,y1},{x2,y2},...}]      ListLogPlot[{l1,...,ln}]
ListLogLinearPlot[{{x1,y1},...}]    ListLogLinearPlot[{l1,...}]
ListLogLogPlot[{{x1,y1},{x2,y2},...}] ListLogLogPlot[{l1,...}]
```

```
al=Table[Random[NormalDistribution[]],{20}]
points=Table[{0.2*i,Exp[0.1*i]+0.1*al[[i]]},{i,1,20}]
g1=ListLogPlot[points,PlotStyle->{PointSize[0.02],
 Hue[0.7]}]; g2=LogPlot[x+Sin[x],{x,0.5,3},PlotStyle->Red];
f[x_]:=x^5+Exp[-x^5]; Show[{g1,g2}]
LogLogPlot[Evaluate[N[f[x],30]],{x,0.1,100}]
```

Plots of piecewise functions

Maple:

```
f:=proc(x)  if cond1  then expr1  else expr2 end if;  end proc;
plot('f(x)',x=a..b,ops);                        plot(f,a..b,ops);
with(plots):    f:=x->piecewise(cond1,expr1,cond2,expr2,expr3);
                                     plot(f(x),x=a..b,ops);
```

```
with(plots): setoptions(thickness=5);
f:=proc(x) if x<0 then 0 elif x<1 then x else 1 fi; end;
plot('f(x)',x=-2..2); plot(f,-2..2);
g:=x->piecewise(x<0,0,x<1,x,1);    plot(g(x),x=-2..2);
```

Mathematica:

```
f1[x_]:=var1/;cond1; ... fn[x_]:=varn/;condn; ...
f2[x_]:=Piecewise[{{val1,cond1},...,{valn,condn}}];
f3[x_]:=UnitStep[x];        Plot[{f1[x],f2[x],f3[x]},{x,a,b}]
```

```
f1[x_]:=0/;x<0; f1[x_]:=x/;0<=x<=1; f1[x_]:=1/;x>1;
f2[x_]:=Piecewise[{{0,x<0},{x,0<=x<=1},{1,x>1}}];
f3[x_]:=x*UnitStep[x]-x*UnitStep[x-1]+UnitStep[x-1];
GraphicsColumn[{Plot[f1[x],{x,-2,2},PlotStyle->Hue[0.5]],
 Plot[f2[x],{x,-2,2},PlotStyle->Hue[0.7]],
 Plot[f3[x],{x,-2,2},PlotStyle->Hue[0.9]]}]
```

Density plots

A density plot of the function $f(x,y)$, $(x,y) \in [x_1,x_2] \times [y_1,y_2]$.

Maple:

```
with(plots):          densityplot(f(x,y),x=x1..x2,y=y1..y2,ops);
```

Mathematica:

```
DensityPlot[f[x,y],{x,x1,x2},{y,y1,y2},ops]
        ListDensityPlot[{{a11,...,a1n},...{an1,...,ann}},ops]
```

Problem 3.5 Construct the density plot of $f(x,y) = xe^{-x^2-y^2}$, $(x,y) \in$ $[-2,2] \times [-2,2]$, with color gradient and the corresponding legend.

Maple:

```
with(plots):
A:=colorstyle=HUE,style=patchnogrid,numpoints=5000,axes=boxed;
G1:=densityplot((x,y)->x*exp(-x^2-y^2),-2..2,-2..2,A):
G2:=densityplot((x,y)->0.2*y,3..3.5,-2..2,A):
G3:=textplot([seq([3.8,-1.95+i/8*3.9,
 sprintf("%.1f",-0.4+i/10)],i=0..8)]):
display({G1,G2,G3},scaling=constrained);
```

Mathematica:

```
f1[x_,y_]:=x*Exp[-x^2-y^2]; f2[x_,y_]:=0.2*y;
g1=DensityPlot[Evaluate[f1[x,y]],{x,-2,2},{y,-2,2},
 ColorFunction->Hue,PlotPoints->100,PlotRange->All];
g2=DensityPlot[Evaluate[f2[x,y]],{x,3,3.5},{y,-2,2},
 ColorFunction->Hue,PlotPoints->100,AspectRatio->Automatic,
 PlotRange->All]; g3=Graphics[Table[Text[StyleForm[
 NumberForm[N[-0.4+i/10],2],FontSize->10],{2.7,-1.95+i/8*3.9}],
 {i,0,8}],PlotRange->All]; Show[{g1,g2,g3}]
list1=Table[Random[Real,{1,10}],{x,1,10},{y,1,10}];
ListDensityPlot[list1,PlotRange->{{1,8},{1,8}}]
```

□

Bar graphs: different types of 2D and 3D bar graphs

Maple:

```
with(stats): with(stats[statplots]):          histogram(data,ops);
with(Statistics):                      Histogram([data1,...],ops);
BarChart([data1,...,],ops);            ColumnGraph([data1,...],ops);
LineChart([data1,...],ops);            PieChart([data1,...],ops);
                                       PieChart['interactive'](data);
```

Mathematica:

```
<<BarCharts`                    <<PieCharts`           <<Histograms`
BarChart[{data1,...},ops]        StackedBarChart[{data1,...},ops]
PercentileBarChart[{data1,...},ops]      PieChart[{y1,...},ops]
Histogram[{x1,...},ops]    GeneralizedBarChart[{data1,...},ops]
                    BarChart3D[{{z11,...},{z21,...},...},ops]
Histogram3D[{{x1,y1},...},ops]
```

Problem 3.6 Compare different bar graphs.

Maple:

```
with(stats): with(stats[statplots]):
list1:=[random[normald](200)]:
histogram(list1); histogram(list1,area=count);
histogram(list1,color=blue);
with(Statistics): NX:=RandomVariable(Normal(0,1));
G1:=DensityPlot(NX,range=-3..3,thickness=3,color=red):
list1:=Sample(Normal(0,1),1500);
G2:=Histogram(list1,range=-3..3,color=blue):
plots[display]({G1,G2});
list2:=[i^2 $ i=1..5]; PieChart(list2,sector=0..360,
 color=[blue,red,green,yellow,white]);
BarChart(list2); ColumnGraph(list2); LineChart(list2);
```

Mathematica:

```
<<BarCharts`; <<PieCharts`; <<Histograms`;
list1=Table[Random[Integer,{1,5}],{i,1,5}];
list2=Table[i^2,{i,1,5}]; list3={"a1","a2","a3","a4","a5"};
g1=BarChart[list1,BarStyle->{Hue[0.7]},BarEdges->True,
 BarOrientation->Horizontal]; g2=BarChart[{list1,list2},
 BarStyle->{Hue[0.5],Hue[0.8]}]; g3=StackedBarChart[
 {list1,list2},BarStyle->{Hue[0.5],Hue[0.8]}];
g4=PercentileBarChart[{list1,list2},BarLabels->list3,
 BarStyle->{Hue[0.5],Hue[0.6]},BarSpacing->0.2];
g5=PieChart[list1,PieLabels->list3,PieExploded->All,
 PieStyle->{Hue[0.5],Hue[0.6],Hue[0.75],Hue[0.85],Hue[0.9]}];
g6=PieChart[list1,PieLabels->list3,PieExploded->{1},
 PieStyle->{Hue[0.5],Hue[0.6],Hue[0.75],Hue[0.85],Hue[0.9]}];
GraphicsGrid[{{g1,g2},{g3,g4},{g5,g6}}]
```

```
list4={{1,2,3},{6,5,4},{9,10,11}}; BarChart3D[list4,
  AxesLabel->{"x","y","z"},BarSpacing->{0.3,0.3},
  BarStyle->Hue[Random[]]]
n=9; list5=Table[PDF[NormalDistribution[n,3],x],{x,0,2*n}];
list6=Table[Hue[0.85],{2*n+1}]; list6[[10]]=Blue;
BarChart[list5,BarLabels->Range[0,2*n],BarStyle->list6]
```

In *Mathematica 7*, several new functions for automated dynamic chart-
ing have been developed and added to the *Mathematica* kernel, e.g.
`BarChart`, `PieChart`, `BubbleChart`, `SectorChart`, `BarChart3D`, `PieChart3D`,
`Histogram`, `Histogram3D`, etc. The packages `BarCharts`, `PieCharts`, and
`Histograms` are no longer needed. We mention here some examples:

```
f[{{x1_,x2_},{y1_,y2_}},___]:=Rectangle[{x1,y1},
  {x2,y2}]; l1=Table[Random[Integer,{1,5}],{i,1,5}];
l2=Table[i^2,{i,1,5}]; BarChart[{l1,l2},
  ChartElementFunction->f, ChartLegends->{"A","B"}]
PieChart[l1,ChartElementFunction->"GlassSector",
  ChartStyle->"Pastel"]
PieChart3D[l1,ChartStyle->"Pastel"]
Histogram[RandomReal[NormalDistribution[0,1],1000]]
```

                                                                            □

3.6   Animations

2D and 3D animations of functions

Maple:

```
with(plots):                       animatecurve(f(x),x=a..b,ops);
                              animate(f(x,t),x=a..b,t=t1..t2,ops);
                     animate3d(f(x,y,t),x=a..b,y=c..d,t=t1..t2,ops);
display([G1,...,GN],insequence=true);
display3d([G1,...,GN],insequence=true);
```

Mathematica:

2D and 3D animations of plot sequences can be produced using the
new function `Animate`, where the speed, the direction, and the pause
of animation can be controlled by the corresponding buttons (also see
Sect. 2.3.7).

```
Animate[exprPlot,{t,t1,t2,tStep},ops]
```

```
f[x_,t_]:=Sin[x+t]+Sin[x-2*t];
Animate[Plot[Sin[-(x-t)^2],{x,-2,2},PlotRange->{-1,1}],
  {t,0.2,0.5}]
Animate[Plot[f[x,t],{x,-10,10},PlotRange->{-10,10}],
  {t,0,30,0.001}]
Animate[Plot3D[Sin[x-y-t],{x,0,Pi},{y,0,Pi},PlotRange->All],
  {t,0,2*Pi}]
Animate[Plot[x*Sin[x*t],{x,-2*Pi,0},PlotRange->{-6,6}],
  {t,0.2,0.5}]
Animate[Plot[Sin[x*i],{x,0,Pi},PlotRange->{-1,1}],{i,1,10}]
```

Problem 3.7 Show that the solutions $u_1(x,t) = \cos(x-2t)$ and $u_2(x,t) = \cos(x+2t)$, of the wave equation are traveling waves.

Maple:

```
with(plots): N:=200;  A:=array(1..2):
u1:=(x,t)->cos(x-2*t); u2:=(x,t)->cos(x+2*t);
setoptions(thickness=3,scaling=constrained,axes=none);
A[1]:=animate(u1(x,t),x=0..4*Pi,t=1..10,frames=N):
A[2]:=animate(u2(x,t),x=Pi/2..9*Pi/2,t=1..10,
      color=green,frames=N):  display(A);
```

Mathematica:

```
u1[x_,t_]:=Cos[x-2*t]; u2[x_,t_]:=Cos[x+2*t];
SetOptions[Plot,Axes->None,Frame->False,
 PlotStyle->Thickness[0.02],AspectRatio->1];
Animate[GraphicsRow[{Plot[u1[x,t],{x,0,4*Pi},PlotStyle->Red],
 Plot[u2[x,t],{x,Pi/2,9*Pi/2},PlotStyle->Blue]}],{t,1,5}]
```

□

Problem 3.8 Observe the Lissajous curves in polar coordinates.

Maple:

```
with(plots):
animatecurve([sin(7*x),cos(11*x),x=0..2*Pi],coords=polar,
   numpoints=300,frames=300,color=blue,thickness=3);
```

Mathematica:

```
m=70; n=3; g={};
l1=Table[{Sin[n*x],Cos[(n+2)*x]},{x,0,2*Pi,1/m}]//N;
k=Length[l1]; Do[g=Append[g,Evaluate[ListPolarPlot[
 Take[l1,{1,i}],PlotMarkers->{"o",10},
 PlotRange->{{-1,1},{-1,1}},AspectRatio->1,
 PlotStyle->Blue]]],{i,1,k}]; ListAnimate[g]
Animate[PolarPlot[{Sin[Prime[n]*x],Cos[Prime[n+1]*x]},
 {x,0,2*Pi},PlotPoints->40,PlotStyle->Blue,Frame->True,
 FrameTicks->False,MaxRecursion->3],{n,1,20,2},
 AnimationRate->2]
```

In *Mathematica* (*ver* $\geq$ 6), some of the functions are located in the *Wolfram Research Website* and are available by downloading the corresponding packages from it. Here we put one example of the function MovieParametricPlot that is available in the legacy standard Add-On package Animation (for *ver* = 5.2 and updated for *ver* = 6). This package can be downloaded from the *Wolfram Research Website* according to the corresponding path in your computer, e.g. Mathematica $\rightarrow$ 6.0 $\rightarrow$ AddOns $\rightarrow$ LegacyPackages $\rightarrow$ Graphics. The first line in the solution we recommend for turning off the obsolescence messages.

```
Off[General::obspkg]; Off[General::newpkg]; Off[General::wrsym];
<<Graphics`Animation`
MovieParametricPlot[{Sin[n*t],Cos[(n+2)*t]},{t,0,2*Pi},
 {n,1,5,2},PlotStyle->Blue,Frame->True,FrameTicks->False]
```

$\square$

Problem 3.9 Animations of two sequences of points in 2D.

Maple:

```
with(plots): n:=100: G:=[]:
L1:=[seq([cos(j*Pi/n),sin(j*Pi/n)],j=0..n)]:
L2:=[seq([-cos(j*Pi/n),-sin(j*Pi/n)],j=0..n)]:
for i from 1 to n do
 G:=[op(G),plot([L1[1..i],L2[1..i]], x=-1..1,y=-1..1,
     symbol=circle,style=point,color=[blue,red])]: od:
display(G,insequence=true);
```

Mathematica:

```
n=100; g={};
l1=Table[{Cos[j*Pi/n],Sin[j*Pi/n]},{j,0,n}]//N;
l2=Table[{-Cos[j*Pi/n],-Sin[j*Pi/n]},{j,0,n}]//N;
Do[g=Append[g,Evaluate[ListPlot[{Take[l1,{1,i}],
 Take[l2,{1,i}]},PlotMarkers->{"o",9},AspectRatio->1,
 PlotStyle->{Blue,Red},PlotRange->{-1,1}]]],{i,1,n}];
ListAnimate[g]
```

☐

Problem 3.10 Animations of two sequences of points in 3D.

Maple:

```
with(plots): n:=250: G:=[]: k:=20:
L1:=[seq([cos(k*j*Pi/n),j,sin(k*j*Pi/n)],j=0..n)]:
L2:=[seq([-cos(k*j*Pi/n),-j,-sin(k*j*Pi/n)],j=0..n)]:
for i from 1 to n do
  G:=[op(G),spacecurve({L1[1..i],L2[1..i]},style=line,
     axes=none,thickness=3,shading=zhue)]: od:
display3d(G,insequence=true);
```

Mathematica:

```
n=25; m=150; g={}; k=20;
Animate[ParametricPlot3D[{{Cos[k*j*t/m],t*j/m,Sin[k*t*j/m]},
 {-Cos[k*j*t/m],-t*j/m,-Sin[k*j*t/m]}},{t,0,2*Pi},
 PlotPoints->70,MaxRecursion->1,Boxed->False,Axes->False,
 ColorFunction->Function[{x,y},Hue[x]],PlotStyle->
 {{Thickness[0.02]},{Thickness[0.02]}},PlotRange->
 {{-1,1},{-1,1},{-1,1}}],{j,0,n},AnimationRate->0.9]
n=250; g={}; k=20;
l1=Table[{Cos[k*j*Pi/n],j/n,Sin[k*j*Pi/n]},{j,0,n}]//N;
l2=Table[{-Cos[k*j*Pi/n],-j/n,-Sin[k*j*Pi/n]},{j,0,n}]//N;
Do[g=Append[g,Evaluate[GraphicsRow[{ListPointPlot3D[
 Take[l1,{1,i}],ColorFunction->Function[{x,y,z},Hue[z]],
 Axes->False,BoxRatios->{1,1,1},Boxed->False,
 PlotStyle->{{PointSize[0.05]},{PointSize[0.05]}},
 PlotRange->{{-1,1},{-1,1},{-1,1}}],ListPointPlot3D[
 Take[l2,{1,i}],ColorFunction->Function[{x,y,z},Hue[z]],
 Axes->False,BoxRatios->{1,1,1},Boxed->False,
 PlotStyle->{{PointSize[0.05]},{PointSize[0.05]}},
 PlotRange->{{-1,1},{-1,1},{-1,1}}]}]]],{i,1,n}];
ListAnimate[g]
```

☐

Problem 3.11 Animations of two structures in 3D.

Maple:

```
with(plots):  n:=40: G:=NULL:
f1:=(x,y)->(x^2-y^2)/9-9: f2:=(x,y)->(x^2-y^2)/9:
G1:=plot3d(f1(x,y),x=-9..9,y=-9..9,style=patchnogrid):
G2:=plot3d(f2(x,y),x=-9..9,y=-9..9,style=patchnogrid):
for i from 1 to n do
 G:=G,display3d([G1,G2],orientation=[i*180/n,50]): od:
display3d([G],scaling=constrained,insequence=true);
```

Mathematica:

```
n=30; g={}; f1[x_,y_]:=(x^2-y^2)/9-9; f2[x_,y_]:=(x^2-y^2)/9;
g1=Plot3D[f1[x,y],{x,-9,9},{y,-9,9},Mesh->False];
g2=Plot3D[f2[x,y],{x,-9,9},{y,-9,9},Mesh->False];
Do[g=Append[g,Show[{g1,g2},SphericalRegion->True,
 ViewCenter->{0.5,0.5,0.5},ViewPoint->{i/10,2,2},
 Boxed->False,Axes->False,PlotRange->{{-10,10},
 {-10,10},{-10,10}}]],{i,-n,n}];
ListAnimate[g,AnimationRate->4]
```

☐

Problem 3.12 Observe the motion of a point rolling along a curve.

Maple:

```
with(plots):
G:=plot(cos(x),x=-Pi..Pi,axes=boxed,thickness=2,color=blue):
animate(pointplot,[[[t,cos(t)]],symbol=circle,symbolsize=20,
        color=red],t=-Pi..Pi,frames=100,background=G);
```

Mathematica:

```
g=Plot[Cos[x],{x,-Pi,Pi},AspectRatio->1,Frame->True,
  PlotStyle->{Blue,Thickness[0.02]}];
Animate[Show[g,Graphics[{Purple,Disk[{x,Cos[x]},0.1]}],
  PlotRange->{{-Pi,Pi},{-1.5,1.5}}],{x,-Pi,Pi}]
```

☐

Chapter 4

Algebra

Algebra (or abstract algebra) is one of the important areas of the mathematics that studies various *algebraic structures*, such as sets, groups, fields, spaces and other.

4.1 Finite Sets

Any algebraic structure can be described as a set or a collection of distinct objects and a finite number of operations that can act on them.

Maple:

The three concepts are defined: *sequences, lists, and sets.*

```
Sequence1:=a1,a2,a3,a4,a5;          Sequence2:=seq(a||i,i=6..10);
List1:=[Sequence1];                            Set1:={Sequence1};
```

Note. The order and repetition of entries are preserved in sequences and lists and are not preserved in sets. The order in the sets can be changed within a *Maple* session.

Mathematica:

A single concept, a *list* is defined, that is an ordered set of objects separated by commas and enclosed in braces {elements}, or can be defined with the function List[elements].

```
{list1={a1,a2,a3,a4,a5},          list2=List[a6,a7,a8,a9,a10],
 set1=list1,                      seq1=Sequence[x,y], f[seq1]}
```

Note. The *list* is a fundamental concept in *Mathematica*. Others concepts (for example, sets, matrices, tables, vectors, arrays, tensors, etc.) are represented as lists (see Sect. 2.3.6). A sequence of arguments for any function is defined with **Sequence**.

I.K. Shingareva, C. Lizárraga-Celaya, *Maple and Mathematica*, 2nd ed.,
DOI 10.1007/978-3-211-99432-0_4, © Springer-Verlag Vienna 2009

Problem 4.1 Let A be the set of elements $\{m, a, t, h, e, m, a, t, i, c, a\}$. Show that the number of all subsets of A is equal to 128 and select all subsets that have a single element.

Maple:

```
with(combinat): A:={m,a,t,h,e,m,a,t,i,c,a}; B:=subsets(A):
i1:=1; while not B[finished] do
 printf("%10d,%20a\n",i1,B[nextvalue]()); i1:=i1+1; end:
choose(A,1);
```

Mathematica:

```
A={m,a,t,h,e,i,c};{B=Subsets[A],Length[B],Subsets[A,{1}]}
```

Some of the letters are repeated and, therefore, the set $A = \{m, a, t, h, e, i, c\}$ has 7 different elements. First, A is a subset of A. There exist 7 subsets of A that have 6 elements:

$$\{a, c, e, h, i, m\}, \{a, c, e, h, i, t\}, \{a, c, e, h, m, t\}, \{a, c, e, i, m, t\},$$
$$\{a, c, h, i, m, t\}, \{a, e, h, i, m, t\}, \{c, e, h, i, m, t\}.$$

There are 21 subsets of A that have 5 elements, 35 subsets that have 4 elements, 35 subsets that have 3 elements, 21 subsets that have 2 elements, 7 subsets that have one element, and a special set that has no elements, *empty set*, thus making up 128 elements in total. All the subsets that have a single element: $\{a\}, \{c\}, \{e\}, \{h\}, \{i\}, \{m\}, \{t\}$.   □

Problem 4.2 Let $A = \{a_1, a_2, \ldots, a_n\}$ be the set that consists of n different elements. Verify that A has 2^n different subsets.

Maple:

```
1+sum(binomial(n,k),k=1..n);
```

Mathematica:

```
1+Sum[Binomial[n,k], {k,1,n}]
```

The number of subsets of k $(k \geq 1)$ different elements is the number of ways to select k elements from n elements and is the binomial coefficient $\binom{n}{k}$. Therefore, the total number of different subsets (including the empty set $\emptyset$) is $1 + \sum_{k=1}^{n} \binom{n}{k} = 2^n$.   □

Problem 4.3 *Application of set theory to tonal music.* We write tonal music in the mathematical language. We consider a melody for one musical instrument (piano), for example, of Sonata in C-major KV545 by W.A. Mozart (Vienna, 1788):

Let A_T, F_M, and D_M denote the main sets determined by the allocation, respectively, of pitch, musical figures (or musical duration), and musical dynamics of a note in a musical composition for one tonal instrument (with its corresponding timbre). Given the lengths n, m, k of the three sets, we can write $A_T = \{a \mid 1 \leq a \leq n, \ a, n \in \mathbb{N}\}$, $F_M = \{f \mid 1 \leq f \leq m, \ f, m \in \mathbb{N}\}$, and $D_M = \{d \mid 1 \leq d \leq k, \ d, k \in \mathbb{N}\}$. Construct two important sets in the area of musical composition: the set C of all possible musical compositions and a set M representing a particular melody (of our problem). So we have $C = A_T \times F_M \times D_M$ and $M = \{(a, f, d) \mid a \in A_T \wedge f \in F_M \wedge d \in D_M\}$, $M \subseteq C$.

Maple:

```
with(combinat): A_T:={i $ i=1..89}; F_M:={i $ i=1..20};
D_M:={i $i=1..8}; C:=cartprod([A_T,F_M,D_M]): k:=0:
while not C[finished] do C[nextvalue](); k:=k+1: end do: k;
```

Mathematica:

```
<<Combinatorica`
{aT=Table[i,{i,1,89}], fM=Table[i,{i,1,20}],dM=Table[i,{i,1,8}]}
{c=CartesianProduct[CartesianProduct[aT,fM],dM], Length[c]}
```

In order to construct A_T we consider the pitches that can be produced by a piano. Let $\{-4, -3, -2, -1, 0, 1, 2, 3, 4\}$ denote the set of all octaves. The pitches within each octave belong to the set

$$\{C, \ C\sharp/D\flat, \ D, \ D\sharp/E\flat, \ E, \ F, \ F\sharp/G\flat, \ G, \ G\sharp/A\flat, \ A, \ A\sharp/B\flat, \ B\},$$

where, e.g. $C\sharp/D\flat$ means the equivalence of pitches $C\sharp$ and $D\flat$. Therefore, we can describe all the pitches in the form of the following table (the equivalent pitches such as $C\sharp$ and $D\flat$ are excluded), for example, $A[-4] = 1$, $A\sharp[-4] = 2$, $B[-4] = 3$, $C[-3] = 4$, $\ldots$, $C[4] = 88$, Silence $= 89$

(for more details see Problem 4.22). We assume that the maximum number of the musical figures that describe the musical duration of a note is equal to 20, for example:

We form the set of the musical figures, where each musical figure has its natural number, $1, \ldots, 20$. Let the set of the musical dynamics of a note in a musical composition consist of 8 elements, for example, $\{ppp, pp, p, mp, mf, f, ff, fff\}$. Then we form the set $D_M = \{d \mid 1 \le d \le 8, d \in \mathbb{N}\}$, where $ppp = 1$, $pp = 2$, etc. As result we can, for example, find 14240 single-voice melodies. With this it is possible to form a huge set C. The set M of our problem has the form:

$$M = \{(52, 2, 4), (56, 3, 4), (59, 3, 4), (51, 17, 5), (52, 5, 4), (54, 5, 4), (52, 3, 3),$$
$$(89, 10, 1), (61, 2, 5), (59, 3, 4), (64, 3, 4), (59, 3, 4), (57, 20, 4), (56, 5, 3),$$
$$(57, 5, 3), (56, 3, 3), (89, 10, 1)\}.$$

□

4.2   Infinite Sets

Let us denote the main infinite sets as follows:
the set of *natural numbers* (or strictly positive integers), $\mathbb{N} = \{1, 2, \ldots\}$,
the set of *integers*,[1] $\mathbb{Z} = \{0, \pm 1, \pm 2, \ldots\} = \{x \mid x \text{ is an integer}\}$,
the set of *rational numbers*,[2] $\mathbb{Q} = \{x \mid x = p/q, \quad p, q \in \mathbb{Z}, q \ne 0\}$,
the set of *irrational numbers*, $\mathbb{I}$, consists of nonperiodic infinite decimal numbers that cannot be expressed as a fraction p/q ($p, q \in \mathbb{Z}, q \ne 0$),
the set of real numbers, $\mathbb{R} = \mathbb{Q} \cup \mathbb{I}$,
the set of *complex numbers*, $\mathbb{C} = \{z \mid z = a + ib, a, b \in \mathbb{R}, i^2 = -1\}$, an extension of the real numbers, used to represent all roots of polynomials.
the set of *algebraic numbers*, $\mathbb{A} = \{\alpha \in \mathbb{C} \mid P(\alpha) = a_0 + a_1 \alpha + \cdots + a_n \alpha^n, a_i \in \mathbb{Z}\}$ consists of roots of a nonzero polynomial $P(x)$ with integer coefficients.

We are interested in studying how these sets can be represented in *Maple* and *Mathematica*. In general, one of the advantages of computer algebra systems for scientific computation is the case in which

---

[1] $\mathbb{Z}$ for "Zahl", the German word for "number"
[2] $\mathbb{Q}$ for "Quotient", the English word

traditional approximate numerical computations can be combined with symbolic computations and with high-precision approximate numerical computations. Therefore, it is important to understand the following results for both systems *Maple* and *Mathematica*:

1. In general, computers represent a real number in the following form that is called *scientific notation*:

$$x=(-1)^s \cdot (0.a_1 a_2 \ldots a_t) \cdot b^E = (-1)^s \cdot M \cdot b^{E-t}, \; a_1 \neq 0, \; 0 \leq a_i \leq b-1, \quad (4.1)$$

where the *sign* $s = 0$ or 1, the *basis* $b \in \mathbb{N}$, $b \geq 2$, that admits a particular computer, the *mantissa* $M \in \mathbb{N}$, whose length is the maximum number of digits a_i and is equal to t, and the *exponent* $E \in \mathbb{N}$, $E_{\min} \leq E \leq E_{\max}$. The digits $a_1 a_2 \ldots a_q$ ($q \leq t$) are called the first q *significant digits* of x. The condition $a_1 \neq 0$ means that a number always has a unique representation.

2. *Maple* and *Mathematica* have *two floating-point number systems*: *hardware-floats*, which are the *binary IEEE double-precision* numbers supported by the computer architecture, and *software-floats*, which are the numbers that support *arbitrary precision arithmetic* (that depends on the computer characteristics) in exact and approximate computations.

3. In *Maple* and *Mathematica*, at any moment of time, the floating-point number systems are controlled automatically.

In *Maple*, the global variable `Digits` controls a number of decimal digits included in the mantissa of any floating-point number. *Maple* can handle a very large number of digits, the maximum number of digits in our computer, `Maple_floats(MAX_DIGITS)`, is 268435448, the predefined value is 10. The result of the function `D_H:=evalhf(Digits)` is the number of decimal digits that the corresponding hardware and software platform (in one particular computer) can reliably deliver to *Maple*, for example, in our computer (with operating system Windows) `D_H:=evalhf(Digits)=14`. In applications, e.g., for evaluating and plotting various functions, where `Digits` $\leq$ `evalhf(Digits)`, the computations can be performed as hardware-floats, since hardware floating-point arithmetic is usually much faster.

Mathematica, by definition, can handle any number of digits. The number of digits in the *hardware floating-point number system* (hardware-floats of double precision) in one computer is given by the value of variable $MachinePrecision. The number of digits in the *software floating-point number system* (software-floats of arbitrary precision) is essentially arbitrary. The practical limit is determined by the amount of memory in the computer and the time of calculations. In *Mathematica*, there exists several functions and options that control the precision and exactitude in various numerical operations, for example, Precision, Accuracy, SetPrecision, SetAccuracy, AccuracyGoal, WorkingPrecision. By definition, in *Mathematica* most of calculations are done using the hardware-floats or *machine numbers*. If for a given number a, the number of given digits is less than $MachinePrecision and the value of $a \in$ [$MinMachineNumber, $MaxMachineNumber], then *Mathematica* determines that the number is a *machine number*. In many *Mathematica* functions the number of digits is defined arbitrary and the numerical error will propagate correctly during arithmetical operations.

Let us consider in more detail the basic algebraic structures of modern mathematics, the sets of natural, integer, rational, irrational, complex, and algebraic numbers.

Natural, integer, and rational numbers.

We recall that one of the important features of symbolic mathematics is exact computations or computations with high (arbitrary) precision. Both computer algebra systems provide exact results in the arithmetical expressions with integers and automatically simplify rational numbers m/n $(m, n \in \mathbb{Z})$ so that numerator and denominator are relatively prime numbers, i.e. $\gcd(m, n) = 1$ (gcd is the *greatest common divisor*).

Maple:

```
n:=27;  80000^90000;  5^(6^7); -55;   Maple_floats(MAX_DIGITS);
kernelopts(maxdigits);  ithprime(n);  isprime(n); nextprime(n);
prevprime(n); ifactor(n);  r:=5/6;  numer(r);denom(r);gcd(5,6);
```

Mathematica:

```
{n=27,  80000^90000,  5^(6^7),  -55,  $MaxNumber,  Prime[n],
 PrimeQ[n],NextPrime[n],RandomPrime[{10,20}], FactorInteger[n],
 r=5/6, Rational[5,6], Numerator[r], Denominator[r],  GCD[5,6]}
```

Problem 4.4 Let primes belong to a given interval $[n_1, n_2]$. Generate a set of primes Sp and a set of random primes Srp of length N.

Maple:

```
with(RandomTools): randomize(): n1:=10; n2:=20; N:=7;
Srp:=NULL; f:=x->x^x: z1:=rand(3..109);
Sp:=(k1,k2)->select(isprime,[$k1..k2]); Sp(n1,n2);
AddFlavor(randprime=proc(k1,k2) local r1,r2:
 r1:=rand(k1..k2); r2:=nextprime(r1()); end proc):
SS:=x->SetState(state=f(x)*round(10^(z1())*time()));
for i from 1 to N do
 SS(i); Srp:=Srp,Generate(randprime(n1,n2)); od: Srp;
```

Mathematica:

```
SeedRandom[]; f[x_]:=x^x; {n1=10, n2=20, n=7, srp={},
 sp=Select[Range[n1,n2],PrimeQ]}
SS[x_]:=SeedRandom[f[x]*10^RandomInteger[{3,109}]];
Do[SS[i]; srp=Append[srp,RandomPrime[{n1,n2}]],{i,1,n}]; srp
```

Since in *Maple* there is no a direct random function for primes (as `RandomPrime` in *Mathematica*), we have to write a procedure that calls some random tools, then we add it to the random "flavors" as a new internal structure (the function `AddFlavor`) so that it can be simply called from the package `RandomTools`. In both systems, generating N primes, in each step we have to reset a "random" seed for the random number generator. □

Irrational and real numbers

If the result is an irrational number, the systems represent it in an unevaluated form. To obtain the value of such numbers we have to approximate using floating-point arithmetic. Real numbers are represented as floating-point numbers.

Maple:

```
Pi;      evalf(Pi);      evalf(Pi,50);    3.4; -2.3E-4; -2.3e4;
HFloat(23,-45);          SFloat(23,-45);          Float(23,-45);
```

Note. The functions `HFloat` (hardware-floats) and `SFloat`, `Float` (software-floats or *Maple*-floats) represent, respectively, the double-precision and arbitrary-precision real numbers. *Maple* uses a *decimal-based number system* for representing *software-floats*.

Mathematica:

```
{Pi,    Pi//N,    N[Pi,50], -2.3*10^(-4), ScientificForm[239.45],
FortranForm[2.0*10^(-50)],    Sin[1.5], Sin[1.5`20],
N[{Sin[1.5], Sin[1.5`20]}],  InputForm[{Sin[1.5],Sin[1.5`20]},
NumberMarks->False], FullForm[Sin[1.5`20],NumberMarks->False]}
```

Note. In *Mathematica*, the arbitrary-precision software-floats are represented in the following form: **number`nnn** (with a single opening apostrophe ` and an integer number **nnn**), the double-precision hardware-floats are represented in the form: **number**.

We know that there exist differences between the algebraic structures of real numbers and computer model numbers. These two classes of numbers are not isomorphic.

Maple and *Mathematica* have two floating-point number systems, double-precision hardware-floats and arbitrary-precision software-floats. Therefore the set $\mathbb{R}$ can be represented only as the subsets of finite dimension, $\mathbb{R}_H$, $\mathbb{R}_S$. Since any real number x is rounded or truncated by a computer, we can denote the new numbers, *floating-point numbers in both subsets*, $fl_H(x) \in \mathbb{R}_H$, $fl_S(x) \in \mathbb{R}_S$ that can not necessarily coincide with the original real number x.

Subsets $\mathbb{R}_H$ and $\mathbb{R}_S$ are characterized by properties that are different from those of the set $\mathbb{R}$. Since $\mathbb{R}_H$ and $\mathbb{R}_S$ are the proper subsets of $\mathbb{R}$, the elementary algebraic operations on floating-point numbers do not coincide with all the properties of similar operations on $\mathbb{R}$.

Since the set of real numbers has a field structure (the complete ordered field), we consider some important differences between $\mathbb{R}_H$, $\mathbb{R}_S$ and the field of real numbers $\mathbb{R}$:

1. Commutative laws are valid for addition and multiplication that we denote by $\circ$:

$$fl_H(a \circ b) = fl_H(b \circ a), \quad fl_S(a \circ b) = fl_S(b \circ a).$$

2. Associative laws are not valid:

$$fl_H(a) \circ fl_H(b \circ c) \neq fl_H(a \circ b) \circ fl_H(c),$$
$$fl_S(a) \circ fl_S(b \circ c) \neq fl_S(a \circ b) \circ fl_S(c).$$

This situation can occur, for example, if we add long signed numbers that cause overflow (or underflow), that is, whose values are greater (or less) than the maximum (or minimum) machine numbers admitted by the computer in the floating-point number systems.

In *Maple*, for the set $\mathbb{R}_H$, we have the following maximum and minimum machine numbers: $\mathrm{MaxR_H} = 0.99999999000000002 \cdot 10^{308}$, $\mathrm{MinR_H} = 0.10000001000000001 \cdot 10^{-306}$. Therefore, if we choose $a = 0.1 \cdot 10^{308}$, $b = 0.11 \cdot 10^{309}$, $c = -0.1001 \cdot 10^{309}$, we obtain $fl_H(a) + fl_H(b + c) = 0.19900000000000008 \cdot 10^{308}$, but $fl_H(a+b) + fl_H(c) =$ indefinite. For the set $\mathbb{R}_S$, the maximum and minimum machine numbers (in a specific computer): $\mathrm{MaxR_S} = 0.1 \cdot 10^{2147483647}$, $\mathrm{MinR_S} = 0.1 \cdot 10^{-2147483645}$. Similarly, if we choose $a = 0.9 \cdot 10^{2147483646}$, $b = 0.91 \cdot 10^{2147483647}$, $c = -0.1 \cdot 10^{2147483647}$, we obtain $fl_H(a) + fl_H(b + c) = \infty$, but $fl_H(a + b) + fl_H(c) = -0.1 \cdot 10^{2147483647}$.

```
useHardwareFloats:=true: Digits; MinRH:=evalhf(DBL_MIN);
MaxRH:=evalhf(DBL_MAX); a:= 0.1e+308; b:=0.11e+309;
c:=-0.1001e+309; HFloat(a)+HFloat(b+c); HFloat(a+b)+HFloat(c);
useHardwareFloats:=false: MaxDigits:=Maple_floats(MAX_DIGITS);
MaxRS:=Maple_floats(MAX_FLOAT); MinRS:=Maple_floats(MIN_FLOAT);
a:=0.9e+2147483646; b:=0.91e+2147483647; c:=-0.1e+2147483647;
SFloat(a)+SFloat(b+c); SFloat(a+b)+SFloat(c);
```

In *Mathematica*, for the set $\mathbb{R}_H$, the maximum and minimum machine numbers are: $\mathrm{MaxR_H} = 1.79769 \cdot 10^{308}$, $\mathrm{MinR_H} = 2.22507 \cdot 10^{-308}$. But if we choose $a > \mathrm{MaxR_H}$ or $b < \mathrm{MinR_H}$, *Mathematica* automatically converts the hardware-floats $\mathbb{R}_H$ to the software-floats $\mathbb{R}_S$. For the set $\mathbb{R}_S$, the maximum and minimum machine numbers are (in the same computer): $\mathrm{MaxR_S} = 1.920224672692357 \cdot 10^{646456887}$, $\mathrm{MinR_S} = 5.207723940958924 \cdot 10^{-646456888}$. If we choose $a = 1.920224672692357 \cdot 10^{646456887}$, $b = 1.920224672692357 \cdot 10^{646456887}$, $c = -1.920224672692357 \cdot 10^{646456887}$, we have $fl_S(a) + fl_S(b+c) = 1.920224672692357 \cdot 10^{646456887}$, but $fl_S(a + b) + fl_S(c) = $Overflow[].

```
Off[General::unfl]; Off[General::ovfl]; Off[General::stop];
{$MinMachineNumber,$MaxMachineNumber,$MaxNumber,$MinNumber}
N[Table[10.^j,{j,1,311}]]
Table[{$MinNumber/j,$MaxNumber*j},{j,1,10}]//TableForm
{a=$MaxNumber+0.1,b=$MaxNumber+0.11,c=-$MaxNumber+0.1001}
{a+(b+c),(a+b)+c}
```

3. Distributive laws are not valid:

$$fl_H(a+b)fl_H(c) \neq fl_H(ac)+fl_H(bc), \quad fl_S(a+b)fl_S(c) \neq fl_S(ac)+fl_S(bc),$$
$$fl_H(c)fl_H(a+b) \neq fl_H(ca)+fl_H(cb), \quad fl_S(c)fl_S(a+b) \neq fl_S(ca)+fl_S(cb).$$

Similarly, this situation can occur if we add and multiply long signed numbers whose values are greater (or less) than the maximum (or minimum) machine numbers that admits a computer in the floating-point systems ($\mathbb{R}_H$, $\mathbb{R}_S$).

4. The subsets $\mathbb{R}_H$ and $\mathbb{R}_S$ are completely characterized by the basis b, the number of significant digits t, and the range [MinE,MaxE] (MinE<0 and MaxE>0) of variation of the exponent E. Therefore, we can denote these subsets in the general form $\mathbb{R}_H$(b, t, MinE, MaxE). In both systems we have the following results:

Maple:

$$\mathbb{R}_H(2, 53, -1021, 1024), \quad \mathbb{R}_S(10, 268435448, -2147483646, 2147483646)$$

Mathematica:

$$\mathbb{R}_H(2, 53, -1021, 1024), \quad \mathbb{R}_S(10, \infty, -646456888, 646456887)$$

Note. For the decimal-based number system, the parameters of the subset $\mathbb{R}_H$ are transformed to $\mathbb{R}_H(10, 15, -307, 308)$.
In *Mathematica*, the global variable $MaxPrecison provides the maximum number of precision digits for the numbers of $\mathbb{R}_S$, its predefined value is Infinity. For a specific hardware and software platform, the value can be evaluated approximately as Log[10,$MaxNumber].

5. The element 0 does not belong to $\mathbb{R}_H$, $\mathbb{R}_S$, i.e., $0 \notin \mathbb{R}_H$ and $0 \notin \mathbb{R}_S$, since otherwise $a_1 = 0$ in Eq. (4.1). Therefore it is a special case.

6. The element 0 is not unique. The sets $\mathbb{R}_H$, $\mathbb{R}_S$ consist of isolate numbers. If we add two floating-point numbers, for example, $fl_H(x) + fl_H(y)$, where we assume $fl_H(y) < fl_H(x)$ and $fl_H(y) < \epsilon_M$, where ϵ_M is the *machine epsilon*, we obtain $fl_H(x) + fl_H(y) = fl_H(x)$. The machine epsilon $\epsilon_M = b^{1-t}$ usually provides the distance between 1 and its closest floating-point number different from 1. Generally, it is possible to say that ϵ_M is the *unit roundoff* or the *relative machine precision* that is the distance between a real number and the corresponding floating-point number represented in the systems $\mathbb{R}_H$ and $\mathbb{R}_S$, i.e., $|x - fl_H(x)|/|x| \leq$

$0.5\,\epsilon_M$, $|x - fl_S(x)|/|x| \leq 0.5\,\epsilon_M$. We note that ϵ_M depends on the basis b and the number of digits t. In *Maple* and *Mathematica* we can obtain the result $\epsilon_M = 2^{1-53} = .222044604925031308 \cdot 10^{-15}$ as follows:

Maple:

```
useHardwareFloats:=true: Digits; epsilon_M:=evalhf(DBL_EPSILON);
x1:=HFloat(1.0); for i from 1 while HFloat(1.0+x1)>1.0 do
  eps_M:=x1; x1:=HFloat(x1/2); end do: eps_M;
eps:=NextAfter(SFloat(0.),SFloat(1.)); 1.+eps; n:=100; 1+eps/n;
Digits:=300; eps:=NextAfter(SFloat(0.),SFloat(1.)); 1+eps;
Digits:=10; NextAfter(HFloat(1.),HFloat(2.));
NextAfter(SFloat(1.),SFloat(2.));
```

Mathematica:

```
{$MachineEpsilon//FullForm,$MachineEpsilon/2+1.}
{n=100,SetPrecision[$MachineEpsilon/n+1.,100],
 $MachineEpsilon+1.===1.}
x1=1.0; While[(1.0+x1)-1.>0.,epsM=x1;x1=0.5*x1]; epsM//FullForm
```

7. The *roundoff error* is generated in subsets $\mathbb{R}_H$, $\mathbb{R}_S$ when a real number $x \neq 0$ is replaced by its corresponding floating-point value $fl_H(x) \in \mathbb{R}_H$, $fl_S(x) \in \mathbb{R}_S$, and this error is small. There exist two measures of roundoff error, the *relative error*, $|x - fl_H(x)|/|x| \leq 0.5\,\epsilon_M$, $|x - fl_S(x)|/|x| \leq 0.5\,\epsilon_M$, and the absolute error, $|x - fl_H(x)|$, $|x - fl_S(x)|$. In both systems, there exist various concepts or functions related to the roundoff error, for example, in *Maple*, the global variables Digits, Rounding, the function fnormal; in *Mathematica*, the functions Round, Precision, Accuracy:

Maple:

```
Digits; csc(0.91^191); Rounding:=0; csc(0.91^191);
Rounding:=infinity; csc(0.91^191); a:=1.; b:=10^(-15);
fnormal(a+b); fnormal(a+I*b);
```

Mathematica:

```
{x1=N[Sqrt[2],20],Round[x1]//FullForm}
{Precision[x1],Accuracy[x1],MachineNumberQ[x1],FullForm[x1]}
```

8. In $\mathbb{R}_H$ and $\mathbb{R}_S$, there are no indeterminate forms. In both systems, values are not defined for the indeterminate forms, for example, $\infty + (-\infty)$, $\infty - (+\infty)$, $-\infty - (-\infty)$, $\pm\infty/\pm\infty$, $\pm 0 \cdot \pm\infty$, $\pm 0/\pm 0$.

Maple:

```
Inf:=infinity; Inf+(-Inf); Inf-(+Inf); -Inf-(-Inf); Inf/(+Inf);
Inf/(-Inf); -Inf/(+Inf); -Inf/(-Inf); +0*(+Inf); +0*(-Inf);
-0*(+Inf); -0*(-Inf); 0/0;
```

Mathematica:

```
Off[Infinity::indet];Off[Power::infy];inf=Infinity; {inf+(-inf),
 inf-(+inf),-inf-(-inf),inf/(+inf),inf/(-inf),-inf/(+inf),
 -inf/(-inf),+0*(+inf),+0*(-inf),-0*(+inf),-0*(-inf),0/0}
```

Algebraic and complex numbers

Algebraic numbers are defined as roots of the univariate polynomial $P(x) = a_0x^0 + a_1x + \cdots + a_nx^n$ (see in more detail Sect. 4.10) over the field $\mathbb{Q}$ of rational numbers. For example, $\sqrt{5}$ is an algebraic number. The algebraic numbers form the *algebraic number field* $\mathbb{A}$.

In both systems, *Maple* and *Mathematica*, we can determine the roots of $P(x) \in \mathbb{Q}[x]$ (for representing algebraic numbers) using the functions RootOf and Root, respectively.

All rational numbers are algebraic numbers of the first degree, the number i (the imaginary unit) is an algebraic number of the second degree (it is a root of $x^2 + 1$), the number $5^{1/n}$ ($n \in \mathbb{N}$) is an algebraic number of degree n (it is a root of the irreducible polynomial $x^n - 5$).

Algebraic numbers (which are a generalization of rational numbers) form subfields of algebraic numbers in the fields of real and complex numbers with special algebraic properties.

Complex numbers are considered as an ordered pair of real numbers $(a; b)$ with the equality notation, $(a; b) = (c; d), a = c; b = d$; the four fundamental operations: addition, $(a; b) + (c; d) = (a+c; b+d)$, subtraction, $(a; b) - (c; d) = (a-c; b-d)$, multiplication, $(a; b) \cdot (c; d) = (ac-bd; ad+bc)$, division, $(a; b)/(c; d) = ((ac + bd)/(c^2 + d^2); (bc - ad)/(c^2 + d^2))$.

All complex numbers can be written in *Cartesian form* $(a; b) = a + bi$, where the imaginary unit $i = (0; 1)$, $i^2 = (0; 1)^2 = (-1; 0)$ ($i^2 = -1$), the real part $a = \Re(a + bi)$ and the imaginary part $b = \Im(a + bi)$.

In *Maple* and *Mathematica*, the complex numbers are represented in the form of ordered pairs of real numbers, $\mathbb{C} = \mathbb{R} \times \mathbb{R} = \{(a, b) \mid a \in \mathbb{R}, b \in \mathbb{R}\}$, where we know how to represent the real numbers. The imaginary unit i is denoted by I in both computer algebra systems. In order to introduce a complex number (in both systems) it is necessary to type, for example, 1+I*2 or the function Complex(1,2) (in *Maple*); 1+I*2 or the function Complex[1,2] (in *Mathematica*). With more detail, we consider the theory of complex functions in Chapter 8.

Maple:

```
sqrt(5);        p:=x^2-5;       alias(a=RootOf(x^2-5)); subs(x=a,p);
evala(a^2);     allvalues(a);   convert(a,radical);           evalf(a);
r1:=RootOf(x^2-5,index=1);  r2:=RootOf(x^2-5,index=2);evalf(r1);
evalf(r2);      z:=3-4*I;       Complex(2/3,3);          Re(z); Im(z);
abs(z);         argument(z);    evalc(sin(z));           conjugate(z);
```

Mathematica:

```
{Sqrt[5],   p=x^2-5,   a={Root[#1^2-5&,1],      Root[#1^2-5&,2]},
 p/.x->a,  ComplexExpand[a^2],     a//N, {Sqrt[5],Sqrt[-5]}//N,
 r1=Root[#1^2-5&,1],   r2=Root[#1^2-5&,2],           {r1,r2}//N,
 z=3-4*I,  Complex[2/3,3],  Re[z],Im[z], Sign[z],Abs[z],Arg[z],
 ComplexExpand[Sin[z]],   Conjugate[z],  Element[z,Complexes]}
```

Note. The square root function is multivalued for complex numbers, computer algebra systems cannot recognize this (in particular, the functions sqrt and Sqrt), but it is possible to select the branches of complex multivalued functions with indexed RootOf and Root.

The function RootOf has two forms: *indexed* and *non indexed*. The non indexed form of RootOf(P(x),x) represents a root or all the roots of $P(x)$ (depending of the context). The indexed form RootOf(P(x),x,index=n) represents the n-th root of $P(x)$.

To simplify the expressions that include RootOf and Root, we can apply various functions (e.g., evalf and N, respectively).

The function Root in *Mathematica* has the *indexed* form and represents the n-th root of $P(x)$.

Problem 4.5 For a given $n \in \mathbb{N}$, generate an n-degree univariate polynomial $P(x)$ and determine the exact and approximate values of the algebraic numbers.

Maple:

```
a:=[1,-10,-10,20,-50,12];
p:=convert(['a[i]*x^(i-1)'$'i'=1..6],`+`); A_n:=[solve(p,x)];
for i from 1 to nops(A_n) do evalf(A_n[i]); od;
```

Mathematica:

```
{a={1,-10,-10,20,-50,12},c=Table[a[[i]],{i,1,6}],
u=Table[x^(i-1),{i,1,6}], p=c.u,
aN1=HoldForm[aN1]==Table[Root[p/.{x->#1}&,j],{j,1,5}],
aN2=HoldForm[aN2]==Table[Root[p,x,j],{j,1,5}],N[aN1],N[aN2]}
```

□

Floating-point approximations in $\mathbb{N}$, $\mathbb{Z}$, $\mathbb{Q}$, $\mathbb{I}$, $\mathbb{R}$, $\mathbb{C}$, $\mathbb{A}$

Maple:

---

1. Numerical approximation of **expr** using arbitrary-precision software-floats (up to 10 significant digits) `evalf(expr)`, and double-precision hardware-floats, `evalhf(expr)`.
2. The global and local definitions of an arbitrary-precision software-floats, respectively, with the global variable `Digits:=n;` `evalf(expr)` and with the function `evalf(expr,n)`.

Note. By default, the value of the environmental variable `Digits=10` (see ?Digits, ?environment), and the working precision is 10 decimal places. The number of displayed decimal places is controlled by the interface variable `displayprecision` (see ?displayprecision, ?interface).

```
135+638; -13*77; 467/31; (3+8*4+9)/2; (-5)^(22); -5^(1/3);
evalf((-5)^(1/3)); evalf(sqrt(122)); evalf((1+sqrt(5))/2);
evalf(exp(1),50);Digits:=50;evalf(exp(1));evalf(0.9);evalhf(0.9);
sin(evalf(Pi)); interface(displayprecision=5): sin(evalf(Pi));
```

Mathematica:

---

1. Numerical approximation of **expr** using double-precision hardware-floats (up to 6 significant digits), `N[expr]` or `expr//N`, and arbitrary-precision software-floats (up to n significant digits), `N[expr,n]`, `NumberForm[expr,n]`.
2. Scientific notation of the numerical approximation of **expr** up to n significant digits, `ScientificForm[expr,n]`.

```
{135+638,-13*77,467/31,(3+8*4+9)/2,(-5)^(22),-5^(1/3)}
{Sqrt[17],Pi+2*Pi*I,1/3+1/5+1/7,N[E,30],N[Pi],Pi//N}
{ScientificForm[N[Sin[Exp[Pi]]],9],NumberForm[N[Sin[Exp[Pi]]],9]}
```

In *Maple* and *Mathematica*, the machine code of procedures and functions can be reproduced to increase the speed in numerical calculations (see the functions Compile in both systems).

Problem 4.6 Let us define a finite set $A = \{x_1, x_2, \ldots, x_k\}$ of real numbers by the formula $x_{i+1} = (x_i^n \exp(-x_i^n))^{1/n}$ (for given n and k). Calculate the real numbers of A and compare the computation time in both systems, *Maple* and *Mathematica*, for the two cases: compiling numerical procedures and functions and without compiling.

Maple:

```
n:=9; f1:=proc(x::float)::float; (x^n*exp(-x^n))^(1/n) end proc;
ti1:=time(): C_f1:=Compiler:-Compile(f1);
for x from 0 to 10 by 0.001 do C_f1(x); od: tt1:=time()-ti1;
f2:=x->(x^n*exp(-x^n))^(1/n); ti2:=time():
for x from 0 to 10 by 0.001 do f2(x); od: tt2:=time()-ti2;
```

Mathematica:

```
n=9; f1=Compile[{x},(x^n*Exp[-x^n])^(1/n)];
f2[x_]:=(x^n*Exp[-x^n])^(1/n);
Table[f1[x],{x,0.,10.,0.001}]//Timing
Table[f2[x],{x,0.,10.,0.001}]//Timing
```

□

Problem 4.7 A finite set of real numbers $A = \{x_1, x_2, \ldots, x_n\}$ is defined by $x_{i+1} = ax_i(1-x_i)$, $0 < a < 10$, $i = 0, \ldots, n$, where a is a given parameter. Define a list of coordinates $[i, x_i]$ such that $x_0 = 0.1$, $a = 3$, $n = 100$. Plot the graph of A.

Maple:

```
a:=3; x[0]:=0.1; n:=100; for i from 1 to n do
x[i]:=a*x[i-1]*(1-x[i-1]): od: Seq1:=seq([i,x[i]],i=2..n):
plot([Seq1],style=point,symbol=circle,symbolsize=5);
```

Mathematica:

```
n=100; l=Array[x,n,0]; a=3; l[[1]]=0.1;
Do[l[[i]]=a*l[[i-1]]*(1-l[[i-1]]),{i,2,n}]
l1=Table[{i,l[[i]]},{i,2,n}]
ListPlot[l1,PlotStyle->{PointSize[0.02],Hue[0.7]}]
```

□

4.3   Operations on Sets

Maple:

The union and intersection of sets, the elimination of elements of sets
are defined with the functions union, intersect, minus, remove, has:

```
Set3:=Set1 union Set2;              Set3:=Set1 intersect Set2;
Set3:=Set1 minus Set2;             remove(has,Set1,Set2);
```

Mathematica:

The sets are represented in the form of lists. The union of different
elements of sets, the intersection of sets, the elimination of elements of
sets are defined with the functions Union, Intersection, Complement.

```
Set3=Union[set1,set2]              Set3:=Intersection[set1,set2]
Set3:=Complement[set1,set2]   Take[set1,{m,n}] Select[set1,crit]
```

Note. The package DiscreteMath`Combinatorica` (for $ver < 6$) or the package
Combinatorica` (for $ver \geq 6$) contains a variety of functions of set theory.

Problem 4.8 Let A, B, C the subsets of $\mathbb{R}$ defined by $A = \{x \mid x \in \mathbb{R}, 0 \leq x \leq 3\}$, $B = \{x \mid x \in \mathbb{R}, -1 \leq x \leq 2\}$, $C = \{x \mid x \in \mathbb{R}, -2 \leq x \leq 1\}$. Demonstrate
that $A \cap (B \cup C) = (A \cap B) \cup (A \cap C)$ and $A \cup (B \cap C) = (A \cup B) \cap (A \cup C)$.

Maple:

```
with(RealDomain); A:=x>=0 and x<=3; B:=x>=-1 and x<=2;
C:=x>=-2 and x<=1; solve(evalb(A and (B or C)));
solve(evalb((A and B) or (A and C))); solve(evalb(A or
      (B and C))); solve(evalb((A or B) and (A or C)));
```

Mathematica:

```
{a=x>=0&&x<=3, b=x>=-1&&x<=2, c=x>=-2&&x<=1}
{Reduce[a&&(b||c)], Reduce[(a&&b)||(a&&c)], Reduce[a||(b&&c)],
 Reduce[(a||b)&&(a||c)]}
```

Let us solve the problem step by step. $B \cup C = \{x \mid -2 \leq x \leq 2\}$, $A \cap B = \{x \mid 0 \leq x \leq 2\}$, $A \cap C = \{x \mid 0 \leq x \leq 1\}$. Therefore, we have $A \cap (B \cup C) = \{x \mid 0 \leq x \leq 2\}$
and $(A \cap B) \cup (A \cap C) = \{x \mid 0 \leq x \leq 2\}$. Then, $B \cap C = \{x \mid -1 \leq x \leq 1\}$,
$A \cup B = \{x \mid -1 \leq x \leq 3\}$, $A \cup C = \{x \mid -2 \leq x \leq 3\}$. Therefore, $A \cup (B \cap C) = \{x \mid -1 \leq x \leq 3\}$ and $(A \cup B) \cap (A \cup C)) = \{x \mid -1 \leq x \leq 3\}$.   □

4.4   Equivalence Relations and Induction

Binary relations. Let X be a nonempty arbitrary set. We say that R is a binary relation in X if $R \subseteq X \times X$. Also we can represent a binary relation as a correspondence from X to X, $f : X \to X$, considering a binary relation as a particular case of the correspondence.

Let X be a given nonempty set and R be a binary relation defined on X between some pairs of elements of X. We say that R is an equivalence relation ($\sim$) if the following conditions hold:

(i) *reflexivity:* $x \sim x \ \forall x \in X$;

(ii) *symmetry:* if $x_1 \sim x_2 (x_1, x_2 \in X)$, then $x_2 \sim x_1$;

(iii) *transitivity:* if $x_1 \sim x_2$ and $x_2 \sim x_3 (x_1, x_2, x_3 \in X)$, then $x_1 \sim x_3$.

Problem 4.9 Let us define a binary relation $\sim$ on $\mathbb{C}$ as follows: $z_1 \sim z_2$ ($z_1 \in \mathbb{C}$, $z_2 \in \mathbb{C}$) if and only if $z_1 - z_2 \in \mathbb{R}$. Prove that $\sim$ is an equivalence relation.

Maple:

```
interface(showassumed=0); z1:=x1+I*y1; z2:=x2+I*y2;
z3:=x3+I*y3;is(z1-z1,real);assume(z1-z2,real,z2-z3,real);
is(z1-z2,real); is(z2-z1,real); evalb(z2-z1=-(z1-z2));
is(-(z1-z2),real); is(z1-z3,real); is(z1-z2,real);
is(z2-z3,real);
```

Mathematica:

```
{{z1=x1+I*y1,z2=x2+I*y2,z3=x3+I*y3},FreeQ[z1-z1,Reals]}
Assuming[{z1-z2,z2-z3} \[Element] Reals,
{FreeQ[z1-z2,Reals],FreeQ[z2-z1,Reals],z2-z1===-(z1-z2),
 FreeQ[-(z1-z2),Reals],FreeQ[z1-z3,Reals],
 FreeQ[z1-z2,Reals],   FreeQ[z2-z3,Reals]}]
```

Let $z_1, z_2 \in \mathbb{C}$. Clearly, $z_1 \sim z_1$, since $z_1 - z_1 = 0 \in \mathbb{R}$. If $z_1 - z_2 \in \mathbb{R}$, then $z_2 - z_1 \in \mathbb{R}$, since $-(-z_2 + z_1) = -(z_1 - z_2) \in \mathbb{R}$. Therefore, $z_1 \sim z_2$ implies that $z_2 \sim z_1$. If $z_1 \sim z_2$ and $z_2 \sim z_3$, then $z_1 - z_2 \in \mathbb{R}$, $z_2 - z_3 \in \mathbb{R}$, and, therefore, we have $z_1 - z_3 \in \mathbb{R}$, since $z_1 - z_3 = (z_1 - z_2) + (z_2 - z_3)$. $\square$

Let us recall that the *Principle of Well-ordering* in $\mathbb{N}$ reads that every non-empty subset of $\mathbb{N}$ has a least element in $\mathbb{N}$. Frequently, if we have some proposition $Pr(n)$, $n \in \mathbb{N}$, we are interested to know whether $Pr(n)$ is true for $\forall\, n \in \mathbb{N}$ or $\forall\, n > N$, $n \in \mathbb{N}$, $N \in \mathbb{N}$. So we are led to the

Principle of Induction: Suppose that $Pr(1)$ is true. If $Pr(k)$ is true, then $Pr(k+1)$ is true. Then $Pr(n)$ is true for $\forall n \in \mathbb{N}$.

Problem 4.10 Let $Pr(n)$ be a proposition $Pr(n) = 1 + 2 + \ldots + n = n(n+1)/2$, $n \in \mathbb{N}$. Prove by induction that $Pr(n)$ is true for $\forall n \in \mathbb{N}$.

Maple:

```
PrL:=n->Sum(i, i=1..n);     PrR:=n->1/2*n*(n+1);
Eq1:=value(PrL(1))=PrR(1); Eq2:=PrL(k)=PrR(k);
Eq3:=PrL(k)+(k+1)=PrR(k)+(k+1);Eq4:=PrL(k+1)=factor(rhs(Eq3));
Eq5:=algsubs(k+1=n,Eq4);   Eq6:=evalb(subs(k=n,Eq2)=Eq5);
```

Mathematica:

```
prL[n_]:=Sum[i,{i,1,n}]//HoldForm; prR[n_]:=1/2*n*(n+1);
{eq1=ReleaseHold[prL[1]]==prR[1], eq2=prL[k]==prR[k],
 eq3=prL[k]+(k+1)==prR[k]+(k+1),   eq4=prL[k+1]==
 Factor[eq3[[2]]], eq5=eq4/.{k+1->n,2+k->n+1},
 eq6=(eq2/.{k->n})===eq5}
```

□

4.5   Mathematical Expressions

The concept of a *mathematical expression* is a general concept for any well-formed combination of mathematical symbols, for example, arithmetic and algebraic expressions. Expressions and their evaluations were defined in a formal system, called λ-*calculus* and developed by A. Church and S.C. Kleene (in 1930), for investigating definitions of functions, computable functions, applications of functions, the recursions. The λ-calculus has been an important influence on development of modern mathematics and, in particular, on symbolic mathematics.

In general, one of the fundamental principles of computer algebra systems is that everything (formulas, graphics, data, programs, etc.) can be represented as *symbolic mathematical expressions*. In *Maple* and *Mathematica*, we consider *symbolic mathematical expressions* as combinations of constants, variables, functions, etc., that can be evaluated according to the particular rules of symbolic systems.

Maple:

---

1. *Symbolic mathematical expression*, `expr`, is a valid declaration, and it is generated as a combination of constants, variables, operators, and functions.
2. Each expression is represented in the form of a tree in which each node (and the leaf) has a particular data type.
3. To analyze any element of the tree, various functions can be used, e.g., `type, whattype, nops, op`.

Note. Algebraic expressions are constructed with variables and algebraic numbers using addition, multiplication, and rational powers, for example, `expr1:=2*x+1`.
Boolean expressions, `bexpr`, are formed using the *logical and relation operators*, for example, `bexpr1:=evalb(not(p and q)=(not p) or (not q));`

Mathematica:

---

1. *Symbolic mathematical expression*, `expr`, is generated as a combination of a small number of *basic* or *atomic* types of objects: *Symbol, String, Integer, Real, Rational, Complex.*
2. Each expression is represented in the form of a tree in which each node (and the leaf) has a particular data type.
3. To analyze any element of the tree, various functions can be used, e.g., `Head, Length, Part, Extract, Map, Level, Depth, LeafCount`, etc.

Note. There exists a rich collection of *symbolic functions* for manipulating symbolic expressions, e.g., the head of `expr` can be obtained with `Head[expr]`, a group of functions, ending in `Q` (`AtomQ, DigitQ, IntegerQ`, etc.), allow us to test an expression (they return `True` for an explicit true answer, and `False` otherwise).

The structure and various forms of `expr` can be analyzed with the symbolic functions `FullForm[expr]`, `InputForm[expr]`, `TreeForm`:

```
l1={5, 1/2, 9.1, 2+3*I, x, {A, B}, a + b, a*b}
{Head /@ l1, FullForm[l1], InputForm[l1], TreeForm[l1]}
```

The symbolic function `SameQ` or its equivalent operator `lhs===rhs` returns `True` if `lhs` is identic to `rhs`, and `False` otherwise.

Boolean expressions, `bexpr`, are formed using the *logical and relational operators*, e.g., `LogicalExpand[!(p&&q)===LogicalExpand[!p||!q]]`

Function expressions and expressions

It is known that the mathematical concept of a function expresses a dependence between two quantities, the *independent variable* or *argument of the function* (which is given) and the other, the *dependent variable* or *value of the function* (which is determined). There are many ways to define a function, e.g., by formulas, tables of values, plots, computational algorithms, descriptions of properties, etc. In computer algebra systems, as in formal mathematics, there exists a distinction between function expressions and expressions.

For example, in *Maple*, a function written in the form f:=x->2*x+1 is different from the expression f:=2*x+1. The function expression in this example is f(x).

In *Mathematica*, a function of the form f[x_]:=2*x+1 is different from the expression f=2*x+1. The function expression in this example is f[x].

The expressions may be evaluated, plotted, etc., but in a formal sense, they are not functions. To define a function we have to define a "mapping". Mapping notation denotes a correspondence between an independent variable and a *function expression*.

Evaluation of mathematical expressions (formulas)

Maple:

The *Maple* functions subs, algsubs, eval, applyrule, macro:

```
expr:=x^2+y^2;        subs(x=a,expr);          subs({x=a,y=b},expr);
eval(expr,x=a);                     algsubs(y=b,algsubs(x=a,expr));
eval(expr,[x=a,y=b]);          applyrule(y=b,applyrule(x=a,expr));
macro(newName=functionComposition);          newName(args,expr);
```

Note. The macro function is an abbreviation mechanism (similar to alias), but in this case we apply evalf at the results of subs,

```
macro(evalNew=evalf@subs); evalNew(x=Pi,sin(x));
```

Mathematica:

The *Mathematica* replacement operator /. or the corresponding function ReplaceAll, the functions Replace, ReplaceRepeated (or //.),

```
expr=x^2+y^2; {expr/.x->a,expr/.{x->a,y->b},expr//.{x->a,y->b},
Replace[expr,{x->a},{-1}],          Replace[expr,{x->a,y->b},{-1}]}
```

4.6 Simplifying Mathematical Expressions

In general, symbolic mathematical expressions, contain a combination of alphabetical symbols, numerical values, mathematical operations and operators, and parenthesis. According to the language and the laws of algebra, the simplification method generally consists in elimination of parentheses, regrouping terms that have the same format.

Frequently simplifications of symbolic mathematical expressions can be sufficiently tedious or difficult, therefore computer algebra systems allow us to automate and to facilitate the symbolic simplification (e.g., operations for regrouping terms, reducing to to the simplest or standard forms, cancelling terms, changing the form of expressions, etc.).

Maple:

```
factor(expr); factor(expr,F);        eval(expr); evala(expr);
expand(expr); combine(expr,ops);              simplify(expr,ops);
normal(expr); numer(expr); denom(expr);  frontend(func,[args]);
convert(expr,form,var); map(f,expr);  match(expr=patt,var,'s');
```

factor(expr), factoring polynomials with numeric ($\in \mathbb{Z}, \mathbb{Q}, \mathbb{R}, \mathbb{C}$) or algebraic number coefficients; factor(expr,F), factoring polynomials over the algebraic number field F,

expand(expr), distributing products over sums,

combine(expr,ops), combining terms into a single term,

eval(expr), evaluating, evala(expr), evaluating in an algebraic number field, other types of evaluations, e.g., evalb(expr), evalc(expr), etc.,

simplify(expr,ops), applying simplification rules to expressions,

normal(expr), finding factored normal form,

numer(expr), denom(expr), extracting the numerator/denominator of expr,

convert(expr,form,var), changing the form of an expression,

map(function,expr), manipulating large expressions,

match(expr=patt,var,'s'), pattern matching,

frontend(func,[args]), extension of the computation domain for many *Maple* functions.

```
f:=(x^3+2*x^2-x-2)/(x^3+x^2-4*x-4);   F:=(x+y+1)^3*(x+3*y+3)^2;
N1:=factor(numer(f)); N2:=subs(x=2,N1);  D1:=factor(denom(f));
D2:=subs(x=3,D1);      F1:=simplify(f);       F2:=subs(x=4,F1);
```

```
F3:=subs(x=-3,F1);    convert(f,parfrac,x);    B:=collect(F,x);
C:=map(factor,B);assume(x>0,y>0,t>0); expand([ln(x*y),ln(x^y),
 (x*y)^z]);         combine([exp(t)*exp(p),t^p*t^q,ln(t)+ln(p)]);
```

Mathematica:

Factor[expr]	Factor[expr,Extension->a]	FactorTerms[expr,x]
Expand[expr]	ExpandAll[expr] PowerExpand[expr]	Apart[expr]
Collect[expr,x]	Together[expr] Cancel[expr]	Evaluate[expr]
Numerator[expr]	Denominator[expr]Hold[expr]	HoldForm[expr]
Short[expr]	Thread[f[args]]	Thread[f[args],head]
Map[f,expr]	f/@expr Apply[f,expr]	f@@expr
Simplify[expr]	FullSimplify[expr]	Simplify[expr,{assum}]

Factor, Factor[expr,Extension->a], factoring polynomials over the integers
   or an algebraic number field,

Short, Map or /@, Apply or @@, Thread, manipulating large expressions,

Expand, PowerExpand, distributing products over sums,

Apart, converting expr into a sum of partial fractions,

Together, combining terms in a sum over a common denominator and cancelling
   factors in the result,

Evaluate, evaluating an expression,

Hold, maintaining expr in unevaluated form,

Cancel, common-factor cancellation in the rational expression expr,

Numerator/Denominator, extracting the numerator/denominator of expr,

Simplify, FullSimplify, applying simplification rules to expressions or doing
   simplifications using assumptions, assum.

```
{expr1=(x^3+2*x^2-x-2)/(x^8-21*x^4-100),expr2=(x+y)^2*(3*x-y)^3,
 expr3=(x+y+1)^3/(x+3*y+3)^2, Factor[Numerator[expr1]]/.x->2,
 Factor[Denominator[expr1],GaussianIntegers->True],
 Factor[Denominator[expr1],Extension->{I,Sqrt[5]}]}
{ExpandNumerator[expr3], ExpandDenominator[expr3],
 expr4=ExpandAll[expr3], Expand[expr3], Collect[expr2,x]}
{Log[x*y],Log[x^y],(x*y)^z}//PowerExpand
{Exp[t]*Exp[p],t^p*t^q}//Together
{Cancel[expr1*(x+1+y)/expr2],Apart[expr1],FullSimplify[expr4],
 Map[Factor,expr4],Expand[expr4^10]//Short, Simplify[(x^m)^n],
 Simplify[(x^m)^n,{{m,n}\[Element]Integers}],
 Simplify[Log[x-y]+Log[x+y],{x>y}]}
```

Note. The function `Element[x,domain]` specifies that $x \in$ domain. The element operator ($\in$) can be entered with the sequence `Esc elem Esc`. Assumptions can be employed with some other functions, for example, `FullSimplify`, `FunctionExpand`, `Refine`, `Limit`, `Integrate`.

Problem 4.11 Let $\tan(x) = x + \frac{1}{3}x^3 + \frac{2}{15}x^5 + \frac{17}{315}x^7 + \frac{62}{2835}x^9$ be the Maclaurin series expansion for $\tan(x)$. Demonstrate that the inverse series of the given series about $x = 0$ is $x = y - \frac{1}{3}y^3 + \frac{1}{5}y^5 - \frac{1}{7}y^7 + \frac{1}{9}y^9$.

Maple:

```
Order:=10; Ec1:=y=series(tan(x),x=0,Order); x:=solve(Ec1,x);
x:=convert(x,polynom);
```

Mathematica:

```
ord=10; f[t_]:=Normal[InverseSeries[Series[
  Tan[x],{x,0,ord}]]]/.x->t; f[z]
```

Since we would like to obtain the inverse series up to $O(x^{10})$, we have to set the environment variable `Order=10`. Then if we find the Maclaurin series expansion for the function $\tan(x)$ and solve the result with respect to the variable x, we obtain the inverse series. For performing further operations with this series, it is recommended to transform the series into a polynomial. □

4.7 Trigonometric and Hyperbolic Expressions

It is known that trigonometry and hyperbolic trigonometry have an enormous variety of applications in a number of academic fields (in mathematics, natural sciences, engineering, in the theory of music) and in practice (navigation, land surveying, building, architecture, measurement problems). Frequently, in solving scientific problems, we have to apply trigonometric and hyperbolic identities to simplify trigonometric and hyperbolic expressions and solve more complicated equations.

In symbolic mathematics, automated simplification of trigonometric and hyperbolic expressions is an important problem that has not been completely solved by current computer algebra systems. But with the aid of computer algebra systems, we can simplify many trigonometric and hyperbolic expressions, which are difficult to do with pencil and paper, and can achieve much better results for many hard problems, e.g. constructing approximate analytical solutions of nonlinear standing wave problems (for a more detail discussion, see Problem 11.17). We consider some important functions for both systems.

Maple:

```
simplify(expr,trig);    combine(expr,trig);   expand(expr,expr1);
convert(expr,exp);      convert(expr,trig); convert(expr,trigh);
testeq(expr1=expr2);frontend(func,[args]);  type(expr,'trigh');
with(RealDomain);    func(expr);              RealDomain:-func(expr);
```

`simplify(expr,trig)`, simplifying trigonometric and hyperbolic expressions,

`combine(expr,trig)`, combining trigonometric and hyperbolic expressions according to the trigonometric and hyperbolic transformations being eliminated products and powers,

`expand(expr,expr1)`, expanding an expression `expr`, avoiding expanding `expr1`,

`frontend(func,[args])`, extending the domain of computation for many *Maple* functions allowing to work, for example, with trigonometric and hyperbolic expressions as with rational expressions,

`type(expr,'trig')`, `type(expr,'trigh')`, verifying whether `expr` is trigonometric and hyperbolic expression,

`RealDomain` package, computing under the assumption that the basic number system is the field of real numbers. Since the implementation of `RealDomain` is a module, the equivalent form for computing one function in `RealDomain` is `RealDomain:-func(expr)`,

`convert(expr,trig)`, `convert(expr,trigh)`, rewriting trigonometric and hyperbolic expressions to other forms,

`testeq(expr1=expr2)`, testing (with a probability) if two expressions (including trigonometric and hyperbolic terms) are identical.

Note. The result `false` of the function `testeq(expr1=expr2)` is always correct, but the result `true` may be incorrect with very low probability.

```
expr1:=c1*sin(x)+c2*x; expr2:=c1*cos(x+y)+c2*sin(x+y);
expr3:=c3*cosh(x)^2-sin(y)^2; frontend(coeff,[expr1,x]);
RealDomain:-solve(cos(I*x)^2-1,x); with(RealDomain);
solve(cos(I*x)^2-1,x); expand(expr2,cos(x+y));
expand(expr2,op(indets(expr2,function)));
factor(convert(simplify(convert(expr3,exp)),trig));
factor(convert(RealDomain:-simplify(convert(expr3,exp)),trig));
```

Mathematica:

Setting the option `Trig->True` within various polynomial manipulation functions, *Mathematica* will treat trigonometric and hyperbolic functions as polynomial elements and will use standard trigonometric identities for manipulating and simplifying expressions.

`Cancel[expr,Trig->True]`		`Together[expr,Trig->True]`
`Expand[expr,Trig->True]`	`TrigToExp[expr]`	`ExpandAll[expr,patt]`
`TrigReduce[expr]`	`TrigFactor[expr]`	`TrigExpand[expr]`
`PowerExpand[expr]`	`ExpToTrig[expr]`	`FunctionExpand[expr]`
`ComplexExpand[expr]`	`ComplexExpand[expr,{x1,...,xn}]`	

`Expand[expr,patt]`, expanding an expression `expr` and avoiding expanding parts of `expr` that do not contain terms matching the pattern `patt`,

`ComplexExpand[expr]`, `ComplexExpand[expr,{x1,..,xn}]`, expanding `expr` assuming that, respectively, all variables are real and variables matching any of the `{x1,..,xn}` are complex,

`TrigExpand`, expanding trigonometric and hyperbolic functions in `expr`,

`TrigReduce`, combining trigonometric and hyperbolic expressions according to the trigonometric and hyperbolic transformations being eliminated products and powers,

`TrigFactor`, factoring trigonometric and hyperbolic functions in `expr`,

`TrigToExp`, `ExpToTrig`, converting trigonometric and hyperbolic functions to exponentials and viceversa.

```
{expr1=c1*Sin[x]+c2*x, expr2=c1*Cos[x+y]+c2*Sin[x+y],
expr3=c3*Cosh[x]^2-Sin[y]^2}
{Coefficient[expr1,x],Solve[Cos[I*x]^2==1,x,
InverseFunctions->True], Reduce[Cos[I*x]^2==1,x,Reals],
TrigExpand[expr2],ExpandAll[expr2,Sin[x+y],Trig->True]//Factor,
ExpandAll[expr2,_Function,Trig->True]//Factor,
ComplexExpand[TrigToExp[expr3]]//FullSimplify, Factor[
Simplify[TrigToExp[expr3],Element[{x,y},Reals]]//ExpToTrig]}
```

Problem 4.12 Let $A_1 = (\csc^4 x - 1)/\cot^4 x$ and $A_2 = \sin^4 x + \cos^4 x$ be trigonometric expressions. Show that A_1 and A_2 can be reduced, respectively, to the expressions $2/\cos^2 x - 1$ and $2\cos^4 x + 1 - 2\cos^2 x$. Transform the expression A_2^4 in a sum of trigonometric terms with the combined arguments $4x$, $8x$, $12x$, and $16x$.

Maple:

```
A1:=(csc(x)^4-1)/cot(x)^4; A2:=sin(x)^4+cos(x)^4;
expand(simplify(A1,trig)); simplify(convert(A2,sincos));
combine(A2^4,trig); expand(simplify(expand(
 algsubs(sin(x)^2+cos(x)^2=1,normal(convert(A1,sincos))))));
```

Mathematica:

```
a1=(Csc[x]^4-1)/Cot[x]^2; a2=Sin[x]^4+Cos[x]^4;
{FullSimplify[TrigToExp[a1]]//ExpToTrig, Factor[
 TrigExpand[a2]/.{Sin[x]^2->1-Cos[x]^2,Sin[x]^4->
 (1-Cos[x]^2)^2}],Together[Expand[a2^4],Trig->True]}
```

Applying the well-known trigonometric identities to the expressions $A1$ and $A2$, we obtain the required results. Also, it is possible to apply the *Maple* functions `convert` and `algsubs`.   □

Problem 4.13 Let $A_1 = \sin^5 t / \cot^2 t - \cot^2 t + 1/\tan^2 t + 1/\csc^2 t - \cos^2 t$ be a trigonometric expression. Rewrite A_1 in terms of the sine function with combined arguments.

Maple:

```
A1:=sin(t)^5/cot(t)^2-cot(t)^2+1/tan(t)^2+1/csc(t)^2-cos(t)^2;
A2:=convert(A1,sin); A3:=convert(A2,parfrac,sin(t)); factor(A3);
```

Mathematica:

```
subs={Tan[x_]->2*sin[x]^2/sin[2*x],Cot[x_]->sin[2*x]/2/sin[x]^2,
      Sec[x_]->2*sin[x]/sin[2*x],Csc[x_]->1/sin[x],
      Cos[x_]->sin[2*x]/2/sin[x],Sin[x_]->sin[x]}
{a1=Sin[t]^5/Cot[t]^2-Cot[t]^2+1/Tan[t]^2+1/Csc[t]^2-Cos[t]^2,
 a2=a1/.subs, a3=Factor[a2]}
```

Since in *Maple* there exist numerous conversion forms allowing us to rewrite an original expression in terms of another functions, we can apply this useful *Maple* function `convert`. In *Mathematica*, we can write the corresponding trigonometric substitutions (with another function name, e.g., `sin`).   □

Problem 4.14 Show that the trigonometric expressions $\sin(a - b) = \sin a \cos b - \cos a \sin b$ and $\sin(4a) = 4 \sin a \cos^3 a - 4 \sin a \cos a$ are identical.

Maple:

```
testeq(sin(a-b)=sin(a)*cos(b)-cos(a)*sin(b));
testeq(sin(4*a)=4*sin(a)*cos(a)^3-4*sin(a)*cos(a));
evalb(expand(sin(a-b)=sin(a)*cos(b)-cos(a)*sin(b)));
evalb(expand(sin(4*a) = 4*sin(a)*cos(a)^3-4*sin(a)*cos(a)));
```

Mathematica:

```
(Sin[a-b]//TrigExpand)===Sin[a]*Cos[b]-Cos[a]*Sin[b]
(Sin[4*a]//TrigExpand)===4*Sin[a]*Cos[a]^3-4*Sin[a]*Cos[a]
```

Additionally, it is possible to apply the *Maple* functions `evalb` and `expand`. □

4.8   Defining Functions or Mappings

Frequently in mathematics and other sciences we are interested in relating a set to another one. Let X and Y be non empty sets and there be a rule (or relation), denoted by f, that determines a correspondence between the elements of X with the elements of Y. We say that the rule f defines a *function* from X to Y or a *mapping on sets* if
(i) all the elements of set X are related with the elements of Y;
(ii) to each element of X there is assigned a unique element of Y.
      We denote the function from X to Y in the form $f : X \to Y$. For $x \in X$, $f(x)$ is called the *image* of x under f, and X and Y are called, respectively, the *domain* and *codomain* of f, the subset of Y given as $f(X) = \{f(x) \mid x \in X\}$ is called the *image* of X or the *range* of f.
      In symbolic mathematics functions can also be defined by an *algorithm*, which gives a description for computing the value of the function. Functions defined by an algorithm are called *computable* or *procedural functions*.
      A *Maple* function expression is an expression of type function, e.g., if we define a function as an arrow operator `F1:=x->2*x+1;`, we have `type('F1(a)','function'); op(0,'F1(a)');`.
      In *Mathematica*, if we define a function as a formalized, named transformation rule (for more details on transformation rules see Chapter 2), e.g., `f1[x_]:=2*x+1`, we have `FreeQ[f[x],_Function]`, where `f[x]` is an expression of type function. The function name is also an expression and can be, for example, extracted as `Head[Sin[x]]` or `Sin[x][[0]]`. Note

that f[x] is a standard form, f@x is a prefix form for f[x], and x//f is
a postfix form for f[x].

Classes of functions

Maple:

> Predefined and user-defined functions, mathematical and procedural func-
> tions, active and inert functions, anonymous and named functions.

Mathematica:

> Predefined and user-defined functions, mathematical and procedural func-
> tions, pure functions and functions defined in terms of a variable, anonymous
> and named functions.

Function names

It is a usual practice in mathematics to introduce functions with names.
However, in symbolic mathematics anonymous functions can be useful,
e.g., in functional operations.

Maple:

> The function names are generated as the expression names (see Chapter 1).
> *Abbreviations* for long function names or any expressions, alias, macro, e.g.,
> alias(H=Heaviside); diff(H(t),t); macro(sq=sqrt); sq(4);
> In order to eliminate this abbreviation, alias(H=H); macro(sq=sq);

Mathematica:

> The *names* of *predefined functions* are usually complete English words (e.g.,
> FullSimplify) or traditional abbreviations of words (for very common func-
> tions), e.g., DSolve, Mod.
> Mathematical functions with names of famous mathematicians, are named
> with PersonSymbol, e.g., the Legendre polynomials $P_n(x)$, LegendreP[n,x].
> Function names can be changed with the operator /., e.g., f[x]/.f->F.

Definition and evaluation of functions or mappings

Scalar-valued functions of a single variable, $f_1 : X \to Y$,

$f_1(x) = \text{expr}, \, x \in X, \, \text{expr} \in Y$,

Scalar-valued functions of several variables, $f_2 : X^n \to Y$,

$f_2(x_1, \ldots, x_n) = \text{expr}, \, x_i \in X^n, \, i = 1, \ldots, n, \, \text{expr} \in Y$,

Vector-valued functions of a single variable, $f_3 : X \to Y^n$,

$f_3(t) = \langle x_1(t), \ldots, x_n(t) \rangle, \, t \in X, \, (x_1(t), \ldots, x_n(t)) \in Y^n$,

Vector-valued functions of several variables, $f_4 : X^n \to Y^m$,

$f_4(x_1, \ldots, x_n) = \langle f_1(x_1, \ldots, x_n), \ldots, f_m(x_1, \ldots, x_n) \rangle$,

$x_i \in X^n$ (i=1, ..., n), $\big(f_1(x_1, \ldots, x_n), \ldots, f_m(x_1, \ldots, x_n) \big) \in Y^m$,

Vector-valued functions of a vector argument, $f_5 : X^n \to Y^m$,

$f_5(V) = \langle f_1(V), \ldots, f_m(V) \rangle, \, V = <V_1, \ldots, V_n> \in X^n, \, \big(f_1(V), \ldots, f_m(V) \big) \in Y^m$.

The *Maple* functional operator (->),

```
f1:=x->expr; f2:=(x1,...,xn)->expr;   f3:=t-><x1(t),...,xn(t)>;
f4:=(x1,...,xn)-><f1(x1,...,xn),...,fm(x1,...,xn)>;
f5:=(V::Vector)-><f1(V[1],...V[n]),...,fm(V[1],...V[n])>;
f1(x); f2(x1,...xn);  f3(t);  f4(x1,...xn); f5(<V[1],...V[n]>);
```

The *Maple* function unapply, the result is functional operator ->,

```
f1:=unapply(expr,x); f2:=unapply(expr,x1,...,xn);
f3:=unapply(<x1(t),...,xn(t)>,t);
f4:=unapply(<f1(x1,...,xn),...,fm(x1,...,xn)>,x1,...,xn);
f5:=unapply(<f1(V[1],...,V[n]),...,fm(V[1],...,V[n])>,
                                               V::Vector);
f1(x);   f2(x1...xn);   f3(t);  f4(x1...xn); f5(<V[1]...V[n]>);
```

The *Maple* procedure proc,

```
f1:=proc(x) expr end; f2:=proc(x1,...,xn) expr end;
f3:=proc(t) <x1(t),...,xn(t)> end;
f4:=proc(x1,...xn) <f1(x1,...xn),...,fm(x1,...,xm)> end;
f5:=proc(V::Vector)<f1(V[1],...V[n]),...,fm(V[1],...,V[n]>end;
f1(x);  f2(x1,...xn);   f3(t); f4(x1,...xn); f5(<V[1],...V[n]>);
```

The *Mathematica* transformation rule f[x_]:=expr,

```
f1[x_]:=expr;
f2[x1_,...,xn_]:=expr;
f3[t_]:={x1[t],...,xn[t]};
f4[x1_,...,xn_]:={f1(x1,...,xn),...,fm(x1,...,xn)};
f5[V_?Vector]:={f1[V[[1]],...V[[n]]],...,fm[V[[1]],...V[[n]]]};
{f1[x], f2[x1,...xn], f3[t], f4[x1,...xn],   f5[{x1,...xn}]}
```

The *Mathematica* pure functions, Function[x,expr],

```
f1=Function[x,expr];
f2=Function[{x1,...,xn},expr];
f3=Function[t,{x1[t],...,xn[t]}];
f4=Function[{x1,...,xn},{f1[x1,...,xn],...,fm[x1,...xn]}];
f5=Function[V,{f1[V[[1]],...,V[[n]]],...,
                                    fm[V[[1]],...,V[[n]]]}];
{f1[x], f2[x1,...,xn], f3[t], f4[x1,...,xn],   f5[{x1,...,xn}]}
```

The *Mathematica* pure functions without names of variables,
Function[expr], where expr includes the symbol # instead of a name of
variable,

```
f1=Function[#];
f2=Function[#1,...#n];
f3=Function[{x1[#],...,xn[#]}];
f4=Function[{f1[#1,...,#n],...,fm[#1,...,#n]}];
f5=Function[{f1[#[[1]],...#[[n]]],..., fm[#[[1]],...,#[[n]]]}];
{f1[x], f2[x1,...,xn], f3[t],  f4[x1,...,xn],   f5[{x1,...,xn}]}
```

The *Mathematica* pure functions without names of variables, with #&
notation,

```
f1=#&;
f2=#1...#n&;
f3={x1[#],...,xn[#]}&;
f4={f1[#1,...,#n],...,fm[#1,...,#n]}&;
f5={f1[#[[1]],...#[[n]]],...,fm[#[[1]],...,#[[n]]]}&;
{f1[x], f2[x1,...,xn],  f3[t],  f4[x1,...,xn],   f5[{x1,...,xn}]}
```

The *Mathematica* Module,

```
f1[x_]:=Module[{},expr]; f2[x_,...,xn_]:=Module[{},expr];
f3[t_]:=Module[{},{x1[t],...,xn[t]}];
f4[x1_,...,xn_]:=Module[{},{f1[x1,...,xn],...,fm[x1,...,xn2]}];
f5[V_?VectorQ]:=Module[{}, {f1[V[[1]],...,V[[n]]],...,
                            fm[V[[1]],...,V[[n]]]}];
{f1[x], f2[x1,...,xn], f3[t], f4[x1,...,xn], f5[{x1,...,xn}]}
```

Problem 4.15 Let $f_1 : \mathbb{R} \to \mathbb{R}$, $f_1(x) = x^2$, be a real function of real variable, $f_2 : \mathbb{R}^2 \to \mathbb{R}$, $f_2(x_1, x_2) = x_1 + x_2$, be a real function of two real variables, $f_3 : \mathbb{R} \to \mathbb{R}^3$, $f_3(t) = \langle t^2, t^3, t^4 \rangle$, be a real vector-valued function of real variable, $f_4 : \mathbb{R}^2 \to \mathbb{R}^3$, $f_4(x_1, x_2) = \langle \sin(x_1 + x_2), \cos(x_1 - x_2), \tan(x_1 + x_2) \rangle$, be a real vector-valued function of two real variables, $f_5 : \mathbb{R}^2 \to \mathbb{R}^3$, $f_5(V) = \langle \sin(V_1 + V_2), \cos(V_1 - V_2), \tan(V_1 + V_2) \rangle$ $(V = < V_1, V_2 >)$, be a real vector-valued function of real vector argument. Define and evaluate these functions in both computer algebra systems.

Maple:

```
f11:=x->x^2;  f12:=(x1,x2)->x1+x2; f13:=t-><t^2,t^3,t^4>;
f14:=(x1,x2)-><sin(x1+x2),cos(x1-x2),tan(x1+x2)>;
f15:=(V::Vector)-><sin(V[1]+V[2]),cos(V[1]-V[2]),tan(V[1]+V[2])>;
f11(1); f12(1,1); f13(1); f14(1,1); f15(<1,1>);
f21:=unapply(x^2,x); f22:=unapply(x1+x2,x1,x2);
f23:=unapply(<t^2,t^3,t^4>,t); f24:=unapply(<sin(x1+x2),
  cos(x1-x2),tan(x1+x2)>,x1,x2);
f25:=unapply(<sin(V[1]+V[2]),cos(V[1]-V[2]),tan(V[1]+V[2])>,
  V::Vector); f21(1); f22(1,1); f23(1); f24(1,1); f25(<1,1>);
f31:=proc(x) x^2 end; f32:=proc(x1,x2) x1+x2 end;
f33:=proc(t) <t^2,t^2,t^4> end;
f34:=proc(x1,x2) <sin(x1+x2),cos(x2-x2),tan(x1+x2)> end;
f35:=proc(V::Vector) <sin(V[1]+V[2]),cos(V[1]-V[2]),tan(V[1]+
  V[2])> end; f31(1); f32(1,1); f33(1); f34(1,1); f35(<1,1>);
```

Mathematica:

```
f11[x_]:=x^2; f12[x1_,x2_]:=x1+x2; f13[t_]:={t^2,t^3,t^4};
f14[x1_,x2_]:={Sin[x1+x2],Cos[x1-x2],Tan[x1+x2]};
f15[V_?VectorQ]:={Sin[V[[1]]+V[[2]]],Cos[V[[1]]-V[[2]]],
  Tan[V[[1]]+V[[2]]]};
```

```
{f11[1], f12[1,1], f13[1], f14[1,1], f15[{1,1}]}
f21=Function[x,x^2]; f22=Function[{x1,x2},x1+x2];
f23=Function[t,{t^2,t^3,t^4}];
f24=Function[{x1,x2},{Sin[x1+x2],Cos[x1-x2],Tan[x1+x2]}];
f25=Function[V,{Sin[V[[1]]+V[[2]]],Cos[V[[1]]-V[[2]]],
 Tan[V[[1]]+V[[2]]]}];
{f21[1],f22[1,1],f23[1],f24[1,1],f25[{1,1}]}
f31=Function[#^2]; f32=Function[#1+#2];
f33=Function[{#^2,#^3,#^4}];
f34=Function[{Sin[#1+#2],Cos[#1-#2],Tan[#1+#2]}];
f35=Function[{Sin[#[[1]]+#[[2]]],Cos[#[[1]]-#[[2]]],
 Tan[#[[1]]+#[[2]]]}];
{f21[1],f22[1,1],f23[1],f24[1,1],f25[{1,1}]}
f41=#^2&; f42=#1+#2&; f43={#^2,#^3,#^4}&;
f44={Sin[#1+#2],Cos[#1-#2],Tan[#1+#2]}&;
f45={Sin[#[[1]]+#[[2]]],Cos[#[[1]]-#[[2]]],Tan[#[[1]]+#[[2]]]}&;
{f41[1],f42[1,1],f43[1],f44[1,1],f45[{1,1}]}
f51[x_]:=Module[{},x^2];    f52[x1_,x2_]:=Module[{},x1+x2];
f53[t_]:=Module[{},{t^2,t^3,t^4}]; f54[x1_,x2_]:=Module[{},
 {Sin[x1+x2],Cos[x1-x2],Tan[x1+x2]}];
f55[V_?VectorQ]:=Module[{}, {Sin[V[[1]]+V[[2]]],
 Cos[V[[1]]-V[[2]]], Tan[V[[1]]+V[[2]]]}];
{f51[1], f52[1,1], f53[1], f54[1,1], f55[{1,1}]}
```

For a more detail discussion about the *Maple* and *Mathematica* functions proc, Module, Function, see Chapter 1 and Chapter 2. □

Piecewise functions

Maple:

The piecewise function,
The *Maple* procedure proc and the control structure if,

```
f:=x->piecewise(cond1,expr1,expr2);
g:=x->piecewise(cond1,expr1,cond2,expr2,expr3);
f:=proc(x) if cond1 then expr1 else expr2 end if: end proc:
g:=proc(x) if cond1 then expr1 elif cond2 then expr2
                          else expr3 end if: end proc:
f:=x->Heaviside(x-a)*Heaviside(b-x);    convert(f(x),piecewise);
```

Mathematica:

The conditional operator /;
The `Piecewise` and `UnitStep` functions,

```
f[x_]:=expr/;cond              f[x_]:=UnitStep[x-a]*UnitStep[b-x];
f[x_]:=Piecewise[{{cond1,val1},{cond2,val2},...}];
f[x_]:=Which[cond1,val1,...,condn,valn];
```

Recursive functions

Recursive functions are the functions that are defined in terms of themselves.

Maple:

The recursive procedures (e.g., as a procedure `proc` with the `option remember`),

```
Rpr(1)=n0; Rpr:=proc(n) option remember; Rpr(n-1); end; Rpr(i);
```

Mathematica:

The recursive functions (e.g., as a transformation rule) with the global variable `$RecursionLimit=Infinity`,

```
$RecursionLimit=Infinity; Rpr[1]=n0; Rpr[n_]:=Rpr[n-1]; Rpr[i]
```

Problem 4.16 Let $X = \{x \mid x \in \mathbb{R}, x \geq 0\}$ and the function $f : X \to \mathbb{R}$ be defined by $f(x) = x/(x + n)$, $(n \in \mathbb{N})$. Demonstrate that $f(X) = \{y \mid y \in \mathbb{R}, 0 \leq y < 1\}$. Observe the behavior of the function f, when $n = \{1, \ldots, 20\}$.

Maple:

```
with(plots): interface(showassumed=0);
A1:=convert(x/(x+n),parfrac,x); limit(A1, x=infinity);
subs(x=0,x/(x+n)); solve(q/(q+n)=p, q); f:= x-> x/(x+n);
simplify(f((-p*n)/(p-1))); assume(n,'natural'); G:=NULL; N:=20;
for i from 1 to N do n:=i; G:=G,plot(f(x),x=0..100, 0..1.2); od:
G:=[G]: display(G, insequence=true);
```

Mathematica:

```
f[x_]:=x/(x+n); f1[x_,i_]:=x/(x+i);
{a1=Apart[x/(x+n)],Limit[a1,x->Infinity],x/(x+n)/.x->0,
 Solve[q/(q+n)==p,q], f[(-p*n)/(p-1)]//Simplify}
n=20; Animate[Plot[f1[x,i],{x,0,100}, PlotRange->{0,1.2},
               PlotStyle->Hue[0.7]],{i,1,n}]
```

At first, we demonstrate that $f(X) \subseteq \{y \mid y \in \mathbb{R}, 0 \le y \le 1\}$. Let $x \in X$, then $|x| \ge 0$, and decomposing $x/(x+n)$ $(n \in \mathbb{N})$ into partial fractions, we obtain $0 \le x/(x+n) = 1 - n/(x+n) < 1$, then $f(X) \subseteq \{y \mid y \in \mathbb{R}, 0 \le y < 1\}$. Now we prove that $f(X) = \{y \mid y \in \mathbb{R}, 0 \le y < 1\}$, i.e., for $\forall p \in \mathbb{R} : 0 \le p < 1 \; \exists \, q \ge 0 \Rightarrow f(q) = p$. Therefore, solving the equation $q/(q+n) = p$ $(0 \le p < 1)$, we obtain $q = (-pn)/(p-1)$ $(q \ge 0)$, therefore $f((-pn)/(p-1)) = p$ $((-pn)/(p-1) \ge 0)$, i.e. $\text{Im} f = \{y \mid y \in \mathbb{R}, 0 \le y < 1\}$. $\qquad \square$

Problem 4.17 Let $X = \{x \mid x \in \mathbb{R}, -\pi \le x \le \pi\}$ and the function $f : X \to \mathbb{R}$ be defined by $f(x) = \cos(6(x - a\sin x))$. Observe the behavior of the function f, where $a \in (\frac{1}{2}, \frac{3}{2})$.

Maple:

```
with(plots): y:=x->cos(6*(x-a*sin(x))); G:=NULL; N:=20;
for i from 0 to N do a:=1/2+i/N; G:=G,plot(y(x),x=-Pi..Pi); od:
G:=[G]: display(G,insequence=true);
```

Mathematica:

```
n=20; y[x_,i_]:=Cos[6*(x-(1/2+i/n)*Sin[x])];
Animate[Plot[y[x,i],{x,-Pi,Pi},
  PlotRange->{-1,1},PlotStyle->Hue[0.7]],{i,0,n}]
```

$\qquad \square$

Problem 4.18 Let $Y = \{y \mid y \in \mathbb{R}, y \ge 0\}$ and the function $f : R \to Y$ be defined by $f(x) = x^2$ $(x \in \mathbb{R})$. Prove that $f(R) = Y$ and that the mapping f is not one-to-one.

Maple:

```
interface(showassumed=0); f:=x->x^2; evalb(f(sqrt(y))=y);
assume(a,'real'); x_1:=-a; x_2:=a;  evalb(f(x_1)=f(x_2));
plot(f(x),x=-10..10,color=blue);
```

Mathematica:

```
f[x_]:=x^2; {f[Sqrt[y]]===y, x1=-a, x2=a}
Assuming[a\[Element]Reals, f[x1]===f[x2]]
Plot[f[x],{x,-10,10},PlotStyle->Hue[0.7]]
```

Since $f(\sqrt{y}) = y$, $\mathrm{Im} f = Y$. Let $x_1 = -a$, $x_2 = a$, $x_1 \neq x_2$ ($a \in \mathbb{R}$), then $f(x_1) = (-a)^2 = a^2 = f(x_2)$, and therefore the mapping f is not one-to-one. $\Box$

4.9   Operations on Functions

Let us consider the most important results describing various operations on functions in both systems.

Maple:

```
(f1@f2)(x);    ((f1@f2)@f3)(x);    (f1@f2@...@fn)(x);   (f@@n)(x);
f1:=(f::function)->apply(op(0,f),x);                   f1(f2());
apply(f,expr); apply(f,x1,...,xn);      apply(f,[x1,...,xn]);
map(f,expr);   map(f,[x1,...,xn]);       map2(f,g,[x1,...,xn]);
map(x->expr,expr);                       map(x->expr,[x1,...,xn]);
```

The *Maple* composition operator @,

Applying functions repeatedly, the repeated composition operator @@,

Defining a function which takes a function name as an argument, ->, apply,

Applying functions to other expressions and objects (e.g., lists), apply,

Applying functions to each elements of other expressions and objects, map, map2,

Functions without names (or anonymous functions), ->, map.

```
n:=5; (f@g)(x); ((f@g)@h)(x); (f@@n)(x);
f1:=(f::function)->apply(op(0,f),x+a)+apply(op(0,f),x-a);
f1(F()); apply(g,x^2+y^2); apply(g,x^2,y^2);
apply(g,[x^2,y^2]); map(g,x^2+y^2); map(g,[x^2,y^2]);
map2(f,g,[x^2,y^2]); map(x->x^2,x+y);
l1:=map(x->x^2,['x||i'$'i'=1..9]); convert(l1,`+`);
```

Mathematica:

```
Composition[f1,...,fn][x,y] Composition[f,f,f][x] f1@f2@{x,y,z}
x//f1//f2  f@f@f@x        Nest[f,expr,n]   NestList[f,expr,n]}
f1[f_,x_]:=f[x+x1]+f[x+x2];f1[f2,x]            list1={x1,...,xn}
Apply[f,{expr}] Apply[f,{x1,...,xn}]   f@@{expr} f@@{x1,...,xn}
Map[f,expr]     Map[f,{x1,...,xn}]       f/@expr   f/@{x1,...,xn}
MapAll[f,list1] f//@list1 Scan[f,expr] Function[#]      #1+#2&
```

The Composition function or functional operators @, //, where the arguments are pure functions f1,f2,...,

Applying functions repeatedly, Nest, NestList,

Defining a function which takes a function name as an argument, f1[f_,x_]:=expr,

Applying functions to other expressions and objects (e.g., lists), Apply or f@@expr,

Applying functions to each elements of other expressions and objects (e.g., lists), Map or f/@expr, MapAll or f//@expr, Scan,

Pure functions (or anonymous functions), Function, the #& notation.

```
{n=5,list1={x,y}, Composition[f1,f2][x,y], Composition[f,f,f][x],
 f1@f2@{x,y,z},x//f1//f2,f@f@f@x,Nest[f,x+y,n],NestList[f,x+y,n]}
f1[f_,x_]:=f[x+a]+f[x-a];  Scan[Print,list1]
{f1[F,x], Apply[g,{x^2+y^2}], Apply[g,{x^2,y^2}], f@@{x^2+y^2},
 f@@{x^2,y^2}, Map[g,x^2+y^2],Map[g,{x^2,y^2}], g/@x^2+y^2,
 g/@{x^2,y^2},MapAll[g,list1],g//@list1,Function[#^2][x],#^2&[x]}
Total[Map[#^2&,Table[Subscript[x,i],{i,1,9}]]]
```

Note. The *Mathematica* function Scan applies a function to each element without constructing a new expression.

Inverse functions

Maple:

The inverse function tables eval(invfunc), eval(Invfunc), the invfunc, Invfunc, unprotect functions:

```
eval(invfunc);  eval(Invfunc);   invfunc[sin];  (sin@@(-1))(1);
unprotect('invfunc');  invfunc[f]:=g;   simplify((f@@(-1))@@n);
```

Mathematica:

The *Mathematica* function InverseFunction, or we can create a specific
function, for example, invFunc:

```
{InverseFunction[f][x],
 Composition[Sin,InverseFunction[Sin]][x]}
invFunc[f_,x_]:=Module[{y},Solve[f==y,x,InverseFunctions->
 True][[1,1,2]]/.y->x]; invFunc[a*x^2+b*x+c,x]
{InverseFunction[Sin],InverseFunction[Sin][1]}
```

Problem 4.19 Let $g_1 : \mathbb{R} \to \mathbb{R}$, $g_2 : \mathbb{R} \to \mathbb{R}$, $g_3 : \mathbb{R} \to \mathbb{R}$ be functions defined
by the formulas $g_1(x) = (x+1)^2$, $g_2(x) = x - 1$, $g_3(x) = x + 1$. Evaluate
the following compositions of functions: $g_1 \circ g_2$, $g_2 \circ g_3$, $g_1 \circ g_3$, $g_2 \circ g_2$.
Verify that $(g_1 \circ g_2) \circ g_3 = g_1 \circ (g_2 \circ g_3)$, $g_1 \circ g_2 \neq g_2 \circ g_1$, $g_1 \circ g_3 \neq g_3 \circ g_1$,
$g_2 \circ g_3 = g_3 \circ g_2$.

Maple:

```
g1:=x->(x+1)^2; g2:=x->x-1; g3:=x->x+1;
map(expand,[(g1@g2)(x),(g2@g3)(x),(g1@g3)(x),(g2@g2)(x)]);
A1:=((g1@g2)@g3)(x); A2:=(g1@(g2@g3))(x); A3:=(g1@g2)(x);
A4:=expand((g2@g1)(x)); A5:=expand((g1@g3)(x));
A6:=factor((g3@g1)(x)); A7:=expand((g2@g3)(x));
A8:=factor((g3@g2)(x)); evalb(A1=A2); evalb(A3=A4);
evalb(A5=A6); evalb(A7=A8);
```

Mathematica:

```
g1[x_]:=(x+1)^2; g2[x_]:=x-1; g3[x_]:=x+1; Map[Expand,
{g1@g2[x],g2@g3[x],g1@g3[x],g2@g2[x]}]
{a1=g1@g2@g3[x],a2=g1@g2@g3[x]}
{a3=g1@g2[x],a4=g2@g1[x],a5=g1@g3[x],a6=g3@g1[x],
 a7=g2@g3[x],a8=g3@g2[x]}//Expand
{a1===a2,a3===a4,a5===a6,a7===a8}
```

□

Problem 4.20 Let g_1, g_2, g_3 be the functions from $\mathbb{R}$ to $\mathbb{R}$. Prove that
$(g_1 \circ g_2) \circ g_3 = g_1 \circ (g_2 \circ g_3)$.

Maple:

```
interface(showassumed=0); assume(g1(x),real,
g2(x),real,g3(x),real); is(g3(x),real); is(g1(x),real);
L1:=[((g1@g2)@g3)(x),(g1@g2)(g3(x)),g1(g2(g3(x))),
        g1((g2@g3)(x)),(g1@(g2@g3))(x)];
for k from 1 to nops(L1)-1 do evalb(L1[k]=L1[k+1]); od;
```

Mathematica:

```
l1=(g1@(g2@g3))[x]==g1[g2@g3[x]]==g1[g2[g3[x]]]
l2=((g1@g2)@g3)[x]==g1@g2@g3[x]==g1[g2[g3[x]]]
{n=Length[l1],l1[[n]]===l2[[n]]}
```

Let $x \in \mathbb{R}$. We can apply several times the definition of the composition of functions: $(f \circ g)(x) = f(g(x))$, $x \in \mathbb{R}$. Then we obtain: $[(g_1 \circ g_2) \circ g_3](x) = [g_1 \circ g_2](g_3(x)) = g_1[g_2(g_3(x))] = g_1[(g_2 \circ g_3)(x)] = [g_1 \circ (g_2 \circ g_3)](x)$. $\quad\square$

Problem 4.21 Let $f : X \to Y$ and $g : Y \to X$ be inverse functions. Prove that f and g are bijections.

Maple:

```
`1_X`[x]:=x;`1_X`[x1];A1:=x1=x2; A2L:=x->`1_X`[x]=(g@f)(x);
A2R:=x->(g@f)(x)=`1_X`[x]; A2L(x1); A3:=f(x1)=f(x2);
A4:=A2L(x1)=subs(A3,rhs(A2L(x1))); A5:=A4=(g@f)(x2);
A6:=A5=A2R(x2); `1_Y`:=x->x; A22R:=x->(f@g)(x)=`1_Y`(x);
A7:=A22R(y); subs(g(y)=x,A7);
```

Mathematica:

```
X1[x_]:=x;a2L[x_]:=X1[x]==(g@f)[x];a2R[x_]:=(g@f)[x]==X1[x];
{X1[x1], a1=x1==x2, a2L[x1],a3=f[x1]==f[x2],a4=a2L[x1]
==(a2L[x1][[2]])/.{x1->x2}),a5=a4==(g@f)[x2],a6=a5==a2R[x2]}
Y1[x_]:=x; a22R[x_]:=(f@g)[x]==Y1[x];
{a7=a22R[y],a7/.{f[g][y]->f[x]}}
```

We prove that f is a bijection (the demonstration is similar for g). Let $x_1, x_2 \in X$, $f(x_1) = f(x_2)$. Then we obtain that the mapping f is one-to-one: $x_1 = 1_X(x1) = (g \circ f)(x_1) = g(f(x_1)) = g(f(x_2)) = (g \circ f)(x_2) = 1_X(x_2) = x_2$.
Let $y \in Y$, we prove that $\exists x \in X : f(x) = y$. Since $g : Y \to X$, $g(y) \in X$, and $f(g(y)) = (f \circ g)(y) = 1_Y(y) = y$. If $x = g(y)$, we have $f(x) = f(g(y)) = y$, therefore, y is surjective. $\quad\square$

Problem 4.22 Let us now consider Problem 4.3.

Let $X = \{-4, -3, -2, -1, 0, 1, 2, 3, 4\}$ be the set of the piano octaves, $f = \{C, C1, D, D1, E, F, F1, G, G1, A, A1, B\}$, the set of all pitches for each octave, and $Y = \{n \mid 1 \leq n \leq 89\}$, the subset of natural numbers of length 89. Let $f : X \to Y$ be the function defined by the rule: $A[-4] = 1$, $A1[-4] = 2$, $B[-4] = 3$, $C[-3] = 4$, ..., $C[4] = 88$, Silence $= 89$. Let us construct the set A_T of all pitches that can produce a piano:

$$
\begin{aligned}
AT = \{&A[-4] = 1, A1[-4] = 2, B[-4] = 3, C[-3] = 4, C1[-3] = 5, D[-3] = 6, \\
&D1[-3] = 7, E[-3] = 8, F[-3] = 9, F1[-3] = 10, G[-3] = 11, G1[-3] = 12, \\
&A[-3] = 13, A1[-3] = 14, B[-3] = 15, C[-2] = 16, C1[-2] = 17, \\
&D[-2] = 18, D1[-2] = 19, E[-2] = 20, F[-2] = 21, F1[-2] = 22, \\
&G[-2] = 23, G1[-2] = 24, A[-2] = 25, A1[-2] = 26, B[-2] = 27, \\
&C[-1] = 28, C1[-1] = 29, D[-1] = 30, D1[-1] = 31, E[-1] = 32, \\
&F[-1] = 33, F1[-1] = 34, G[-1] = 35, G1[-1] = 36, A[-1] = 37, \\
&A1[-1] = 38, B[-1] = 39, C[0] = 40, C1[0] = 41, D[0] = 42, \\
&D1[0] = 43, E[0] = 44, F[0] = 45, F1[0] = 46, G[0] = 47, G1[0] = 48, \\
&A[0] = 49, A1[0] = 50, B[0] = 51, C[1] = 52, C1[1] = 53, D[1] = 54, \\
&D1[1] = 55, E[1] = 56, F[1] = 57, F1[1] = 58, G[1] = 59, G1[1] = 60, \\
&A[1] = 61, A1[1] = 62, B[1] = 63, C[2] = 64, C1[2] = 65, D[2] = 66, \\
&D1[2] = 67, E[2] = 68, F[2] = 69, F1[2] = 70, G[2] = 71, G1[2] = 72, \\
&A[2] = 73, A1[2] = 74, B[2] = 75, C[3] = 76, C1[3] = 77, \\
&D[3] = 78, D1[3] = 79, E[3] = 80, F[3] = 81, F1[3] = 82, \\
&G[3] = 83, G1[3] = 84, A[3] = 85, A1[3] = 86, B[3] = 87, \\
&C[4] = 88, \text{Silence} = 89\}
\end{aligned}
$$

Maple:

```
f:=[C,C1,D,D1,E,F,F1,G,G1,A,A1,B]; X:=[-3,-2,-1,0,1,2,3];
Y:=[seq(i,i=1..89)]; fn:=nops(f); Xn:=nops(X); L:=NULL;
Yn:=nops(Y);LN:=[A[-4],A1[-4],B[-4]];RN:=[C[4],Silence];
Pr:=proc(j) local i; global L;
 for i from 1 to fn do L:=L,f[i][X[j]]; od: end proc;
for i from 1 to Xn-1 do print(Pr(i)); od:
MN:=[Pr(Xn)]; LT:=[op(LN),op(MN),op(RN)]; LF:=NULL;
for i1 from 1 to Yn do LF:=LF,LT[i1]=i1; od: AT:=[LF];
```

Mathematica:

```
{f={CC,C1,DD,D1,EE,F,F1,G,G1,A,A1,B},x={-3,-2,-1,0,1,2,3},
 y=Range[1,89],fn=Length[f],xn=Length[x],yn=Length[y],
 ln={A[-4],A1[-4],B[-4]},rn={CC[4],Silence},l={},lf={}}
Pr[j_]:=Module[{i},Do[l=Append[l,f[[i]][x[[j]]]],
 {i,1,fn}];l]; Do[Print[Pr[i]],{i,1,xn-1}]; mn=Pr[xn]
lt=Append[Append[ln,mn],rn]//Flatten
Do[lf=Append[lf,lt[[i1]]==i1],{i1,1,yn}]; at=lf
```

Here we replace C, D, E pitches by its equivalents CC, DD, EE (because of the predefined values of *Mathematica* symbols C, D, E). The pitches $C1$ and $C\sharp$, $D1$ and $D\sharp$, $F1$ and $F\sharp$, $G1$ and $G\sharp$, $A1$ and $A\sharp$ are equivalent. $\square$

4.10   Univariate and Multivariate Polynomials

Polynomials are one of the basic concepts of algebra, mathematics, and other sciences. Polynomials are an important class of smooth functions that have derivatives of all finite orders. In general, a *polynomial* is an expression constructed from one or more variables and constants, with the operations of addition, subtraction, multiplication, and raising to powers (that are natural numbers). In algebra, a univariate polynomial p with respect to one indeterminate (or variable) x over a ring R is defined as a formal expression: $f = a_0 + a_1x + \cdots + a_nx^n$, where $n \in \mathbb{N}$, the coefficients $a_i \in R$. A multivariate polynomial p with respect several variables over a ring R can be defined in a similar manner, the important difference is that the order of monomials in a multivariate polynomial p gets more complicated.

We denote by $\mathbb{Q}[x]$ the set of polynomials in x over $\mathbb{Q}$. Similarly $\mathbb{Z}[x]$, $\mathbb{R}[x]$, $\mathbb{C}[x]$, $\mathbb{D}[x]$ are denoted by choosing the correspondent coefficients for the polynomials.

Polynomial algebra is the base of symbolic mathematics, the core of computer algebra systems. Nowadays solution of polynomial systems is one of the most active areas of investigation in computer algebra. We consider most important functions of polynomial algebra in both systems.

Defining polynomials

Maple:

Univariate and multivariate polynomials, and polynomials over number rings and fields can be defined in different manners (see `?polynomials`).

Moreover, there exist various specific packages, e.g., PolynomialTools, Groebner, SDMPolynom, PolynomialIdeals, RegularChains, for working with polynomials. Also, various special polynomials can be defined with the package orthopoly and using the bernoulli, bernstein, euler, cyclotomic, fibonacci functions.

Defining univariate polynomials of the form $p = a_0 + a_1 x + \cdots + a_n x^n$ ($n = 10$),

```
n:=10; p:=add(a||i*x^i, i=0..n);    p:=add(cat(a,i)*x^i,i=0..n);
p:=a0+add(cat(a,i)*x^i, i=1..n);
p:=a0+sum('cat(a,i)*x^i','i'=1..n);
p:=convert(['a||i*x^i'$'i'=0..n],`+`);          A:=array(0..n);
S:=A[0]; for i from 1 to n do S:=S+A[i]*x^i; od: p:=S;
RandomTools[Generate](polynom(integer(range=-n..n),x,
                            degree=n)); randpoly(x,degree=n);
```

Defining multivariate polynomials of the form $p = a_{00} + a_{10}x + a_{20}x^2 + a_{01}y + a_{11}xy + a_{21}x^2y + a_{02}y^2 + a_{12}xy^2 + a_{22}x^2y^2$ ($n = 2$, $m = 2$),

```
n:=2; m:=2;          p:=add(add(a||i||j*x^i*y^j,i=0..n),j=0..m);
p:=add(add(cat(a,i,j)*x^i*y^j,i=0..n),j=0..m);
p:=sum(sum('a[i,j]*x^i*y^j','i'=0..n),'j'=0..m);
A:=Array(0..n,0..m): S:=0: k:=10;
for i from 0 to n do for j from 0 to m do
 S:=S+A[i,j]*x^i*y^j; od: od: p:=S;    randpoly({x,y},degree=n);
RandomTools[Generate](polynom(integer(range=-k..k),{x,y},
                            degree=n));
```

Mathematica:

Polynomials in *Mathematica* are represented beginning with the constant term. Univariate and multivariate polynomials, and polynomials over number rings and fields can be defined in different manners (see ?*Polynomial*).

Defining univariate polynomials of the form $p = a_0 + a_1 x + \cdots + a_n x^n$ ($n = 10$),

```
n=10; {p=Sum[Row[{a,i}]*x^i,{i,0,n}],  p=a0+Sum[Row[{a,i}]*x^i,
 {i,1,n}], p=Sum[Subscript[a,i]*x^i,{i,0,n}],
 c=Table[a[i],{i,0,n}],                u=Table[x^i,{i,0,n}], c.u}
f[x_,k_]:=Table[a[i],{i,0,k}].x^Range[0,k];            p=f[x,n]
{Cyclotomic[n,x],  deg=RandomInteger[{-n,n}], cR=Table[
 RandomInteger[{-n,n}],{i,0,n}],  u=Table[x^i,{i,0,n}],  cR.u}
```

Defining multivariate polynomials of the form $p = a_{00} + a_{10}x + a_{20}x^2 + a_{01}y + a_{11}xy + a_{21}x^2y + a_{02}y^2 + a_{12}xy^2 + a_{22}x^2y^2$ $(n = 2, m = 2)$,

```
n=2; m=2;    {p=Sum[Sum[Row[{a,i,j}]*x^i*y^j,{i,0,n}],{j,0,m}],
 p=Sum[Sum[a[i,j]*x^i*y^j,{i,0,n}],{j,0,m}],
 p=Sum[Sum[Subscript[a,{i,j}]*x^i*y^j,{i,0,n}],{j,0,m}],
 c=Table[a[i,j],{i,0,n},{j,0,m}]//Flatten,
 u=Table[x^i*y^j,{i,0,n},{j,0,m}]//Flatten, c.u//Sort}
f[x_,y_,k_,t_]:=Table[a[i,j],{j,0,t},
 {i,0,k}].x^Range[0,k].y^Range[0,t]; p=f[x,y,n,m]//Expand
{r=10, deg=RandomInteger[{-r,r}],
 cR=Table[RandomInteger[{-r,r}],{i,0,n},{j,0,m}]//Flatten,
 u=Table[x^i*y^j,{i,0,n},{j,0,m}]//Flatten, cR.u//Sort}
SymmetricPolynomial[2,Table[x[i],{i,1,r}]]]
```

Operations on polynomials

The basic arithmetical operations on polynomials (addition, subtraction, multiplication, and raising to powers) are defined in a direct manner. We consider other important operations.

Maple:

```
divide(p1,p2);    gcd(p1,p2);    lcm(p1,p2);        quo(p1,p2,x);
rem(p1,p2,x);     sort(p,x)      coeff(p,x,n);        coeffs(p,x);
lcoeff(p,x);      tcoeff(p,x);   degree(p,x);       ldegree(p,x);
collect(p,x);     expand(p);     normal(p);          sqrfree(p,x);
roots(p,x);       realroot(p);   psqrt(p);            proot(p,n);
fsolve(p,x);      compoly(p,x);  content(p,x);      primpart(p,x);
type(p,polynom(domain,x));       convert(series,polynom);
discrim(p,x);     resultant(p1,p2,x);                Factor(p) mod m;
factor(p);        evala(Factor(p,a));       Factor(p,a) mod m;
```

divide, **quo**, **rem**, performing exact polynomial division,

gcd, lcm, finding the greatest common divisor and least common multiple of polynomials,

coeff, coeffs, lcoeff, tcoeff, extracting polynomial coefficients,

degree, ldegree, determining the degree and the lowest degree of a polynomial,

collect, grouping the coefficients of like power terms together,

roots, fsolve, finding exact and numerical solutions of polynomial equations,

sort, a polynomial sorting operation,

type(p,polynom(domain,x)), testing if p is a polynomial with respect to x with
coefficients in the domain,

Factor(p) mod m, Factor(p,a) mod m, evala(Factor(p,a)), factorizing p,
respectively, over the *integers modulo* m, the *finite field* defined by the algebraic extension a over the integers modulo m (specified by RootOf), and an
algebraic number field defined by the extension a (specified by RootOf or a set
of RootOfs). Here m is a prime integer,

compoly, decomposing a polynomial into a composition of simpler polynomials,

discrim, resultant, computing the discriminant of the polynomial and the resultant of the two polynomials with respect to the variable x.

Mathematica:

```
PolynomialQuotientRemainder[p,q,x]          PolynomialQ[expr,x]
PolynomialQuotient[p,q,x]               PolynomialRemainder[p,q,x]
PolynomialGCD[p1,...,pn]                   PolynomialLCM[p1,...,pn]
Coefficient[p,x]  Coefficient[p,x,n]          CoefficientList[p,x]
Variables[p]      Exponent[p,x]            Factor[p,Extension->a]
Expand[p,x]       Collect[p,x]             FactorList[p,x] Sort[p]
Decompose[p,x]    Roots[eq,x]                        NRoots[eq,x]
Solve[eq,x]       Reduce[p,x]              Eliminate[{eqs},{vars}]
Root[p,n]         RootReduce[expr]            Resolve[expr,domain]
ToRadicals[expr]  Discriminant[p,x]              Resultant[p1,p2,x]
PolynomialMod[p,m]       GroebnerBasis[{p1,...,pn},{x1,...,xn}]
```

PolynomialQuotient, PolynomialRemainder, performing exact polynomial division,

PolynomialGCD, PolynomialLCM, finding the greatest common divisor and least
common multiple of polynomials,

PolynomialQ, testing if expr is a polynomial with respect to x,

Coefficient, CoefficientList, extracting polynomial coefficients,

Variables, determining a list of all independent variables,

Exponent, determining the degree of a polynomial,

Root, Roots, NRoots, Solve, Resolve, Reduce, finding exact and numerical solutions of polynomial equations,

Eliminate, eliminating variables between a set of simultaneous equations,

RootReduce, ToRadicals, reducing expr to a single Root object, and, viceversa,
reducing all Root objects in terms of radicals,

Sort, a polynomial sorting operation,

Collect, grouping the coefficients of like terms together,

Factor, and other polynomial and algebraic functions with the option Extension,
specifying the field $\mathbb{Q}[a]$ consisting of the rational numbers extended by the
algebraic number a,

Decompose, decomposing a polynomial into a composition of simpler polynomials,

Discriminant, Resultant, computing the discriminant of the polynomial and the
resultant of the two polynomials with respect to the variable x,

GroebnerBasis, generating a list of polynomials that form a Gröbner basis for the
set of polynomials,

PolynomialMod, finding the polynomial reduced modulo m.

Problem 4.23 Find all the solutions of the polynomial equation $x^3 - 5x^2 - 2x + 10 = 0$ which are less than 1.

Maple:

```
sols:=[solve(x^3-5*x^2-2*x+10=0,x)];
numsols:=[fsolve(x^3-5*x^2-2*x+10=0,x)];
selectSol:= proc(list) local i, l; l:=NULL;
 for i from 1 to nops(list) do if evalf(list[i])<1 then
   l:=l,list[i]; fi; od; RETURN(l); end;
l1:=selectSol(sols); l2:=selectSol(numsols);
```

Mathematica:

```
sols=Roots[x^3-5*x^2-2*x+10==0,x]; Reduce[sols&&x<1]
numsols=NRoots[x^3-5*x^2-2*x+10==0,x]; Reduce[numsols&&x<1]
```

We find all the solutions of the polynomial equation with the *Maple* solve
function. For choosing solutions that are less than 1, we construct a procedure
selectSol. In *Mathematica* we apply the functions Roots (or NRoots) and the
logical operator &&.   □

Problem 4.24 Let $p(x) = x^4 + a_1x^3 + a_2x^2 + a_3x + a_4$ be a univariate
polynomial with the real coefficients ($a_i \in \mathbb{R}$, $i = 1, \ldots, 4$). Find all the
polynomials that have the roots $\{a_1, a_2, a_3, a_4\}$.

Maple:

```
n:=4; p1:=(x-a1)*(x-a2)*(x-a3)*(x-a4); p2:=expand(p1);
for i from 1 to n do Eq||i:=coeff(p2,x,i-1)=a||(n-(i-1)); od;
Sol:=allvalues({solve({'Eq||i'$'i'=1..n},{'a||i'$'i'=1..n})}):
for i from 1 to 6 do
 s||i:=evalf(Sol[i]); s||i||f:=remove(has,s||i,I); od:
for i from 1 to 6 do s||i||f od;
```

Mathematica:

```
{n=4, p1=(x-a[1])*(x-a[2])*(x-a[3])*(x-a[4]), p2=Expand[p1]}
{eqL=Table[eq[i]=Coefficient[p2,x,i-1]==a[n-(i-1)],{i,1,n}],
 varL=Table[a[i],{i,1,n}]}
sol=Solve[eqL,varL]; s1=sol//N; s2=Select[s1,FreeQ[#,Complex]&]
{s3=Intersection[s2[[3]],s2[[5]]],
 sf=Union[{s2[[1]]},{s2[[2]]},{s2[[4]]},{s2[[6]]},{s3}]}
```

Let us represent a polynomial with the roots $\{a_i\}$ $(i = 1, \ldots, 4)$ in the form $p(x) = (x - a_1)(x - a_2)(x - a_3)(x - a_4)$. To obtain all the polynomials of the required form, we have to find all the real solutions of the system of linear equations (Eq1, ... Eq4). Then solving this system, we obtain the desired solutions: $\{a_1 = 0, a_2 = 0, a_3 = 0, a_4 = 0\}$, $\{a_1 = 1, a_2 = -2, a_3 = 0, a_4 = 0\}$, $\{a_1 = 1, a_2 = -1, a_3 = -1, a_4 = 0\}$, $\{a_1 = 0.565198, a_2 = -1.76929, a_3 = 0.638897, a_4 = 0\}$, $\{a_1 = 1, a_2 = -1.75488, a_3 = -0.56984, a_4 = 0.324718\}$.   □

4.11 Groups

Group theory is a powerful formal and general method for analyzing complex abstract and real-world systems (for example, multi-particle quantum mechanical systems) in which *symmetry* is present. A group[3] is one of the most fundamental structures in modern mathematics.

Abstract algebra studies various algebraic structures (e.g., magmas, quasigroups, loops, semigroups, monoids, groups, rings, fields).

A *magma*[4] is an elementary algebraic structure, $(M, \circ)$, where M is a nonempty set in which a *composition law* $\circ$ (or a *binary operation*) is defined: $\forall$ $(m_1, m_2) \in M$ a third element of M can be determined, denoted by $m_1 \circ m_2$. A binary operation $\circ$ is closed by definition (and

---

[3]The origins of the concept of a group can be observed in the works of P. Ruffini, E. Galois, L. Kronecker, the abstract concept of a finite group was first formulated by A. Cayley.

[4]The term magma was introduced by N. Bourbaki.

no other axioms are imposed on the operation): $\forall\ m_1, m_2 \in M, m_1 \circ$ $m_2 \in M$. Other algebraic structures can be considered as different kinds of magmas depending on the required axioms of the operation $\circ$. We mention the most important algebraic structures.

A *semigroup*, $(S, \circ)$ is a magma where the operation $\circ$ is associative: $\forall\ s_1, s_2, s_3 \in S, (s_1 \circ s_2) \circ s_3 = s_1 \circ (s_2 \circ s_3)$. Additionally, if for any two elements of S the operation is commutative, we say that $(S, \circ)$ is a *commutative semigroup* or an *Abelian semigroup*.

A semigroup S with an element $e \in S$: $es = se = s\ \forall s \in S$ is called *monoid*. The element e (if there exists) is unique and is called the *identity element*[5] or *neutral element* of S. If S is a monoid with the identity element e, the unique element $\tilde{s} \in S$ (if there exists): $\tilde{s}s = s\tilde{s} = e$ is called the *inverse element* of s.

A *group* is a monoid G in which there exists an inverse element for each element of G. If G is multiplicative, the inverse element of $g \in G$ is denoted by g^{-1}. A commutative group is called an *Abelian group*.

The group theory studies various types of groups, for example, permutation and symmetry groups, torsion and free-torsion Abelian groups, finite and infinite groups, free groups and semigroups, Galois and Lie groups, p-groups and Sylow subgroups, etc.

In symbolic mathematics there exists a special interest on the computational group theory (these algebraic structures are well appropriate for symbolic computing). Computational group theory is a branch of modern mathematics, actually active for the investigation, that studies groups by means of computers, i.e., develops and analyzes algorithms and data structures for computations with groups. Frequently, for many interesting groups it is impossible to obtain new results without a computer. The important and most frequently used representations of groups are, for example, permutation groups and groups defined by generators and relations, matrix groups, solvable groups, and the representation theory.

There are two specialized computer algebra systems which are well appropriate for computations with groups: GAP (Groups, Algorithms, Programming) and Magma (after the name of the elemental algebraic structure). However, in our book we consider the general purpose computer algebra systems, *Maple* and *Mathematica*, and solve some problems on group theory using various functions and packages, e.g., `group` and

---

[5]derived from the German word "einheit"

combinat, in *Maple*, and Combinatorica, in *Mathematica*.

We show how to solve some computational problems in finite group theory with both systems (e.g., permutation groups, the construction of groups using generators and relations, operations with the group elements, calculations of subgroups, normal subgroups, and factor groups). With these symbolic solutions, we can observe that the process of problem solving and teaching in finite group theory becomes more and more easier and interactive now, due to the visual interface of both computer algebra systems, *Maple* and *Mathematica*. Now we mention the most important functions that help us in solving problems.

Maple:

```
permgroup(n,gens);      grelgroup(gens,rels);        galois(f,var);
subgrel(gens,grelgroup);       algcurves[monodromy](f,x,y,group);
with(combinat);         permute(elems);             numbperm(elems);
randperm(elems);        with(group);                  elements(gr);
elements({elems});      grouporder(gr);              invperm(perm);
center(gr);             cosets(gr,sg);              orbit(gr,point);
core(sg,gr);            inter(gr1,gr2);           normalizer(gr,sg);
mulperms(p1,p2);        RandElement(gr);        groupmember(perm,gr);
isabelian(gr);          isnormal(gr,sg);         issubgroup(sg,gr);
convert([perm],'disjcyc');          convert(perm,'permlist',n);
```

permgroup, representing a permutation group of order n by a set of generators, gens,

grelgroup, representing a group by a set of generators, gens, and a set of relations, rels,

galois, computing the Galois group of an irreducible polynomial,

subgrel, representing a subgroup of a group by a set of generators of the subgroup and the group for which this is a subgroup, grelgroup,

monodromy,group (of the package algcurves), with the option group, computing the monodromy group G,

permute, numbperm, randperm (of the package combinat), constructing and working with the permutations of a list of elements,

elements, grouporder, invperm (of the package group), constructing and working with permutation groups, etc.

Mathematica:

```
Permutations[{elems}]          <<Combinatorica`   ToCycles[perm]
FromCycles[{{c1},...{cn},}]     PermutationWithCycle[n,{cycle}]
Permute[p1,p2]     PermutationGroupQ[{perms}] SymmetricGroup[n]
MultiplicationTable[{gr},op]   CyclicGroup[n]  DihedralGroup[n]
AlternatingGroup[n] OrbitRepresentatives[gr,x]     Orbits[gr,x]
NonCommutativeMultiply[x,y,z]    x**y**z
```

Permutations generating a list of all possible permutations of the elements elems,

ToCycles, FromCycles, PermutationWithCycle, representing the permutation perm in the cycle notation, and viceversa, representing the permutation that has the given cycle structure in the list notation,

Permute, finding the multiplication of two permutations,

PermutationGroupQ, verifying if the list of permutations forms a permutation group,

SymmetricGroup, generating the symmetric group on n symbols,

MultiplicationTable, constructing the Cayley table defined by the binary operation *op* on the elements of the group gr,

CyclicGroup, DihedralGroup, AlternatingGroup, constructing, respectively, the cyclic, dihedral, and alternating groups of permutations on n symbols,

Orbits, OrbitRepresentatives, finding the orbits of x induced by the action of the group on x and a representative of each orbit,

NonCommutativeMultiply, the form of multiplication, non-commutative multiplication.

Problem 4.25 Let us consider bijective mappings of a non-empty set X onto X, which can be multiplied under the circle-composition of mappings. Such mappings are called the *permutations* of X. Any set of permutations of X forming a group is called a *permutation group*. The set $S(X)$ of all the permutations on X is a group is called the *symmetric group* on X.

Let S_n be the symmetric group of all the permutations on n symbols, $1, 2 \ldots, n$ (n is given).

1) Find all the permutations $s_i \in S_4$.

2) Determine the number of permutations.

3) Obtain all the disjoint cycles.

4) For any two permutations $p1, p2$, compute the product of permutations.

5) Find all the elements of the group S_4 generated by the disjoint cycles $(1, 2)$ and $(1, 2, 3, 4)$.

6) Display all the elements in the standard mathematical notation for permutations.

Maple:

```
with(group): with(combinat):  S4:=permute(4); nops(S4);
numbperm(4); p1:=convert(S4[3],'disjcyc');
p2:=convert(S4[4],'disjcyc');mulperms(p1,p2);mulperms(p2,p1);
L4:=NULL: for i from 1 to nops(S4) do
L4:=L4,convert(S4[i],'disjcyc'); od: L4;
G4:=permgroup(4,{[[1,2]],[[1,2,3,4]]});
EG4:=elements(G4);nops(EG4); for i from 1 to nops(EG4) do
   print(Array([[1,2,3,4],convert(EG4[i],'permlist',4)])); od;
```

Mathematica:

```
<<Combinatorica`
{s4=Permutations[Range[4]], ls4=Length[s4],p1=ToCycles[s4[[3]]],
 p2=ToCycles[s4[[4]]], p11=FromCycles[p1],p21=FromCycles[p2],
 m1=Permute[p11,p21],m2=Permute[p21,p11],Map[ToCycles,{m1,m2}]}
{l4={}, l4=Append[l4, Table[ToCycles[s4[[i]]],{i,1,ls4}]],
 HoldForm[l4]==l4, p12=FromCycles[{{1,2},{3},{4}}],
 p22=FromCycles[{{1,2,3,4}}], g4=SymmetricGroup[4],
 PermutationGroupQ[g4], lg4=Length[g4]}
Map[MatrixForm,Table[{{1,2,3,4},g4[[i]]},{i,1,lg4}]]
```

It should be noted that in *Maple* and *Mathematica* there are some small differences in representing permutations in the disjoint cycle notation and in the process of multiplying permutations. In *Maple* the most important operations on permutations are performed in the disjoint cycle notations, in *Mathematica*, in the list notation.   □

Problem 4.26 Let p_1 and p_2 be any two permutations $\in S_7$. Obtain p_1^{-1} and verify that $p_1^{-1} p_2 p_1 \neq p_2$.

Maple:

```
with(group): with(combinat):
p1:=convert([4,6,7,3,5,1,2],'disjcyc');
p2:=convert([3,6,7,1,5,4,2],'disjcyc'); p1Inv:=invperm(p1);
convert(p1Inv,'permlist',7); p2p1:= mulperms(p1,p2);
convert(p2p1,'permlist',7); p1Invp2p1:=mulperms(p2p1,p1Inv);
evalb(convert(p1Invp2p1,'permlist',7)=p2);
```

Mathematica:

```
<<Combinatorica`
{p1={4,6,7,3,5,1,2}, p2={3,6,7,1,5,4,2},
 p1Inv=InversePermutation[p1], p2p1=Permute[p2,p1],
 p1Invp2p1=Permute[p1Inv,p2p1],p1Invp2p1===p2}
```

□

Problem 4.27 Let S_3 be the symmetric group of 3 elements (we denote it by $G3$).
    1) Find all the elements of $G3$.
    2) Determine the order of each element of the group.
    3) Determine the order of the group $G3$.

Maple:

```
with(group): with(combinat): G3:=permgroup(3,{[[1,2]],[[2,3]]});
grouporder(G3); EG3:=elements(G3); nops(EG3);
for i from 1 to nops(EG3) do
    print(Array([[1,2,3],convert(EG3[i],'permlist',3)])); od;
S31:=permute(3); nops(S31);
for i from 1 to nops(S31) do convert(S31[i],'disjcyc'); od;
mulperms(EG3[2],EG3[2]); mulperms(EG3[3],EG3[3]);
mulperms(EG3[6],EG3[6]); mulperms(mulperms(EG3[4],EG3[4]),
 EG3[4]); mulperms(mulperms(EG3[5],EG3[5]),EG3[5]);
```

Mathematica:

```
<<Combinatorica`
{n=3,PermutationGroupQ[SymmetricGroup[n]],
 eg3=SymmetricGroup[n], grouporder=n!, leg3=Length[eg3]}
Map[MatrixForm,Table[{{1,2,3},eg3[[i]]},{i,1,leg3}]]
{s31=Permutations[Range[3]],ls31=Length[s31]}
Table[ToCycles[s31[[i]]],{i,1,ls31}]
{Permute[eg3[[2]],eg3[[2]]],Permute[eg3[[3]],eg3[[3]]],
 Permute[eg3[[6]],eg3[[6]]],
 Permute[Permute[eg3[[4]], eg3[[4]]], eg3[[4]]],
 Permute[Permute[eg3[[5]], eg3[[5]]], eg3[[5]]]}
```

The first permutation in the list EG3 (or eg3) is the identity permutation $(1, 2, 3)$ and is denoted in *Maple* by [ ]. The permutations $(2, 1, 3)$, $(1, 3, 2)$, and $(3, 2, 1)$ are of order 2, and the other two permutations, $(2, 3, 1)$ and $(3, 1, 2)$, are of order 3.

□

Problem 4.28 An important method for studying properties of finite groups is the construction of their *abstract group tables*. Let G be a multiplicative group of 6 elements $\{e, a, b, c, d, f\}$. We recall that Cayley's theorem states that every finite group G is isomorphic to a permutation group (or a subgroup SG of the symmetric group S_n on G). Let $f : G \to S_3$ be an isomorphism, where $f(e) = (1,2,3)$, $f(a) = (2,3,1)$, $f(b) = (3,1,2)$, $f(c) = (2,1,3)$, $f(d) = (3,2,1)$, $f(f) = (1,3,2)$.

Construct the multiplication table (or the *Cayley table*) of the elements of G.

Maple:

```
with(group): with(combinat): G:=[e,a,b,c,d,f];
EG3:=[[],[[1,2,3]],[[1,3,2]],[[1,2]],[[1,3]],[[2,3]]];
n:=nops(EG3); print(seq(G[i]=convert(EG3[i],'permlist',3),
   i=1..n)); GT:=Matrix(1..n+1,1..n+1):
for i from 1 to n do for j from 1 to n do for k from 1 to n do
   if EG3[k]=mulperms(EG3[j],EG3[i]) then
   GT[i+1,j+1]:=G[k]; fi; od: od: od:  GT[1,1]:=` `:
for j from 1 to n do GT[1,j+1]:=G[j]; GT[j+1,1]:=G[j]; od: GT;
```

Mathematica:

```
<<Combinatorica`
eg3={{1,2,3},{2,3,1},{3,1,2},{2,1,3},{3,2,1},{1,3,2}}
{n=3, g={e,a,b,c,d,f},leg3=Length[eg3]}
subs=Table[eg3[[i]]->g[[i]],{i,1,leg3}]
multiplication[p1_,p2_]:=Permute[p1,p2];
{mTable1=MultiplicationTable[eg3,multiplication]//MatrixForm,
 mTable2=MultiplicationTable[eg3,multiplication]/.
 {m_Integer:>eg3[[m]]}//MatrixForm, mTable2/.subs}
```

The isomorphism is described in the list EG3 or eg3 according to the statement of the problem. However, it is possible to generate a permutation group with the function permgroup (in *Maple*) or with the function SymmetricGroup (in *Mathematica*), and construct the Cayley tables. In this case we modify the previous solution as follows:

Maple:

```
with(group): with(combinat): G:=[e,a,b,c,d,f]; G3:=permgroup(3,
  {[[1,3]],[[1,2,3]]}); EG31:=[op(elements(G3))]; n:=nops(EG31);
```

```
EG3:=[EG31[1],op(EG31[5..6]),op(EG31[2..4])]; print(seq(G[i]=
convert(EG3[i],'permlist',3),i=1..n));GT:=Matrix(1..n+1,1..n+1):
for i from 1 to n do for j from 1 to n do for k from 1 to n do
 if EG3[k]=mulperms(EG3[j],EG3[i]) then GT[i+1,j+1]:=G[k]; fi;
 od: od: od: GT[1,1]:=` `:
for j from 1 to n do GT[1,j+1]:=G[j]; GT[j+1,1]:=G[j]; od: GT;
```

Mathematica:

```
<<Combinatorica`
mult[p1_,p2_]:=Permute[p1,p2]; {n=3,g={e,a,b,c,d,f},
 tH={"e","a","b","c","d","f"},eg31=SymmetricGroup[n],
 eg3={eg31[[1]],eg31[[4]],eg31[[5]],eg31[[3]],eg31[[6]],
 eg31[[2]]},leg3=Length[eg3]}
subs=Table[eg3[[i]]->g[[i]],{i,1,leg3}]; {mTable1=
MultiplicationTable[eg3,mult]//MatrixForm,mTable2=
MultiplicationTable[eg3,mult]/.{m_Integer:>eg3[[m]]},
mTable2//MatrixForm, mTable2/.subs}
TableForm[mTable2/.subs,TableHeadings->{tH,tH}]
```

□

Problem 4.29 If SG is a subgroup of the group G and if $g \in G$, then $g \in gSG$, where $gSG = \{g\,sg, \; sg \in SG\}$, and we say that gSG is a *left coset* (or *left lateral class*) of SG in G determined by g. Similarly, $g \in SGg$, $SGg = \{sg\,g, \; sg \in SG\}$ is called a *right coset* of SG in G determined by g. In a normal subgroup N any left coset is a right coset, and viceversa. Let G be a group that has a normal subgroup N. We denote by G/N a set of cosets of N in G, then G/N is a group and this group is called *quotient group* or *factor group* of G. Let S_3 be the symmetric group on 3 symbols.

1) Determine all the elements of S_3 using the right cosets of the trivial subgroup $\{e\}$ of the symmetric group S_3.

2) For the group G of the previous problem and the subgroup $SG = \{e, c\}$ of G, represent G as a union of the right cosets of SG in G.

Maple:

```
with(group): n:=3; G||n:=permgroup(n,{[[3,1]],[[3,2,1]]});
SGe:=permgroup(n,{[]}); L||n:=cosets(G||n,SGe);
for i from 1 to nops(L||n) do
 convert(L||n[i],'permlist',n); od; G:=[e,a,b,c,d,f];
EG||n:=[[],[[1,2,3]],[[1,3,2]],[[1,2]],[[1,3]],[[2,3]]];
```

```
m:=nops(EG||n); print(seq(G[i]=convert(EG||n[i],'permlist',n),
 i=1..m)); GP||n:=permgroup(n,{op(EG||n)});
SG:=permgroup(n,{[],[[1,2]]}); issubgroup(SG,GP||n);
SGa:=permgroup(n,cosets(GP||n,SG) minus {[]});
SGb:=permgroup(n,{mulperms([[1,2]],[[1,2,3]]),
 mulperms([[1,2]],[[2,3]])}); GP1:=SGe union SGa union SGb;
```

Mathematica:

```
<<Combinatorica`
mult[p1_,p2_]:=Permute[p1,p2]; {g3={{1,2,3},{2,1,3},{3,2,1},
 {1,3,2},{2,3,1},{3,1,2}},n=3,sge={1,2,3},lg3=Length[g3],
 cosetsR=Table[mult[sge,g3[[i]]],{i,1,lg3}]}
{g={e,a,b,c,d,f},sg={{1,2,3},{2,1,3}},lsg=Length[sg]}
eg3={{1,2,3},{2,3,1},{3,1,2},{2,1,3},{3,2,1},{1,3,2}}
subs=Table[eg3[[i]]->g[[i]],{i,1,lg3}]
sga=Table[mult[sg[[i]],eg3[[2]]],{i,1,lsg}]
sgb=Table[mult[sg[[i]],eg3[[3]]],{i,1,lsg}]
sgd=Table[mult[sg[[i]],eg3[[5]]],{i,1,lsg}]
sgf=Table[mult[sg[[i]],eg3[[6]]],{i,1,lsg}]
{gp1={{sge},sga,sgb,sgd,sgf}/.{subs}, gp2=gp1[[1]],
 gp3=Map[Sort,gp2], Union[gp3,SameTest->SameQ]}
```

The group G is isomorphic to a subgroup of S_3, and we denote this subgroup as $G3$ (or g3). The right cosets are: $SG = SGe$, $SGa = \{ea, ca\} = \{a, f\}$, $SGb = \{eb, cb\} = \{b, d\}$. Therefore $G = SGe \cup SGa \cup SGb$. It should be noted that *Maple* does not display all the representatives of the cosets. So, for obtaining all the representatives, we have to perform multiplications of permutations. To find all results in *Mathematica*, we have to construct our own solution. □

Problem 4.30 Let S_3 and S_4 be the symmetric groups of degree 3 and 4. We can observe that the group S_4 contains every element of S_3, and S_3 and S_4 are groups.

  1) Show that S_3 is not a subgroup of S_4.
  2) Construct a subgroup SG of the symmetric group S_4.

Maple:

```
with(group): G3:=permgroup(3,{[[3,1]],[[3,2,1]]});
G4:=permgroup(4,{[[1,2]],[[1,2,3,4]]}); elements(G3);
elements(G4); issubgroup(G3,G4); m:=3; n:=4;
SG||m:=permgroup(n,{[[1,2]],[[seq (i,i=1..m)]]});
issubgroup(SG||m,G||n); elements(SG||m); elements(G||n);
```

Mathematica:

```
<<Combinatorica`
mult[p1_,p2_]:=Permute[p1,p2]; {m=3,n=4,g3=SymmetricGroup[m],
 g4=SymmetricGroup[n],lg3=Length[g3],lg4=Length[g4]}
{MemberQ[g3,IdentityPermutation[n]], mTable1=
 MultiplicationTable[g3,mult]/.{m_Integer:>g3[[m]]}//MatrixForm,
 mTable2=MultiplicationTable[g3,mult]/.{m_Integer:>g3[[m]]}}
{Map[MemberQ[g3,#]&,mTable2,{2}],
 Map[MemberQ[g3,InversePermutation[#]]&,g3,{1}]}
sg3={{1,2,3,4},{2,1,3,4},{3,2,1,4},{1,3,2,4},{2,3,1,4},{3,1,2,4}}
{MemberQ[sg3,IdentityPermutation[n]],mTable1=MultiplicationTable[
 sg3,mult]/.{m_Integer:>sg3[[m]]}//MatrixForm,
 mTable2=MultiplicationTable[sg3,mult]/.{m_Integer:>sg3[[m]]}}
{Map[MemberQ[sg3,#]&,mTable2,{2}],
 Map[MemberQ[sg3,InversePermutation[#]]&,sg3,{1}]}
```

$\square$

Problem 4.31 Let SG be a subgroup of the group G and $g \in G$, then $g^{-1}SGg = \{g^{-1}\,sg\,g,\ s \in SG\}$ is a subgroup of G and $g^{-1}SGg$ is called the *conjugate subgroup* of SG in G with respect to g. A subgroup that coincides with all the conjugate subgroups is called a *normal subgroup* N of G. In an Abelian group any subgroup is a normal subgroup.

Let us consider **Problem 4.28**.

1) Determine all the subgroups of the group G.
2) Find the normal and conjugate subgroups.
3) Determine the center of the group G.

Maple:

```
with(group): n:=3; G:=[e,a,b,c,d,f];
EG||n:=[[],[[1,2,3]],[[1,3,2]],[[1,2]],[[1,3]],[[2,3]]];
m:=nops(EG||n); print(seq(G[i]=convert(EG||n[i],'permlist',n),
    i=1..m)); G||n:=permgroup(n,{op(EG||n)}); center(G||n);
SGs:=map(x->permgroup(n,x),[{[]},{[[1,2,3]]},
    {[[1,2]]},[[1,2,3]]},{[[1,2]]},{[[1,3]]},{[[2,3]]}]);
map(grouporder,SGs); SG_elems:=map(elements,SGs);
for i from 1 to nops(SGs) do isnormal(G||n,SGs[i]); od;
for i from 1 to nops(SGs) do issubgroup(SGs[i],G||n); od;
for i from 1 to nops(SG_elems) do
   A||i:=map(convert,SG_elems[i],'permlist',3); od;
x:=op(op(2,SGs[4]));y:=[[1,2,3]];
mulperms(mulperms(y,x),invperm(y)); SGs[5];
```

```
x:= op(op(2,SGs[4]));y:=[[1,3]];
mulperms(mulperms(y,x),invperm(y)); SGs[6];
x:=op(op(2,SGs[5])); y:=[[1,2,3]];
mulperms(mulperms(y,x),invperm(y)); SGs[6];
```

Mathematica:

```
<<Combinatorica`
mult[p1_,p2_]:=Permute[p1,p2]; centerGroupQ[z_,g_]:=
  (mult[mult[InversePermutation[z],g],z])===g;
rules={a1___,a2_,a3___,a4_,a5___}:>{a1,a2,a3,a5}/;a2==a4
eg3={{1,2,3},{2,3,1},{3,1,2},{2,1,3},{3,2,1},{1,3,2}}
unionSubgrs[sg_List]:=sg//.rules;
genSubgrs[g_,G_List]:=(unionSubgrs[FoldList[mult[g,#]&,g,G]]);
normalQ[g1_,sg1_List]:=(Complement[Map[mult[g1,#]&,sg1],
  Map[mult[#,g1]&,sg1]])==={};
conjSubgr[g1_,sg1_List]:=Module[{x,y},x=InversePermutation[g1];
    y=Map[mult[#,g1]&,sg1]; Map[mult[x,#]&,y]];
{n=3, g={e,a,b,c,d,f},m=Length[eg3],g3=PermutationGroupQ[eg3],
  idP=IdentityPermutation[n], k=Array[w,m,0],MemberQ[eg3,idP]}
Map[centerGroupQ[idP,#]&,eg3]
Do[k[[i]]=Map[centerGroupQ[eg3[[i]],#]&,eg3];
  If[MemberQ[k[[i]],False],,Print["the center of the group is"];
  Print[eg3[[i]]]],{i,1,m}]
{sgs1=Map[genSubgrs[#,eg3]&,eg3], sgs2=Map[Sort,sgs1],
  sgs3=Union[sgs2,{eg3},SameTest->SameQ],
  sgOrder=Map[Length[#]&,sgs3]}
Do[s1=Map[normalQ[#,sgs3[[j]]]&,eg3];Print[s1],{j,1,m}]
subGroupQ[sg_List,G_List]:=Module[{},
  Complement[{Depth[sg]},Map[Length[#]&,G]]==={};
  mTable=MultiplicationTable[sg,mult]/.{m_Integer:>sg[[m]]};
  Map[MemberQ[sg,#]&,mTable,{2}];
  Map[MemberQ[sg,InversePermutation[#]]&,sg,{1}]]
Map[subGroupQ[#,eg3]&,sgs3]
{conjSubgr[eg3[[2]],sgs3[[3]]],conjSubgr[eg3[[3]],sgs3[[3]]],
  conjSubgr[eg3[[5]],sgs3[[3]]],conjSubgr[eg3[[6]],sgs3[[3]]],
  conjSubgr[eg3[[2]],sgs3[[4]]],conjSubgr[eg3[[2]],sgs3[[2]]]}
```

In the group G there are 6 subgroups: $\{\}$, $\{e,a,b,c,d,f\}$, $\{e,a,b\}$, $\{e,c\}$, $\{e,d\}$, $\{e,f\}$, the first three of which are normal subgroups, the other three ones are subgroups conjugate to each other, the center of the group G is $\{e\}$.

□

Problem 4.32 We say that a group is a *cyclic group* if every element is a power of an element g that generates the group, this element is called the *generator* of G. Any cyclic group is an Abelian group.
1) Generate a cyclic group G of degree 4, that is, $x^4 = 1$.
2) Determine the degree of the group.
3) Find all the elements of the group G.

Maple:

```
with(group): n:=4; G:=grelgroup({x},{[x,x,x,x]}); grouporder(G);
SG:=subgrel({y=[]},G); G_elems:=cosets(SG);
```

Mathematica:

```
<<Combinatorica`
{n=4, g=CyclicGroup[4], Length[g]}
```
□

Problem 4.33 Generate the Klein four-group G, represented by the generators $\{x, y, z\}$ and the relations $xz = y$, $yz = x$, $xy = z$, $yx = z$.
1) Determine: the degree of the group.
2) Find all the elements of the group G.
3) Construct the Cayley table (see **Problem 4.28**).

Maple:

```
with(group): G:=grelgroup({x,y,z},
 {[x,z,1/y],[y,z,1/x],[x,y,1/z],[y,x,1/z]});
grouporder(G); SG:=subgrel({e=[]},G);
PSG:=permrep(SG); m:=grouporder(PSG);
G_elems:=elements(PSG);
EG3:=[[],op((G_elems minus{[]}))]; n:=nops(EG3);
for i from 1 to n do
 convert(EG3[i],'permlist',n); od; G:=[e,x,y,z];
print(seq(G[i]=convert(EG3[i],'permlist',n),
 i=1..n)); GT:=Matrix(1..n+1,1..n+1):
for i from 1 to n do
 for j from 1 to n do
  for k from 1 to n do
   if EG3[k]=mulperms(EG3[j],EG3[i]) then
    GT[i+1,j+1]:=G[k]; fi; od: od: od:
GT[1,1]:=` `:
for j from 1 to n do
 GT[1,j+1]:=G[j]; GT[j+1,1]:=G[j]; od: GT;
```

Mathematica:

```
<<Combinatorica`
m[a1___,a2___]:=NonCommutativeMultiply[a1,a2];
rules={m[a1__,e,a2___]:>m[a1,a2],m[a1___,e,a2__]:>m[a1,a2],
m[a1___,x,x,a2___]:>m[a1,e,a2],m[a1___,y,y,a2___]:>m[a1,e,a2],
m[a1___,z,z,a2___]:>m[a1,e,a2],m[a1___,x,z,a2___]:>m[a1,y,a2],
m[a1___,x,y,a2___]:>m[a1,z,a2],m[a1___,y,z,a2___]:>m[a1,x,a2],
m[a1___,y,x,a2___]:>m[a1,z,a2],m[a1___,z,x,a2___]:>m[a1,y,a2],
m[a1___,z,y,a2___]:>m[a1,x,a2],m[a1_]:>a1};
{gr4Klein={{e**e,e**x,e**y,e**z},{x**e,x**x,x**y,x**z},
{y**e,y**x,y**y,y**z},{z**e,z**x,z**y,z**z}}//.rules,
 Length[gr4Klein],tH={"e","x","y","z"}}
mTable=TableForm[gr4Klein,TableHeadings->{tH,tH}]
```

□

4.12   Rings, Integral Domains, and Fields

A generalization of the concept of a group, is a *ring*. A *ring* is an algebraic structure, $(R,+,*)$, consisting of a set and the two binary operations, *addition* and *multiplication*, for which the set of axioms hold:

1. Axioms of addition: $(R, +)$ is an Abelian group, and the five axioms hold (closure, associativity, the existence of the zero element, the existence of an inverse element, commutativity);

2. Axioms of multiplication: $(R, *)$ is a monoid, and the two axioms hold (closure and associativity);

3. Axioms of distributivity: the two distributive laws of $*$ with respect to $+$.

Rings may have an *identity (or unity) element*, 1, (with respect to multiplication), the *proper divisors of zero*, or can be commutative rings.

A commutative ring with an identity, $1 \neq 0$ and no proper divisors of zero is called an *integral domain*, $\mathbb{D}$, e.g., $\mathbb{Z}, \mathbb{Q}, \mathbb{R}, \mathbb{C}$ are integral domains. We denote the polynomial rings in x, $\mathbb{Z}[x], \mathbb{Q}[x], \mathbb{R}[x], \mathbb{C}[x], \mathbb{D}[x]$, in which the polynomials have coefficients, respectively, from $\mathbb{Z}, \mathbb{Q}, \mathbb{R}, \mathbb{C}, \mathbb{D}$.

In ring theory, the important concept is a special subset I of a ring R, an additive subgroup of R, that is called *ideal*. A set I is called a *right ideal* of R if for all $i \in I, r \in R$, it follows that $ir \in I$. Similarly, A set I is called a *left ideal* of R if for all $i \in I, r \in R$, it follows that $ri \in I$. Moreover, a set I is called an *ideal* or a *two-sided ideal* of R if for all $i \in I, r \in R$, it follows that $ir \in I$ and $ri \in I$.

The concept of a *quotient ring* or *factor ring* is similar to the concept of a factor group in group theory. We start with a ring R and a two-sided ideal I of R, and construct a new ring, the quotient ring R/I, essentially by requiring that all elements of I be zero.

A commutative ring with an identity element $1 \neq 0$ in which every non-zero element is *invertible* (with respect to multiplication) is called a *field*. If $(R, +, *)$ is a non-empty set with two binary operations of addition and multiplication such that R is an additive Abelian group, $R \setminus \{0\}$ is a multiplicative Abelian group, the distributive laws are hold, then R is a field. Every field is an integral domain, but a finite integral domain is a field.

Field theory studies various classes of fields, e.g., the field of complex numbers $\mathbb{C}$, the field of rational numbers $\mathbb{Q}$, the field of algebraic numbers $\mathbb{A}$, the field of real numbers $\mathbb{R}$, a field K (where $K(X)$ is a set of rational functions with coefficients in K), *finite fields* or *Galois fields* containing finitely many elements (for every prime number p and positive integer n, there exists a finite field with p^n elements).

If, for example, we consider such fields where the size is a prime (i.e., $n = 1$), the ring $\mathbb{Z}/p\mathbb{Z}$, or $F_p = \{0, 1, \ldots, p - 1\}$, or $\mathrm{GF}(p)$ is a finite field with p elements, where arithmetic is performed modulo p. The prime number p is called the *characteristic of a field*. If $p = 2$, we obtain the field F_2 that consists of two elements, 0 and 1. Finite fields are very important in the number theory, algebraic geometry, Galois theory, coding theory, and cryptography.

In symbolic mathematics, computational commutative and noncommutative algebra and the field theory are of central importance due to various applications in cryptography, coding theory, digital signal processing, error detection and correction, etc. In *Maple* and *Mathematica* there exist various functions and packages for operations in quotient rings, finite fields, integral domains. We consider the most important functions.

Maple:

`expr mod m;`	`modp(expr,m);`	`mods(expr,m);`
`msolve(eqs,m);`	`Roots(f) mod m;`	`Roots(f,F) mod m;`
`evala(Factors(f,F));`	`Factors(f) mod m;`	`Factors(f,F) mod m;`
`ifactor(n);`	`Expand(f) mod m;`	`evala(Expand(f));`
`with(numtheory);`	`primroot(n,p);`	`GF(p,k); GF(p,k,f);`
`Irreduc(f,F) mod p;`	`Irreduc(f) mod p;`	`Rem(f1,f2,x) mod m;`

mod, modp, mods, computing algebraic expressions expr over the integers modulo m,

msolve, solving the equations over the integers modulo m,

Roots(f) mod m, Roots(f,F) mod m, finding the roots of the polynomial f modulo m and the roots over the finite field F (F is defined as RootOf),

Factors(f) mod m, Factors(f,F) mod m, evala(Factors(f,F)), factorizing the polynomial f, respectively, over the integers modulo m, over the finite field defined by F, and over an algebraic number field defined by the extension F,

Rem(f1,f2,x) mod m, Quo(f1,f2,x) mod m, finding remainder and quotient of f_1/f_2 over the integers modulo m,

ifactor, factorizing the integer n,

Expand(f) mod m, expanding products in the polynomial f over the integers modulo m,

GF(p,k,f), *Galois Field* package for performing arithmetic in the Galois Field, $GF(p^k)$, with p^k elements,

primroot(n,p) (in the package numtheory), finding the first primitive root of p that is greater than the integer n,

Irreduc(f) mod p, Irreduc(f,F) mod p, verifying whether the polynomial f irreducible over the integers modulo p and over the finite field defined by F.

Mathematica:

Mod[a,m]	PowerMod[a,n,m]	PowerMod[a,-1,m]
Solve[eqs&&Modulus==m,{vars}]		Roots[eq,x,Modulus->m]
Factor[f,Modulus->m]		Factor[f,Extension->{a1,...,an}]
FactorInteger[n]		Expand[f,Modulus->n]
PolynomialQuotient[f1,f2,x, Modulus->m]		<<FiniteFields`
fF=GF[p] GF[p,k] GF[p][{k}] Characteristic[fF]		EulerPhi[n]
MultiplicativeOrder[a,m]		IrreduciblePolynomial[x,p,d]

Mod[a,m], PowerMod[a,n,m], computing, respectively, the remainder on division of a by m and a^n modulo m,

Solve, solving equations over the integers modulo m, including the equation of the form Modulus==m,

performing various algebraic operations over the integers modulo m, including the option Modulus->m to various algebraic functions, e.g., Roots, Factor, Expand, PolynomialQuotient, PolynomialRemainder,

performing various algebraic operations over the algebraic number fields, including the option Extension->a to various algebraic functions, e.g., Factor,

FactorInteger, factorizing the integer n,

the package `FiniteFields`, for performing arithmetic in the Galois field, $GF(p^k)$, with p^k elements:

`GF[p,k]`, constructing the Galois field, $GF(p^k)$, `GF[p][{a}]`, the element a in the Galois field $GF(p)$, with p elements (p is prime),

`Characteristic`, finding the characteristic of the Galois field,

`MultiplicativeOrder[a,m]`, computing the multiplicative order of a modulo m, defined as the smallest integer n such that $a^n \equiv 1 \bmod m$,

`EulerPhi`, the Euler totient function $\phi(n)$, i.e., the number of positive integers less than or equal to n which are relatively prime to n,

`IrreduciblePolynomial`, finding an irreducible polynomial in x of degree d over the integers modulo p (p is prime).

Problem 4.34 Let $\bar{m} = \{x \equiv m \,(\bmod\, n),\ x \in \mathbb{Z}\} = \{m + nk,\ k \in \mathbb{Z}\}$ be an equivalence class of integers containing an integer m for a given non-zero integer n ($n > 1$). We know that the equivalent classes modulo n are the cosets of $n\mathbb{Z}$ in $\mathbb{Z}$, we can form the factor-group $\mathbb{Z}/n\mathbb{Z}$ (or $\mathbb{Z}_n$), the elements of $\mathbb{Z}_n$ are these cosets. Then $\mathbb{Z}_n$ is a commutative ring under addition and multiplication modulo n with an identity.

1) Generate various equivalent classes of integers for $n = 12$.

2) Construct the addition and multiplication tables of $\mathbb{Z}_{12}$ and determine whether $\mathbb{Z}_{12}$ is an integral domain.

Maple:

```
with(LinearAlgebra): interface(rtablesize=15); n:=12; n1:=100;
Z||n:=[seq(i, i=0..n-1)]; for i from 0 to n-1 do
EqCl||i:=[seq(i+12*k,k=-n1..n1)]; od; EqCl0; EqCl1;
AddT||n:=Matrix(1..n+1,1..n+1): AddT||n[1,1]:=`+`:
for i from 1 to n do AddT||n[i+1,1]:=Z||n[i]; od:
for j from 1 to n do AddT||n[1,j+1]:=Z||n[j]; od:
for i from 1 to n do for j from 1 to n do
AddT||n[i+1,j+1]:=modp(Z||n[i]+Z||n[j],n); od: od: AddT||n;
MuT||n:=Matrix(1..n+1,1..n+1): MuT||n[1,1]:=`*`:
for i from 1 to n do MuT||n[i+1,1]:=Z||n[i]; od:
for j from 1 to n do MuT||n[1,j+1]:=Z||n[j]; od:
for i from 1 to n do for j from 1 to n do
MuT||n[i+1,j+1]:=modp(Z||n[i]*Z||n[j],n); od: od: MuT||n;
```

Mathematica:

```
{n=12,n1=100,z12=Table[i,{i,0,n-1}],
 eqCl=Table[Table[i+12*k,{k,-n1,n1}],{i,0,n-1}];
 eqCl[[1]],eqCl[[2]]}
tH={"0","1","2","3","4","5","6","7","8","9","10","11"}
addT1=Table[Table[Mod[z12[[i]]+z12[[j]],n],{i,1,n}],{j,1,n}]
TableForm[addT1,TableHeadings->{tH,tH}]
multT1=Table[Table[Mod[z12[[i]]*z12[[j]],n],{i,1,n}],{j,1,n}]
TableForm[multT1,TableHeadings->{tH,tH}]
```

Although $\mathbb{Z}_{12}$ is a commutative ring with $\bar{1}$ as an identity element, $\mathbb{Z}_{12}$ is not an integral domain since there exist proper divisors of zero, e.g., $\bar{2} * \bar{6} = \bar{0}$. □

Problem 4.35 *Application of ring and group theory to the western tonal music system.* We know that this system has the 12-note chromatic scale and is a system of equal temperament. Suppose that a piano has 12 octaves and there is no difference between tones such as $C\sharp/D\flat$, etc. Frequently, for mathematical and computational investigations in music theory, the difference between octaves can be ignored.

   1) Construct an enumeration for all notes of a given piano.
   2) Define a function describing musical transposition.
   3) Define a function describing musical inversion.

Maple:

```
with(group): n:=12; n1:=9; Z||n:=[seq(i, i=0..n-1)];
for m from 0 to n-1 do EqCl||m:=[seq(m+12*k,k=0..n-1)]; od;
EqCl0; EqCl1; MT:=(x::list,N)->map(x->x+N mod 12,x);
MT(EqCl0,n1); p1:=MT(Z12,n1); L1:=NULL:
for i from 1 to n do L1:=L1,MT(Z12,i); od:
L2:=subs(0=12,[L1]); p11:=subs(0=12,p1);
convert(p11,'disjcyc'); map(convert,L2,'disjcyc');
MI:=(x::list,N)->map(x->(2*N-x) mod 12,x);
L1[1];MI(Z12,0); L1[2];MI(L1[1],1); L1[3];MI(L1[2],2);
```

Mathematica:

```
<<Combinatorica`
{n=12,n1=9, z12=Table[i,{i,0,n-1}], eqCl=Table[Table[
 m+12*k,{k,0,n-1}],{m,0,n-1}], eqCl[[1]], eqCl[[2]]}
mT[x_List,k_]:=Map[Mod[#+k,12]&,x];
```

```
{mT[eqCl[[1]],n1],p1=mT[z12,n1], l1={},
 Do[l1=Append[l1,mT[z12,i]],{i,1,n}]; l2=l1/.{0->12},
 p11=p1/.{0->12}, ToCycles[p11],Map[ToCycles,l2]}
mI[x_List,k_]:=Map[Mod[2*k-#,12]&,x];
{l1[[1]],mI[z12,0],l1[[2]],mI[l1[[1]],1],l1[[3]],mI[l1[[2]],2]}
```

First, we choose an arbitrary principal tone a_1 and set $a_1 = 0$, then we can easily enumerate the next tones constructing a bijection between the set of all tones and the set of all integers. Since we are not interested in difference between octaves, we can classify all tones into equivalence classes $\mathbb{Z}_{12} = \{m + 12k, \ k \in \mathbb{Z}, m \in [0, 11]\}$. Therefore, the enumeration of the western tonal music system, described as a commutative ring $\mathbb{Z}_{12}$ with the two operations $(+, *)$ over the integers modulo 12 and the identity $1 \neq 0$, can be written in the following form:

Let us recall that the musical transposition by N tones higher or lower is equivalent to replacing each tone by the corresponding tone (that is N tones higher or lower). So, the transposition function can be written in the form:

$$\text{MT}(x, N) = x + N \bmod 12, \quad x = \{m + 12k, \ k \in \mathbb{Z}, \ m \in [0, 11]\}, \ N \in [0, 11],$$

This function describes all possible permutations of $\mathbb{Z}_n$. It should be noted that for working with permutations, we replace the 0-element of $\mathbb{Z}_n$ by 12. The inversion function can be written as

$$\text{MI}(x, N) = 2N - x \bmod 12, \quad x = \{m + 12k, \ k \in \mathbb{Z}, \ m \in [0, 11]\}, \ N \in [0, 11].$$

$\square$

Problem 4.36 Let us consider a field $\mathbb{Z}_p$ (for a given prime p). Let $f_1 = x^2 + x + 2$, $f_2 = x^6 + 1$ be two univariate polynomials. We know that this field has some common algebraic properties with $\mathbb{Q}, \mathbb{R}$. Let us perform some of them.

1) Solve the system of equations over $\mathbb{Z}_p$.

2) Find the roots of the polynomial f_2 in $\mathbb{Z}_p$.

3) Factorize the polynomial f_2 over $\mathbb{Z}_p$ and over an algebraic number field defined by α (where α are the roots of f_1).

4) Show that $(a + b)^p = a^p + b^p$, $(a + b)^{p^k} = a^{p^k} + b^{p^k}$ for any $a, b \in \mathbb{Z}_p, k \in \mathbb{N}$.

5) Find the inverses of all non-zero elements of $\mathbb{Z}_p$.

Maple:

```
p:=5; k:=4; f1:=x^2+x+2; f2:=x^6+1;
alias(alpha=RootOf(f1)):
sys:={x+2*y=1,2*x+3*y=1}; msolve(sys,p);
Roots(f2) mod p;
Factors(f2) mod p; evala(Factors(f2,alpha));
Expand((a+b)^p) mod p; Expand((a+b)^(p^k)) mod p;
Z||p0:=convert({seq(i,i=0..p-1)} minus {0},list);
Z||pInv:=map(x->x^(-1) mod p, Z||p0);
```

Mathematica:

```
{p=5,k=4,f1=x^2+x+2,f2=x^6+1,sys=x+2*y==1&&2*x+3*y==1,
  alpha={Root[f1,1],Root[f1,2]}}
{Solve[sys&&Modulus==p,{x,y}],Roots[f2==0,x,Modulus->p],
  Factor[f2,Modulus->p],Factor[f2,Extension->alpha],
  Expand[(a+b)^p,Modulus->p],Expand[(a+b)^(p^k),Modulus->p],
  z50=Complement[Table[i,{i,0,p-1}],{0}],
  z5Inv=PowerMod[z50,-1,p]}
```

$\square$

Problem 4.37 Let p be a given prime number, F be the finite field with p elements, and α_1, α_2 be any two elements of F.

1) Define the finite field F.

2) Perform all arithmetic operations in the field F for the elements α_1 and α_2 and compare the results with the corresponding operations using modular arithmetic operations (mod and Mod).

3) Verify whether α_1 is a primitive element or find the corresponding primitive root of p.

4) Perform some algebraic operations (e.g., the division, polynomial expansion, verifying irreducibility, factorization) on the given two polynomials, e.g., $f_1 = (x^2 + x + 1)$, $f_2 = (x^3 + x^2 + x + 1)$, in the field F.

Maple:

```
with(numtheory); p:=127; GF||p:=GF(p,1);
alpha1:=GF||p:-input(23); alpha2:=GF||p:-input(121);
a1:=23; a2:=121; a3:=10;
L1:=[GF||p:-`+`(alpha1,alpha2), GF||p:-`-`(alpha1,alpha2),
     GF||p:-`*`(alpha1,alpha2), GF||p:-`^`(alpha1,a3),
     GF||p:-`/`(alpha1,alpha2), GF||p:-inverse(alpha1)];
```

```
use GF127 in
 L11:=[alpha1+alpha2,alpha1-alpha2,alpha1*alpha2,
       alpha1^10,alpha1/alpha2,inverse(alpha1)]
end use;
 L2:=[a1+a2 mod p,a1-a2 mod p,a1*a2 mod p,a1&^a3 mod p,
      a1/a2 mod p,1/a1 mod p];
GF||p:-isPrimitiveElement(alpha1); primroot(22,p);
f1:=(x^2+x+1);                f2:=(x^3+x^2+x+1);
f3:=Quo(f1^3,f2,x) mod p; f4:=Rem(f1^3,f2,x) mod p;
f5:=sort(expand(f1^3+f2^2*f3^3+f4^2) mod p);
Irreduc(f1) mod p; Factor(f1) mod p; ifactor(p-1);
```

Mathematica:

```
<<FiniteFields`
{p=127, fieldF=GF[127], Characteristic[fieldF],
 alpha1=GF[p][{23}], alpha2=GF[p][{121}],a1=23, a2=121, a3=10}
l1={alpha1+alpha2,alpha1-alpha2,alpha1*alpha2,alpha1^a3,
    alpha1/alpha2,alpha1^(-1)}
l2={Mod[a1+a2,p],Mod[a1-a2,p],Mod[a1*a2,p],PowerMod[a1,a3,p],
    Mod[a1*PowerMod[a2,-1,p],p],PowerMod[a1,-1,p]}
primitiveRootQ[x_,n_]:=Module[{list1},list1=Select[Range[n],
 MultiplicativeOrder[#,n]==EulerPhi[n]&];MemberQ[list1,x]];
primitiveRoot[k_,n_]:=Module[{l1},
 l1=Table[PowerMod[i,1,n],{i,k,n}];First[Select[l1,#>k&]]];
irreduciblePolynomialQ[f_,n_,x_]:=(f===IrreduciblePolynomial[
 x,n,Exponent[f,x]]);
{primitiveRootQ[a1,p],primitiveRoot[22,p]}
{f1=(x^2+x+1),f2=(x^3+x^2+x+1),
 f3=PolynomialQuotient[f1^3,f2,x,Modulus->p],
 f4=PolynomialRemainder[f1^3,f2,x,Modulus->p],
 f5=Expand[f1^3+f2^2*f3^3+f4^2,Modulus->p],
 irreduciblePolynomialQ[f1,p,x],Factor[f1,Modulus->p],
 FactorInteger[p-1]}
```

☐

Chapter 5

Linear Algebra

Linear algebra is the algebra of linear spaces and linear transformations between these spaces. Linear algebra is one of the fundamental areas of modern mathematics that has extensive applications in the natural and social sciences and in engineering, since nonlinear models can be studied by using approximated linear models.

In symbolic mathematics, we consider symbolic linear algebra that studies algorithms for performing symbolic linear algebra computations on computers. Symbolic linear algebra develops and implements many sophisticated algorithms on polynomials, vectors, matrices, combinatorial structures and other mathematical objects. Various algorithms are well known (e.g., Buchberger's Gröbner basis reduction algorithm, Wiedemann's sparse linear system solver for scalars from a finite field, etc.), they form the core of any computer algebra system, and their improvement is a continuous task for researchers. In addition, various algorithms for new important problems are the subject of current investigations (e.g., Diophantine linear system solutions, nonlinear system solutions using homotopic deformations, etc.).

If we apply the standard linear algebra methods to a matrix with numerical exact elements (e.g., the integers or rational numbers), we can obtain the correct results, but if a matrix has symbolic elements, we can obtain new mathematical features or surprising results. The basic feature is discontinuity of standard matrix functions (e.g., the rank function, the reduced row-echelon form of a matrix, the Jordan normal form of a matrix, etc.). On the other hand, the evaluation in exact arithmetic has some bounds, e.g. in some step we can obtain a matrix whose elements are very large intractable expressions, in this case it should apply some compression methods (e.g., see the `LargeExpressions`

I.K. Shingareva, C. Lizárraga-Celaya, *Maple and Mathematica*, 2nd ed.,
DOI 10.1007/978-3-211-99432-0_5, © Springer-Verlag Vienna 2009

package in *Maple*) or, large expressions will display in a shortened form in *Mathematica* (see also the `Short` function).

In *Maple*, there are the large packages, `LinearAlgebra`, `linalg`, and the `LinearAlgebra` subpackage of the `Student` package, that contain functions for working with linear algebra. But the `linalg` package is older and is considered obsolete. In the present book we describe only the more relevant functions of the `LinearAlgebra` package.

In *Mathematica*, vectors, matrices, tensors are represented with lists (see Chapter 2, Sect. 2.3.6). *Mathematica* has a variety of functions for solving problems in linear algebra.

Note. In *Maple*, the functions beginning with the lower-case letters belong to the obsolete `linalg` package, while the functions beginning with the upper-case letters belong to the new `LinearAlgebra` package. In *Mathematica*, the package `LinearAlgebra`MatrixManipulation` is obsolete for *ver* ≥ 6.

5.1   Vectors

In mathematics, a *vector* is an element of the mathematical structure called *vector space* (or *linear space*) on a scalar field $\mathbb{F}$ (e.g., $\mathbb{R}$ or $\mathbb{C}$) that belongs to a nonempty set **V** with the two operations (the vector addition and the scalar multiplication) and certain eight axioms hold (see [34]). In another aspect, the vector, is an ordered set of real or complex numbers (or elements of a scalar field). A two-dimensional vector can be represented with an arrow from the origin, that has determined length and direction from a horizontal axis. In physics, a vector always has some physical meaning and it is represented in terms of a *physical magnitude* and the *direction*. In symbolic mathematics, a vector can be considered as a *mathematical object* or a *data object* (that is a set of variables of the same type, where the elements of the set are indexed).

In *Maple*, a *vector* is the *mathematical object*, `Vector`, represented as a list of elements. However, `Vector` can be converted to the *data object*, `Array` and viceversa, but `Vector` and `Array` are not the same concepts in *Maple*.

In *Mathematica*, a *vector* is a *data object* represented as a list of elements.

Vector representations and components

Maple:

```
with(LinearAlgebra):              Vector[column]([x1,...,xn]);
Vector([x1,...,xn]);                  Vector[row]([x1,...,xn]);
<x1,...,xn>; <x1|...|xn>;     Vector(n,i->f(i)); map(x->f(x),V);
Vector(n,fill=a);            Vector(n,symbol=a); print(V);   op(V);
V[i]; V[i..j]; V[p..-1];   convert(V,Matrix); convert(V,Vector);
type(V,Vector); whattype(V);      convert(V,Array); Dimension(V);
```

Vector, Vector[column], Vector[row], constructing a column and a row vector;
   Maple distinguishes between column and row vectors, these two types of vectors
   are different,

<x1,...,xn>, <x1|...|xn>, constructing a column vector and a row vector, where
   the elements can be represented as another column and row vectors,

Vector(n,i->f(i)), map(x->f(x),V), constructing vectors defining their elements as a univariate function $f(x)$,

Vector(n,fill=a), Vector(n,symbol=a), constructing vectors, where, respectively, every element is a and the symbolic elements are a_i,

print, op, displaying vectors,

V[i], V[i..j], V[i..-1], extracting elements of vectors,

convert(V,Matrix), convert(V,Vector), converting a Vector into a Matrix
   and viceversa; convert(V,Array), converting a Vector into an Array (data
   structure object),

type(V,Vector), whattype(V), verifying and determining the type of an object,

Dimension, determining the dimension of a Vector.

```
with(LinearAlgebra): n:=4; f:=x->x^2; V1:=Vector([x1,x2,x3,x4]);
V2:=Vector[column]([x1,x2,x3,x4]); V3:=Vector[row]([x1,x2,x3,
  x4]); U1:=<x1,x2,x3,x4>; U2:=<x1|x2|x3|x4>;
U3:=Vector(n,i->f(i)); U4:=map(x->f(x),V1); Vector(n,fill=a);
Vector(n,symbol=a); print(V1); op(U1); V2[2]; U1[1]:=w1;
U1[2]:=w2; V1[1..3]; V1[2..-1]; Vm:=convert(V1,Matrix);
Vv:=convert(Vm,Vector); Equal(Vv,V1); print(Vm);
type(Vm,Matrix); whattype(Vm); whattype(Vv); Dimension(V1);
```

Mathematica:

```
v1=Table[expr,{i,i1,in}] v2=Array[f,n]      v3=ConstantArray[c,n]
v4=Range[i1,in]           v5=Array[Subscript[a,#]&,n]  Column[v]
ColumnForm[v] v[[i]]      v[[i]]=c      v1=v2       Take[v,{i1,i2}]
Take[v,{i1,-1}]           Length[v]    Dimensions[v]  VectorQ[v]
```

Table, Array, ConstantArray, Range, vector representations, *Mathematica* does
   not distinguish between row and column vectors,

ColumnForm, Column, displaying the elements of a list in a column,

[[]], Take, extracting a particular vector element or column,

VectorQ, determining whether or not the expression expr has the structure of a
   vector,

Dimensions, Length, determining the dimension of a vector.

```
f1[i_]:=(-1)*i^3; {v1={1,2,3,4,5},
  v2=Table[f1[i],{i,5}], v3=Array[f1,5], v4=ConstantArray[c,5],
  v5=Range[2,5], v6=Array[Subscript[a,#]&,5]}
{v1//ColumnForm, Column[v3], v3//MatrixForm, v4, Length[v4],
  VectorQ[v2], v1[[3]], Take[v2,{1,4}], Take[v2,{2,-1}]}
```

Problem 5.1 Construct various vectors (with real and complex elements), select their elements, and determine the vector dimensions.

Maple:

```
with(LinearAlgebra): V1:=<1,2,3,4>; V2:=[1,2,3,4]; V3:=<4+I,5-I>;
V4:=<I,2>; Vector(5,symbol=v); U:=Vector(3,i->u||i);
Vector([1,2,3]);Vector[row]([1,2,3,4]);Vector[column]([1,2,3,4]);
print(V1,U); V2[2]; U[1]:=w1; V1[1..3]; V1[-3..-1]; V1[-3..-3];
Dimension(V1);type(V1,Vector);VA:=convert(V1,Array);whattype(VA);
```

Mathematica:

```
f1[i_]:=(-1)*i^3; {v1={1,2,3,4,5},
  v2=Table[f1[i],{i,5}], v3=Array[f1,5], v4={4+I,5,I,1+2*I},
  ConstantArray[c,5], Range[5]}
{v1//ColumnForm, Column[v3], v3//MatrixForm, VectorQ[v2],
  v1[[3]], Take[v2,{1,4}], Length[v4]}
{v1=Table[i^2,{i,1,5}], v2=Array[f,5], v1=v2, v1===v2,
  Map[VectorQ,{v1,v2}]}
```

□

Mathematical operations on vectors

Let us consider the mathematical operations on vectors: addition, subtraction, linear combination, inner (or dot) product, vector cross product, vector transpose operation, conjugating transpose operation (or Hermitian transpose operation), the vector norm.

Maple:

```
with(LinearAlgebra):    V1+V2; V1-V2;        a*V1;      a*V1+b*V2;
VectorAdd(V1,V2); Add(V1,V2); Add(V1,V2,a,b);       Add(V1,-V2);
DotProduct(V1,V2);        V1.V2; V1 &x V2;   CrossProduct(V1,V2);
OuterProductMatrix(V1,V2);    Norm(V1,2);              Norm(V1,p);
Transpose(V1);          V1^%T; V1^%H;    HermitianTranspose(V1);
```

Mathematica:

```
v1+v2 v1-v2 a*v1 a*v1+b*v2   v1.v2     Dot[v1,v2]     Cross[v1,v2]
Inner[Times,v1,v2,Plus]       Outer[Times,v1,v2]]        Norm[v1,p]
transpV=v1            transpHV=Conjugate[v1]   <<VectorAnalysis`
DotProduct[v1,v2,coordsys]         CrossProduct[v1,v2,coordsys]
```

Note. Since *Mathematica* does not distinguish between row and column vectors, then $v^T = v$, and conjugating transpose operation (or Hermitian transpose operation) is Conjugate[v].

Problem 5.2 Perform various mathematical operations on vectors.

Maple:

```
with(LinearAlgebra): V1:=<1|2|3>; V2:=<I|1+I|3-I>;
V3:=<a1,a2,a3>; V4:=<a4,a5,a6>; W:=Vector(3,symbol=c);
V3+V4; Add(V3,V4); V1+V2; 2*V1; c1*V1+c2*V2; V1-V2;
V1.V2; V1 &x V2; V1^%T; Transpose(V1);
V2^%H; HermitianTranspose(V2); Norm(W,2); VectorNorm(W,2);
```

Mathematica:

```
{v1={1,2,3},v2={I,1+I,3-I},Map[ColumnForm,{v3={a1,a2,a3},
  v4={a4,a5,a6}}],w1=Array[Subscript[c,#]&,3],
  w2={ConstantArray[c,3]}}
Map[MatrixForm,{v3+v4, v1+v2, 2*v1, c1*v1+c2*v2, v1-v2, v1.v2,
  Dot[v1,v2], Cross[v1,v2], transpV=v1, transpHV=Conjugate[v2],
  Inner[Times,v1,v2,Plus], Outer[Times,v1,v2], Norm[w1,2]}]
```

□

Vector visualization

Let us consider various visual representations of vectors in 2-D and 3-D vector spaces.

Maple:

```
with(plots):                           pointplot({v1,...,vn},ops);
pointplot3d({v1,...,vn},ops);          plottools[arrow](v0,v1},ops);
```

Mathematica:

```
ListPointPlot3D[{v,...,xn},ops]   ListLinePlot[{v1,...,vn},ops]
ListPlot[{v1,...,vn},ops]              Graphics[Arrow[{v1,...,vn]]
              Show[Graphics3D[{Point[v1],...,Point[vn]}]]
         Graphics[{Arrowheads[{spec}],Arrow[{v1,...,vn}]}]
```

Problem 5.3 Graph various vectors and their linear combinations in $\mathbb{R}^2$ and $\mathbb{R}^3$.

Maple:

```
with(plots): with(plottools): V1:=<1,2,3>; V2:=<4,5,6>;
V3:=<1,2>; V4:=<3,4>; pointplot({V3+V4,V3-2*V4},color=blue,
 symbol=circle,symbolsize=20,axes=boxed,view=[-9..9,-9..9]);
pointplot3d({V1+V2,V2-2*V1},color=blue,symbol=circle,
  symbolsize=20,axes=boxed,view=[-9..9,-9..9,-9..9]);
A1:=arrow(V1,V2,.2,.4,.1,color=blue): A2:=arrow(V2,V2*2,
 .2,.4,.1,color=blue): A3:=arrow(V3,V4,.2,.4,.1,color=blue):
A4:=arrow(V4,V4+V3,.2,.4,.1,color=blue): display({A1,A2},
 orientation=[-50,100],axes=boxed); display({A3,A4},axes=framed);
```

Mathematica:

```
{v1={1,2,3}, v2={4,5,6}, v3={1,2}, v4={3,4}}
ListPointPlot3D[{v1+v2,v2-2*v1},PlotStyle->{Blue,PointSize[
 Large]},Boxed->True,PlotRange->{{-9,9},{-9,9},{-9,9}}]
ListPlot[{v3+v4,v3-2*v4},PlotStyle->{Blue,PointSize[0.05]},
 Frame->True,PlotRange->{{-9,9},{-9,9}}]
ListLinePlot[{v3,v4},Filling->Axis]
Show[Graphics3D[{PointSize[0.03],Hue[0.9],Point[v1],Point[v2],
 Point[(v2-v1)/2]}],BoxRatios->{1,1,1}]
```

```
Show[Graphics[{PointSize[0.03],Hue[0.9],Point[v3],Point[v4],
 Point[v3-v4]}],Frame->True]
Graphics[{Blue,Arrow[{{0,0},v3,v4,v3+v4}]},Frame->True]
g1=Graphics[{Blue,Arrow[{{0,0},v3}]}];
g2=Graphics[{Blue,Arrow[{v3,v4}]}];
g3=Graphics[{Blue,Arrow[{v4,v3+v4}]}];
Show[g1,g2,g3,Frame->True]
Graphics[{Arrowheads[{-.1,.1}],Blue,Arrow[{v3,v4}]}]
```

□

Problem 5.4 Let z_1, z_2 be two complex numbers ($z_1, z_2 \in \mathbb{C}$) which can be represented as vectors in the complex plane. Construct the graphical representation of addition $w = z_1 + z_2$ in the complex plane $\mathbb{C}$.

Maple:

```
with(plots): with(plottools): z1:=9+I; z2:=2+7*I; w:= z1+z2;
G1:=arrow([0,0],[Re(z1),Im(z1)],.2,.4,.1,color=blue):
G2:=arrow([0,0],[Re(z2),Im(z2)],.2,.4,.1,color=blue):
G3:=arrow([Re(z1),Im(z1)],[Re(w),Im(w)],.2,.4,.1,color=blue):
G4:=arrow([0,0],[Re(w),Im(w)],.2,.4,.1,color=blue):
display({G1,G2,G3,G4},axes=framed);
```

Mathematica:

```
{z1=9+I, z2=2+7*I, w=z1+z2}
g1=Graphics[{Blue,Arrow[{{0,0},{Re[z1],Im[z1]}}]}];
g2=Graphics[{Blue,Arrow[{{0,0},{Re[z2],Im[z2]}}]}];
g3=Graphics[{Blue,Arrow[{{Re[z1],Im[z1]},{Re[w],Im[w]}}]}];
g4=Graphics[{Blue,Arrow[{{0,0},{Re[w],Im[w]}}]}];
Show[g1,g2,g3,g4,Frame->True]
```

□

Special types of vectors

Let us consider the most important special types of vectors, e.g., zero vectors, unit vectors, constant vectors, random vectors, sparse vectors, scalar multiple of a unit vector, etc.

Maple:

```
with(LinearAlgebra): ZeroVector(n);                    UnitVector(j,n);
ConstantVector(x,n); ScalarVector(x,j,n); Vector(n,shape=zero);
Vector(n,shape=unit[j]);                 Vector(n,shape=constant[x]);
Vector(n,shape=scalar[j,x]);                  Vector(n,storage=sparse);
RandomVector[column](n,generator=G);
```

Note. These types of vectors can be constructed with the predefined **shape** and **storage** parameters in the **Vector** object (see **?shape** or **?storage**). New parameters can be constructed using the *indexing functions* for **Vector** objects (see **?rtable_indexfcn**).

Mathematica:

```
UnitVector[n,j]   ConstantArray[x,n]   RandomSample[{x1,...xn}]
zeroVector[m_]:=Table[0,{i,1,m}]                        zeroVector[n]
unitVector[m_,k_]=Table[If[i==k,1,0],{i,1,m}]    unitVector[n,j]
constantVector[x_,m_]=Table[x,{i,1,m}]         constantVector[x,n]
         randomVector=Table[Random[Integer,{a,b}]/num,{i,1,m}]
                 scalarVector[x_,j_,m_]:=x*UnitVector[m,j]
           SparseArray[{{i1,j1}->val1,{i2,j2}->val2,...}]
```

Problem 5.5 Generate various special types of vectors (with real, complex, or symbolic entries): zero vector, unit vector, constant vector, random vector, sparse vector, scalar multiple of a unit vector.

Maple:

```
with(LinearAlgebra): n:=5: j:=2: x:=1+I: a:=0: b:=10: num:=50:
ZeroVector(n);  UnitVector(j,n); ConstantVector(x,n);
ScalarVector(x,j,n);Vector(n,shape=zero);Vector(n,shape=unit[j]);
Vector(n,shape=constant[x]); Vector(n,shape=scalar[j,x]);
RandomVector[row](n,generator=evalf(rand(a..b)/num));
V2:=Vector(n,storage=sparse); V2[j]:=C; V2;
```

Mathematica:

```
zeroVector[m_]:=ConstantArray[0,n];
unitVector[m_,k_]:=Table[If[i==k,1,0],{i,1,m}];
constantVector[y_,m_]:=Table[y,{i,1,m}];
randomVector[m_]:=Table[Random[Integer,{a,b}]/num,{i,1,m}];
scalarVector[y_,k_,m_]:=y*UnitVector[m,k];
```

```
{n=5,j=2,x=1+I,a=0,b=10,num=50,zeroVector[n],unitVector[n,j],
 UnitVector[n,j],constantVector[x,n],ConstantArray[x,n],
 scalarVector[x,j,n],RandomSample[Range[n]/num],randomVector[n],
 SparseArray[{{a+1,a+1}->7+9*I,{n,a+1}->9*I}]//MatrixForm}
```

☐

5.2 Matrices

In matrix theory, a *matrix* is a rectangular table (*rows* and *columns*) of elements (numbers or symbols) and for which certain axioms of algebra hold. Matrices are used for describing equations and linear transformations, analyzing data that depend on multiple parameters and more. In our work we consider matrices whose elements are symbols and real and complex numbers. In symbolic mathematics, there exist differences in matrix representations, similar to the concept of a vector.

In *Maple*, a matrix is a *mathematical object*, but also a matrix can be defined as a *data object* represented as a *two-dimensional array*, and these objects are different. For example, one-column matrices and vectors, matrices and two-dimensional arrays are different objects in *Maple*.

In *Mathematica*, a matrix is a *data object* and can be represented as a list, a table, and an array.

Matrix representations and components

Maple:

```
with(LinearAlgebra): Matrix([[a11,...,a1m],...,[an1,...anm]]);
<<a11|...|a1m>,...,<an1|...|anm>>;                map(x->f(x),M);
<<a11,...,an1>|...|<a1m,...,anm>>;    Matrix(n,m,(i,j)->f(i,j));
Matrix(n,m,ops);      Matrix(n,m,fill=c); Matrix(n,m,symbol=a);
with(Student[LinearAlgebra]):       A:=MatrixBuilder(); M[i,j];
M[i,1..-1];M[1..-1,j];M[a..b,c..d]; SubMatrix(M,[a..b],[c..d]);
Dimension(M);        ColumnDimension(M);       RowDimension(M);
```

`Matrix([[a11,...,a1m],...,[an1,...anm]])`, constructing a matrix row-by-row (as a list of lists),

<<a11|...|a1m>,...>>, <<a11,...,an1>|...>>, constructing a matrix, respectively row-by-row and column-by-column (as vectors), these forms are sufficiently slow for large matrices,

MatrixBuilder (of the package Student[LinearAlgebra]), an interactive construction of a matrix (up to 5×5),

Matrix(n,m,(i,j)->f(i,j)), map(x->f(x),M), constructing a matrix $n \times m$ with elements that are defined, respectively, by a function $f(i,j)$ and a function $f(x)$ applied to each element of matrix M (for more details see Chapter 1),

Matrix(n,m,fill=c), Matrix(n,m,symbol=a), constructing a matrix $n \times m$ where, respectively, each element is c and the symbolic elements are a_{ij}. In general form, matrices can be constructed defining various options in the function Matrix (for more details see ?Matrix),

M[i,j], M[i,1..-1], M[1..-1,j], M[a..b,c..d], extracting elements, rows, columns, and submatrices of the matrix M,

Dimension, RowDimension, ColumnDimension, determining the dimension, row dimension, and column dimension of a matrix.

Mathematica:

```
m1={{a11,...,a1m},...,{an1,...,anm}}          m3=Array[f,{m,n}]
m2=Table[expr,{i,i1,im},{j,j1,jn}]    m4=ConstantArray[c,{m,n}]
m5=Array[Subscript[a,##]&,{n,m}]    m5//MatrixForm MatrixQ[m3]
ArrayQ[m3] Dimensions[m1] ArrayDepth[m2]      m3[[i,j]] m3[[i]]
m3[[All,j]] m1[[{i1,j1},{i2,j2}]]         Take[m1,{i1,j1},{i2,j2}]
```

{{...},{...}}, Table, Array, ConstantArray, constructing matrices as lists, tables, arrays,

the menu Insert->Table/Matrix, an interactive construction of matrices and tables,

MatrixForm, displaying the elements of lists, arrays in the matrix form,

[[]], Take, extracting elements, rows, columns, submatrices of a matrix,

MatrixQ, ArrayQ, determining whether or not the expression expr is a matrix or an array,

Dimensions, ArrayDepth, Length, determining, respectively, the dimensions, the depth, and the number of rows of matrices and arrays.

Problem 5.6 Generate matrices of various forms and select their elements, rows, columns, and submatrices.

Maple:

```
with(LinearAlgebra): f1:=(i,j)->(-1)*(i*j)^3; f:=x->x^2;
M1:=Matrix([[1,2,3],[4,5,6],[7,8,9]]); map(x->f(x),M1);
M2:=<<3|4|5>,<4|5|6>,<5|6|7>,<6|7|8>>; M3:=Matrix(3,3);
M4:=<<1,2,3>|<2,3,4>|<3,4,5>|<4,5,6>>; M5:=Array(1..3,1..3,[]);
M6:=Matrix(3,3,(i,j)->a||i||j); M7:=Matrix(3,3,(i,j)->f1(i,j));
Matrix(2,2,fill=c); Matrix(2,2,symbol=a); op(M1); M2[2,2]:=-1;
print(M3); M4[2,2]; M2[2,1..-1]; M2[1..-1,3]; M1[2..2,1..2];
SubMatrix(M1,[2..2],[1..2]); Dimension(M3); ColumnDimension(M2);
```

Mathematica:

```
f1[i_,j_]:=(-1)*(i*j)^3; {m1={{3,3,4,5},{3,4,5,6},{4,5,6,7},
  {5,6,7,8}}, m2=Table[i*j,{i,3},{j,5}],m3=Array[f1,{3,5}],
  m4=ConstantArray[c,{3,3}],m5=Array[Subscript[a,##]&,{3,3}]}
Map[MatrixForm,{m1,m2,m3,m4,m5}]
{MatrixForm[m1],m5//MatrixForm,MatrixQ[m3],ArrayQ[m3],
  Dimensions[m2],ArrayDepth[m1],m3[[2,1]],m5[[1]]}
Map[MatrixForm,{m2[[All,2]],m2[[{1,2},{3,5}]],
                m2[[{1,2},All]],Take[m2,{1,3},{2,4}]}]
```

$\square$

Data types of matrices

Both systems are suitable for exact and floating-point numerical linear algebra computations. Therefore, the elements of matrices may consist of different types of numbers defined in both systems (e.g., real or complex hardware-floats, real or complex software-floats, integers and rational numbers, and in general case, symbolic values).

Maple:

	MatrixOptions(M,datatype=val,shape=val);
Matrix(n,m,ops);	Matrix(n,m,datatype=val);

MatrixOptions, displaying or setting matrix options,

Matrix, Matrix,datatype, constructing a numerical $n \times m$ matrix and performing numerical linear algebra operations, where the option datatype=val indicates a data type of a matrix,

datatype=float[8], datatype=complex[8], the elements of a matrix are real and complex hardware-floats (for more details see Sect. 4.2),

datatype=sfloat, datatype=complex(sfloat), the elements of a matrix are
real and complex software-floats (here the integer 8 means the number of bytes
for storing the floating-point real and complex numbers,

n., an integer with a decimal point belongs to software-floats (software-floats are
sufficiently slower than hardware-floats).

```
n:=2; m:=3; L:=[float[8],complex[8],sfloat,complex(sfloat)];
for i in L do Matrix(n,m,datatype=i); od;
```

Mathematica:

> Numerical matrices can be constructed by explicitly introducing the data
> types of matrix entries (integers and rational numbers, machine-precision
> and arbitrary-precision real and complex floating-point numbers).

```
f1[i_,j_]:=i/(i+j); Map[MatrixForm,
{{{1,2},{3,4}},Array[f1,{2,2}],{{1.1,1.2},{1.3,1.4}},
 {{1.1+1.2*I,1.3+1.4*I},{1.5+1.6*I,1.7+1.8*I}},
 {{1.1`30,1.2`30},{1.3`30,1.4`30}}}]
```

Problem 5.7 Let M be a symmetric $n \times n$ matrix whose elements are
real hardware-floats or software-floats. Determine the eigenvalues of the
matrix M and observe the difference.

Maple:

```
with(LinearAlgebra): L:=[[1,9],[9,-5]]; O1:=datatype=float[8];
O2:=shape=symmetric; O3:=datatype=sfloat;
M1:=Matrix(2,2,L,O1,O2); M2:=Matrix(2,2,L,O1);
M3:=Matrix(2,2,L,O2,O3); M4:=Matrix(2,2,L,O3);
Map(Eigenvalues,{M1,M2}); Digits:=30; Map(Eigenvalues,{M3,M4});
```

Mathematica:

```
Map[MatrixForm,
{m1={{1.,9.},{9.,-5.}},m2={{1`30,9`30},{9`30,-5`30}}}]
{Eigenvalues[m1], Eigenvalues[m2]}
```

A difference in solution methods and results can occur in *Maple*, depending on
the declarations of matrices. In this problem, this difference is caused by the
symmetric declaration of the matrix.   □

Mathematical operations on matrices

Maple:

```
with(LinearAlgebra): M1+M2; M1-M2; a*M; c1*M1+c2*M2; M1.M2;
M1^n; MatrixPower(M1,n); MatrixAdd(M1,M2); Add(M1,M2);
Add(M1,M2,c1,c2); Add(M1,-M2); Multiply(M1,M2);
MatrixScalarMultiply(M1,a); MatrixVectorMultiply(M1,V1);
MatrixMatrixMultiply(M1,M2); M1^(-1); MatrixInverse(M1);
M1^%T; Transpose(M1); M1^%H; Trace(M1); HermitianTranspose(M1);
Norm(M1,2); Norm(M1,p); Determinat(M1); factor(M1^5);
```

Mathematica:

```
m1+m2 m1-m2 a*m1       c1*m1+c2*m2 m1.m2          m1.v1    m1^n
MatrixPower[m1,n]      MatrixPower[m1,-1]          Dot[m1,m2]
Inner[f,m1,m2,g]       Outer[Times,m1,m2]]         Norm[m1,p]
Transpose[m1]         ConjugateTranspose[m1]     Det[m1] Tr[m]
```

Problem 5.8 Perform various mathematical operations on matrices.

1) Find the sum, the difference, the product of a matrix and a scalar, the product of these matrices.

2) Obtain the inverse matrix, the transpose matrix, the Hermitian transpose matrix and perform other mathematical operations on matrices.

Maple:

```
with(LinearAlgebra): M1:=<<1,2,9>|<2,3,9>|<5,6,9>>;
M2:=Matrix(3,3); W:=Vector(3); M3:=Matrix([[1,2,1],
[4,5,4],[7,8,9]]); M4:=Matrix(3,3,(i,j)->a||i||j);
M1+M3; M1-M3; M1.M3; M3^(-1); M1^%T, M1^%H, 2*M1;
M3*(-1); M3^(-3); M3^0; MatrixAdd(M3,M4); Add(M3,M4);
evalm(M4&*W); Multiply(M3,M4); MatrixMatrixMultiply(M3,M4);
MatrixScalarMultiply(M1,9); ScalarMultiply(M1,C);
factor(M3^5); Determinant(M3); Norm(M3,2);
```

Mathematica:

```
Map[MatrixForm,{m1=Table[Random[Real,{0,1}],{i,1,3},{j,1,3}],
    m2=Table[If[i<=j,0,1],{i,1,3},{j,1,3}]}]
```

```
Map[MatrixForm,{m1+m2,m1-m2,m1.m2,m1^(-1),2*m1,c1*m1+c2*m2,
    MatrixPower[m1,-3],Dot[m1.m2],Norm[m1,2],Transpose[m1],
    ConjugateTranspose[m1],Det[m1],Tr[m1]}]
Map[MatrixForm,{Inner[f,m1,m2,g],Outer[Times,m1,m2]}]
```

$\square$

Problem 5.9 Let M_1, M_2, and W be two matrices and a vector.

1) Find the mixed products: the vector-matrix product and the matrix-vector product.

2) Perform other operations on matrices: remove and insert rows and columns, construct submatrices, etc.

Maple:

```
with(LinearAlgebra): M1:=<<1,2,9>|<2,3,9>|<3,4,9>|<5,6,9>>;
m:=4; n:=3; Vector[row](n,symbol=v).Matrix(n,m,symbol=a);
Matrix(n,m,symbol=a).Vector[column](m,symbol=v);
Matrix(1,n,symbol=v).Matrix(n,m,fill=a);
Matrix(n,m,fill=b).Matrix(m,1,symbol=v);
MatrixVectorMultiply(M1,<0,K,0,K>); <<M1,M1[1,1..4]>>;
<<M1|M1>>; DeleteRow(M1,1..2); DeleteColumn(M1,[1,2]);
DeleteRow(M1,1); SubMatrix(M1,[1,2],[1,2]);
```

Mathematica:

```
{m1={{1,2,3,5},{2,3,4,6},{9,9,9,9}},
 m=4, n=3, m2=Table[Random[Integer,{0,10}],{i,1,4},{j,1,4}],
 m3=Table[If[i<j,Sin[Pi*(i+j)],1],{i,1,4},{j,1,4}]}
vInd[s_,n_]:=Array[Subscript[s,#]&,n];
mInd[s_,n_,m_]:=Array[Subscript[s,##]&,{n,m}];
Map[MatrixForm,{vInd[v,n].mInd[a,n,m],
 mInd[a,n,m].vInd[v,m],mInd[v,1,n].ConstantArray[a,{n,m}],
 ConstantArray[b,{n,m}].mInd[v,m,1]}]
Map[MatrixForm,{m1,m2,,m3,m1.{{0},{K},{0},{K}}}]
Map[MatrixForm,{Append[m1,m1[[1]]],Join[m1,m1,2],
 Take[m1,{3,3},All],Take[m1,All,{3,4}],Take[m1,{2,3},All],
 Take[m1,{1,2},{1,2}]}]
```

$\square$

Note. In *Mathematica* instead of the functions `AppendRows`, `TakeRows`, `Append Columns`, `TakeColumns`, `TakeMatrix`, `SubMatrix` (for *ver* < 6) the new functions `Append`, `Join`, and `Take` should be applied respectively (for *ver* ≥ 6).

Elementary row and column operations on matrices

Let us consider the elementary row and column operations on matrices:
    a) interchanging two rows or columns,
    b) multiplying any row or column by a nonzero constant,
    c) adding a multiple of a row or column to another row or column.

Maple:

```
RowOperation(A,[i1,i2]);              ColumnOperation(A,[j1,j2]);
RowOperation(A,i,s);                    ColumnOperation(A,j,s);
RowOperation(A,[i1,i2],s);          ColumnOperation(A,[j1,j2],s);
```

Mathematica:

```
a[[{i1,i2}]]=a[[{i2,i1}]]       a[[All,{j1,j2}]]=a[[All,{j2,j1}]]
a[[i]]=s*a[[i]]                               a[[All,j]]=s*a[[All,j]]
a[[i1]]=a[[i1]]+a[[i2]]*s a[[All,j1]]=a[[All,j1]]+a[[All,j2]]*s
```

Problem 5.10 Let A be a matrix. Perform the elementary row and column operations on the matrix A.

Maple:

```
with(LinearAlgebra):         A:=<<1,2,9>|<2,3,9>|<3,4,9>>;
RowOperation(A,[3,2]);             ColumnOperation(A,[1,3]);
RowOperation(A,3,3);                 ColumnOperation(A,1,2);
RowOperation(A,[1,2],9);       ColumnOperation(A,[2,3],9);
```

Mathematica:

```
{a={{1,2,3},{2,3,4},{9,9,9}}, a//MatrixForm}
{a1=a,a1[[{3,2}]]=a1[[{2,3}]],a1//MatrixForm}
{a2=a,a2[[All,{1,3}]]=a2[[All,{3,1}]],a2//MatrixForm}
{a3=a,a3[[3]]=3*a3[[3]],a3//MatrixForm}
{a4=a,a4[[All,1]]=2*a4[[All,1]],a4//MatrixForm}
{a5=a,a5[[1]]=a5[[1]]+a5[[2]]*9,a5//MatrixForm}
{a6=a,a6[[All,2]]=a6[[All,2]]+a6[[All,3]]*9,a6//MatrixForm}
```

□

Matrix visualization

Maple:

```
with(plots):   matrixplot(M,ops);                    pointplot3d(M,ops);
               sparsematrixplot(M,ops);  listdensityplot(M,ops);
```

Mathematica:

```
ListPlot3D[m,ops]    MatrixPlot[m,ops]    ListPointPlot3D[m,ops]
       Show[Graphics3D[{Point[m[[1]]],...,Point[m[[n]]]}],ops]
```

Problem 5.11 Let M be an $n \times n$ matrix. Observe various visual forms of the matrix in $\mathbb{R}^2$ and $\mathbb{R}^3$.

Maple:

```
with(plots): M:=Matrix([[1,-2,3],[-4,-5,6],[7,-8,9]]);
matrixplot(M,heights=histogram,gap=0.3,style=patchnogrid);
pointplot3d(M,color=blue,axes=boxed,symbolsize=30,symbol=cross);
sparsematrixplot(M,color=blue);
listdensityplot(M,smooth=false,colorstyle=RGB);
```

Mathematica:

```
m={{1,-2,3},{-4,-5,6},{7,-8,9}}
ListPlot3D[m,Mesh->All,Filling->Axis]
MatrixPlot[m,ColorFunction->(RGBColor[1,#,1]&)]
ListPointPlot3D[m,PlotStyle->PointSize[Large]]
Show[Graphics3D[{PointSize[0.03],Hue[0.9],Point[m[[1]]],
     Point[m[[2]]],Point[m[[3]]]}]]
```

□

Special types of matrices

Let us consider the most important special types of matrices, e.g., zero, identity, constant, and random matrices; Hilbert and Hankel matrices; sparse and band matrices; diagonal and triangular matrices; scalar multiple of an identity matrix; symmetric and skew-symmetric matrices, etc.

Maple:

```
with(LinearAlgebra):       ZeroMatrix(n,m); IdentityMatrix(n,m);
ConstantMatrix(C,n,m);     RandomMatrix(n,m,generator=a..b,ops);
Matrix(n,m,shape=zero);          Matrix(n,m,shape=constant[C]);
Matrix(n,m,shape=identity);          HilbertMatrix(n,m,var);
HankelMatrix([elems],n);       DiagonalMatrix([elems],n,m);
BandMatrix([elems],k,n,m);            ScalarMatrix(C,n,m);
Matrix(n,m,[elems],storage=val); Matrix(n,n,[elems],shape=val);
```

Note. These types of matrices can be constructed with the predefined **shape** and **storage** parameters in the **Matrix** object (see ?shape or ?storage). New parameters can be constructed using the *indexing functions* for **Matrix** objects (see ?rtable_indexfcn).

Mathematica:

```
zeroM=ConstantArray[0,{n,m}] IdentityMatrix[n]  RandomMatrix[n]
C*IdentityMatrix[n]  DiagonalMatrix[m1]  ConstantArray[c,{n,m}]
 HilbertMatrix[n]          blockM=ArrayFlatten[{{m1,m2...},...}]
HankelMatrix[n]       SparseArray[{{i1,j1},...}->{a1,...},{n,n}]
    SparseArray[{Band[{i1,j1}]->a1,Band[i2,j2]->a2,...},{n,n}]
```

Note. The **UpperDiagonalMatrix**, **LowerDiagonalMatrix**, **TridiagonalMatrix** functions are not available (for *ver* $\geq$ 6), but it is easy to construct the corresponding matrices using, e.g. **SparseArray** function with **Band** construction. Instead of the **BlockMatrix** function, it should be applied the **ArrayFlatten** function. The **ZeroMatrix** function has been replaced by the **ConstantArray** function.

Problem 5.12 Construct and observe various special types of matrices.

Maple:

```
with(LinearAlgebra): n:=5: f:=(i,j)->(i+j);
g:=(i,j)->(x||i||j+I*y||i||j); ZeroMatrix(n,n);
IdentityMatrix(n,n); ConstantMatrix(C,n);
RandomMatrix(n,n,generator=0..10,outputoptions=
  [shape=triangular[upper]]); Matrix(n,n,shape=zero);
Matrix(n,n,shape=identity); Matrix(n,n,shape=constant[C]);
A:=Matrix(n,n,f); A1:=Matrix(n,n,g); L1:=[1,2,3,4,5];
L2:=[1,9,-1]; L3:=<a,b,c,d,e>; L4:=[[1],[2,3],[2,3],[4,5],
  [6,7]]; A2:=Matrix(n,n,storage=sparse); A2[2,3]:=a23; A2;
DiagonalMatrix(L1,n,n); BandMatrix(L2,1,n,n);
ScalarMatrix(C,n,n); HilbertMatrix(n,n,x); HankelMatrix(L3);
```

```
M1:=Matrix(n,n,shape=symmetric); M1[2,3]:=4; M1;
M2:=Matrix(n,n,shape=skewsymmetric); M2[2,3]:=5; M2;
Matrix(n,L4,shape=triangular[lower]);
```

Mathematica:

```
n=5; m1=Array[Sqrt,n];
f[i_,j_]:=i+j; triDiagonalMatrix[d1_,d2_,d3_]:=Table[
 If[i-j==1,d2,If[i-j==-1,d3,If[i==j,d1,0]]],{i,n},{j,n}];
a1=Table[f[i,j],{i,1,n},{j,1,n}]; a2=Array[
 Subscript[x,##]&,{n,n}]+I*Array[Subscript[y,##]&,{n,n}];
Map[MatrixForm,{m1,a1,a2}]
Map[MatrixForm,{m2=ConstantArray[0,{n,n}],m3=IdentityMatrix[n],
 ConstantArray[c,{n,n}],Table[c,{i,1,n},{j,1,n}],
 Table[Random[Integer,{0,10}],{i,1,n},{j,1,n}],
 HilbertMatrix[n],HankelMatrix[n],C*IdentityMatrix[n],
 DiagonalMatrix[m1],SparseArray[{{i_,i_},{1,n}}->{9,17},{n,n}],
 ArrayFlatten[{{m2,m3},{m3,m2}}],triDiagonalMatrix[1,2,3],
 SparseArray[{Band[{1,1}]->1,Band[{2,1}]->2,Band[{1,2}]->3},
 {n,n}],SparseArray[{{i_,i_},{n,1},{n-1,1}}->{2,17,9},{n,n}]}]
```

                                                                          □

Standard matrix operations

Let us consider the most important standard matrix operations, e.g., determinants, ranks, coranks, adjoint matrices, cofactors, minors, inverse matrices.

Maple:

```
with(LinearAlgebra):   Determinant(M);   Rank(M);    Adjoint(M);
                       Minor(M,i,j);               MatrixInverse(M);
```

Mathematica:

```
Det[m]         MatrixRank[m]        Inverse[m]       Minors[m,k]
```

Problem 5.13 Let A be an $n \times n$ matrix with symbolic entries a_{ij}. Find
a) the determinant of A, b) the rank and corank of the matrix, c) the adjoint matrix, d) cofactors, e) minors, and f) inverse matrix.

Maple:

```
with(LinearAlgebra): n:=3; IM:=IdentityMatrix(n,n);
f:=(i,j)->a||i||j; A:=Matrix(n,n,f); Determinant(A); Rank(A);
MatrixInverse(A); coRank:=M->ColumnDimension(M)-Rank(M);
coRank(A); Equal(map(factor,A.Adjoint(A)),Determinant(A).IM);
Cofactor:=Matrix(n,n,(i,j)->(-1)^(i+j)*Minor(A,i,j));
Equal(Adjoint(A),Transpose(Cofactor));
map(simplify,MatrixInverse(A)-Adjoint(A)/Determinant(A));
Equal(MatrixInverse(A),map(simplify,Adjoint(A)/Determinant(A)));
```

Mathematica:

```
n=3; a1=IdentityMatrix[n]; a2=Array[Subscript[a,##]&,{n,n}]
rank[m_]:=Length[m]-Length[NullSpace[m]];
coRank[m_]:=Dimensions[m,1]-rank[m];
{Det[a2],rank[a1],rank[a1]===MatrixRank[a1],coRank[a1]}
Map[MatrixForm,{Minors[a2,1],Minors[a2,2]}]
{k=Length[a2],k1=Length[Minors[a2,1]],k2=Length[Minors[a2,2]]}
mIJ[m_,n_]:=Table[m[[n-i+1,n-j+1]],{i,1,n},{j,1,n}];
cofactor[m_,n_]:=Table[(-1)^(i+j)*mIJ[Minors[m,n-1],n][[i,j]],
  {i,1,n},{j,1,n}]; adjointM[m_,n_]:=Transpose[cofactor[m,n]];
Map[MatrixForm, {mIJ[Minors[a2,1],k1],mIJ[Minors[a2,2],k2],
    adjointM[a2,k],cofactor[a2,k],Inverse[a2]}]
(Inverse[a2].a2//FullSimplify)===IdentityMatrix[n]
(a2.adjointM[a2,k]//FullSimplify)===
    (Det[a2]*IdentityMatrix[n]//FullSimplify)
adjointM[a2,k]===Transpose[cofactor[a2,k]]
(Inverse[a2])===(adjointM[a2,k]/Det[a2])//FullSimplify
```

□

5.3 Functions of Matrices

In general, a function of a matrix denotes a function that maps a matrix to a matrix and can be defined in several ways [16] (e.g., Jordan canonical form definition, polynomial interpolation definition, Cauchy integral definition).

There are various methods for evaluating functions of a square matrix. We consider the basic functions of matrices that are valid in both systems of symbolic algebra (e.g., matrix powers and, polynomials, matrix exponential, trigonometric and hyperbolic functions of a matrix, etc.).

Maple:

```
with(LinearAlgebra):   MatrixInverse(M);        MatrixPower(M,n);
MatrixExponential(M);  MatrixFunction(M,f,x); M^n;        M^(-1);
```

Mathematica:

```
Inverse[m]              MatrixPower[m,n]              MatrixExp[m]
```

Problem 5.14 Let A and B be $n \times n$ nonsingular matrices.

1) Evaluate matrix powers of A and verify the basic properties of matrix powers.

2) Verify for an *involutive matrix* A_{in} that $A_{\text{in}} = A_{\text{in}}^{-1}$ or $A_{\text{in}}^2 = I$.

3) Evaluate the exponentials of A and A^{-1}.

4) Verify that $e^A e^B \neq e^B e^A$ in general case and $e^A e^B = e^B e^A$ for the commutative matrices A and B.

5) Evaluate $e^{-A_{\text{in}}t}$.

Maple:

```
with(LinearAlgebra): n:=3; AIn:=Matrix(2,2,[[1,0],[1,-1]]);
A:=Matrix(n,n,[[1,0,0],[1,1,0],[0,2,1]]);
B:=Matrix(n,n,[[-1,0,1],[-1,-1,0],[0,2,1]]);
Ac:=Matrix(2,2,[[1,0],[0,0]]);Bc:=Matrix(2,2,[[0,0],[0,1]]);
Equal(IdentityMatrix(n),A.A^(-1)); Equal(A^2,A.A);
Equal((A^2).(A^3),A^(2+3)); Equal((A^2)^3,A^(2*3));
Equal(AIn,MatrixInverse(AIn));Equal(AIn^2,IdentityMatrix(2));
MatrixExponential(A); AExp:=evalf(MatrixExponential(A));
evalf(MatrixInverse(AExp)); evalf(MatrixExponential(-A));
Z1:=evalf(MatrixExponential(A).MatrixExponential(B));
Z2:=evalf(MatrixExponential(B).MatrixExponential(A));
Equal(Z1,Z2); Equal(A.B,B.A);
Z1:=MatrixExponential(Ac).MatrixExponential(Bc);
Z2:=MatrixExponential(Bc).MatrixExponential(Ac); Equal(Z1,Z2);
Equal(Ac.Bc,Bc.Ac); Ac.Bc; Bc.Ac; MatrixExponential(-AIn,t);
```

Mathematica:

```
n=3; Map[MatrixForm, {aIn={{1,0},{1,-1}},
a={{1,0,0},{1,1,0},{0,2,1}},b={{-1,0,1},{-1,-1,0},{0,2,1}},
ac={{1,0},{0,0}},bc={{0,0},{0,1}}}]
```

```
{IdentityMatrix[n]===a.Inverse[a], MatrixPower[a,2]===a.a,
 MatrixPower[a,2].MatrixPower[a,3]===MatrixPower[a,(2+3)],
 MatrixPower[MatrixPower[a,2],3]===MatrixPower[a,(2*3)],
 aIn===Inverse[aIn],MatrixPower[aIn,2]===IdentityMatrix[2]}
Map[MatrixForm,{MatrixExp[a],aExp=N[MatrixExp[a]],
   N[Inverse[aExp]],N[MatrixExp[-a]],MatrixExp[-aIn*t]}]
{z1=N[MatrixExp[a].MatrixExp[b]],z2=N[MatrixExp[b].MatrixExp[a]],
 z1===z2,a.b===b.a}
{z1=MatrixExp[ac].MatrixExp[bc],z2=MatrixExp[bc].MatrixExp[ac],
 z1===z2,ac.bc===bc.ac,ac.bc, bc.ac}                                    □
```

Problem 5.15 Let A be an $n \times n$ nonsingular matrix.

1) Evaluate various matrix functions of A, e.g., $\sqrt{A}$, $\sin(A)$, $\cos(A)$, $\sinh(A)$, $\cosh(A)$.

2) The *Pascal matrix* is an infinite-dimensional matrix in which the entries are binomial coefficients. Generate an $n \times n$ Pascal matrix and evaluate various matrix functions of P.

3) For $n \times n$ arbitrary matrices A_i $(i = 0, \ldots, n - 1)$ construct a polynomial with matrix coefficients.

Maple:

```
with(LinearAlgebra): n:=3; A:=Matrix(2,2,[[1+I,0],[1-I,1]]);
PascalM:=Matrix(n,n,(i,j)->binomial(i+j-2,j-1));
f:=x->add(a||i*x^i,i=0..n-1); MatrixFunction(A,sin(x),x);
MatrixFunction(A,cos(x),x);   MatrixFunction(A,sinh(x),x);
MatrixFunction(A,cosh(x),x);  MatrixFunction(A,sqrt(x),x);
MatrixFunction(PascalM,exp(x*t),x); f(x); f(PascalM);
evalf(MatrixFunction(PascalM,LambertW(x),x));
CharacteristicMatrix(PascalM,x);
for i from 0 to n-1 do A||i:=Matrix(n,n,symbol=a||i); od;
F:=x->add(A||i*x^i,i=0..n-1); F(lambda);
```

Mathematica:

```
f[x_]:=Sum[a[i]*MatrixPower[x,i],{i,0,n-1}]; f1[i_,j_]:=
 Binomial[i+j-2,j-1]; {n=3,a1={{1+I,0},{1-I,1}},
 mPascal=Array[f1,{n,n}],f[x]}
Map[MatrixForm,{MatrixExp[mPascal*t],f[mPascal],
 charMatrix=x*IdentityMatrix[n]-mPascal,
 Table[A[i]=Array[Subscript[a[i],##]&,{n,n}],{i,0,n-1}]}]
fun[x_]:=Sum[A[i]*x^i,{i,0,n-1}]; fun[\[Lambda]]//MatrixForm
Map[MatrixForm,{Sin[a1],Cos[a1],Sinh[a1],Cosh[a1],Sqrt[a1]}]         □
```

5.4   Vector Spaces

A *vector space* (or linear space) is a set of vectors, defined over a field F, where the two operations (the vector addition and scalar multiplication) are defined and certain axioms hold (see [34]). Vector spaces are the basic objects of study in linear algebra, mathematics, science, and engineering. Various classes of mathematical objets can be considered as vectors (e.g., real numbers, matrices, polynomials, continuous functions, solutions of differential equations, etc.). The most familiar vector spaces are the *Euclidean spaces* $\mathbb{R}^2$ and $\mathbb{R}^3$ over the field $\mathbb{R}$. Vectors in these spaces can be represented by ordered pairs and triples of real numbers, and are isomorphic to *geometric vectors* (see Sect. 5.1). The behavior of geometric vectors under the two operations (vector addition and scalar multiplication) represents a simple model that can help in understanding the behavior of vectors in more abstract vector spaces (e.g., the vector space of polynomials over $\mathbb{R}$).

We know that two arbitrary vectors of a vector space can be parallel or *linearly dependent* (we can write each vector as a linear combination of the others), and non-parallel or *linearly independent* (we cannot write each vector as a linear combination of the others). An ordered set B is called a basis of a vector space V if the following conditions hold:

  (i) all the elements of the basis B are linearly independent,

  (ii) all the elements of the basis B belong to the vector space V,

  (iii) each vector of V can be uniquely written as a linear combination of the basis vectors.

We consider the finitely generated vector space V, i.e., if there exists a finite set of vectors span $S = \{\mathbf{v}_1, \ldots, \mathbf{v}_n\} \in V$ such that span $S = V$. If the vectors in S are also linearly independent, then the set S is a *basis* for V. In symbolic mathematics, there exist various methods for constructing bases of vector spaces. We consider the most important functions in both systems for generating vector spaces, bases, bases for row spaces or column spaces of a matrix, bases for the nullspace (or kernel) of a matrix, for determining dimensions and linear dependency of vectors.

Maple:

```
with(LinearAlgebra):  n,m:=Dimension(M);        RowDimension(M);
ColumnDimension(M);   Basis(V);     RowSpace(V); ColumnSpace(V);
IntersectionBasis(VS);SumBasis(VS);Dimension(V);  NullSpace(V);
```

Mathematica:

NullSpace[m]	Dimensions[m]	Length[m]

Problem 5.16 Let VS_1, VS_2, A, and V be, respectively, two Euclidean finite-dimensional vector spaces over the field $\mathbb{R}$, an $n \times n$ matrix, and an n-dimensional vector.

1) Find bases for the vector spaces.
2) Determine the bases for row space and column spaces of A.
3) Find the basis for the null space of the matrix A.
4) Determine the dimensions of the vector spaces.
5) Determine the basis for the direct sum of the vector spaces.
6) Find the basis for the intersection of the vector spaces.
7) Determine the dimension, the row dimension, and column dimension of the matrix A.

Maple:

```
with(LinearAlgebra):
A:=<<1,3,1>|<1,3,1>|<2,4,1>>; V:=<1,1,1>;
VS1:={<1,2,3>,<4,5,6>,<7,8,9>}; VS2:={Vector([1,1,3]),
  Vector([2,4,6]),Vector([3,5,9])};  for i from 1 to 2 do
  Rank(convert(convert(VS||i,list),Matrix));od; B1:=Basis(VS1);
B2:=Basis(VS2); RowSpace(A); ColumnSpace(A); NullSpace(A);
for i from 1 to 2 do Dim_VS||i:=nops(B||i); od;
SumBasis([VS2,V]); IntersectionBasis([VS2,V]); Dimension(V);
Dimension(A); RowDimension(A); ColumnDimension(A);
```

Mathematica:

```
{a={{1,1,2},{3,3,4},{1,1,1}},b={0,0,0},v={1,1,1},
 vs1={{1,2,3},{4,5,6},{7,8,9}},vs2={{1,1,3},{2,4,6},{3,5,9}}}
rowSpace[m_]:=DeleteCases[RowReduce[m],_?(Union[#]=={0}&)];
columnSpace[m_]:=DeleteCases[RowReduce[
  Transpose[m]],_?(Union[#]=={0}&)];
Map[MatrixForm,{rowSpace[a],columnSpace[a],NullSpace[a],
    MatrixRank[vs1],MatrixRank[vs2]}]
Map[Dimensions[#]&,{rowSpace[a],columnSpace[a],NullSpace[a]}]
{augM1=MapThread[Append,{Transpose[vs1],b}],
 rr=RowReduce[augM1],c1={C1=k,C2=-2*k,C3=k}/.{k->1},
 Sum[c1[[i]]*vs1[[i]],{i,1,2}]===-c1[[3]]*vs1[[3]]}
```

```
{augM2=MapThread[Append,{Transpose[vs2],b}],
 rr=RowReduce[augM2],c2={C1=-k,C2=-k,C3=k}/.{k->1},
 Sum[c2[[i]]*vs2[[i]],{i,1,2}]===-c2[[3]]*vs2[[3]],
 b1=vs1[[1;;2]],b2=vs2[[1;;2]],Map[Dimensions[#,1]&,{b1,b2}]]}
{nullSp1=NullSpace[Transpose[vs1]],
 linDependence1=Solve[Thread[nullSp1.{v1,v2,v3}=={0}],{v3}]}
{nullSp2=NullSpace[Transpose[vs2]],
 linDependence2=Solve[Thread[nullSp2.{v1,v2,v3}=={0}],{v3}]}
{dims=Dimensions[a],rowDim=dims[[1]],colDim=dims[[2]],
 m=Length[a],n=Length[First[a]]}
```
                                                                                □

5.5 Normed and Inner Product Vector Spaces

A vector space on which a *vector norm* is defined is called a *normed vector space* (e.g., Euclidian space E^2 over the field R with the norm that coincides with the vector length). If a normed vector space is *complete* with respect to the metric (induced by the norm) we say that it is a *Banach space*. An *inner product vector space* is a vector space with the additional structure of inner product. If an inner product space is *complete* with respect to the metric (induced by the inner product) we say that it is a *Hilbert space*.

We consider in both systems the most important concepts of normed and inner product vector spaces: norms of vectors and matrices, inner products, orthogonal vectors and matrices, the angle between two vectors, orthogonal and orthonormal sets of vectors in the Gram–Schmidt orthonormalization process.

Maple:

```
with(LinearAlgebra):        VectorNorm(V,p);              Norm(V,p);
Normalize(V,p); V.U;        DotProduct(V,U);        MatrixNorm(M,p);
IsOrthogonal(M);            IsUnitary(M);          VectorAngle(V,U);
GramSchmidt([V,U],normalized=true);           GramSchmidt([V,U]);
```

Mathematica:

```
Orthogonalize[{u,v}]   Orthogonalize[{u,v},f]      Normalize[v]
Normalize[v,f]         Projection[u,v,f]        Projection[u,v]
Norm[m,p] v/Norm[v]    Inner[Times,v,u,Plus]            v.u
```

Note. The package LinearAlgebra`Orthogonalization` is obsolete for $ver \geq 6$.

Problem 5.17 Let V, U, and M be two n-dimensional vectors and an $n \times n$ matrix (with symbolic entries).
1) Find Euclidean norms of the vectors and the matrix.
2) Obtain the inner product.
3) Determine orthogonal vectors and matrices.
4) Compute the angle between the vectors.
5) Find the vector projection of V onto U.
6) Determine an orthonormal set of vectors in the Gram–Schmidt orthonormalization process.

Maple:

```
with(LinearAlgebra): n:=3; M:=<<1,0>|<0,-1>>;
V:=Vector(n,i->v||i); U:=Vector(n,i->u||i); VectorNorm(V,2);
MatrixNorm(M,2); Norm(U,2); DotProduct(V,U);
Equal(V/Norm(V,2),Normalize(V,Euclidean)); VectorAngle(V,U);
vectorProj:=(v,u)->DotProduct(v,u)/DotProduct(u,u)*u;
vectorProj(V,U); IsOrthogonal(M); IsUnitary(M);
GramSchmidt([V,U]); GramSchmidt([V,U],normalized=true);
```

Mathematica:

```
angle[x_,y_]:=ArcCos[x.y/(Norm[x]*Norm[y])];
{m1={{1,0},{0,-1}}, n=3, v1=Array[Subscript[v,#]&,n],
 u1=Array[Subscript[u,#]&,n]}
Map[Simplify,{Norm[v1,2], Norm[u1,2], Norm[m1,2]}]
{v1/Norm[v1,2]=== Normalize[v1], Normalize[v1],v1.u1,
 v1.u1===Inner[Times,v1,u1,Plus]}
{angle[v1,u1], Projection[v1,u1], gS=Orthogonalize[{v1,u1}]}
gS//TraditionalForm
{v2={1,2,3}, u2={3,4,5}, g=Orthogonalize[{v2,u2}],
Table[{g[[i]].g[[i]], g[[i]].g[[3-i]]},{i,1,2}]}
```

□

Problem 5.18 Let U and V be two column matrices. Find the first step in the Gram–Schmidt orthonormalization process.

Maple:

```
u:=<<1,2,9>>; v:=<<2,1,5>>; p:=v^%T.u; q:= u^%T.u;
map(whattype,{p,q}); map(op,{p,q}); w_1:=v-u*p[1,1]/q[1,1];
map(whattype,{u[1.. 1,1],v[1..-1,1]});
w_2:=v-u*(v[1..-1,1]^%T.u[1..-1,1])/(u[1..-1,1]^%T.u[1..-1,1]);
evalm(w_1-w_2);
```

Mathematica:

```
{u={1,2,9}, v={2,1,5}, p=v.u, q=u.u, Map[NumberQ,{p,q}]}
{w1=v-u*p/q, w2=v-u*(v.u)/(u.u), w1-w2}
```

Since *Mathematica* does not distinguish between row and column vectors, then
$v^T = v$ and all operations become more simpler than in *Maple*, where we have
to operate carefully with different mathematical objects.   □

5.6   Systems of Linear Equations

A *linear system of algebraic equations* is a set of linear equations over
a field or a commutative ring. The problem of solving linear systems
of algebraic equations is one of oldest problems in mathematics and it
has a great number of applications in various branches of sciences and
engineering.

We consider the most important concepts and methods in both systems of symbolic algebra (e.g., generation of linear equations from a
coefficient matrix, generation of a coefficient matrix from linear equations, augmented matrices, exact solutions, approximate solutions using
least-squares method, homogenous linear systems, the nullspace or kernel of a matrix).

Maple:

```
with(LinearAlgebra): Sol:=LinearSolve(A,B);                A.Sol=B;
Determinant(A);        NullSpace(A); Rank(A);          augM:=<A|B>;
Eqs:=GenerateEquations(A,[x],B);(A,B):=GenerateMatrix(Eqs,[x]);
A1:=GenerateMatrix(Eqs,[x],augmented=true);     LinearSolve(A1);
     solParameterized :=LinearSolve(A,B,method='QR',free='k');
     solHomogeneousPar:=LinearSolve(A,ZeroVector(n),free='k');
     LeastSquares(A,B,free='k');   LeastSquares(A,B,optimize);
```

Mathematica:

```
NullSpace[a] MatrixRank[a] Det[a]   s=LinearSolve[a,b]   a.s===b
genEqs=Thread[a.x==b]  genM=Normal[CoefficientArrays[genEqs,x]]
augM=Transpose[Append[Transpose[a],b]]        rr=RowReduce[augM]
srr=Flatten[Take[Transpose[rr],-1]]}          bs=NullSpace[a]
sPart=LinearSolve[a,b]      sParam=k*Flatten[bs]+Flatten[sPart]
sPart=LinearSolve[a,vec0]      sHom=k*Flatten[bs]+Flatten[sPart]
LeastSquares[a,b]===PseudoInverse[a].b         PseudoInverse[a]
```

Problem 5.19 Let A_i and B_i $(i = 1, 2)$ be, respectively, an $n \times n$ non-singular and singular matrix and n-dimensional vectors over the field $\mathbb{Z}$.

1) Generate the nonhomogeneous system of linear equations that corresponds to the coefficient matrix A_1 and the vector B_1, and viceversa, generate the coefficient matrix that corresponds to the non-homogeneous system of linear equations.

2) Construct the corresponding augmented matrix.

3) Determine the determinants and the nullspaces of the matrices A_i.

4) Find the exact and parameterized solutions of the corresponding nonhomogeneous systems of linear equations.

5) Find the approximate solution of the nonhomogeneous system of linear equations (for singular case) using the least squares method.

6) Find the parameterized solution of the linear homogeneous system (for singular case).

Maple:

```
with(LinearAlgebra): n:=3; X1:=Vector(n,symbol=v);
A1:=<<6,3,3>|<1,1,2>|<1,-1,-6>>; B1:=<3,2,-1>;
A2:=<<6,3,3>|<1,1,1>|<1,-1,-1>>; B2:=<3,2,2>;
map(NullSpace,{A1,A2}); map(Determinant,{A1,A2});
X:=LinearSolve(A1,B1);  Equal(A1.X,B1);
generateEqs:=GenerateEquations(A1,convert(X1,list),B1);
generateMat:=GenerateMatrix(generateEqs,convert(X1,list));
generateMat[1].X1=-generateMat[2]; augmentedMat:=<A1|B1>;
sol1:=LinearSolve(augmentedMat); map(Rank,{A1,A2});
coRank:=M->ColumnDimension(M)-Rank(M); map(coRank,{A1,A2});
aM:=GenerateMatrix(generateEqs,convert(X1,list),
  augmented=true); LinearSolve(aM);
solParameterized:=LinearSolve(A2,B2,method='QR',free='k');
solHomogeneous:=LinearSolve(A2,ZeroVector(n),free='k');
LeastSquares(A2,B2,free='k'); LeastSquares(A2,B2,optimize);
```

Mathematica:

```
rank[m_]:=Length[m]-Length[NullSpace[m]];
corank[m_]:=Dimensions[m,1]-rank[m];
Map[MatrixForm,{a1={{6,1,1},{3,1,-1},{3,2,-6}},b1={3,2,-1},
   n=Length[a1],a2={{6,1,1},{3,1,-1},{3,1,-1}},b2={3,2,2},
   n=Length[a2],x1=Array[Subscript[v,#]&,n]}]
Map[MatrixForm,{NullSpace[a1],Det[a1],rank[a1],
  rank[a1]===MatrixRank[a1],corank[a1],x=LinearSolve[a1,b1]}]
```

```
{a1.x===b1,generateEqs=Thread[a1.x1==b1],
generateMat=Normal[CoefficientArrays[generateEqs,x1]],
Thread[generateMat[[2]].x1==-generateMat[[1]]]}
{augmentedMat=Transpose[Append[Transpose[a1],b1]],
rr=RowReduce[augmentedMat],sol1=Flatten[Take[Transpose[rr],-1]]}
{rank[a2],rank[a2],corank[a2],basis=NullSpace[a2],Det[a2]}
{solPart=LinearSolve[a2,b2],solParameterized=k*Flatten[basis]
 +Flatten[solPart],test=a2.x2/.x2->solParameterized//Simplify}
{solPartHom=LinearSolve[a2,ConstantArray[0,n]],
 solParameterizedHom=k*Flatten[basis]+Flatten[solPartHom],
 test=a2.x2/.x2->solParameterizedHom//Simplify}
{LeastSquares[a2,b2]===PseudoInverse[a2].b2,LeastSquares[a2,b2]}
```

To obtain the same parameterized solutions of the nonhomogeneous system of
linear equations (for singular case) in *Maple* and *Mathematica*, we have included
the *Maple* options `method='QR'` and `free='k'` into the `LinearSolve` function.

$\square$

5.7   Linear Transformations

A *linear transformation* T (or a linear operator) is a function $T:V \to W$
between two vector spaces V and W over the same field F that preserves
the two operations, the vector addition and the scalar multiplication, i.e.,
$T(x+y)=T(x)+T(y)$, $T(ax)=aT(x)$ ($\forall x, y \in V$, $a \in F$). We can combine
the two axioms (additivity and homogeneity) $T(ax+by)=aT(x)+bT(y)$
($\forall x, y \in V$, $a, b \in F$) and say that the linear transformation preserves
linear combinations. In abstract algebra, a linear transformation is a
homomorphism of the vector spaces.

If V and W are finite-dimensional vector spaces and if we choose the
bases in these spaces, then every linear transformation $T:V \to W$ can be
represented as the *matrix associated with the linear transformation* T.
A linear transformation can be represented by different matrices (since
the values of the entries of the matrix associated with T depend on the
basis that we chose).

We consider the the most relevant concepts and operations concerning linear transformations in symbolic mathematics.

Maple:

```
with(LinearAlgebra):      T:=v->expr;   X:=Matrix(n,m,symbol=x);
Y:=Matrix(n,m,symbol=y);  T(a*X+b*Y);             a*T(X)+b*T(Y);
```

Mathematica:

```
x=Array[Subscript[X,##]&,{n,m}]  y=Array[Subscript[Y,##]&,{n,m}]
T[v_]:=expr;                     T[a*x+b*y]         a*T[x]+b*T[y]
```

Problem 5.20 Let $\mathcal{M}_{n\times m}$ and $\mathcal{M}_{n\times 1}$ be, respectively, vector spaces of $n \times m$ and $n \times 1$ arbitrary matrices over the field $\mathbb{R}$ and let $T:\mathcal{M}_{n\times m} \rightarrow \mathcal{M}_{n\times 1}$ be a function defined by the formula:

$$T \begin{pmatrix} a_{11} & \cdots & a_{1m} \\ \cdots & \cdots & \cdots \\ a_{n1} & \cdots & a_{nm} \end{pmatrix} = \begin{pmatrix} a_{11} \\ \cdots \\ a_{n1} \end{pmatrix}$$

Show that T is the linear transformation.

Maple:

```
with(LinearAlgebra): n:=3; m:=4;
T:=v->Matrix(n,1,[seq(v[i,1],i=1..n)]); X:=Matrix(n,m,symbol=x);
Y:=Matrix(n,m,symbol=y); Z1:=T(a*X+b*Y); Z2:=a*T(X)+b*T(Y);
Equal(Z1,Z2); expand(T(a*X+b*Y)-(a*T(X)+b*T(Y)));
```

Mathematica:

```
{n=3,m=4,v=Array[Subscript[V,##]&,{n,1}],
 x=Array[Subscript[X,##]&,{n,m}],y=Array[Subscript[Y,##]&,{n,m}]}
T[v_]:=Table[v[[i,1]],{i,1,n}]; {Map[MatrixForm,{T[a*x+b*y],
 a*T[x]+b*T[y]}], T[a*x+b*y]===a*T[x]+b*T[y]}
```
□

Problem 5.21 Let V and $\mathbb{Q}[x]$ be n-dimensional vector spaces of vectors and polynomials over the field $\mathbb{R}$ and let $T:V \rightarrow \mathbb{Q}[x]$ be a function defined by the formula:

$$T \begin{pmatrix} a_1 \\ \vdots \\ a_n \end{pmatrix} = \sum_{i=1}^{n} a_i x^{i-1}.$$

Show that T is the linear transformation.

Maple:

```
with(LinearAlgebra): n:=9; v:=Vector(n);
T:=v->add(v[i]*x^(i-1),i=1..n); P:=Vector(n,symbol=p);
Q:=Vector(n,symbol=q); Z1:=T(a*P+b*Q); Z2:=a*T(P)+b*T(Q);
expand(T(a*P+b*Q)-(a*T(P)+b*T(Q))); evalb(expand(Z1)=expand(Z2));
```

Mathematica:

```
{n=9,v=Array[Subscript[V,#]&,n],
 p=Array[Subscript[P,#]&,n], q=Array[Subscript[Q,#]&,n]}
T[v_]:=Sum[v[[i]]*x^(i-1),{i,1,n}]; {T[a*p+b*q],a*T[p]+b*T[q],
 Expand[T[a*p+b*q]]===Expand[a*T[p]+b*T[q]]}
```

□

5.8 Eigenvalues and Eigenvectors

Eigenvectors[1] of a linear transformation T are vectors ($\neq 0$) that satisfy the equation $Tx = \lambda x$ for a scalar λ, called *eigenvalue* of T, that corresponds to the eigenvector x. An *eigenspace* of a linear transformation for a particular eigenvalue is a space formed by all the eigenvectors associated with this eigenvalue.

A linear transformation can be considered as an operation on vectors that usually changes its magnitudes and directions. The direction of the eigenvector of T does not change (for positive eigenvalues) or changes in the opposite direction (for negative eigenvalues). The eigenvalue of an eigenvector is a scaling factor by which it has been multiplied. The *spectrum* of a linear transformation defined on finite-dimensional vector spaces is the set of all eigenvalues.

These concepts are of great importance in various areas of mathematics, especially in linear algebra, functional analysis, nonlinear mathematical equations, etc.

We consider the most important concepts and methods, related to eigenvalues and eigenvectors, in both systems of symbolic algebra: characteristic matrices and polynomials, eigenvalues, eigenvectors, minimal polynomials, the diagonalization and diagonal factorization of a matrix, the trace of a matrix, the Cayley–Hamilton theorem.

Maple:

```
with(LinearAlgebra):Trace(M);z:=Eigenvectors(M);Eigenvalues(M);
CharacteristicMatrix(M,x);          CharacteristicPolynomial(M,x);
MinimalPolynomial(M,x); p:=z[2];          d:=DiagonalMatrix(z[1]);
Equal(d,MatrixInverse(p).M.p);      Equal(M,p.d.MatrixInverse(p));
```

---

[1]from the German term "eigen"

Mathematica:

```
Eigenvalues[m]  Eigenvectors[m]          z=Eigensystem[m]  Tr[m]
charMatrix=m-x*IdentityMatrix[n]         Det[m-x*IdentityMatrix[n]]
CharacteristicPolynomial[m,x]                    p=Transpose[z[[2]]]
d=DiagonalMatrix[z[[1]]] d===Inverse[p].m.p  m===p.d.Inverse[p]
```

Note. Since it is not possible to obtain explicit expressions in exact arithmetic for the eigenvalues of a matrix of dimension ≥ 5, *Maple* and *Mathematica* represent eigenvalues implicitly, with the **RootOf** function (these expressions can be simplified and evaluated numerically).

Problem 5.22 Let A be an $n \times n$ matrix with integer entries.
   1) Obtain characteristic polynomials and matrices.
   2) Find eigenvalues and eigenvectors.
   3) Construct the minimal polynomial.
   4) Obtain the trace of the matrix.
   5) Verify the Cayley–Hamilton theorem.
   6) Perform the diagonalization and diagonal factorization of A.

Maple:

```
with(LinearAlgebra): A:=<<1,0,0>|<2,5,0>|<0,6,9>>;
CharacteristicMatrix(A,x); CharacteristicPolynomial(A,x);
Eigenvalues(A); Eigenvectors(A,output='list');
MinimalPolynomial(A,x); Trace(A); CayleyHamiltonT:=subs(x=A,
CharacteristicPolynomial(A,x)); simplify(CayleyHamiltonT);
z:=Eigenvectors(A);    p:=z[2]; d:=DiagonalMatrix(z[1]);
Equal(d,MatrixInverse(p).A.p); Equal(A,p.d.MatrixInverse(p));
```

Mathematica:

```
{a1={{1,2,0},{0,5,6},{0,0,9}},n=Length[a1],Tr[a1]}
characteristicM=a1-x*IdentityMatrix[n]//MatrixForm
{CharacteristicPolynomial[a1,x],Det[a1-x*IdentityMatrix[n]],
 Det[x*IdentityMatrix[n]-a1]}
t=Solve[Det[a1-x*IdentityMatrix[n]]==0,x]
eVec1=Table[NullSpace[a1-t[[i,1,2]]*IdentityMatrix[n]],{i,1,3}]
{eVals=Eigenvalues[a1],z=Eigensystem[a1],eVec2=Eigenvectors[a1]}
Do[Print["EigVal ", i, " is ", z[[1,i]], " with EigVec",
 z[[2,i]]],{i,1,3}]; minimalPolynomial[m_,x_]:=Module[
 {n=Length[m],minP},minP=Det[x*IdentityMatrix[n]-m]/Apply[
 PolynomialGCD[##]&,Flatten[Minors[x*IdentityMatrix[n]-m,n-1]]]];
```

```
{minimalPolynomial[a1,x],p=CharacteristicPolynomial[a1,x],
 l=CoefficientList[p,x],k=Length[l],CayleyHamiltonT=
 Sum[l[[i+1]]*MatrixPower[a1,i],{i,0,k-1}]//MatrixForm}
Map[MatrixForm,{p=Transpose[z[[2]]],d=DiagonalMatrix[z[[1]]]}]
{Map[MatrixForm,{a1,p,d,Inverse[p]}],d===Inverse[p].a1.p,
 a1===p.d.Inverse[p]}
```

$\square$

5.9   Matrix Decompositions and Equivalence Relations

A *decomposition of a matrix* A of rank $r > 0$ is a factorization (or reduction) of the matrix A (using elementary transformations) into some *canonical form* or *normal form* that is a block diagonal matrix with the $(r \times r)$ identity matrix I_r in the upper left corner.

In general, a canonical form of a mathematical object is a standard way to represent it. The canonical forms (in a set of mathematical objects with an equivalence relation) represent the equivalence classes.

We say that two matrices A and B are *equivalent*, $A \sim B$, if one of them can be converted to the other using elementary transformations. Equivalent matrices have the same order and the rank r. The necessary and sufficient condition for the two $m \times n$ matrices A and B to be equivalent is that there exist two nonsingular matrices P $(m \times m)$ and Q $(n \times n)$ so that the equivalence relation $B = PAQ$ holds.

A set of $m \times n$ matrices forms a *canonical set* $\mathcal{C}_\mathcal{S}$ of matrices with respect to an *equivalence relation* if any matrix $M \notin \mathcal{C}_\mathcal{S}$ is equivalent to one and only one matrix of $\mathcal{C}_\mathcal{S}$.

There are different equivalence relations of matrices, e.g., the congruence relations $B = X^T AX$ and $B = X^H AX$, the similarity relation $B = X^{-1}AX$, the relation of orthogonal similarity $B = Q^T AQ$ (Q is orthogonal, i.e., $Q^{-1} = Q^T$), the relation of unitary similarity $B = Q^H AQ$ (Q is unitary, i.e., $Q^{-1} = Q^*$).

In symbolic mathematics, there are several methods of matrix decompositions. Each matrix decomposition relates to a particular class of problems.

Decompositions related to solving systems of linear equations

We consider the most important matrix decompositions in both systems: the LU decomposition, Gaussian elimination, Gauss–Jordan elimination or reduced row-echelon form, fraction-free elimination, the QR decomposition, and Cholesky decomposition.

Maple:

```
with(LinearAlgebra):     AM:=<A|B>;          LUDecomposition(M,ops);
              LUDecomposition(M,method='GaussianElimination');
LUDecomposition(M,method='RREF');        ReducedRowEchelonForm(M);
LUDecomposition(M,method=FractionFree); GaussianElimination(M);
LU1:=LUDecomposition(AM,output='U');     BackwardSubstitute(LU1);
LU2:=LUDecomposition(AM,output='R');     ForwardSubstitute(LU2);
(q,r):=QRDecomposition(M);BackwardSubstitute(r,Transpose(q).B);
```

Mathematica:

```
    elimGauss[m_,b_]:=Last/@RowReduce[Flatten/@Transpose[{m,b}]];
   id=IdentityMatrix[n]    mb=Transpose[Insert[Transpose[m],b,n+1]]
  mid=Join[m,id,2]    {q,r}=QRDecomposition[m]]  LinearSolve[r,q.b]
  gj=RowReduce[mb]  rr=RowReduce[mid]  inv=Take[rr,{1,n},{n+1,2*n}]
  m.inv===id        {lu,p,c}=LUDecomposition[m]          {l,u,l.u}
  l=lu*SparseArray[{i_,j_}/;j<i->1,{n,n}]+id       u=lu*SparseArray[
  {i_,j_}/;j>=i->1,{n,n}]  luBack1=LinearSolve[a]     x=luBack1[b]
  a.x===b          luBack2=LinearSolve[u,LinearSolve[l,b[[p]]]]
```

Problem 5.23 Let A and B be an $n \times n$ matrix and an n-dimensional vector over the field $\mathbb{C}$.

Solve the linear system of equations $Ax = B$ performing several matrix decompositions: 1) Gaussian elimination. 2) Gauss–Jordan elimination or reduced row-echelon form. 3) LU-decompositions. 4) the QR-decomposition.

Maple:

```
with(LinearAlgebra): n:=4; B:= <1,0,2,-1>; Id:=IdentityMatrix(n);
A:=<<1,0,2,3>|<2,9,6,7>|<3,6,5,7>|<2,7,5,6>>; Rank(A);AB1:=<A|B>;
G1:=GaussianElimination(AB1);BackwardSubstitute(G1); AId:=<A|Id>;
GJ1:=ReducedRowEchelonForm(AB1); GJ2:=LUDecomposition(AId,
 method='RREF'); LUD1:=LUDecomposition(AB1,output='U');
BackwardSubstitute(LUD1); LUD2:=LUDecomposition(AB1,output='R');
ForwardSubstitute(LUD2); RR:=ReducedRowEchelonForm(AId);
Inv:=SubMatrix(RR,1..n,n+1..2*n); Equal(A.Inv,Id);
(q,r):=QRDecomposition(A); BackwardSubstitute(r,Transpose(q).B);
Equal(q.r,A); GE:=LUDecomposition(AB1,
 method='GaussianElimination'); BackwardSubstitute(GE[3]);
```

Mathematica:

```
{a1={{1,2,3,2},{0,9,6,7},{2,6,5,5},{3,7,7,6}},
 b1={1,0,2,-1}, n=MatrixRank[a1],id=IdentityMatrix[n]}
elimGauss[a_,b_]:=Last/@RowReduce[Flatten/@Transpose[{a,b}]];
Map[MatrixForm,{a1b1=Transpose[Insert[Transpose[a1],b1,n+1]],
 g1=elimGauss[a1,b1],gj1=RowReduce[a1b1],gj1[[All,n+1]],
 a1.gj1[[All,n+1]]===b1,aid=Join[a1,id,2]}]
{luBack1=LinearSolve[a1],x=luBack1[b1],a1.x===b1}
{lu,p,c}=LUDecomposition[a1]; l=lu*SparseArray[{i_,j_}/;j<i->1,
 {n,n}]+id; u=lu*SparseArray[{i_,j_}/;j>=i->1,{n,n}];
Map[MatrixForm,{l,u,l.u,l[[p]],l[[p]].u}]
luBack2=LinearSolve[u,LinearSolve[l,b1[[p]]]]
{rr=RowReduce[aid],inv=Take[rr,{1,n},{n+1,2*n}],a1.inv===id}
Map[MatrixForm,{q,r}=QRDecomposition[a1]]
LinearSolve[r,q.b1]
```
                                                                                          □

Matrix decompositions related to eigenvalues

Let us consider matrix decompositions related to *eigenvalues* in both systems: spectral decomposition, Jordan decomposition, Schur decomposition, Hermite decomposition, singular value decomposition in *Mathematica*, and Smith form in *Maple*.

Maple:

```
with(LinearAlgebra):                    z:=Eigenvectors(M); p:=z[2];
d:=DiagonalMatrix(z[1]);         Equal(d,MatrixInverse(p).M.p);
Equal(M,p.d.MatrixInverse(p)); HermiteForm(M,output=['H','U']);
SmithForm(M,output=['U','V']);    SchurForm(M,output=['T','Z']);
JordanBlockMatrix([[x1,n1],...]);   JordanForm(M,output=['Q']);
```

Mathematica:

```
z=Eigensystem[m]    p=Transpose[z[[2]]] d=DiagonalMatrix[z[[1]]]
d===Inverse[p].m.p m===p.d.Inverse[p]    JordanDecomposition[m]
SchurDecomposition[m]                 SingularValueDecomposition[m]
SchurDecomposition[m]                       HermiteDecomposition[m]
```

Problem 5.24 Let A_i $(i = 1, \ldots, 5)$ be nonsingular matrices with symbolic, integer, real, and complex entries.

1) Construct a block diagonal matrix and a Jordan block matrix.
2) Perform Jordan decomposition (for symbolic case).
3) Perform Schur decomposition (for real and complex cases).
4) Find the spectral and Hermite decompositions (for integer case).
5) Construct Smith form and perform singular value decomposition (for complex case).

Maple:

```
with(LinearAlgebra): A1:=Matrix(2,2,(i,j)->a||i||j);
A2:=Matrix(3,3,(i,j)->b||i||j); A3:=Matrix(3,3,(i,j)->i+j);
A4:=Matrix(3,[[1,2+2*I,3+3*I],[4*I,-3,0],[I,0,9]],
 datatype=complex(sfloat)); A5:=Matrix(3,[[1,2,3],[4,-3,0],
 [0,0,9]],datatype=sfloat); DiagonalMatrix([A1,A2]); n:=5;
JordanBlockMatrix([[lambda,n]]); (J,Q):=JordanForm(A1,
 output=['J','Q']); map(simplify,J-Q^(-1).A1.Q);
(T,Z):=SchurForm(A4,output=['T','Z']); Z1:=map(fnormal,A4,2);
Z2:=map(fnormal,Z.T.Z^%H,2); evalf(Z1-Z2);
(T,Z):=SchurForm(A5,output=['T','Z']); Z1:=map(fnormal,A5,2);
Z2:=map(fnormal,Z.T.Z^%T,2); evalf(Z1-Z2);
z:=Eigenvectors(A3); p:=z[2]; d:=DiagonalMatrix(z[1]);
map(simplify,d-p^(-1).A3.p); map(simplify,A3-p.d.p^(-1));
HermiteForm(A3,output=['H','U']); map(simplify,H-U.A3);
(S,U,V):=SmithForm(A4,output=['S','U','V']);
map(fnormal,S-U.A4.V,2);
```

Mathematica:

```
f[i_,j_]:=i+j; n=5; blockJordan[x_,n_]:=Table[If[i-j==-1,1,
 If[i==j,x,0]],{i,n},{j,n}]; Map[MatrixForm,{a1=Array[
 Subscript[a,##]&,{2,2}], a2=Array[Subscript[b,##]&,{3,3}],
 a3=Array[f,{3,3}], a4=N[{{1,2+2*I,3+3*I},{4*I,-3,0},
 {I,0,9}}], a5=N[{{1,2,3},{4,-3,0},{0,0,9}}],
 m023=ConstantArray[0,{2,3}], m032=ConstantArray[0,{3,2}]}]
Map[MatrixForm, {ArrayFlatten[{{a1,m023},{m032,a2}}],
    blockJordan[\[Lambda],n]}]
{j,q}=JordanDecomposition[a1]; Map[MatrixForm,
 {j,q,a1-j.q.Inverse[j]//FullSimplify}]
{t,z}=SchurDecomposition[a4]; Map[MatrixForm, {t,z,z1=a4,
 z2=Chop[t.z.ConjugateTranspose[t]],Chop[z1-z2]}]
```

```
{t,z}=SchurDecomposition[a5];  Map[MatrixForm,{t,z,z1=a5,
 z2=Chop[t.z.ConjugateTranspose[t]],Chop[z1-z2]}]
Map[MatrixForm,{zE=Eigensystem[a3],p=Transpose[zE[[2]]],
    d=DiagonalMatrix[zE[[1]]]}]
Map[FullSimplify,{d-Inverse[p].a3.p,a3-p.d.Inverse[p]}]
{h,u}=HermiteDecomposition[a3]; Map[MatrixForm,
 {h,u,u-h.a3//FullSimplify}]
{s,v,u}=SingularValueDecomposition[a4]; Map[MatrixForm,
{s,v,u,Chop[a4-s.v.Conjugate[Transpose[u]]]//FullSimplify}]
```

□

5.10   Bilinear and Quadratic Forms

Let V be a vector space over a scalar field F. A *linear form* or a *linear functional* is a linear transformation $T : V \to F$ of the vector space V to the scalar field F. If $\dim V = n$, a linear form of n variables, x_i $(i = 1, \ldots, n)$, defined over F is a polynomial $\sum_{i=1}^{n} a_i x_i$ in which the coefficients are elements of F.

A *bilinear form B* is a bilinear transformation $B : V \times V \to F$, i.e., is a linear and homogenous expression with respect to each set of variables x_i, y_j $(i = 1, \ldots, m, \, j = 1, \ldots, n)$. Bilinear forms can be expressed in the matrix form: $B(X,Y) = X^T A Y = \sum_{i=1}^{m} \sum_{j=1}^{n} a_{ij} x_i y_j$, where A is an $n \times n$ matrix of the bilinear form $B(X,Y)$, and vectors $X, Y \in V$. A *quadratic form* is a homogenous polynomial of degree two in a finite number of variables.

Quadratic and bilinear forms are basic mathematical objects in various branches of mathematics, physics, chemistry (e.g., in number theory, Riemannian geometry, Lie theory, theoretical mechanics, etc.).

Maple:

```
with(LinearAlgebra):    DotProduct(U,V);    BilinearForm(U,V);
BilinearForm(U,V,IdentityMatrix(n));           BilinearForm(U,V,A);
sort(expand(U^%T.A.V));        BilinearForm(U,V,conjugate=false);
Equal(A,A^%H);          IsUnitary(A);  sort(expand(U.A.V^%H));
```

Mathematica:

```
n=Length[a]          Dot[u,v]     u.v      u.IdentityMatrix[n].v
Sort[Expand[u.a.v]]  Dot[u,Conjugate[v]]         u.Conjugate[v]
v.a.Conjugate[u]     HermitianMatrixQ[a]            unitaryQ[a]
unitaryQ[a_]:= (Conjugate[Transpose[a]].a===IdentityMatrix[n]);
```

Note. Since *Mathematica* does not distinguish between row and column vectors, then $v^T = v$, and conjugating transpose operation (or Hermitian transpose operation) is `Conjugate[v]`.

Problem 5.25 Let U, V, and A be, respectively, n-dimensional vectors and an $n \times n$ matrix with symbolic entries over the fields $\mathbb{R}$ and $\mathbb{C}$.

1) Verify that if A is the $n \times n$ identity matrix, the real and Hermitian bilinear forms are equivalent to the corresponding inner products.

2) Generate the real and Hermitian bilinear forms as the triple product of U, A, and V (i.e., a generalization of the corresponding inner products).

Maple:

```
with(LinearAlgebra): n:=3; U:=Vector(n,symbol=u);
V:=Vector(n,symbol=v); A:=Matrix(n,n,symbol=a);
DotProduct(U,V,conjugate=false);
BilinearForm(U,V,conjugate=false);
BilinearForm(U,V,IdentityMatrix(n),conjugate=false);
BF1:=sort(expand(BilinearForm(U,V,A,conjugate=false)));
DotProduct(V,U); BilinearForm(V,U); BilinearForm(V,U,
IdentityMatrix(n)); BF2:=sort(expand(BilinearForm(V,U,A)));
```

Mathematica:

```
n=3; Map[MatrixForm,{u=Array[Subscript[U,#]&,n],
 v=Array[Subscript[V,#]&,n],a=Array[Subscript[A,##]&,{n,n}]}]
{Dot[u,v],u.v,u.IdentityMatrix[n].v,bf1=Sort[Expand[u.a.v]]}
{Dot[u,v\[Conjugate]], u.v\[Conjugate],
 u.IdentityMatrix[n].v\[Conjugate],
 bf2=Sort[Expand[u.a.v\[Conjugate]]]}//TraditionalForm
```

□

Problem 5.26 Let U, V, and A be, respectively, n-dimensional vectors and an $n \times n$ matrix with symbolic entries over the field $\mathbb{R}$. For a given bilinear form f_1 generate the matrix notation, i.e., $f_1 = U^T.A.V$.

Maple:

```
with(LinearAlgebra): n:=3; U:=Vector(n,symbol=u);
V:=Vector(n,symbol=v); A:=Matrix(n,n,symbol=a);
f1:=sort(2*u[1]*v[1]-3*u[1]*v[2]+4*u[2]*v[1]+2*u[2]*v[2]
-3*u[2]*v[3]+3*u[3]*v[2]-2*u[3]*v[3]);f2:=sort(expand(U^%T.A.V));
L1:=NULL:L2:=NULL:L3:=NULL:L4:=NULL: c1:=nops(f1);c2:=nops(f2);
for i from 1 to c2 do w2:=op(i,f2); L1:=L1,select(has,w2,a);
 L2:=L2,remove(has,w2,a); od: for i from 1 to c1 do
 w1:=op(i,f1); L3:=L3,remove(has,w1,indets(w1));
 L4:=L4,select(has,w1,indets(w1)); od:
L1:=[L1]; L2:=[L2]; L3:=[L3]; L4:=[L4]; k:=nops(L4); m:=nops(L2);
B:=Matrix(n,n,symbol=b); L5:=convert(Vector(m,0),list);
for i from 1 to k do for j from 1 to m do if L4[i]=L2[j] then
 L5[j]:=L3[i]; fi; od; od; L5; L1; t:=1:
for i from 1 to n do for j from 1 to n do b[i,j]:=L5[t]: t:=t+1:
 od: od: A; B; f1Mat:=sort(expand(U^%T.B.V)); evalb(f1=f1Mat);
```

Mathematica:

```
n=3; Map[MatrixForm,{u=Array[Subscript[U,#]&,n],
 v=Array[Subscript[V,#]&,n],a=Array[Subscript[A,##]&,{n,n}]}]
{f1=Sort[2*Subscript[U,1]*Subscript[V,1]-3*Subscript[U,1]*
 Subscript[V,2]+4*Subscript[U,2]*Subscript[V,1]+2*Subscript[U,2]*
 Subscript[V,2]-3*Subscript[U,2]*Subscript[V,3]+3*Subscript[U,3]*
 Subscript[V,2]-2*Subscript[U,3]*Subscript[V,3]],
 f2=Sort[Expand[u.a.v]]}
{l1={},l2={},l3={},l4={},c1=Length[f1],c2=Length[f2]}
Do[l1=Append[l1,Take[f2[[i]],-1]],{i,1,c2}];
Do[l2=Append[l2,Take[f2[[i]],2]],{i,1,c2}];
Do[l3=Append[l3,Take[f1[[i]],1]],{i,1,c1}];
Do[l4=Append[l4,Take[f1[[i]],-2]],{i,1,c1}];
{l1,l2,l3,l4,k=Length[l4],m=Length[l2],
 b=Array[Subscript[B,##]&,{n,n}],l5=ConstantArray[0,m]}
Do[If[l4[[i]]==l2[[j]],l5[[j]]=l3[[i]]],{j,1,m},{i,1,k}];
{l5=Flatten[l5],l1,t=1}
Do[b[[i,j]]=l5[[t]];t=t+1,{i,1,n},{j,1,n}]; Map[MatrixForm,{a,b}]
{f1Mat=Sort[Expand[u.b.v]],f1===f1Mat}
```
                                                                        □

Problem 5.27 Let A be an $n \times n$ arbitrary matrix over a field F and let X, Y be n-dimensional arbitrary vectors (with symbolic entries). Verify that the transformation $B : V \times V \to F$ defined by $B(X,Y) = X^{\mathrm{T}}AY$ is a bilinear form ($\forall~X,Y \in V$).

Maple:

```
with(LinearAlgebra): n:=3; X:=Vector(n,symbol=x);
X1:=Vector(n,symbol=x1); X2:=Vector(n,symbol=x2);
Y:=Vector(n,symbol=y); Y1:=Vector(n,symbol=y1);
Y2:=Vector(n,symbol=y2); A:=Matrix(n,n,symbol=a);
B11:=BilinearForm(a*X1+b*X2,Y,A,conjugate=false);
B12:=collect(B11,[a,b]); B13:=a*BilinearForm(X1,Y,A,
 conjugate=false)+b*BilinearForm(X2,Y,A,conjugate=false);
evalb(B12=B13); B21:=BilinearForm(X,a*Y1+b*Y2,A,conjugate=false);
B22:=collect(expand(B21),[a,b]); B23:=a*BilinearForm(X,Y1,A,
 conjugate=false)+b*BilinearForm(X,Y2,A,conjugate=false);
B24:=collect(expand(B23),[a,b]); evalb(B22=B24);
```

Mathematica:

```
n=3; Map[MatrixForm,{x=Array[Subscript[X,#]&,n],
 x1=Array[Subscript[X1,#]&,n],x2=Array[Subscript[X2,#]&,n],
 y=Array[Subscript[Y,#]&,n],y1=Array[Subscript[Y1,#]&,n],
 y2=Array[Subscript[Y2,#]&,n],a=Array[Subscript[A,##]&,{n,n}]}]
{b11=(A*x1+B*x2).a.y, b12=Collect[b11,{A,B}]}
{b13=A*x1.a.y+B*x2.a.y,Expand[b12]===Expand[b13]}
{b21=x.a.(A*y1+B*y2),b22=Collect[b21,{A,B}]}
{b23=A*x.a.y1+B*x.a.y2,b24=Collect[b23,{A,B}],
 Expand[b22]===Expand[b24]}
```

□

5.11 Linear Algebra with Modular Arithmetic

Let $m \neq 0$ be an integer. We say that two integers a and b are congruent modulo m if there is an integer k such that $ab = km$, and in this case we write $a = b \bmod m$ or $a \equiv b \pmod m$. Modular arithmetic is the arithmetic of congruences.

In symbolic mathematics, there are various methods and algorithms that allow us to perform modular arithmetic (e.g., in *Maple*, with the functions mod, modp1, modp2 and in *Mathematica*, with the function Mod and the options Mode->Modular, Modulus-> n.

Now we are interested to perform modular arithmetic with matrices and vectors. Modular arithmetic can be performed in $\mathbb{Z}/m$, the integers modulo m $(m > 0)$.

In Maple, there is the `LinearAlgebra[Modular]` subpackage for performing linear algebra operations and functions in $\mathbb{Z}/p$ (p is a prime number and the vector or matrix entries are in the range $0..p-1$).

In Mathematica, we can generate a list of functions that accept the option `Modulus`, apply the `SetOptions` function for each element of the list defining modulo m operation, and perform modular arithmetic (for more details, see the next problem).

Maple:

```
isprime(m);      mod(expr,m);    modp(LinearAlgebra:-function,m);
with(LinearAlgebra[Modular]);                      Mod(m,M,datatype);
Create(m,nR,nC,datatype);                            Copy(m,M);
Transpose(m,M); Inverse(m,M);  Determinant(m,M);      Rank(m,M);
Multiply(m,M1,M2);             LinearSolve(m,M,k,inplace=false);
X:=LUDecomposition(m,M,pV,'det');             LUApply(m,M,pV,X);
RowReduce(m,M,nR,nC,nVars,'det','pd','rank','sig','incR',flag);
```

Note. The efficiency of computations can be improved if we define data types of matrices and vectors (e.g., `float[8]`, `integer[]`, `integer[4]`, `integer[8]`)

Mathematica:

```
Mod[m,p]                              func[vars,Mode->Modular]
func[vars,Modulus->p]                 MatrixRank[m,Modulus->p]
Det[m,Modulus->p]                     LinearSolve[m,b,Modulus->n]
RowReduce[m,Modulus->p]               LUDecomposition[m,Modulus->p]
```

Note. Including the `Modular` option in some functions (for example, `Eliminate`, `AlgebraicRules`, `MainSolve`, `Solve`, `SolveAlways`), we can find a modulus for which a system of equations has a solution.

Problem 5.28 Let n be a prime number and let M and B be an $n \times n$ nonsingular matrix and an n-dimensional vector with the integer entries that belong to the set $\{0, \ldots, n-1\}$.

     1) Introduce the matrix M and the vector B modulo n.

     2) Solve the linear system of equations $Mx = B$ modulo n constructing the reduced row-echelon form and verify the solution.

     3) Obtain the determinant and the rank of M modulo n.

4) Solve the linear system of equations modulo n using the function LinearSolve and verify the solution.

5) Solve the linear system of equations modulo n applying the LU-decomposition method and verify the solution.

Maple:

```
with(LinearAlgebra[Modular]); n:=5; isprime(n);
M:=Matrix(n,n,[[1,2,3,3,0],[2,1,2,1,3],[1,0,2,3,4],[3,4,1,0,2],
  [1,0,3,2,4]]); B:=Vector(n,[1,4,2,1,3]);
MM:=Mod(n,M,integer[4]); BM:=Mod(n,B,integer[4]); AM:=<MM|BM>;
modp(LinearAlgebra:-Determinant(Matrix(M,datatype=integer)),n);
M2:=Copy(n,AM); RowReduce(n,M2,n,n+1,n,'det',0,'rank',0,0,true);
M2; det; rank; Multiply(n,MM,M2,1..n,n+1)=Copy(n,AM,1..n,n+1);
M3:=Copy(n,AM): B3:=LinearSolve(n,M3,1,inplace=false); M3;
Multiply(n,MM,B3)=BM;    M4:=Copy(n,MM): B4:=Copy(n,BM):
pv:=Vector(n-1): LUDecomposition(n,M4,pv,'det'): M4,pv;
LUApply(n,M4,pv,B4): Multiply(n,MM,B4)=BM;
```

Mathematica:

```
{n=5,PrimeQ[n],m={{1,2,3,3,0},
  {2,1,2,1,3},{1,0,2,3,4},{3,4,1,0,2},{1,0,3,2,4}},b={1,4,2,1,3}}
funcsMod=Map[ToExpression,Select[DeleteCases[Names["System`*"],
  _?(StringTake[#,1]=="$"&)],!FreeQ[Options[Symbol[#]],Modulus]&]]
functionsModp[x_]:=Scan[SetOptions[#,Modulus->x]&,funcsMod];
functionsModp[n]
Map[MatrixForm,{mm=Mod[m,n],bm=Mod[b,n],
                am=Transpose[Append[Transpose[mm],bm]]}]
{Det[m,Modulus->n],MatrixRank[m,Modulus->n]}
{m1=RowReduce[am,Modulus->n],
 m2=Mod[mm.Flatten[Take[m1,{1,n},-1]],n],bm,m2===bm}
{b3=LinearSolve[mm,bm,Modulus->n],m3=Mod[mm.b3,n],bm,m3===bm}
{lu,p,c}=LUDecomposition[mm,Modulus->n]; lu//MatrixForm
l=Table[If[j<i,lu[[i,j]],0],{i,1,n},{j,1,n}]+IdentityMatrix[n]
u=Table[If[j>=i,lu[[i,j]],0],{i,1,n},{j,1,n}]
b4=LinearSolve[u,LinearSolve[l,bm[[p]],Modulus->n],Modulus->n]
{m4=Mod[mm.b4,n], bm, m4===bm}
```

$\Box$

5.12   Linear Algebra over Rings and Fields

As we know the methods of linear algebra can be applied over arbitrary rings and fields, i.e., if we can perform arithmetic in a given ring R
or field F, then we can, for example, add and multiply matrices and vectors, compute determinants and inverse matrices, solve linear systems
over R and F. In symbolic mathematics, for example, in the *Magma*
and *Axiom* computer algebra systems, it is possible to construct a ring
or field and apply linear algebra methods over it.

In *Maple*, working with the `LinearAlgebra` package we can only apply linear algebra methods over specific rings and can obtain incorrect results for a matrix with arbitrary entries. Working with
the `LinearAlgebra[Generic]` subpackage, we can apply linear algebra methods over fields, Euclidean domains, integral domains and
rings.

In *Mathematica*, since there is no a special package for linear algebra computations over arbitrary fields, we can solve problems by
different ways, e.g., applying the `Mod` function and the `Modulus`
option, constructing a finite field object with the `FiniteFields`
package and applying the linear algebra functions (e.g., `RowReduce`,
`LinearSolve`), or defining explicitly the matrix and vector entries
that belong to an infinite field (e.g, $\mathbb{Q}$, $\mathbb{R}$, $\mathbb{C}$).

Maple:

```
with(LinearAlgebra[Generic]);        (F[`0`],F[`1`],F[`+`],F[`-`],
           F[`*`],F[`/`],F[`=`]):=(0,1,`+`,`-`,`*`,`/`,`=`);
Determinant[F](M);    AM:=<M|B>;    ReducedRowEchelonForm[F](AM);
LinearSolve[F](M,B); RREF[F](AM); MatrixVectorMultiply[F](M,X);
p:=primeNum;         F:=GF(p,1);          M1:=map(F:-ConvertIn,M);
B1:=map(F:-ConvertIn,B);                  AM1:=map(F:-ConvertIn,AM);
ReducedRowEchelonForm[F](AM1);                  Determinant[F](M1);
LinearSolve[F](M1,B1);RREF[F](AM1);         MatrixInverse[F](M1);
```

Note. To work with the `LinearAlgebra[Generic]` subpackage first it is necessary
to construct (using the table or module representations) a commutative ring, field,
Euclidean domain, or integral domain defining the operations $+, -, *, /$, the equality
notion $=$, and the constants 0 and 1 or defining a finite field with the `GF` package (for

more details see Sect. 4.12). Then the constructed field F can be used as an index in the linear algebra function of the subpackage.

Mathematica:

```
Mod[m,p]   LinearSolve[m,b,Modulus->p]   RowReduce[m,Modulus->p]
<<FiniteFields`          ff=GF[p]               Characteristic[ff]
SetFieldFormat[GF[p],   FormatType->FunctionOfCoefficients[Nff]]
    ffMat[x_List]:=Table[Table[Nff[x[[i,j]]],{j,1,n}],{i,1,n}];
    ffVec[x_List]:=Table[Nff[x[[i]]],{i,1,n}];        m1=ffMat[m]
b1=ffVec[b]         am1=Transpose[Append[Transpose[m1],b1]]}]
RowReduce[am1]      LinearSolve[m1,b1]   Det[m1]   Inverse[m1]
```

Problem 5.29 Let M and B be an $n \times n$ matrix and an n-dimensional vector with the rational number entries.

1) Solve the linear system of equations $Mx = B$ over the *infinite field* of rational numbers $\mathbb{Q}$ constructing the reduced row-echelon form and verify the solution.

2) Obtain the determinant of M over the field $\mathbb{Q}$.

3) Solve the linear system of equations over the filed $\mathbb{Q}$ using the function `LinearSolve` and verify the solution.

Maple:

```
with(LinearAlgebra[Generic]); n:=3; (Q['0'],Q['1'],Q['+'],
Q['-'],Q['*'],Q['/'],Q['=']):= (0,1,'+','-','*','/','=');
M:=Matrix(n,n,[[1/2,2/3,4/5],[3/7,1/3,2/5],[1/2,0,2/7]]);
B:=Vector(n,[2/3,3/5,4/7]); AM:=<M|B>;
X1:=ReducedRowEchelonForm[Q](AM);
MatrixVectorMultiply[Q](M,X1[1..n,n+1])=B; Determinant[Q](M);
X2,dim:=LinearSolve[Q](M,B); MatrixVectorMultiply[Q](M,X2)=B;
X3:=RREF[Q](AM); MatrixVectorMultiply[Q](M,X3[1..n,n+1])=B;
X4:=RREF[Q](AM,method=GaussianElimination);
MatrixVectorMultiply[Q](M,X4[1..n,n+1])=B;
LinearAlgebra[LinearSolve](M,B)=X2;
```

Mathematica:

```
n=3; Map[MatrixForm,{m={{1/2,2/3,4/5},{3/7,1/3,2/5},{1/2,0,2/7}},
          b={2/3,3/5,4/7},am=Transpose[Append[Transpose[m],b]]}]
{x1=RowReduce[am],m.x1[[All,n+1]]===b,Det[m],x2=LinearSolve[m,b],
 m.x2===b}
```

In *Maple*, to work with the `LinearAlgebra[Generic]` subpackage we first construct the field of rational numbers $\mathbb{Q}$ defining the operations $+, -, *, /$, the equality notion $=$, and the constants 0 and 1. In *Mathematica*, we define explicitly the corresponding matrix and vector entries.   □

Problem 5.30 Let p be a prime number and let M and B be an $n \times n$ matrix and an n-dimensional vector with the integer entries.

1) Solve the linear system of equations $Mx = B$ over the *finite field* $\mathbb{Z}_p$ constructing the reduced row-echelon form and verify the solution.

2) Obtain the determinant of M over the field $\mathbb{Z}_p$.

3) Solve the linear system of equations over the filed $\mathbb{Z}_p$ using the function `LinearSolve` and verify the solution.

Maple:

```
with(LinearAlgebra[Generic]); n:=3; p:=5; GF||p:=GF(p,1);
print(GF||p); M:=Matrix(n,n,[[5,-1,2],[6,2,3],[9,6,7]]);
B:=Vector(n,[8,5,9]); AM:=<M|B>; M1:=map(GF||p:-ConvertIn,M);
B1:=map(GF||p:-ConvertIn,B); AM1:=map(GF||p:-ConvertIn,AM);
X1:=ReducedRowEchelonForm[GF||p](AM1);
MatrixVectorMultiply[GF||p](M1,X1[1..n,n+1])=B1;
Determinant[GF||p](M1); X2,dim:=LinearSolve[GF||p](M1,B1);
MatrixVectorMultiply[GF||p](M1,X2)=B1; X3:=RREF[GF||p](AM1);
MatrixVectorMultiply[GF||p](M1,X3[1..n,n+1])=B1;
X4:=RREF[GF||p](AM1,method=GaussianElimination);
MatrixVectorMultiply[GF||p](M1,X4[1..n,n+1])=B1;
```

Mathematica:

```
<<FiniteFields`
{n=3,p=5,ff=GF[p], Characteristic[ff],
 m={{5,-1,2},{6,2,3},{9,6,7}},b={8,5,9},
 am=Transpose[Append[Transpose[m],b]]}
SetFieldFormat[GF[p],FormatType->FunctionOfCoefficients[Z5]]
ffMat[x_List]:=Table[Table[Z5[x[[i,j]]],{j,1,n}],{i,1,n}];
ffVec[x_List]:=Table[Z5[x[[i]]],{i,1,n}];
Map[MatrixForm,{m1=ffMat[m],b1=ffVec[b],
 am1=Transpose[Append[Transpose[m1],b1]]}]
{x1=RowReduce[am1],m1.x1[[All,n+1]]===b1,Det[m1],
 x2=LinearSolve[m1,b1],m1.x2===b1,m1.x2,b1}
```

In *Maple*, in order to work with the `LinearAlgebra[Generic]` subpackage we first construct the finite field $\mathbb{Z}_p$ using the `GF` package (for more details see

Chapter 4, Sect. 4.12). Since the GF package has a special representation for the elements of the field, we have to convert the matrix and vector entries into the special format using ConvertIn. Then we perform linear algebra computations over the finite field.

In *Mathematica*, the idea is similar. We first construct the finite field $\mathbb{Z}_p$ using the FiniteFields package (for more details see Sect. 4.12) and the GF function. Since the FiniteField package has a special representation for the elements of the field, we have to convert the matrix and vector entries into the special format (we have written the ffMat and ffVec functions). Then we perform linear algebra computations over the finite field. □

Problem 5.31 Let us consider the finite field $GF(2^4)$ as polynomials in x over the integers modulo 2 and represent elements of the field as polynomials modulo $1 + x^3 + x^4$, the irreducible polynomial of degree 4 in x. Let M be an $n \times n$ matrix whose entries are elements of the field $GF(2^4)$.

1) Construct the finite field $GF(2^4)$ as polynomials in $\mathbb{Z}_2[x]$ modulo $1 + x^3 + x^4$.

2) Find all the elements in $GF(2^4)$. 3) Find the inverse matrix and the determinant of M over the field $GF(2^4)$.

Maple:

```
with(LinearAlgebra[Generic]); n:=3; GF16:=GF(2,4,1+x^3+x^4);
print(GF16); Es:={}:
for i from 1 to 100 do Es:=Es union {GF16:-random(x)}; od:
Es; nops(Es); Els:=map(GF16:-ConvertOut,Es);
M:=Matrix(n,n,[[x^3,x^2,x],[x^2,x,0],[x,0,x^2]]);
M1:=map(GF16:-ConvertIn,M); M2:=MatrixInverse[GF16](M1);
MatrixMatrixMultiply[GF16](M1,M2); Determinant[GF16](M1);
```

Mathematica:

```
<<FiniteFields`
{mn=3,ff=GF[2,4],n=FieldSize[ff],
 FieldIrreducible[ff,x], Characteristic[ff]}
 m={{x^3,x^2,x},{x^2,x,0},{x,0,x^2}}
{f16=Table[FromElementCode[ff,j],{j,0,n-1}], f16x=Map[
 ElementToPolynomial[#,x]&,f16]/.ElementToPolynomial[0,x] >0]
convertIn[x_List,d_,nf_]:=Module[{lk,k,v,i,j,lk1,lk2},lk={};
Do[lk=Append[lk,Table[If[f16x[[v]]===x[[i,j]],k=v,0],
 {v,1,nf}]],{j,1,d},{i,1,d}]; lk1=DeleteCases[Flatten[lk],0];
```

```
lk2=Table[f16[[lk1[[v]]]],{v,1,Length[lk1]}];
lk3=Partition[lk2,d]]; convertOut[x_List,s_]:=Map[
ElementToPolynomial[#,s]&,x,{2}]/.ElementToPolynomial[0,s]->0;
m1=convertIn[m,mn,n]
{m2=Inverse[m1],m1m2=m1.m2//FullSimplify,m1m2=convertOut[m1m2,x],
 m2x=convertOut[m2,x],d1=Det[m1]//FullSimplify,
 ElementToPolynomial[d1,x]}
Map[MatrixForm,{m1m2,m2x}]
```

In *Maple*, the `ConvertIn` and `ConvertOut` functions, specify, respectively, the representation of the matrix entries as the field elements (in a special format) and the polynomial representation of the field elements with higher degree monomials to the left. Solving this problem in *Mathematica*, we have written the corresponding functions `convertIn` and `convertOut`. □

5.13 Tensors

A tensor is a mathematical object that generalizes the notion of scalar, vector, and matrix. In linear algebra and differential geometry, a tensor is a multilinear function. In physics and engineering, a tensor is associated with each point of a geometric space, varying continuously with position and represents a physical quantity. Physical laws are independent of any particular non inertial coordinate systems describing a physical motion mathematically. Therefore, changes of representation must satisfy some conditions that ensure that the physical quantity does not change. A study of these consequences leads to tensor analysis.

In symbolic mathematics, there are various methods for defining and evaluating tensors. We consider the basic concepts in both systems describing the built-in important functions and showing how to define the new functions for working with tensors (creation of tensors, elementary operations of tensor algebra and calculus, general relativity calculations).

In *Maple*, the functions for working with tensors are contained in the packages: `tensor`, `Physics`, and `DifferentialGeometry`.

The `DifferentialGeometry` package contains various functions and subpackages for specialized operations in curved spaces. The most important subpackage for performing tensor analysis in curved spaces is the `Tensor` subpackage.

The `tensor` package also permits operations in curved spaces, but it's main purpose is to work with *tensor components*.

The new `Physics` package contains various functions for performing *abstract tensor algebra* in flat spaces.

In *Mathematica*, a tensor is represented as a set of nested lists. The nesting level is the rank of the tensor.

Defining tensors and analyzing the structure

Maple:

```
with(tensor):      t1_compts:=array(symmetric,sparse,1..n,1..n);
t1_compts[i,j]:=t1ij;                 t1:=create([1,-1],t1_compts);
invert(t1,'det_t1'); get_compts(t1);                    get_char(t1);
get_rank(t1);                                type(t1,tensor_type);
with(Physics);                       Define(T1(x,y,z),antisymmetric);
with(DifferentialGeometry);     with(Tools);           with(Tensor);
DGsetup([vars],M);          T1:=DGtensor([vars],[indTypes],M);
            T2:=CanonicalTensors(type,indTypes,signature,M);
```

`create`, `entermetric`, `invert`, defining the metric tensor components. The index character is defined as a list of positive and negative indices, 1 or −1, which correspond, respectively, to the contravariant and covariant indices of the tensor,

`type`, analyzing the tensor-type object,

`get_compts`, `get_char`, `get_rank`, determining the tensor components, the index character, and the rank.

To obtain essential information on the `Physics` package, see `?Physics[Setup]` (on the computational environment of the package), `?Physics[Define]` (on definitions of tensors), `?Physics[Parameters]` (on definition of the parameters).

For the `DifferentialGeometry` package, we can recommend to read the following: `?DifferentialGeometry[DGsetup]` (on definition of coordinate systems, frames, Lie algebras), `?DifferentialGeometry[Preferences]` (on worksheet preferences for the package).

```
with(tensor): T1_compts:=array(symmetric,sparse,1..3,1..3);
T1_compts[1,1]:=a; T1_compts[2,2]:=b; T1_compts[3,3]:=c;
```

```
T1:=create([-1,-1],eval(T1_compts));  T2:=invert(T1,'det_T1');
det_T1; type(T1,tensor_type); type(T2,tensor_type);
T1Components:=get_compts(T1); get_char(T1); get_rank(T1);
with(Physics): Define(R[mu,nu,alpha,beta](x,y,z),
 symmetric={mu,nu}); with(DifferentialGeometry): with(Tools):
with(Tensor): DGsetup([x,y,z,w],M); CanonicalTensors(
 "SymplecticForm","bas"); DGtensor([1,2,3,4],
 [["cov_bas","con_bas","cov_bas","con_bas"],[]]);
```

Mathematica:

```
t1=Table[expr,{i1,1,n},{i2,1,m},...]              t1=Array[f,{n,m}]
t1=SparseArray[{{i1,j1}->t11,...}]     Det[t2]    ArrayDepth[t1]
t1={{{t11,t12,...},{t21,t22,...}},...}}}          Dimensions[t1]
TreeForm[t1]    MatrixForm[t1]       Length[t1]     t1[[i,j,k,l]]
TensorQ[t1] Inverse[t1] epsilon=Array[Signature[{##}]&,{3,3,3}]
```

Table, Array, creating a tensor,

TreeForm, MatrixForm, visualizing tensors as a tree or a matrix,

Length, Dimensions, ArrayDepth, TensorQ, analyzing tensors,

[[ ]], extracting tensor components, subtensors,

Array, Signature, defining antisymmetric tensors.

```
{Table[i1*i2,{i1,2},{i2,3}], Array[(#1*#2)&,{2,3}]}
 t1={{{a,b,c},{d,e,f}},{{g,h,i},{j,k,l}}}
{TreeForm[t1], MatrixForm[t1], Length[t1],
 Dimensions[t1],TensorRank[t1],ArrayDepth[t1],TensorQ[t1]}
{n=3,t2=SparseArray[{{1,1}->a,{2,2}->b,{3,3}->c}],
 MatrixForm[t2], Inverse[t2]//MatrixForm, Det[t2],
 TensorQ[t2], MatrixForm[t2[[All,2]]]}
epsilon=Array[Signature[{##}]&,{3,3,3}]
```

Problem 5.32 *Tensor notation.* We would like to display covariant and contravariant components of a tensor in the mathematical notation (as in textbooks), e.g., $\Gamma_{ij}^k(z_1, z_2)$, R_{lki}^q, R^{ji}, Γ_{kji}. Construct the tensor notation in both systems.

Maple:

```
with(tensor); coord:=[z1,z2]; n:=3;
Gamma_compts:=array(symmetric,sparse,1..n,1..n,1..n);
Gamma_compts[1,1,1]:=1; Gamma_compts[2,2,2]:=z1^2;
Gamma_compts[3,3,2]:=z2;
Gamma:=create([-1,-1,1],eval(Gamma_compts));
with(DifferentialGeometry): with(Tensor): DGsetup([x,y,z],M);
K1:=KroneckerDelta("bas",1); K2:=KroneckerDelta("bas",2);
with(Physics): Define(R[mu,nu,alpha,beta]); Define(R,query);
```

Mathematica:

```
{Subsuperscript[\[CapitalGamma],{i,j},k][z1,z2],
 Subsuperscript[R,{l,k,i},q],Superscript[R,{j,i}],
 Subscript[\[CapitalGamma],{k,j,i}]}
```

In *Maple*, within the `tensor` package, the index character is defined as a list of the form $[-1, -1, 1]$ (for indicating covariant and contravariant indices of a tensor). But in the `Physics` package (for performing abstract tensor algebra) there is no distinction between free *covariant* and *contravariant* indices and both indices are represented the same way as if they were covariant, e.g., `R[mu,nu,alpha,beta]` (for more details see `?Physics,conventions`). So, all the indices (free and contracted) are displayed as subscripts. In the `Tensor` subpackage, there are the special types of tensors, e.g., `"cov_bas"`, `"con_bas"`, etc. (for more details see `?DifferentialGeometry[Tools][DGinfo]`).   □

Tensor algebra

Maple:

```
with(tensor): t1+t2; t1-t2; prod(t1,t2); prod(t1,t2,[t1i,t2i]);
lin_com(c1,t1,c2,t2);                           contract(t1,[i1,i2]);
with(Physics);A[i,j]*B[i,j];Setup(op={A,B},%Commutator(A,B)=0);
A*B; Bra(A,n).Ket(A,m);    LinearAlgebra:-KroneckerProduct(C,D);
```

+, -, the sum and the difference of two tensors of the same rank and type (i.e., the same number of contravariant and covariant indices),

  prod, inner and outer tensor products,

lin_com, linear combination of tensors,

contract, the contraction operation, i.e., contract a tensor over one or more pairs
   of indices (with opposite index character),

*, a generalized tensor product (for a commutative, anticommutative, or noncom-
   mutative product),

., a scalar product between Bras, Kets vectors (representing the quantum state of
   a system in Dirac's notation), and quantum operators,

KroneckerProduct, the Kronecker product of two matrices (a special case of tensor
   product).

```
with(tensor): e1:=create([-1],array(1..3,[1,0,0]));
e2:=create([-1],array(1..3,[0,1,0]));
e3:=create([-1],array(1..3,[0,0,1]));
E1:=lin_com(1,prod(e1,e1),-1,prod(e2,e2));
T1:=create([1],array([x,y,z])): for i from 1 to 2 do
T2[i]:=create([1],array([i,5*i,9*i])) od:
C:=lin_com(x,T1,eval(T2[1]),eval(T2[2])); with(Physics);
Define(T3(x,y,z),symmetric); Define(T4(x,y,z),antisymmetric);
Simplify(T3[i,j]*T4[i,j]);Setup(op={A,B},%Commutator(A,B)=0);
A*B; Bra(A,n).Ket(A,m); Setup(op=H); Bra(A,n).H.Ket(B,m);
with(LinearAlgebra): A:=Matrix(3,3,symbol=a);
B:=Matrix(2,2,symbol=b); KroneckerProduct(A,B);
```

Mathematica:

t1+t2 t1-t2	Inner[f,t1,t2,g]	Outer[f,t1,t2] t1.t2	
c1*t1+c2*t2	Transpose[t1] Tr[t1,f]	KroneckerProduct[m1,m2]	

Inner, Outer, generalized inner and outer products of the tensors t_1 and t_2 (where
   f and g are, respectively, multiplication and addition operators,

., the dot product of t_1 and t_2 (the last index of t_1 contracted with the first index
   of t_2),

Transpose, transposing the first two indices in a tensor,

Tr, a generalized trace of a tensor,

KroneckerProduct, the Kronecker product of matrices m_i (a special case of tensor
   product).

```
{e1={1,0,0},e2={0,1,0},e3={0,0,1},v1={x,y,z}}
t1=Outer[Times,e1,e1]-Outer[Times,e2,e2]//MatrixForm
{t2=Table[{i,5*i,9*i},{i,1,2}],t2//MatrixForm,
 linearCombination=x*v1+t2[[1]]+t2[[2]]}
```

```
{t3=Array[f,{2,3,3}],Dimensions[t3]}
{t4=Transpose[t3],Dimensions[t4]}
KroneckerProduct[Array[Subscript[a,##]&,{3,3}],
 Array[Subscript[b,##]&,{2,2}]]//MatrixForm
```

Problem 5.33 Let A_{ij}^{kl} be a tensor of rank 4 with symbolic entries. Perform various tensor contractions.

Maple:

```
with(tensor): n:=3; A:=create([-1,-1,1,1],convert(Array(1..n,
 1..n,1..n,1..n,(i,j,k,l)->a[i,j,k,l]),array)):
lprint(get_compts(A)); A1:=contract(A,[1,3]);
A2:=contract(A,[1,4]); A3:=contract(A,[2,3]);
A4:=contract(A,[2,3],[1,4]); A5:=contract(A,[1,4],[2,3]);
get_rank(A); for i from 1 to 5 do get_rank(A||i); od;
```

Mathematica:

```
contract[ts_,ind:{_,_}..]:=
 Nest[Tr[#,Plus,2]&,Transpose[ts,Module[{s=Range[
 ArrayDepth[ts]]},s[[Join[ind,Delete[s,Map[List,Join[
 ind]]]]]]]]=s;s]],Length[{ind}]]; a=Array[A,{3,3,3,3}];
Map[TableForm,{a1=contract[a,{1,3}], a2=contract[a,{1,4}],
 a3=contract[a,{2,3}], a4=contract[a,{2,3},{1,4}],
 a5=contract[a,{1,4},{2,3}]}]
Map[ArrayDepth,{a,a1,a2,a3}]
{TensorQ[{a4,a5}],ArrayDepth[a4]-1,ArrayDepth[a5]-1}
```
□

Problem 5.34 Let V be the space of all skew-symmetric Hermitian 2×2 matrices M_i, where $\operatorname{tr} M_i = 0$. We can choose the basis in V, $B = \{e_1, e_2, e_3\}$, and define the Lie product $[v, u] = vu - uv$, where $v, u \in V$. The right-hand side of this commutator can be expressed by the antisymmetric tensor in $\mathbb{R}^3$, Levi-Civita symbol ε_{ik}^j, according to the following relation:

$$[e_i, e_k] = \sum_{j=1}^{3} \varepsilon_{ik}^j e_j, \quad i, k = 1, 2, 3.$$

Here $\varepsilon_{ik}^j = 0$ if any index is repeated, $\varepsilon_{ik}^j = 1$ or -1 if $\{i, j, k\}$ form, respectively, an even or odd permutation of $\{1, 2, 3\}$. Construct the Levi-Civita tensor and verify the commutator relation.

Maple:

```
with(Physics): Setup(dimension=3,signature=`+`);
with(LinearAlgebra): n:=3; E[1]:=1/2*Matrix(2,2,[[0,-I],[-I,0]]);
E[2]:=1/2*Matrix(2,2,[[0,-1],[1,0]]);
E[3]:=1/2*Matrix(2,2,[[-I,0],[0,I]]); LieProduct:=(V,U)->V.U-U.V;
M1:=Matrix(n,n);M2:=Matrix(n,n); LeviCivita[1,2,3];
CommutatorL:=(m,n)->LieProduct(E[m],E[n]);
CommutatorR:=(m,n)->add(LeviCivita[m,n,l]*E[l],l=1..3);
for i from 1 to 3 do for k from 1 to 3 do M1[i,k]:=
 CommutatorL(i,k);M2[i,k]:=CommutatorR(i,k);od;od;  print(M1,M2);
```

Mathematica:

```
Map[MatrixForm, {e1=1/2*{{0,-I},{-I, 0}},
 e2=1/2*{{0,-1},{1,0}},e3=1/2*{{-I,0},{0,I}},b={e1,e2,e3}}]
LieProduct[v_List,u_List]:=v.u-u.v;
LeviCivitaT[i_,j_,k_]:=Module[{lc,ind,l1},ind={i,j,k};
 l1=Union[ind,{1,2,3}]; If[Length[l1]<3||Length[l1]>3,lc=0,
 lc=Signature[ind]]];
Table[LeviCivitaT[i,j,k],{i,1,3},{j,1,3},{k,1,3}]//MatrixForm
CommutatorRel[m_,n_]:=LieProduct[b[[m]],b[[n]]]===
 Sum[LeviCivitaT[m,n,l]*b[[l]],{l,1,3}];
Table[CommutatorRel[i,k],{i,1,3},{k,1,3}]//MatrixForm
CommutatorL[m_,n_]:=LieProduct[b[[m]],b[[n]]];
CommutatorR[m_,n_]:=Sum[LeviCivitaT[m,n,l]*b[[l]],{l,1,3}];
Map[MatrixForm,{Table[CommutatorL[i,k],{i,1,3},{k,1,3}],
 Table[CommutatorR[i,k],{i,1,3},{k,1,3}]}]
```

                                                                                        □

Problem 5.35 Convert a tensor to an operator.

Maple:

```
with(tensor): n:=3; t1_compts:=array(symmetric,sparse,1..n,1..n):
t1_compts[1,1]:=a: t1_compts[2,2]:=b: t1_compts[3,3]:=c:
t1:=create([-1,-1],eval(t1_compts)); t1Comps:=get_compts(t1);
A:=map(unapply,t1Comps,a,b,c); A(1,2,3); convert(A,listlist);
```

Mathematica:

```
{n=3,t1={{a,0,0},{0,b,0},{0,0,c}}}; t1//MatrixForm
f[x_]:=x^2; m1=SparseArray[{i_,i_}:>f[i],{n,n}]; f1={a,b,c};
m2=SparseArray[{i_,i_}:>f1[[i]],Length[f1]];
rules={{1,1}->a1,{2,2}->a2,{3,3}->a3}; m3=SparseArray[rules];
Map[MatrixForm,{m1,m2,m3}]
```

$\square$

Tensor differentiation

Maple:

```
with(tensor):      t1_compts:=Array(symmetric,sparse,1..n,1..n);
coord:=[vars];     t1_compts[i,j]:=t1ij;             t2:=invert(t1);
t1:=create([-1,-1],eval(t1_compts));    D1:=d1metric(t1,coord);
D2:=d2metric(D1,coord);           partial_diff(t1,coord);
Riemann(t2,D2,Cf1); directional_diff(F(vars),VfieldCon,coord);
Cf1:=Christoffel1(D1);            Cf2:=Christoffel2(t2,Cf1);
cov_diff(t1,coord,Cf2);                cov_diff(R1,coord,Cf2);
VfieldCon:=create([1], array([t1_compts[1,k],t1_compts[1,k]]));
```

d1metric, d2metric, the first and second partial derivatives of metric tensors,

partial_diff, the partial derivative of tensors with respect to given coordinates,

cov_diff, the covariant derivative of tensors,

directional_diff, the directional derivative of tensors,

Christoffel1, Christoffel2, the Christoffel symbols of the first and the second kinds (the connection coefficients of an affine connection in a coordinate basis),

Riemann, the covariant Riemann curvature tensor.

```
with(tensor): coord:=[z1,z2]; n:=2:
t1_compts:=array(symmetric,sparse, 1..n,1..n);
t1_compts[1,2]:=exp(rho(z1,z2)); t1:=create([-1,-1],
 eval(t1_compts)); t2:=invert(t1,'det_t1');
D1:=d1metric(t1,coord); D2:=d2metric(D1,coord);
partial_diff(t1,coord); Cf1:=Christoffel1(D1);
Cf2:=Christoffel2(t2,Cf1); R1:=Riemann(t2,D2,Cf1);
t3:=create([1],array([t1_compts[1,2],t1_compts[1,2]]));
cov_diff(t1,coord,Cf2); cov_diff(R1,coord,Cf2);
directional_diff(F(z1,z2),t3,coord);
```

Problem 5.36 In the Einstein theory of general relativity, the important *Schwarzschild exact solution* (or the Schwarzschild vacuum), obtained by Karl Schwarzschild in 1915, describes the gravitational field outside a spherical non-rotating mass. Let us consider a spherically symmetric spacetime with the Schwarzschild coordinates r, θ, ϕ, t and the Schwarzschild metric $g_{\alpha\beta}$.

1) Determine the inverse metric $g^{\lambda\sigma}$.

2) Compute the Christoffel symbols of the first and the second kinds.

Maple:

```
with(tensor): coord:=[r,theta,phi,t]; n:=4;
g_compts:=array(symmetric,sparse, 1..n,1..n);
g_compts[1,1]:=1/(1-m/r); g_compts[2,2]:=r^2;
g_compts[3,3]:=r^2*sin(theta)^2; g_compts[4,4]:=-(1-2*m/r);
g:=create([-1,-1],eval(g_compts)); gInv:=invert(g,'det_g');
D1:=d1metric(g,coord): D2:=d2metric(D1,coord):
Cf1:=Christoffel1(D1): Cf2:=Christoffel2(gInv,Cf1):
lprint(op(D1)); lprint(op(D2)); lprint(op(Cf1));
lprint(op(Cf2)); print(Cf2);
```

Mathematica:

```
{coord={r,theta,phi,t},n=4,
 g={{1/(1-m/r),0,0,0},{0,r^2,0,0},{0,0,r^2*Sin[theta]^2,0},
 {0,0,0,-(1-2*m/r)}}}
Christoffel1[j_,k_,a_,ts_]:=Module[{},FullSimplify[(1/2)*
 (D[ts[[a,j]],coord[[k]]]+D[ts[[a,k]],coord[[j]]]-D[ts[[j,k]],
 coord[[a]]])]]; Christoffel2[j_,k_,i_,ts_,tsInv_]:=Module[{},
 Sum[tsInv[[i,a]]*Christoffel1[j,k,a,ts],{a,1,n}]//FullSimplify];
Map[Simplify,{gInv=Inverse[g],Det[gInv],TensorQ[gInv]}]
Map[MatrixForm,{g,Simplify[gInv]}]
Map[FullSimplify,{cf1=Table[Christoffel1[j,k,i,g],{i,1,n},
 {k,1,n},{j,1,n}],cf2=Table[Christoffel2[j,k,i,g,gInv],{i,1,n},
 {k,1,n},{j,1,n}]}]
Map[TableForm,{cf1,cf2}]
cf1L=Table[{{i,j,k}==cf1[[i,j,k]]},{i,1,n},{j,1,n},{k,1,n}]
cf2L=Table[{{i,j,k}==cf2[[i,j,k]]},{i,1,n},{j,1,n},{k,1,n}]
TableForm[Partition[Flatten[cf1L],2],TableSpacing->{1,1}]
TableForm[Partition[Flatten[cf2L],2],TableSpacing->{1,1}]
```

Since in *Mathematica* there are no functions for computing the Christoffel symbols, we show how to write the corresponding functions and obtain the same results in both systems.   □

General Relativity curvature tensors in a coordinate basis

Maple:

```
with(tensor):
    tensorsGR([coord],cov_metric,con_metric,det_metric,Cf1,Cf2,
              TRiemann,TRicci,RicciSc,TEinstein,TWeyl,flag);
    display_allGR([coord],cov_metric,con_metric,det_met,Cf1,Cf2,
    TRiemann,TRicci,RicciSc,TEinstein,TWeyl);displayGR(name,ts);
```

tensorsGR, computing General Relativity curvature tensors in a coordinate basis. For a given spacetime with coordinates coord (that can be written in the form [t,x,y,z]), and a given covariant metric and the inverse metric (contravariant metric), cov_metric, we can compute the following: the Christoffel symbols of the first and the second kinds, Cf1 and Cf2, the covariant Riemann tensor, Riemann, the covariant Ricci tensor, Ricci, the Ricci scalar, RicciSc, the covariant Einstein tensor, Einstein, the covariant Weyl tensor, Weyl.

display_allGR, displayGR, displaying, the nonzero components of all of the general relativity tensors or a specific general relativity tensor.

```
with(tensor): coord:=[r,theta,phi,t]; n:=4;
t1_compts:=array(symmetric,sparse,1..n,1..n);
t1_compts[1,1]:=f(r); t1_compts[2,2]:=-g(r)/f(r);
t1_compts[3,3]:=-r^2; t1_compts[4,4]:=-r^2*sin(theta)^2;
t1:=create([-1,-1],eval(t1_compts));
tensorsGR(coord,t1,t2,det_t1,Cf1,Cf2,TRiemann,TRicci,RicciSc,
TEinstein,TWeyl); display_allGR(coord,t1,t2,det_t1,Cf1,Cf2,
TRiemann,TRicci,RicciSc,TEinstein,TWeyl);
Christoffel1:=get_compts(Cf1); Christoffel2:=get_compts(Cf2);
RicciT:=get_compts(TRicci); EinsteinT:=get_compts(TEinstein);
```

Problem 5.37 Important solutions of the Einstein field equations include the Schwarzschild exact solution describing the spacetime geometry of empty space surrounding any spherically symmetric uncharged and nonrotating mass.

1) Determine the Schwarzschild metric.

2) Compute the scalar curvature and the Einstein tensor.

Maple:

```
with(tensor): coord:=[t,r,theta,phi]; n:=4;
t1_compts:=array(symmetric,sparse,1..n,1..n);
t1_compts[1,1]:=f(r); t1_compts[2,2]:=-g(r)/f(r);
t1_compts[3,3]:=-r^2; t1_compts[4,4]:=-r^2*sin(theta)^2;
t1:=create([-1,-1],eval(t1_compts));
tensorsGR(coord,t1,t2,det_t1,Cf1,Cf2,TRiemann,TRicci,RicciSc,
 TEinstein,TWeyl); RicciT:=get_compts(TRicci);
dsolve({RicciT[1,1],RicciT[2,2]},{f(r),g(r)});
displayGR(Ricciscalar,RicciSc); R1:=get_compts(RicciSc);
displayGR(Einstein,TEinstein); get_compts(TEinstein);
```

Mathematica:

```
{coord={t,r,theta,phi},n=4,
 t1={{f[r],0,0,0},{0,-g[r]/f[r],0,0},{0,0,-r^2,0},
 {0,0,0,-r^2*Sin[theta]^2}}}
Christoffel1[j_,k_,a_,ts_]:=Module[{},(1/2)*(D[ts[[a,j]],
 coord[[k]]]+D[ts[[a,k]],coord[[j]]]-D[ts[[j,k]],coord[[a]]])];
Christoffel2[j_,k_,i_,ts_,tsInv_]:=Module[{},Sum[tsInv[[i,a]]*
 Christoffel1[j,k,a,ts],{a,1,n}]];
Map[Simplify,{t1Inv=Inverse[t1],Det[t1Inv],TensorQ[t1Inv]}]
Map[MatrixForm,{t1,Simplify[t1Inv]}]
Map[FullSimplify,{cf1=Table[Christoffel1[j,k,i,t1],{i,1,n},
 {k,1,n},{j,1,n}],cf2=Table[Christoffel2[j,k,i,t1,t1Inv],
 {i,1,n},{k,1,n},{j,1,n}]}]
Map[TableForm,{cf1,cf2}]
cf1L=Table[{{i,j,k}==cf1[[i,j,k]]},{i,1,n},{j,1,n},{k,1,n}]
cf2L=Table[{{i,j,k}==cf2[[i,j,k]]},{i,1,n},{j,1,n},{k,1,n}]
TableForm[Partition[Flatten[cf1L],2],TableSpacing->{1,1}]
TableForm[Partition[Flatten[cf2L],2],TableSpacing->{1,1}]
RiemannT=Table[D[cf2[[i,j,l]],coord[[k]]]-D[cf2[[i,j,k]],
 coord[[l]]]+Sum[cf2[[a,j,l]]*cf2[[i,k,a]]-cf2[[a,j,k]]*
 cf2[[i,l,a]],{a,1,n}],{i,1,n},{j,1,n},{k,1,n},{l,1,n}];
RicciT=Table[Sum[RiemannT[[i,j,i,l]]],{i,1,n}],{j,1,n},{l,1,n}];
sys={RicciT[[1,1]]==0,RicciT[[2,2]]==0}//Simplify
DSolve[sys,{f[r],g[r]},r]; RicciSc=Sum[t1Inv[[i,j]]*
RicciT[[i,j]],{i,1,n},{j,1,n}]//FullSimplify
EinshteinT=RicciT-1/2*RicciSc*t1//FullSimplify
```

□

Chapter 6

Geometry

In this chapter we consider the most important geometric concepts in a
Euclidean space $\mathbb{R}^n$, where $\mathbb{R}^2$ represents the two dimensional plane and
$\mathbb{R}^3$, the three dimensional space.

6.1   Points in the Plane and Space

Maple:

```
with(plots):                                pointplot(points,ops);
pointplot3d(points,ops);                    matrixplot(matrix,ops);
listplot(list,ops);                         listplot3d(list,ops);
with(Statistics):                  ScatterPlot(seqX,seqY,ops);
```

Mathematica:

```
ListPlot[{{x1,y1},{x2,y2},...,{xn,yn}},
  PlotRange->val,Filling->val,FillingStyle->val,DataRange->val,
          MaxPlotPoints->val,Joined->True,PlotMarkers->val,
              PlotStyle->Directive[PointSize[val],valColor]]
ListLinePlot[{{x1,y1},{x2,y2},...,{xn,yn}},ops]
ListPlot3D[{{x1,y1,z1},...},MeshShading->val,ops]
ListPointPlot3D[{{x1,y1,z1},...},ops]
Show[Graphics3D[{PointSize[s],Point[{x1,y1,z1}],...}]]
```

Problem 6.1 Determine the points of intersection of the curves $f(x) = x^3/9 + 3$ and $g(x) = x^2/2 + 2$.

I.K. Shingareva, C. Lizárraga-Celaya, *Maple and Mathematica*, 2nd ed.,
DOI 10.1007/978-3-211-99432-0_6, © Springer-Verlag Vienna 2009

Maple:

```
with(plots):
f:=x->x^3/9+3; g:=x->x^2/2+2; L:=[f(x),g(x)];
plot(L,x=-4..4,-1..11,color=[green,blue],thickness=[3,5]);
S:=[fsolve(f(x)=g(x),x)];
for i from 1 to nops(S) do
  Y[i]:=eval(f(x),x=S[i]); eval(g(x),x=S[i]);
od;
P:=[seq([S[i],Y[i]],i=1..nops(S))];
G1:=plot(L,x=-4..4,-1..11,color=[green,blue],
 thickness=[3,5]): G2:=pointplot(P,color=red,
 symbol=circle,symbolsize=30):
display({G1,G2});
```

Mathematica:

```
f[x_]:=x^3/9+3; g[x_]:=x^2/2+2;
SetOptions[Plot,PlotRange->{{-4,4.1},{-1,11}},
 PlotStyle->{{Green,Thickness[0.008]},{Blue,Thickness[0.015]}}]
Plot[{f[x],g[x]},{x,-4,4}]
{s=NSolve[f[x]==g[x],x],n=Length[s],y=f[x]/.s,z=g[x]/.s}
points=Table[{Flatten[s][[i,2]],y[[i]]},{i,1,n}]
g1=Plot[{f[x],g[x]},{x,-4,4}];
g2=ListPlot[points,PlotStyle->{Red,PointSize[0.04]}];
Show[{g1,g2}]
```

□

Problem 6.2 Graph the points $(0,-1,1)$, $(0,-5,0)$, $(0,-1,0)$, $(0,4,2)$.

Maple:

```
plots[pointplot3d]({[0,-1,1],[0,-5,0],[0,-1,0],[0,4,2]},
  axes=boxed,symbol=circle,symbolsize=20,color=blue);
```

Mathematica:

```
Show[Graphics3D[{PointSize[0.05],Hue[Random[]],
 Point[{0,-1,1}],Point[{0,-5,0}],Point[{0,-1,0}],
 Point[{0,4,2}]},BoxRatios->{2,2,1},
 ViewPoint->{5.759,2.606,-1.580}],BoxRatios->1]
```

□

Problem 6.3 Graph points in the plane and the space.

Maple:

```
with(plots): with(Statistics): n:=100:
listplot([seq([i,i^3],i=1..n)],color=blue);
pointplot([seq([i,i^3],i=1..n)],color=blue);
x1:=<seq(1..n)>: y1:=<seq(i^3,i=1..n)>:
ScatterPlot(x1,y1,title="Scatter Plot");
listplot3d([seq(seq([i,sin(2*i)*cos(3*j)],i=1..4),
 j=1..4)],shading=z);
points1:={seq(seq(seq([i,sin(i)+cos(j),sin(i)+cos(k)],
 i=1..3),j=1..3),k=1..3)}:
points2:={seq([cos(Pi*t/10),sin(Pi*t/10),Pi*t/10],t=0..40)}:
pointplot3d(points1,symbol=circle,symbolsize=15,
 shading=z,axes=boxed);
pointplot3d(points2,symbol=circle,symbolsize=15,
 shading=z,axes=boxed);
```

Mathematica:

```
n=100; list1=Table[{i,i^3},{i,1,n}];
SetOptions[ListPointPlot3D,PlotStyle->
 Directive[Magenta,PointSize[0.02]],BoxRatios->{1,1,1},
 ColorFunction->Function[{x,y,z},Hue[z]]];
ListPlot[list1,PlotStyle->Directive[PointSize[0.01],Red]]
ListPlot[list1,Joined->True,PlotStyle->Blue]
ListLinePlot[list1,PlotStyle->Blue]
ListPlot[Table[{i,Random[Real,{-5,5}]},{i,-5,5}],
 Joined->True,Filling->Axis,FillingStyle->LightPurple]
ListPlot3D[Flatten[Table[{i,Sin[2*i]*Cos[3*j]},
 {j,1,4},{i,1,4}],1],BoxRatios->{1,1,1},Mesh->False,
 ColorFunction->Function[{x,y,z},Hue[z]]]
points1=Flatten[Table[{i,Sin[i]+Cos[j],Sin[i]+Cos[k]},
 {k,1,3},{j,1,3},{i,1,3}],2];
points2=Table[{Cos[Pi*t/10],Sin[Pi*t/10],Pi*t/10},{t,0,40}];
ListPointPlot3D[points1]
ListPointPlot3D[points2]
```

□

6.2 Parametric Curves

2D parametric curve $\{x = x(t), y = y(t)\}$.

Maple:

```
plot([x(t),y(t),t=t1..t2],x1..x2,y1..y2,ops);
```

```
plot([t^2*sin(t),t^3*cos(t),t=-10*Pi..10*Pi],axes=boxed);
plot([(-t)^3*cos(t),(-t)^2*sin(t),t=-8*Pi..8*Pi]);
```

Mathematica:

```
ParametricPlot[{x[t],y[t]},{t,t1,t2},ops]
```

```
SetOptions[ParametricPlot,AspectRatio->1,PlotRange->All,
   PlotStyle->{Red,Thickness[0.005]},Frame->True];
ParametricPlot[{t^2*Sin[t],t^3*Cos[t]},{t,-10*Pi,10*Pi}]
ParametricPlot[{(-t)^3*Cos[t],(-t)^2*Sin[t]},{t,-8*Pi,8*Pi}]
```

3D parametric curve $\{x = x(t), y = y(t), z = z(t)\}$, $t \in [t_1, t_2]$.

Maple:

```
with(plots);          spacecurve([x(t),y(t),z(t)],t=t1..t2,ops);
```

```
with(plots):
x:=t->-1/2*cos(3*t); y:=t->-1/4*sin(3*t); z:=t->1/7*t;
spacecurve([x(t),y(t),z(t)],t=0..10*Pi,numpoints=400);
```

Mathematica:

```
ParametricPlot3D[{x[t],y[t],z[t]},{t,t1,t2},ops]
```

```
x[t_]:=-1/2*Cos[3*t]; y[t_]:=-1/4*Sin[3*t];
z[t_]:=1/7*t; Evaluate[ParametricPlot3D[{x[t],y[t],z[t]},
  {t,0,10*Pi},PlotPoints->200,BoxRatios->{1,1,1},
  ColorFunction->Function[{x,y,z},Hue[z]],Axes->None,
  Boxed->False]]
```

Problem 6.4 Let $g(x, y) = \cos(2x - \sin(2y))$. Graph the intersection of $g(x, y)$ with the planes $x = 5$, $y = 0.5$.

Maple:

```
with(plots): g:=(x,y)->cos(2*x-sin(2*y)):
plot3d(g(x,y),x=0..2*Pi,y=0..2*Pi,axes=boxed);
G1:=spacecurve([5,t,0],t=0..2*Pi):
G2:=spacecurve([5,t,g(5,t)],t=0..2*Pi):
G3:=spacecurve([t,0.5,0],t=0..2*Pi):
G4:=spacecurve([t,0.5,g(t,0.5)],t=0..2*Pi):
display3d({G1,G2},axes=boxed);
display3d({G3,G4},axes=boxed);
spacecurve({[5,t,0],[5,t,g(5,t)],[t,0.5,0],
            [t,0.5,g(t,0.5)]},t=0..2*Pi,axes=boxed);
```

Mathematica:

```
g[x_,y_]:=Cos[2*x-Sin[2*y]];
SetOptions[ParametricPlot3D,BoxRatios->{1,1,1},
 ColorFunction->Function[{x,y,z},Hue[z]],
 PlotRange->{All,All,All}];
Plot3D[g[x,y],{x,0,2*Pi},{y,0,2*Pi},Mesh->False]
g1=ParametricPlot3D[{5,t,0},{t,0,2*Pi}];
g2=ParametricPlot3D[{5,t,g[5,t]},{t,0,2*Pi}];
g3=ParametricPlot3D[{t,0.5,0},{t,0,2*Pi}];
g4=ParametricPlot3D[{t,0.5,g[t,0.5]},{t,0,2*Pi}];
GraphicsColumn[{Show[{g1,g2}],Show[{g3,g4}]}]
Show[{g1,g2,g3,g4}]
```

□

Problem 6.5 Graph $f(x,y) = x^2 \sin(4y) - y^2 \cos(4x)$ for (x,y) on the circle $x^2 + y^2 = 1$.

Maple:

```
with(plots): f:=(x,y)->x^2*sin(4*y)-y^2*cos(4*x);
spacecurve({[cos(t),sin(t),0],[cos(t),sin(t),
 f(cos(t),sin(t))]},t=0..2*Pi,axes=boxed,numpoints=200);
```

Mathematica:

```
SetOptions[ParametricPlot3D,BoxRatios->{1,1,1},
 ColorFunction->Function[{x,y,z},Hue[z]],PlotPoints->200];
f[x_,y_]:=x^2*Sin[4*y]-y^2*Cos[4*x];
g1=ParametricPlot3D[{Cos[t],Sin[t],0},{t,0,2*Pi}];
g2=ParametricPlot3D[{Cos[t],Sin[t],f[Cos[t],Sin[t]]},
 {t,0,2*Pi}]; Show[g1,g2]
```

□

6.3   Implicitly Defined Curves

Maple:

```
implicitplot(f(x,y)=c,x=x1..x2,y=y1..y2,ops);
```

```
with(plots): f:=(x,y)->x^2-x+y^2-y+2; R:=-10..10;
implicitplot(f(x,y)=21,x=R,y=R,grid=[50,50]);
implicitplot(sin(x+y)=x*cos(x),x=-10..10,y=-10..10);
```

Mathematica:

```
ContourPlot[f[x,y]==c,{x,x1,x2},{y,y1,y2},ops]
ContourPlot[{f1[x,y]==c1,f2[x,y]==c2,...},{x,x1,x2},{y,y1,y2}]
```

```
f[x_,y_]:=x^2-x+y^2-y+2;
ContourPlot[f[x,y]==21,{x,-10,10},{y,-10,10},
 Frame->True,PlotRange->Automatic]
ContourPlot[Sin[x+y]==x*Cos[x],{x,-10,10},{y,-10,10}]
```

6.4   Curves in Polar Coordinates

Maple:

```
            plot([r(t),theta(t),t=t1..t2],coords=polar,ops);
with(plots): polarplot([r(t),theta(t),theta=theta1..theta2]);
```

Mathematica:

```
PolarPlot[f[t],{t,t1,t2},ops]
            ParametricPLot[{r*Cos[t],r*Sin[t]},{t,t1,t2},ops]
```

Problem 6.6 Lissajous curves are defined by the parametric equations
$x(t) = \sin(nt)$, $y(t) = \cos(mt)$, $t \in [0, 2\pi]$, where (n, m) are different
coprimes. Observe the forms of the *Lissajous curves* for various (n, m).

Maple:

```
with(plots): for i from 1 to 3 do
 n:=ithprime(i): m:=ithprime(i+1): x:=sin(n*t): y:=cos(m*t):
 plot([x,y,t=0..2*Pi],scaling=constrained,colour=red); od;
G:=[seq(plot([sin(ithprime(i)*t),cos(ithprime(i+1)*t),
 t=0..2*Pi],scaling=constrained,colour=blue),i=1..10)]:
display(G,insequence=true);
```

Mathematica:

```
SetOptions[ParametricPlot,PlotStyle->Blue,PlotRange->All,
 Frame->True,FrameTicks->False,AspectRatio->1];
Do[n=Prime[i]; m=Prime[i+1]; x=Sin[n*t]; y=Cos[m*t];
   Print[ParametricPlot[{x,y},{t,0,2*Pi},PlotStyle->Blue,
   PlotRange->All,Frame->True,FrameTicks->False,
   AspectRatio->1]],{i,1,3}];
{ps=Table[Prime[i],{i,1,10}],pN=Length[ps]}
g=Table[ParametricPlot[{Sin[ps[[k]]*t],Cos[ps[[k+1]]*t]},
 {t,0,2*Pi}],{k,1,pN-1}]; ListAnimate[g,AnimationRate->1]
```

□

Problem 6.7 Plot the rose function, $r = \sin(n\theta)$ $(n = 1, \ldots, 5)$, Fermat's spiral, $r^2 = \theta$, Archimedes' spiral, $r = \theta$, the hyperbolic spiral, $r = 1/\theta$, the cardioid, $r = a(1 + \cos\theta)$ $(a > 0)$, the lemniscate, $r^2 = a^2\cos(2\theta)$ $(a > 0)$, the logarithmic spiral, $r = ae^{b\theta}$ $(a > 0, |b| < 1)$.

Maple:

```
with(plots):  A:=array(1..5):  a:=1; b:=0.1;
for i from 1 to 5 do
 A[i]:=polarplot([sin(i*t),1],t=0..2*Pi,thickness=[2,1],
 title=convert(n=i,string),color=[blue,red]) od:
display(A,axes=none,scaling=constrained);
plot([sqrt(t),t,t=0..17*Pi],coords=polar);
polarplot([t,t,t=0..17*Pi]);
polarplot([1/t,t,t=0..17*Pi],-0.4..0.4);
polarplot([a*(1+cos(t)),t,t=0..17*Pi]);
polarplot([a*sqrt(cos(2*t)),t,t=0..17*Pi],numpoints=1000);
polarplot([a*exp(b*t),t,t=0..17*Pi],numpoints=200);
```

Mathematica:

```
a=1; b=0.1; SetOptions[PolarPlot,Frame->True,PlotRange->All,
  ImageSize->150,AspectRatio->1];
GraphicsRow[Table[PolarPlot[{Sin[i*t],1},{t,0,2*Pi},PlotStyle->
  {{Blue,Thickness[0.03]},{Red,Thickness[0.01]}},FrameTicks->
  False,PlotLabel->Row[{"n=",i}]],{i,1,5}]]
GraphicsRow[{PolarPlot[Sqrt[t],{t,0,17*Pi},PlotRange->{-6,6}],
  PolarPlot[t,{t,0,17*Pi}],PolarPlot[1/t,{t,0.0001,10*Pi},
  PlotRange->{{-0.4,0.4},{-1,1}}],PolarPlot[a*(1+Cos[t]),
  {t,0,17*Pi}],PolarPlot[a*Sqrt[Cos[2*t]],{t,0,17*Pi},
  PlotPoints->200]}]
PolarPlot[a*Exp[b*t],{t,0,17*Pi},PlotPoints->200]
```

□

Problem 6.8 Find the area between the graphs, $r_1 = 1$ and $r_2 = \cos(3t)$.

Maple:

```
with(plots): r1:=t->1; r2:=t->cos(3*t); Rt:=0..2*Pi;
plot([r1(t),r2(t)],t=Rt,view=[Rt,-1..1]);
polarplot([r1(t),r2(t)],t=Rt,view=[-1..1,-1..1]);
Circ:=Pi; Rose:=3/2*int(r2(t)^2,t=-Pi/6..Pi/6);
AreaG:=evalf(Circ-Rose);
```

Mathematica:

```
r1[t_]:=1; r2[t_]:=Cos[3*t]; nD=10;
Plot[{r1[t],r2[t]},{t,0,2*Pi},
  PlotRange->{{0,2*Pi},{-1,1}},PlotStyle->Blue]
PolarPlot[{r1[t],r2[t]},{t,0,2*Pi},
  PlotRange->{{-1,1},{-1,1}},PlotStyle->Hue[0.9]]
{circ=Pi, rose=3/2*Integrate[r2[t]^2,{t,-Pi/6,Pi/6}],
  areaG=N[circ-rose,nD]}
```

□

6.5   Secant and Tangent Lines

Problem 6.9 Let $f(x) = x^3 - 4x^2 + 5x - 2$. Plot $f(x)$, the secant line passing through some two points, for instance, $(b, f(b))$, $(b+t, f(b+t))$, for various values of t and $b = 1$; and the tangent line passing through some point, for instance, $(a, f(a))$, for various values of a.

Maple:

```
with(plots): f:=x->x^3-4*x^2+5*x-2; b:=1;
SL:=(x,t)->((f(b+t)-f(b))/t)*(x-b)+f(b);
TL:=(x,t)->D(f)(t)*(x-t)+f(t);
G1:=animate(f(x),x=0..4,t=0.1..2,frames=100,
  view=[0..3,-2..2],thickness=10,color=plum):
G2:=animate(SL(x,t),x=0..4,t=0.1..2,frames=100,
  view=[0..3,-2..2],thickness=5,color=green):
G3:=animate(TL(x,a),x=0..4,a=0.1..2.9,frames=100,
  view=[0..3,-2..2],thickness=5,color=blue):
display([G1,G2,G3]); with(Student[Calculus1]):
for a from 0.1 to 2.9 by 0.5 do Tangent(f(x),x=a); od;
Tangent(f(x),x=0.1,output=plot);
Tangent(f(x),x=0.1,output=plot,showpoint=false,
  tangentoptions=[color=blue,thickness=5]);
```

Mathematica:

```
f[x_]:=x^3-4*x^2+5*x-2; b=1;
secL[x_,t_]:=((f[b+t]-f[b])/t)*(x-b)+f[b];
tanL[x_,t_]:=f[t]+f'[t]*(x-t);
SetOptions[Plot,PlotRange->{{0,3},{-2,2}}];
g=Plot[f[x],{x,0,4},PlotStyle->{Hue[0.85],Thickness[0.01]}];
Animate[Show[g,Plot[secL[x,t],{x,0,4},PlotStyle->
  {Green,Thickness[0.01]}],Plot[tanL[x,t],{x,0,4},PlotStyle->
  {Blue,Thickness[0.01]}]],{t,0.1,2},AnimationRate->0.1]
```

□

6.6   Tubes and Knots

Tubes and knots around 3D parametric curve.

Maple:

```
with(plots):  tubeplot([x(t),y(t),z(t)],t=t1..t2,
                  radius=r,tubepoints=m,numpoints=n,ops);
```

```
with(plots): tubeplot([2*sin(t),cos(t)-sin(2*t),cos(2*t)],
  t=0..2*Pi,axes=boxed,radius=0.25,numpoints=100,
  scaling=constrained);
```

```
tubeplot([-2*cos(t)+2*cos(2*t)+2*sin(2*t),
 -2*cos(2*t)+2*sin(t)-2*sin(2*t),-2*cos(2*t)],t=0..2*Pi,
 axes=boxed,radius=0.25,numpoints=100,color=gold);
```

Mathematica:

```
ParametricPlot3D[{x[t],y[t],z[t]},{t,t1,t2},
                              PlotStyle->{Tube[r]},ops]
```

```
{curve1={2*Sin[t],Cos[t]-Sin[2*t],Cos[2*t]},
 curve2={-2*Cos[t]+2*Cos[2*t]+2*Sin[2*t],-2*Cos[2*t]+
 2*Sin[t]-2*Sin[2*t],-2*Cos[2*t]}}
ParametricPlot3D[curve1,{t,0,2*Pi},PlotStyle->{Tube[0.2],
 Hue[Random[]]},PlotRange->All,BoxRatios->{1,1,1}]
ParametricPlot3D[curve2,{t,0,2*Pi},PlotStyle->{Tube[0.2],
 Hue[Random[]]},PlotRange->All,BoxRatios->{1,1,1}]
```

In *Mathematica 7,* the new Tube graphics primitive has been introduced.
Tubes and knots can be plotted as a parametric space curve as follows:

```
{curve1={2*Sin[t],Cos[t]-Sin[2*t],Cos[2*t]},
 curve2={-2*Cos[t]+2*Cos[2*t]+2*Sin[2*t],-2*Cos[2*t]+
 2*Sin[t]-2*Sin[2*t],-2*Cos[2*t]}}
ParametricPlot3D[curve1,{t,0,2*Pi},PlotStyle->Red,
PlotRange->All]/.Line[ps_,s___]:>Tube[ps,0.2,s]
```

6.7   Surfaces in Space

Parametrically defined surfaces, $x = x(u,v)$, $y = y(u,v)$, $z = z(u,v)$, and
   implicitly defined surfaces, $f(x,y,z) = c$.

Maple:

```
with(plots):   plot3d([x(u,v),y(u,v),z(u,v)],u=u1..u2,v=v1..v2);
        implicitplot3d(f(x,y,z)=c,x=x1..x2,y=y1..y2,z=z1..z2);
```

```
with(plots): F:=(x,y,z)->-x^2-2*y^2+z^2-4*y*z;
implicitplot3d({F(x,y,z)=-100,F(x,y,z)=0,
     F(x,y,z)=100},x=-6..6,y=-6..6,z=-6..6);
```

Mathematica:

```
ParametricPlot3D[{x[u,v],y[u,v],z[u,v]},{u,u1,u2},{v,v1,v2}]
ContourPlot3D[f[x,y,z],{x,x1,x2},{y,y1,y2},{z,z1,z2},
                                    Contours->{c1,c2,...},ops]
```

```
F[x_,y_,z_]:=-x^2-2*y^2+z^2-4*y*z;
ContourPlot3D[F[x,y,z],{x,-6,6},{y,-6,6},{z,-6,6},
  Contours->{-100,0,100}]
```

Problem 6.10 Plot the ellipsoid $\dfrac{x^2}{16}+\dfrac{y^2}{4}+z^2=1$, the hyperboloid of one

sheet $\dfrac{x^2}{16}+\dfrac{y^2}{4}-z^2=1$, and the hyperboloid of two sheets $\dfrac{x^2}{9}-\dfrac{y^2}{4}-\dfrac{z^2}{4}=1$.

Maple:

```
with(plots): x:=(u,v)->3*cos(u)*cos(v);
y:=(u,v)->2*cos(u)*sin(v); z:=(u,v)->sin(u);
plot3d([x(u,v),y(u,v),z(u,v)],u=-Pi/2..Pi/2,v=-Pi..Pi,
  axes=boxed, orientation=[58,60]);
x:=(u,v)->4*sec(u)*cos(v); y:=(u,v)->2*sec(u)*sin(v);
z:=(u,v)->tan(u);
plot3d([x(u,v),y(u,v),z(u,v)],u=-Pi/4..Pi/4,v=-Pi..Pi,
  axes=boxed,orientation=[10,72],color=green);
implicitplot3d(x^2/9-y^2/4-z^2/4=1,x=-10..10,y=-8..8,
  z=-5..5,axes=normal,numpoints=500,orientation=[63,43]);
```

Mathematica:

```
SetOptions[ParametricPlot3D,
  BoxRatios->{1,1,1}]; x[u_,v_]:=3*Cos[u]*Cos[v];
y[u_,v_]:=2*Cos[u]*Sin[v]; z[u_,v_]:=Sin[u];
ParametricPlot3D[{x[u,v],y[u,v],z[u,v]},
  {u,-Pi/2,Pi/2}, {v,-Pi,Pi}]
x[u_,v_]:=4*Sec[u]*Cos[v]; y[u_,v_]:=2*Sec[u]*Sin[v];
z[u_,v_]:=Tan[u]; ParametricPlot3D[
  {x[u,v],y[u,v],z[u,v]},{u,-Pi/4,Pi/4},{v,-Pi,Pi}]
ContourPlot3D[x^2/9-y^2/4-z^2/4,{x,-10,10},{y,-8,8},
  {z,-5,5},Contours->{1}]
```

□

6.8   Level Curves and Surfaces

Maple:

```
with(plots):          contourplot(f(x,y),x=x1..x2,y=y1..y2,ops);
                      contourplot3d(f(x,y,z),x=x1..x2,y=y1..y2,ops);
```

```
with(plots): f1:=(x,y)->x^2-x+y^2-y+2; R:=-10..10;
contourplot(f1(x,y),x=R,y=R,axes=boxed);
contourplot(f1(x,y),x=R,y=R,grid=[50,50],axes=boxed,
   contours=10,filled=true,coloring=[blue,black]);
contourplot3d(f1(x,y),x=R,y=R);
```

Mathematica:

```
ContourPlot[f1[x,y],{x,x1,x2},{y,y1,y2},ops]
    ContourPlot3D[f2[x,y,z], {x,x1,x2},{y,y1,y2},{z,z1,z2},ops]
                      list1=Table[f1[x,y],{x,x1,x2},{y,y1,y2}]
          list2=Table[f2[x,y,z],{z,z1,z2},{y,y1,y2},{x,x1,x2}]
ListContourPlot[list1,ops]          ListContourPlot3D[list2,ops]
```

for more detail, see Options[ContourPlot].

```
f1[x_,y_]:=x^2-x+y^2-y+2;
f2[x_,y_,z_]:=-x^2-2*y^2+z^2-4*y*z;
SetOptions[ContourPlot,ImageSize->200];
GraphicsRow[{ContourPlot[f1[x,y],{x,-10,10},{y,-10,10}],
 ContourPlot[f1[x,y],{x,-10,10},{y,-10,10},Contours->5],
 ContourPlot[f1[x,y],{x,-10,10},{y,-10,10},
 ContourLines->False],ContourPlot[f1[x,y],{x,-10,10},
 {y,-10,10},Contours->{10,20,30},ContourShading->False]}]
Plot3D[f1[x,y],{x,-10,10},{y,-10,10},
 BoxRatios->{1,1,1},Mesh->False]
list1=Table[Random[Real,{1,10}],{x,1,10},{y,1,10}];
ListContourPlot[list1,DataRange->{{-8,8},{-8,8}}]
list2=Table[f1[x,y],{x,1,10},{y,1,10}];
ListContourPlot[list2,DataRange->{{-8,8},{-8,8}}]
ContourPlot3D[Evaluate[f2[x,y,z],{x,-2,2},{y,-2,2},
 {z,-2,2},Contours->{0,8},BoxRatios->{1,1,1}]]
list3=Table[f2[x,y,z],{z,-2,2},{y,-2,2},{x,-2,2}];
ListContourPlot3D[Evaluate[list3,Axes->True,
 DataRange->{{-1,1},{-1,1},{-1,1}},Contours->{0,8}]]
```

Problem 6.11 Plot the function $h(x, y) = \dfrac{x - y}{x^2 + y^2}$ and some level curves, $(x, y) \in [-4, 4] \times [-4, 4]$.

Maple:

```
with(plots);
h:=(x,y)->(x-y)/(x^2+y^2); R:=-4..4; C:=[-0.5,-0.9,-1.5];
plot3d(h(x,y),x=R,y=R,grid=[50,50],orientation=[15,67]);
contourplot(h(x,y),x=R,y=R,grid=[150,150],
  axes=boxed,contours=20); contourplot(h(x,y),x=R,y=R,
  grid=[150,150],axes=boxed,contours=C);
```

Mathematica:

```
h[x_,y_]:=(x-y)/(x^2+y^2);
SetOptions[ContourPlot,AspectRatio->1,ImageSize->300];
Plot3D[h[x,y],{x,-4,4},{y,-4,4},PlotPoints->{20,20},
  BoxRatios->{1,1,1},PlotRange->All]
GraphicsRow[{ContourPlot[Evaluate[h[x,y]],{x,-4,4},{y,-4,4},
  Contours->20,ContourShading->False,PlotPoints->{80, 80}],
  ContourPlot[h[x,y],{x,-4,4},{y,-4,4},Contours->
  {-0.5,-0.9,-1.5},ContourShading->False,ContourStyle->Blue],
  ContourPlot[h[x,y]==-0.5,{x,-4,4},{y,-4,4},ContourShading->
  False,ContourStyle->Blue]}]
```

□

6.9 Surfaces of Revolution

A surface of revolution is a surface obtained by rotating a curve lying on some plane, $z = f(x)$ (or a parametric curve $x = x(t)$, $z = z(t)$), around a straight line (the axis of rotation) that lies on the same plane (z-axis).

Maple:

```
with(Student[Calculus1]):        SurfaceOfRevolution(f(x),x=a..b,
            axis=value,output=value,surfaceoptions=[values],ops);
                           VolumeOfRevolution(f(x),x=a..b,ops);
```

Mathematica:

```
RevolutionPlot3D[fz,{t,t1,t2},ops]
        RevolutionPlot3D[{fz,t},{t,t1,t2},ops]
                RevolutionPlot3D[{fx,fz},{t,t1,t2},ops]
                        RevolutionPlot3D[{fx,fy,fz},{t,t1,t2},ops]
```

Problem 6.12 Plot the surface of revolution created by revolving the region bounded by the graphs of $f(x) = x^2$, $x = 0$, $x = 1$, and the x-axis around the y-axis.

Maple:

```
f:=x->x^2; F:=[x*cos(t),x*sin(t),f(x)];
plot3d(F,x=0..1,t=0..2*Pi);
with(Student[Calculus1]):
SurfaceOfRevolution(f(x),x=0..1,axis=vertical,
 output=plot,surfaceoptions=[shading=Z]);
SurfaceOfRevolution(f(x),x=0..1,axis=vertical,
 output=value);
SurfaceOfRevolution(f(x),x=0..1,axis=vertical,
 output=integral);
VolumeOfRevolution(f(x),x=0..1,axis=vertical,
 output=plot);
VolumeOfRevolution(f(x),x=0..1,axis=vertical,
 output=value);
VolumeOfRevolution(f(x),x=0..1,axis=vertical,
 output=integral);
```

Mathematica:

```
f[x_]:=x^2;
RevolutionPlot3D[f[x],{x,0,1},BoxRatios->{1,1,1}]
RevolutionPlot3D[{f[x],x},{x,0,1},BoxRatios->{1,1,1}]
RevolutionPlot3D[{t,t^2},{t,0,1},BoxRatios->{1,1,1}]
RevolutionPlot3D[{t,t^2},{t,0,1},
 BoxRatios->{1,1,1},ViewVertical->{-1,0,0}]
```

□

6.10   Vector Fields

2D and 3D vector fields and gradient fields

Maple:

```
with(plots):                    gradplot(f(x,y),x=x1..x2,y=y1..y2,ops);
                  fieldplot([f(x,y),g(x,y)],x=x1..x2,y=y1..y2,ops);
fieldplot3d([f(x,y,z),g(x,y,z),h(x,y,z)],x=x1..x2,y=y1..y2,
                                                  z=z1..z2,ops);
             gradplot3d(f(x,y,z),x=x1..x2,y=y1..y2,z=z1..z2,ops);
```

```
with(plots): Rx:=-5..5; Ry:=-4*Pi..4*Pi;
contourplot(x*cos(y)+4*x-sin(y),x=Rx,y=Ry,grid=[50,50]);
fieldplot([x*sin(y)+cos(y),cos(y)+4],x=Rx,y=Ry,
 grid=[30,30],arrows=slim,color=x);
gradplot((1+2*x^2)*(y-1)/(1+2*y^2),x=-1..1,y=-1..1,
 grid=[20,20],arrows=thick,color=x^2);
fieldplot3d([x-20*y+20*z,x-4*y+20*z,x-4*y+20*z],x=-4..4,
 y=-8..8,z=-8..8,grid=[9,9,9],scaling=constrained,axes=boxed);
gradplot3d(10*x^2-5*y^2+2*z^2-10,x=-1..1,y=-1..1,z=-1..1,
 axes=boxed,grid=[7,7,7]);
```

Mathematica:

```
<<VectorFieldPlots`
             VectorFieldPlot[{fx,fy},{x,x1,x2},{y,y1,y2},ops]
VectorFieldPlot3D[{fx,fy,fz},{x,x1,x2},{y,y1,y2},{z,z1,z2},ops]
             GradientFieldPlot[f[x,y],{x,x1,x2},{y,y1,y2},ops]
GradientFieldPlot3D[f[x,y,z],{x,x1,x2},{y,y1,y2},{z,z1,z2},ops]
             ListVectorFieldPlot[{field1,field2,...},ops]
             ListVectorFieldPlot3D[{field1,field2,...},ops]
         HamiltonianFieldPlot[f[x,y],{x,x1,x2},{y,y1,y2},ops]
```

```
<<VectorFieldPlots`
ContourPlot[x*Cos[y]+4*x-Sin[y],{x,-5,5},{y,-4*Pi,4*Pi}]
VectorFieldPlot[{x*Sin[y]+Cos[y],Cos[y]+4},{x,-5,5},
 {y,-4*Pi,4*Pi},BaseStyle->Blue,AspectRatio->1]
VectorFieldPlot3D[{x-20*y+20*z,x-4*y+20*z,x-4*y+20*z},
 {x,-4,4},{y,-8,8},{z,-8,8},PlotPoints->9,VectorHeads->True]
```

```
GradientFieldPlot[(1+2*x^2)*(y-1)/(1+2*y^2),
 {x,-1,1},{y,-1,1},BaseStyle->Blue]
GradientFieldPlot3D[10*x^2-5*y^2+2*z^2-10,{x,-1,1},
 {y,-1,1},{z,-1,1}]
```

In *Mathematica 7*, the new functions `VectorPlot`, `VectorPlot3D` have been introduced and incorporated into the `Mathematica` kernel. The `GradientFieldPlot`, `GradientFieldPlot3D`, `HamiltonianFieldPlot` functions have been incorporated into the `VectorPlot` and `VectorPlot3D` functions.

```
VectorPlot[{x*Sin[y]+Cos[y],Cos[y]+4},{x,-5,5},
 {y,-4*Pi,4*Pi},VectorColorFunction->Hue]
VectorPlot3D[{x-20*y+20*z,x-4*y+20*z,x-4*y+20*z},
 {x,-4,4},{y,-8,8},{z,-8,8},PlotPoints->9,
 VectorColorFunction->Hue]
VectorPlot[Evaluate[D[(1+2*x^2)*(y-1)/(1+2*y^2),
 {{x, y}}]],{x,-1,1},{y,-1,1}]
VectorPlot3D[Evaluate[D[10*x^2-5*y^2+2*z^2-10,
 {{x,y}}]],{x,-1,1},{y,-1,1},{z,-1,1}]
```

6.11   Cylindrical Coordinates

Maple:

```
with(plots):           Rtheta:=theta1..theta2;           Rr:=r1..r2;
              cylinderplot(f(r,theta),theta=Rtheta,r=Rr,ops);
        cylinderplot([r(theta),theta,r],theta=Rtheta,r=Rr,ops);
```

```
with(plots): f:=(r,theta)->cos(r)-(2+sin(4*theta));
cylinderplot(f(r,theta),theta=0..2*Pi,r=0..4*Pi,grid=[50,50]);
```

Mathematica:

```
RevolutionPlot3D[fz,{r,r1,r2},{t,t1,t2},ops]
      RevolutionPlot3D[{fz,r},{r,r1,r2},{t,t1,t2},ops]
            RevolutionPlot3D[{fx,fz},{r,r1,r2},{t,t1,t2},ops]
```

```
f[r_,theta_]:=Cos[r]-(2+Sin[4*theta]);
RevolutionPlot3D[Evaluate[{f[r,theta],r}],{r,0,4*Pi},
 {theta,0,2*Pi},BoxRatios->{1,1,1},PlotPoints->20]
```

6.12   Spherical Coordinates

Maple:

```
with(plots):          Rtheta:=theta1..theta2;          Rr:=r1..r2;
                      sphereplot(f(z,r),z=z1..z2,r=Rr,ops);
        sphereplot([z(theta),theta,z],z=z1..z2,theta=Rtheta,ops);
```

```
with(plots):
sphereplot(2*sin(z)*cos(r)-r,r=0..2*Pi,z=0..Pi,
        axes=boxed,grid=[40,40],orientation=[0,0]);
```

Mathematica:

```
SphericalPlot3D[f[z,r],{z,z1,z2},{r,r1,r2},ops]
```

```
SphericalPlot3D[Evaluate[2*Sin[z]*Cos[r]-r],{z,0,Pi},
 {r,0,2*Pi},BoxRatios->{1,1,1},PlotStyle->
 Directive[LightPurple,Opacity[0.9]],PlotPoints->5]
```

6.13   Standard Geometric Shapes

Standard geometric 2D and 3D shapes and their transformations

Maple:

Standard geometric shapes (e.g., circles, disks, points, lines, rectangles, polygons, cuboids, cylinders, spheres, cones) and their transformations are constructed with the packages geometry, geom3d, plottools.

```
with(geometry): with(plottools): with(plots):
Ops:=color=red,linestyle=3:
G1:=circle([1,1],1,color=blue):
G2:=line([-1,-1],[1,1],Ops):
G3:=line([1,1],[1,0],Ops): G4:=line([1,0],[-1,-1],Ops):
G5:=sphere([0,0,0],1):
display({G1,G2,G3,G4},scaling-constrained);
display({cuboid([0,0,0],[1,2,3]),cylinder([0,0,-1],1,4)},
 style=patchcontour,scaling=constrained,
 orientation=[-60,60]);
```

```
display({sphere([0,0,0],1),cone([0,0,1],1,-2)},
 style=wireframe,scaling=constrained,
 orientation=[-60,60]);
display(rotate(G5,0,Pi,0),axes=boxed);
display([G5,translate(G5,1,2,3)],orientation=[-50,80]);
display([G5,translate(scale(G5,0.5,0.5,0.5),1,1,1)],
 orientation=[-50,80]);
```

Mathematica:

Standard geometric shapes and their transformations can be constructed with the functions Graphics[shape], Graphics3D[shape].

```
SetOptions[Graphics3D,BoxRatios->{1,1,1}];
g1=Graphics[{Blue,Circle[{1,1},1]}];
g2=Graphics[{Red,Dashing[0.01],Line[{{-1,-1},
 {1,1},{1,0},{-1,-1}}]}]; Show[{g1,g2}]
Show[Graphics3D[{{Blue,Cuboid[{0,0,0},{1,2,3}]},
 {Purple,Cylinder[{{0,0,-1},{0,0,3}},0.9]}}]]
g4=Sphere[]; Graphics3D[{g4,Translate[Rotate[
 g4,Pi,{1,1,1}],{2,2,2}]}]
Graphics3D[{g4,Translate[g4,{1,2,3}]}]
Graphics3D[{g4,Translate[Scale[g4,{0.5,0.5,0.5},
 {1,1,1}],{2,2,2}]}]
Graphics3D[GeometricTransformation[
 GeometricTransformation[g4,ScalingMatrix[2,{1,1,1}]],
 ReflectionMatrix[{1,1,1}]]]
SetOptions[Graphics3D,Lighting->"Neutral",
 PlotRange->All,BoxRatios->{1,1,1}];
Graphics3D[{Opacity[0.6],Sphere[{1,1,1},3],
 Cylinder[{{0,0,-1},{1,1,1}},3]}]
```

Note. Some functions (e.g. WireFrame, Helix, DoubleHelix, etc.) of the package Graphics`Shapes` are available on the Wolfram website. The functions Torus, MoebiusStrip are available in the kernel function ExampleData.

Chapter 7

Calculus and Analysis

7.1 Real Functions

In this Chapter we consider real functions of real variables, $f : \mathbb{R}^n \to \mathbb{R}^n$ ($n \in \mathbb{N}$). At first, let $n = 1$ and we consider various forms of definition, visualization, and operations on real functions of the real variable in both systems (in more detail about functions, see Sect. 4.8).

Maple:

```
f:=x->expr;           f:=unapply(expr,x);      f:=proc(x) expr end;
               f:=proc(x) options operator,arrow; expr end proc;
(f1@f2@...@fn)(x);              (f@@n)(x);  plot(f(x),x=a..b,ops);
map(f,expr);           map(x->expr,expr);                apply(f,expr);
```

Mathematica:

```
f[x_]:=expr;          f=Function[x,expr];          f=Function[#];
f=#&;              f[x_]:=Module[{},expr]; Plot[f[x],{x,a,b},ops]
Composition[f1,...,fn][x]        f@f@f@x  Composition[f,f,f][x]
x//f1//f2 Apply[f,{expr}]        f@@{expr}  Map[f,expr]  f/@expr
MapAll[f,list1] f//@list1 Nest[f,expr,n]       NestList[f,expr,n]
```

Problem 7.1 Define the function $f(x) = 1 - \sin(x^2)$ and evaluate $f(1)$, $f(0)$, $f(a^2 + b^2)$.

Maple:

```
f:=(x)->1-sin(x^2); f(x);
evalf(f(1)); f(0); simplify(f(a^2+b^2));
```

I.K. Shingareva, C. Lizárraga-Celaya, *Maple and Mathematica*, 2nd ed.,
DOI 10.1007/978-3-211-99432-0_7, © Springer-Verlag Vienna 2009

Mathematica:

```
f[x_]:=1-Sin[x^2];
{f[t], N[f[1]], f[0], Simplify[f[a^2+b^2]]}
```

☐

Problem 7.2 Graph various elementary functions, for example, linear functions, parabolas, hyperbolas, rational and irrational functions, trigonometric and hyperbolic functions and their inverses, exponential and logarithmical functions.

Maple:

```
with(plots): a:=1; b:=2; c:=3; G:=array(1..4):
for k from -1.5 to 1.5 do G[trunc(k+2.5)]:=plot(k*x,x=-2..2,
  scaling=constrained,thickness=3,color=COLOR(RGB,rand()/10^12,
  rand()/10^12,rand()/10^12)); od: display(G); setcolors();
plot([a*x^2,x^2+c,(x-a)^2,a+(x-b)^2,a*x^2+b*x+c],x=-2..2);
plot([x^3,a+(x-b)^3,a*x^3+b*x+c,x^4,a*x^2+b*x^4],x=-2..2);
plot([a/x,1/(a-x),(a-x)/(c+x),a+b/(x-c)],x=-5..5,
  view=[-5..5,-10..10],discont=true);
plot([a/(x^2+c),b*x/(x^2+c),x^2+1/x,x+1/x^2],x=-5..5,
  view=[-5..5,-10..10],discont=true);
plot([x^(2/3),x*sqrt(x),sqrt(x^2-1),x*sqrt(x/(a-x))],
  x=-2..2,view=[-2..2,0..5]);
plot([sin(x)+cos(x),sinh(x),arcsin(x/c),sqrt(tan(x))],x=-Pi..Pi);
plot([exp(-x^2),b^x,x^x,c+log(x^2),log(b/x)],x=-2..2);
```

Mathematica:

```
a=1; b=2; c=3; SetOptions[Plot,ImageSize->250,PlotStyle->{Red,
  Blue,Magenta,Purple,Green}]; GraphicsRow[Table[
  Plot[k*x,{x,-2,2},PlotStyle->Hue[Random[]]],{k,-1.5,1.5}]]
GraphicsRow[{Plot[{a*x^2,x^2+c,(x-a)^2,a+(x-b)^2,
  a*x^2+b*x+c},{x,-2,2}], Plot[{x^3,a+(x-b)^3,a*x^3+b*x+c,
  x^4,a*x^2+b*x^4},{x,-2,2}], Plot[{a/x,1/(a-x),(a-x)/(c+x),
  a+b/(x-c)},{x,-5,5}, PlotRange->{{-5,5},{-10,10}}],
  Plot[{a/(x^2+c),b*x/(x^2+c),x^2+1/x,x+1/x^2},{x,-5,5},
  PlotRange->{{-5,5},{-10,10}}]}]
GraphicsRow[{Plot[{x^(2/3),x*Sqrt[x],Sqrt[x^2-1],
  x*Sqrt[x/(a-x)]},{x,-2,2},PlotRange->{{-2,2},{0,5}}],
  Plot[{Sin[x]+Cos[x],Sinh[x],ArcSin[x/c],Sqrt[Tan[x]]},
  {x,-Pi,Pi}], Plot[{Exp[-x^2],b^x,x^x,c+Log[x^2],
  Log[b/x]},{x,-2,2}]}]
```

☐

7.2 Limits of Sequences and Functions

The limit of a real function $f(x)$ when x tends to x_0, the limit of a sequence x_n of real numbers when n tends to ∞.

Maple:

```
limit(f(x),x=x0); limit(f(x),x=infinity);limit(f(x),x=x0,left);
limit(f(x),x=x0,right); limit(f(x),x=x0,real);Limit(f(x),x=x0);
limit(f(x),x=x0,complex);        limit(f(x,y),{x=x0,y=y0},left);
```

Mathematica:

```
Limit(f[x],x->x0)            Limit[f[x],x->x0,Assumptions->value]
Limit[f[x],x->x0,Analytic->True]Limit[f[x],x->x0,Direction->-1]
Limit[f[x],x->x0,Direction->1] Limit[Limit[f[x,y],x->x0],y->y0]
Limit[f[x],x->Infinity]        TraditionalForm[Limit[f[x],x->x0]]
```

Problem 7.3 Graph the function $f(x) = \dfrac{x^3 - 3x}{x^3 - x}$, $x \in [-2, 2]$, and evaluate the limits $\lim\limits_{x \to 0} f(x)$, $\lim\limits_{x \to \pm 1} f(x)$, $\lim\limits_{x \to +\infty} f(x)$, $\lim\limits_{x \to 0} \sin(1/x)$.

Maple:

```
f:=x->(x^3-3*x)/(x^3-x); limit(sin(1/x),x=0);
plot(f(x),x=-2..2,-100..100,discont=true);
limit(f(x),x=0); limit(f(x),x=infinity);
limit(f(x),x=1,right); limit(f(x),x=-1,left);
```

Mathematica:

```
f[x_]:=(x^3-3*x)/(x^3-x);
Plot[f[x],{x,-2,2},Exclusions->{x^3-x==0},Frame->True]
{Limit[f[x],x->0], Limit[f[x],x->Infinity],
 Limit[f[x],x->1,Direction->-1],
 Limit[f[x],x->-1,Direction->1],Limit[Sin[1/x],x->0]}
```

□

Problem 7.4 Evaluate the limit of the sequence

$$\lim_{n \to \infty} \left(\frac{1}{1 \cdot 2 \cdot 3 \cdot 4} + \cdots + \frac{1}{n(n+1)(n+2)(n+3)} \right).$$

Maple:

```
L1:=evalf(limit(add(1/(i*(i+1)*(i+2)),i=1..100000),n=infinity));
s:=1/(m*(m+1)*(m+2)); L2:=limit(sum(s,'m'=1..n),n=infinity);
s1:=convert(s,parfrac,m,sqrfree); ex1:=map(sum, s1,'m'=1..n);
limit(ex1, n=infinity); ex2:=expand(map(Sum, s1,'m'=1..n));
ex3:=ex2=op(1,ex2)-Sum(1/m,m=2..n+1)+1/2*Sum(1/m,m=3..n+2);
ex4:=expand(op(2,ex3)); ex41:=1/2+1/4+1/2*Sum(1/m,m=3..n);
ex42:=-1/2-Sum(1/m,m=3..n)-1/(n+1);
ex43:=1/2*Sum(1/m,m=3..n)+1/(2*(n+1))+1/(2*(n+2));
ex5:=ex41+ex42+ex43; limit(ex5,n=infinity);
```

Mathematica:

```
L1=N[Limit[Sum[1/(i*(i+1)*(i+2)),{i,1,100000}],n->Infinity]]
{s=1/(m*(m+1)*(m+2)), L2=Limit[Sum[s,{m,1,n}],n->Infinity]}
{s1=Apart[s], ex1=Map[Sum[#,{m,1,n}]&,s1],
 Limit[ex1,n->Infinity],
 ex2=Expand[Map[HoldForm[Sum[#,{m,1,n}]]&,s1]],
 ex3=ex2==ex2[[3]]-HoldForm[Sum[1/m,{m,2,n+1}]]+
 1/2*HoldForm[Sum[1/m,{m,3,n+2}]]], ex4=Expand[ex3[[2]]]}
{ex41=1/2+1/4+1/2*HoldForm[Sum[1/m,{m,3,n}]]],
 ex42=-1/2-HoldForm[Sum[1/m,{m,3,n}]]-1/(n+1),
 ex43=1/2*HoldForm[Sum[1/m,{m,3,n}]]+1/(2*(n+1))+1/(2*(n+2)),
 ex5=ex41+ex42+ex43, Limit[ex5,n->Infinity]}
```

□

7.3   Continuity of Functions

Maple:

```
f:=x->piecewise(cond1,expr1,cond2,expr2,expr3);          f(x);
discont(f,var);              fdiscont(f,x=a..b,tolerance,var,ops);
iscont(f,x=a..b);                        iscont(f,x=a..b,'closed');
iscont(f,x=a..b,'open');        plot(f,x=a..b,ops,discont=true);
with(Student[Calculus1]):                    Asymptotes(f,x,ops);
```

Mathematica:

```
f[x_]:=expr/;cond        f[x_]:=Which[cond1,val1,...,condn,valn];
              f[x_]:=Piecewise[{{cond1,val1},{cond2,val2},...}];
                      f[x_]:=UnitStep[x-a]*UnitStep[b-x];
Plot[f[x],ops,Exclusions->{eqs}]
```

Problem 7.5 Define and plot various continuous and discontinuous real functions.

Maple:

```
with(plots); with(Student[Calculus1]):
setoptions(plot,color=magenta,thickness=3);
f:=x->piecewise(x>1,-x^2+2,x<=1,x^2);f1:=x->2*x^2/tan(2*x);
plot('f(x)','x'=-2..2); plot(f1(x),x=-Pi..Pi,
  view=[-Pi..Pi,-30..30],discont=true); plot(f1(x),x=-Pi..Pi,
  view=[-Pi..Pi,-30..30]); plot(-2*Heaviside(x-1)*x^2+
  2*Heaviside(x-1)+x^2,x=-3..3); x1:=expand((9*x+12)/(3*x+4));
Asymptotes(op(1,x1)); Asymptotes(op(2,x1)); epsilon:=0.09;
G1:=plot(x1,x=-4..-4/3-epsilon,view=[DEFAULT,-4..4]):
G2:=plot(x1,x=-4/3+epsilon..4,view=[DEFAULT,-4..4]):
G3:=pointplot([-4/3,0],symbol=circle,symbolsize=40):
display({G1,G2,G3}); Asymptotes(f1(x),x=0..2*Pi);
discont(f1(x),x); fdiscont(f1(x),x=0..2*Pi,0.001);
iscont(f1(x),x=0..2*Pi); iscont(f1(x),x=0..2*Pi,'closed');
iscont(f1(x),x=0..2*Pi,'open'); plot(abs(tan(2*x))+
  abs(tan(3*x)),x=-1..1,discont=true,view=[DEFAULT,0..30]);
```

Mathematica:

```
f[x_?NumberQ]:=Which[x>1,-x^2+2,x<=1,x^2];
Plot[f[x],{x,-2,2},PlotStyle->{Hue[0.8],Thickness[0.01]}]
Plot[2*x^2/Tan[2*x],{x,-Pi,Pi},Exclusions->{Tan[2*x]==0}]
Plot[2*x^2/Tan[2*x],{x,-Pi,Pi}]
Plot[-2*UnitStep[x-1]*x^2+2*UnitStep[x-1]+x^2,{x,-3,3},
  PlotStyle->{Hue[0.8],Thickness[0.01]}]
Plot[(9*x+12)/(3*x+4),{x,-4,4},PlotRange->{{-4,4},{-4,4}},
  PlotStyle->{Hue[0.85],Thickness[0.01]},Epilog->{Hue[0.85],
  PointSize[0.02],Disk[{-4/3,0},0.12],White,Point[{-4/3,3}]}]
Plot[Abs[Tan[2*x]]+Abs[Tan[3*x]],{x,-1,1}]
```

Problem 7.6 Graph the function $f(x) = \lim_{n \to \infty} (x-a) \arctan(x^n)$, $a = 1$.

Maple:

```
a:=1; interface(showassumed=0):
L:=limit(expand((x-a)*arctan(x^n)),n=+infinity);
assume(x>a); y1:=factor(L); y2:=subs(x=a,
  (x-a)*arctan(x^n)); assume(x>=0 and x<a); y3:=factor(L);
```

```
assume(x>=-a and x<0); y4:=factor(L);
assume(x<-a); evala(L); y5:=undefined;
f:=x->piecewise(x>a,Pi/2*(x-a),x>-a and x<=a,0,undefined);
plot('f(x)','x'=-5..5,thickness=5,scaling=constrained);
```

Mathematica:

```
a=1; {y1=Limit[Expand[(x-a)*ArcTan[x^n]],
 n->Infinity,Assumptions->{x>a}],
 y2=(x-a)*ArcTan[x^n]/.{x->a}, y3=Limit[Expand[
 (x-a)*ArcTan[x^n]],n->Infinity,Assumptions->{x<a && x>0}],
 y4=Limit[Expand[(x-a)*ArcTan[x^n]],n->Infinity,
 Assumptions->{x>-a && x<0}], y5=Limit[Expand[TrigToExp[
 (x-a)*ArcTan[x^n]]],n->Infinity,Direction->1,
 Assumptions->{x<-a && n>0 && Element[n,Integers]}]}
f[x_?NumberQ]:=Which[x>a,Pi/2*(x-a),x>-a && x<=a,0];
Plot[f[x],{x,-5,5},PlotStyle->{Hue[0.85],Thickness[0.01]}]
```
□

7.4 Differential Calculus

In Maple, most of calculus functions are contained in the packages:
student, Student, VectorCalculus. The Student package contains
the subpackages: Calculus1, LinearAlgebra, Precalculus, Vector
Calculus, and MultivariateCalculus, with functions covering the
basic material of the corresponding course.

With the VectorCalculus package multivariate and vector calculus
operations can be performed.

The derivatives of $f(x)$ with respect to x, the total differential of a
function, the relative minimum/maximum of $f(x)$ near x_0.

Maple:

```
diff(f(x),x);       Diff(f(x),x);D(f)(x);           diff(f(x),x$n);
Diff(f(x),x$n);    (D@@n)(f)(x);              fracdiff(f,var,nu,ops);
PDEtools[declare](x(t),y(t),Dt=t);            diff(f(x(t),y(t)),t);
minimize(f(x),x=a..b,location=true);       maximize(f(x),x=a..b);
with(Student[Calculus1]):     InflectionPoints(f(x),x=a..b,ops);
ExtremePoints(f(x),x=a..b,ops);CriticalPoints(f(x),x=a..b,ops);
with(Student:-Precalculus):          Distance([x1,y1],[x2,y2]);
with(Optimization): Maximize(f(x),cnstr); Minimize(f(x),cnstr);
```

Mathematica:

```
f'[x]     f''[x]   f'''[x]   f''''[x]  ...   Sqrt' Sin'          Dt[f]
D[f[x],x]   D[f[x],{x,n}]      Derivative[n]         Derivative[n][f]
Derivative[n][f][x]     Maximize[f,{x,x0}]       Minimize[f,{x,x0}]
FindMaximum[f[x],{x,x0}]         FindMaximum[{f[x],cnstr},{x,x0}]
FindMinimum[f[x],{x,x0}]         FindMinimum[{f[x],cnstr},{x,x0}]
NMaximize[{f[x],cnstr},var,ops]  NMinimize[{f[x],cnstr},var,ops]
EuclideanDistance[{x1,y1},{x2,y2}]
```

Problem 7.7 Let $g(x)=\dfrac{\cos^2(x^2+1)}{(\sin x+1)^2}$ and $f(x)=\sqrt{x-2}\,(2x^2+1)^3$. Calculate the derivatives of the functions: $g(x)$, $f(x)$, $f(g(x))$, and $g(f(x))$.

Maple:

```
g:=x->cos(x^2+1)^2/(sin(x)+1)^2; f:=x->sqrt(x-2)*(2*x^2+1)^3;
D(f)(x); diff(g(x),x); diff(f(g(x)),x); diff(g(f(x)),x);
```

Mathematica:

```
g[x_]:=Cos[x^2+1]^2/(Sin[x]+1)^2;
f[x_]:=Sqrt[x-2]*(2*x^2+1)^3;
{f'[x], g'[x], D[f[g[x]],x], D[g[f[x]],x]}//Simplify
{Derivative[1][f][x], Derivative[1][g][x]}//Simplify
```
□

Problem 7.8 Let $f(x)=x^3-4x^2+8x-2$. Calculate $f'(x)$, $f'(\frac{7}{3})$ and
find the value of x for which $f'(x)=a$.

Maple:

```
a:=10; f:=x->x^3-4*x^2+8*x-2; D(f)(x); D(f)(7/3);
solve(D(f)(x)=a,x);
```

Mathematica:

```
a=10; f[x_]:=x^3-4*x^2+8*x-2; g=Derivative[1][f];
{f'[x], D[f[x],x]/.x->7/3, g[x], g[7/3]}
Solve[f'[x]==a,x]
```
⊔

Problem 7.9 Let $f(x)=(5x-2)^2(2-5x^2)^3$. Calculate $f'(x)$. Find the
values of x for which the tangent line of $f(x)$ at $(x,f(x))$ is horizontal.

Maple:

```
f:=x->(5*x-2)^2*(2-5*x^2)^3;
df:=diff(f(x),x); factor(df); solve(df=0,x);
```

Mathematica:

```
f[x_]:=(5*x-2)^2*(2-5*x^2)^3; df=f'[x];
Factor[df]; Solve[df==0,x]
```

□

Problem 7.10 Let $f(x) = x^2 \cos^2 x$. Calculate $f'(x)$, $f''(x)$, $f^{(3)}(x)$, and $f^{(4)}(x)$. Graph $f(x)$ and their derivatives for $x \in [-\pi, \pi]$.

Maple:

```
f:=x->x^2*cos(x)^2; df1:=diff(f(x),x); df2:=diff(f(x),x$2);
df3:=diff(f(x),x$3);df4:=diff(f(x),x$4);
plot([f(x),df1,df2,df3,df4],x=-Pi..Pi,color=[blue,green,
 red,plum],linestyle=[SOLID,DOT,DASH,DASHDOT],thickness=5);
```

Mathematica:

```
f[x_]:=x^2*Cos[x]^2; df1=f'[x]; df2=f''[x];
df3=f'''[x]; df4=f''''[x]; Plot[{f[x],df1,df2,df3,df4},
 {x,-Pi,Pi},PlotStyle->{{Hue[0.3],Thickness[0.02]},
  {Hue[0.5],Thickness[0.03]},{Hue[0.7],Thickness[0.02],
  Dashing[{0.03}]},{Hue[0.8],Thickness[0.02]},
  {Hue[0.9],Thickness[0.01],Dashing[{0.01,0.02}]}},
  AspectRatio->1,PlotRange->All,ImageSize->300]
```

□

Problem 7.11 Let $f(x) = 2x^3 - 8x^2 + 10x$. Find $f'(x)$ and $f''(x)$. Graph $f(x)$, $f'(x)$, and $f''(x)$, $x \in [-1, 5]$. Determine the critical and inflection points of $f(x)$.

Maple:

```
with(plots): a:=-1; b:=5; f:=x->2*x^3-8*x^2+10*x;
df1:=diff(f(x),x); df2:=diff(f(x),x$2);
plot([f(x),df1,df2],x=a..b,-15..20,thickness=[5,1,1],
 color=[blue,green,magenta]); factor(df1);
p_cr:=[solve(df1=0,x)]; p_inf:=solve(df2=0,x);
eval(df2,x=p_cr[1]); eval(df2,x=p_cr[2]);
```

```
for i from 1 to nops(p_cr) do Y_cr[i]:=eval(f(x),
  x=p_cr[i]); od; Y_inf:=eval(f(x),x=p_inf);
G1:=plot([f(x),df1,df2],x=a..b,-15..20,
  color=[blue,green,magenta],thickness=[5,1,1]):
P:=[seq([p_cr[i],Y_cr[i]],i=1..nops(p_cr)),[p_inf,Y_inf]];
G2:=pointplot(P,color=red,symbol=circle,symbolsize=20):
display([G1,G2]); with(Student[Calculus1]):
CriticalPoints(f(x),x=a..b); ExtremePoints(f(x),x=a..b);
InflectionPoints(f(x),x=a..b);
```

Mathematica:

```
a=-1; b=5; f[x_]:=2*x^3-8*x^2+10*x; df1=f'[x];
df2=f''[x]; Plot[{f[x],df1,df2},{x,a,b},AspectRatio->1,
  PlotStyle->{{Hue[0.5],Thickness[0.01]},{Hue[0.7],
  Thickness[0.013]},{Hue[0.9],Thickness[0.015]}}]
{Factor[df1], pcr=Solve[df1==0,x], pinf=Solve[df2==0,x],
  xcr={pcr[[1,1,2]],pcr[[2,1,2]]}, Evaluate[df2/.pcr],
  ycr=Map[f,{pcr[[1,1,2]],pcr[[2,1,2]]}],yinf=f[pinf[[1,1,2]]]}
g1=Plot[{f[x],df1,df2},{x,a,b},AspectRatio->1,PlotStyle->
  {{Hue[0.5],Thickness[0.01]},{Hue[0.7],Thickness[0.013]},
  {Hue[0.9],Thickness[0.015]}}]; ps={{xcr[[1]],ycr[[1]]},
  {xcr[[2]],ycr[[2]]},{pinf[[1,1,2]],yinf}}
g2=ListPlot[ps,PlotStyle->{PointSize[0.03],Hue[0.4]}];Show[g1,g2]
{FindMaximum[f[x],{x,a,b},WorkingPrecision->20,
  MaxIterations->200], FindMinimum[f[x],{x,b},
  WorkingPrecision->20,MaxIterations->200]}
```

□

Problem 7.12 Let $f(x)=(x-\pi)(\pi+x)\cos x$. Plot $f(x)$ and the tangent line of $f(x)$ at t($x_0, f(x_0)$) for $x_0 \in [0, 3\pi]$.

Maple:

```
with(plots): f:=x->(x-Pi)*(Pi+x)*cos(x); df:=diff(f(x),x);
G1:=plot(f(x),x=0..3*Pi,color=blue,thickness=5):
tanLine:=proc(x0) local lin,G2: lin:=subs(x=x0,df)*(x-x0)+f(x0);
  G2:=plot(lin,x=0..3*Pi,color=green,thickness=3):
  display({G1,G2}); end proc:
for i from 0 to 10 do tanLine(i*3*Pi/10) end do;
```

Mathematica:

```
f[x_]:=(x-Pi)*(Pi+x)*Cos[x]; df=f'[x];
tanLine[x0_]:=Module[{lin,g1,g2},lin=(df/.x->x0)*(x-x0)+f[x0];
 g1=Plot[f[x],{x,0,3*Pi},PlotStyle->{Blue,Thickness[0.02]}];
 g2=Plot[lin,{x,0,3*Pi},PlotStyle->{Green,Thickness[0.01]}];
 Show[g1,g2]];  Do[Print[tanLine[i*3*Pi/10]],{i,0,10}]
```

□

Problem 7.13 For the function $f(x) = ax + b$ and the point (x_0, y_0) not in $f(x)$ find the value of x such that the distance between (x_0, y_0) and $(x, f(x))$ is minimal.

Maple:

```
with(Student:-Precalculus): f:=x->a*x+b;
d_min:=simplify(Distance([x0,y0],[x,f(x)]));
dd_min:=simplify(diff(d_min,x));
xd_min:=solve(numer(dd_min)=0,x);
xd_min:=simplify(xd_min): yd_min:=simplify(f(xd_min));
```

Mathematica:

```
f[x_]:=a*x+b;
{mindis=ExpandAll[Sqrt[(y0-f[x])^2+(x-x0)^2]],
 dmd=D[mindis,x], xmd=Solve[Numerator[dmd]==0,x],
 ymd=Simplify[f[x]/.xmd]}
```

□

Problem 7.14 Find the dimensions of a cone of the minimal volume that is circumscribed about a sphere of radius R.

Maple:

```
Sol:=[solve((h-R)/R=(sqrt(r^2+h^2))/r,h)];
Sol1:=remove(has,Sol,0); V:=subs(h=op(Sol1),(Pi/3)*r^2*h);
dV:=simplify(diff(V,r)); p_cr:=[solve(dV=0,r)];
p_cr1:=select(has,p_cr,R); for i from 1 to nops(p_cr1) do
 p_cr2[i]:=p_cr1[i]/p_cr1[i]; od:
for i from 1 to nops(p_cr2) do if p_cr2[i]>0 then
 x_cr:=p_cr2[i]*p_cr1[i]; fi; od: x_cr;
ddV:=simplify(diff(V,r$2)); subs(r=x_cr,ddV); subs(r=x_cr,V);
```

Mathematica:

```
sol=Solve[(h-R)/R==Sqrt[r^2+h^2]/r,h]
sol1=Flatten[DeleteCases[sol,{x_->0}]]
{v=Pi/3*r^2*h/.sol1,dv=D[v,r]//Simplify, pCr=Solve[dv==0,r],
 pCr1=Flatten[DeleteCases[pCr,{x_->0}]],
 pCr2=pCr1/pCr1,n=Length[pCr2]}
Do[If[pCr2[[i]]>0,xCr=pCr2[[i]]*pCr1[[i]]],{i,1,n}];
{xCr, ddv=D[v,{r,2}]//Expand, ddv/.xCr, v/.xCr}
```

□

Problem 7.15 Calculate dy/dx if $x^3 + y^3 = 1$.

Maple:

```
ec1:=D(x^3+y^3=1); ec2:=subs(D(x)=1,ec1); isolate(ec2,D(y));
```

Mathematica:

```
Solve[Dt[x^3+y^3==1] /. Dt[x]->1, Dt[y]]
```

□

Problem 7.16 Determine the relative and the global minima and maxima of the function $f(x) = x\cos(2x)$, $x \in [-\pi, \pi]$.

Maple:

```
f:=x->x*cos(2*x);
plot(f(x),x=-Pi..Pi,color=blue,thickness=3);
evalf(minimize(f(x),x=-2..0,location=true));
evalf(maximize(f(x),x=-2..0,location=true));
evalf(minimize(f(x),x=0..2,location=true));
evalf(maximize(f(x),x=0..2,location=true));
with(Optimization): Maximize(f(x),{x>=-3,x<=3});
Minimize(f(x),{x>=-3,x<=3});
```

Mathematica:

```
f[x_]:=x*Cos[2*x];
Plot[f[x],{x,-Pi,Pi},PlotStyle->{Blue,Thickness[0.02]}]
{FindMinimum[f[x],{x,-1}],FindMinimum[f[x],{x,1}]}
{FindMaximum[f[x],{x,-1}],FindMaximum[f[x],{x,1}]}
{NMaximize[{f[x],-2<=x<=2},{x},Method->"DifferentialEvolution"],
 NMinimize[{f[x],-3<=x<=3},{x}]}
```

□

7.5   Integral Calculus

Maple:

```
with(student);        with(Student);        with(Student[Calculus1]);
RiemannSum(f(x),x=a..b,ops);              leftbox(f(x),x=a..b,ops);
rightbox(f(x),x=a..b,ops);                leftsum(f(x),x=a..b,ops);
rightsum(f(x),x=a..b,ops);        ApproximateInt(f(x),x=a..b,ops);
int(f(x),x);        int(f(x),x=a..b,ops);        I1:=Int(f(x),x);
changevar(f(x)=u,I1);        intparts(I1,u);              value(I1);
simpson(f(x),x=a..b,ops);              trapezoid(f(x),x=a..b,ops);
```

leftbox, rightbox, leftsum, rightsum, graphical and numerical approximations of an integral,

RiemannSum, construction of Riemann sums,

int,Int, value, evaluation of indefinite and definite integrals,

changevar, intparts, integration by substitution and by parts,

simpson, trapezoid, approximations of definite integrals, etc.

Note. The items relative to approximating of definite integrals are discussed in more
detail in Chapter 12.

Mathematica:

```
Integrate[f[x],x]                    Integrate[f[x],{x,a,b}]
NIntegrate[f[x],{x,a,b}]             Integrate[f[x],{x,a,b}]//N
```

Integrate, evaluation of indefinite and definite integrals,

NIntegrate, numerical approximation of definite integrals.

Problem 7.17 Let $f(x) = x^3 + 5x^2 - 2x + 1$. Approximate $\displaystyle\int_0^5 f(x)dx$ by

means of Riemann sums corresponding to the regular partition of $[0, 5]$

in n subintervals.

Maple:

```
with(student): with(Student[Calculus1]):
f:=x->x^3+5*x^2-2*x+1; leftbox(f(x),x=0..5,25);
rightbox(f(x),x=0..5,25); S_i:=leftsum(f(x),x=0..5,25);
```

```
evala(value(S_i)); S_s:=value(rightsum(f(x),x=0..5,n));
limit(S_s, n=infinity);
RSl:=RiemannSum(f(x),x=0..5,method=left,partition=25);
RSr:=RiemannSum(f(x),x=0..5,method=right,partition=25);
Rsm:=RiemannSum(f(x),x=0..5,method=midpoint,partition=25);
RSs:=RiemannSum(f(x),x=0..5,method=right,partition=n);
limit(RSs,n=infinity);
```

Mathematica:

```
f[x_]:=x^3+5*x^2-2*x+1;
a=0; b=5; n=25; dx=(b-a)/n; Integrate[f[x],{x,a,b}]
xL[k_]:=a+(i-1)*dx; Sum[f[xL[i]]*dx,{i,1,n}]
xR[k_]:=a+i*dx; Sum[f[xR[i]]*dx,{i,1,n}]
xM[k_]:=a+(i-1/2)*dx; Sum[f[xM[i]]*dx,{i,1,n}]
RiemannSumL[f_,{x_,a_,b_},n_]:=Sum[Evaluate[
  (f/.x->(a+k*(b-a)/n))],{k,0,n-1}]*(b-a)/n;
{RiemannSumL[f[x],{x,a,b},n],
  Limit[RiemannSumL[f[x],{x,a,b},m],m->Infinity]}
```

More details about Riemann sums see in Chapter 12 and Sect. 12.3 □

Problem 7.18 Let $f(x) = 3x^2$. Prove that the functions $F(x) = x^3$ and $F_1(x) = x^3 + 1$ are antiderivatives of $f(x)$.

Maple:

```
f:=x->3*x^2; F:=x->x^3; F1:=x->x^3+1;
diff(F(x),x)=f(x); diff(F1(x),x)=f(x);
```

Mathematica:

```
f[x_]:=3*x^2; F[x_]:=x^3; F1[x_]:=x^3+1;
{D[F[x],x]===f[x], D[F1[x],x]===f[x]}
```

□

Problem 7.19 Prove that $\int 4x^3 \, dx = x^4 + C$.

Maple:

```
f:=x->4*x^3; F:=x->x^4; diff(F(x),x)=f(x); int(4*x^3,x)+C;
```

Mathematica:

```
f[x_]:=4*x^3; F[x_]:=x^4;
{D[F[x],x]===f[x], Integrate[4*x^3,x]+C}
```

☐

Note. The arbitrary constant of integration in the next problems will be omitted.

Problem 7.20 Evaluate the integrals: $\int \dfrac{2x+1}{\sqrt{2x}}\, dx,\ \int (2x-1)\, e^{-2x}\, dx,$

$\int \cot x\, dx,\ \int \dfrac{dx}{\sqrt{2x-x^2}},\ \int \dfrac{4x^2-3x-2}{x^3+27}\, dx,\ \int \dfrac{x^2}{x-3}\, dx.$

Maple:

```
expand(int((2*x+1)/sqrt(2*x),x));
collect(int((2*x-1)*exp(-2*x),x),exp);
int(cot(x),x)+C; int(1/sqrt(2*x-x^2),x);
fr:=(4*x^2-3*x-2)/(x^3+27); simplify(int(fr,x));
Den:=factor(denom(fr)); Fr:=a/op(1,Den)+(b*x+d)/op(2,Den);
fr1:=fr-Fr; fr2:=collect(factor(fr1*Den),x):
c2:=coeff(fr2,x,2): c1:=coeff(fr2,x,1):
c0:=coeff(fr2,x,0): sys:={c0=0,c1=0,c2=0}:
Sol:=solve(sys,{a,b,d}): assign(Sol):
simplify(int(Fr,x),log); Intd:=x^2/(x-3); int(Intd,x);
Intd1:=convert(Intd,parfrac,x); int(Intd1,x);
```

Mathematica:

```
CS[a_.x_^2+b_.x_+c_.]:=
 a*((x+b/(2*a))^2-(b^2-4*a*c)/(4*a^2));
CompleteSquare[X_]:=If[TrueQ[X==Expand[X]],X,CS[Expand[X]]];
{Integrate[(2*x+1)/Sqrt[2*x], x]//Expand,
 Integrate[(2*x-1)*Exp[-2*x],x], Integrate[Cot[x],x],
 Integrate[1/Sqrt[CompleteSquare[2*x-x^2]]*Dt[x]/.{x->t+1}/.
 {Dt[t]->1},t]/.{t->x-1}//FullSimplify}
{fr=(4*x^2-3*x-2)/(x^3+27), Simplify[Integrate[fr,x]]}
{den=Factor[Denominator[fr]], frc=a/den[[1]]+(b*x+d)/
 den[[2]], fr1=fr-frc, fr2=Together[Factor[fr1*den]]}
{c2=Coefficient[fr2,x,2], c1=Coefficient[fr2,x,1],
 c0=Coefficient[fr2,x,0], sys={c0==0,c1==0,c2==0},
 sol=Solve[sys,{a,b,d}], frc/.sol[[1]]}
i1=Simplify[Integrate[frc/.sol[[1]],x]]
Expand[i1/.{Denominator[fr]->Factor[Denominator[fr]]}]
{intd=Apart[x^2/(x-3)], Map[Integrate[#,x]&,intd]}
```

☐

Problem 7.21 Evaluate the integrals: $\int x^2\sqrt{x-2}\,dx,\ \int\sqrt{6+4x-2x^2}\,dx.$

Maple:

```
with(student): f:=x->x^2*sqrt(x-2); I1:=Int(f(x),x);
Cv:=changevar(sqrt(x-2)=u,I1); I2:=value(Cv);
I3:=simplify(subs(u=sqrt(x-2),I2));
f1:=x->sqrt(2)*completesquare((6+4*x-2*x^2)/2)^(1/2);
I1:=Int(f1(x),x);
Intd1:=subs(x-1=2*sin(t),f1(x));
I2:=int(combine(sqrt(2)*2*cos(t)*2*cos(t),trig),t);
I3:=expand(subs(t=arcsin((x-1)/2),I2));
I4:=simplify(completesquare(I3,x));
```

Mathematica:

```
f[x_]:=x^2*Sqrt[x-2]; f[x]
{i1=Integrate[f[x],x]//Expand,
 i2=Integrate[f[x]*Dt[x]/.{x->u^2+2}/.{Dt[u]->1},u],
 i3=Expand[i2/.{u->Sqrt[x-2]}]}
CS[a_.x_^2+b_.x_+c_.]:=a*((x+b/(2*a))^2-(b^2-4*a*c)/(4*a^2));
CompleteSquare[X_]:=If[TrueQ[X==Expand[X]],X,CompleteSquare[
 Expand[X]]]; f1[x_]:=Sqrt[2]*CompleteSquare[(6+4*x-
 2*x^2)/2]^(1/2); {f1[x], i1=Integrate[f1[x],x],
 intd1=TrigReduce[f1[x]*Dt[x]/.{x->2*Sin[t]+1}/.{Dt[t]->1}],
 intd2=TrigReduce[Map[Simplify[#1]&,intd1]], i2=Factor[
 Integrate[intd2,t]], i3=Expand[i2/.{t->ArcSin[(x-1)/2]}],
 i4=TrigExpand[Map[Simplify[#1]&,i3]]}
```

□

Problem 7.22 Evaluate the integrals: $\int\frac{e^{2x}dx}{e^x+3},\ \int\frac{dx}{2+\sin x},\ \int\sin^5 x\,dx.$

Maple:

```
with(student): int(exp(2*x)/(exp(x)+3),x);
f:=x->exp(2*x)/(exp(x)+3);
I1:=Int(f(x),x); Cv:=changevar(exp(x)=z,I1);
I2:=value(Cv); I3:=simplify(subs(z=exp(x),I2));
int(1/(2+sin(x)),x); f1:=x->1/(2+sin(x));
I1:=Int(convert(f1(x),tan),x); Cv:=changevar(tan(x/2)=t,I1);
I2:=value(Cv); I3:=simplify(subs(t=tan(x/2),I2));
f2:=x->sin(x)^5; I1:=Int(convert(f2(x),cos),x);
Cv:=changevar(cos(x)=z,I1); I2:=value(Cv);
I3:=simplify(subs(z=cos(x),I2));
```

Mathematica:

```
Integrate[Exp[2*x]/(Exp[x]+3),x]
f[x_]:=Exp[2*x]/(Exp[x]+3); f[x]
intd=Apart[f[x]*Dt[x]/.{x->Log[z]}/.{Dt[z]->1}]
{i1=Map[Integrate[#1,z]&,intd], i2=i1/.z->Exp[x]}
f1[x_]:=1/(2+Sin[x]); f2[x_]:=Sin[x]^5;
{f1[x],f2[x],Integrate[1/(2+Sin[x]),x]//FullSimplify}
{intd1=f1[x]/.{Sin[x]->2*Tan[x/2]/(1+Tan[x/2]^2)},
 i1=Integrate[intd1,x]}
{intd2=Factor[intd1*Dt[x]/.{x->2*ArcTan[t]}/.{Dt[t]->1}],
 i2=Integrate[intd2,t],i3=ExpandAll[i2/.{t->Tan[x/2]}]}
{intd3=f2[x]/.{Sin[x]->-Cos[x+pi/2]},i4=Integrate[intd3,x]}
{i5=Integrate[intd3*Dt[x]/.{x->ArcCos[z]}/.{Dt[z]->1},z]/.
 {pi->Pi}, i6=FullSimplify[i5/.{z->Cos[x]}]}
```
□

Problem 7.23 Evaluate the integrals: $\int \ln x\, dx,\ \int e^{-2x} \cos 2x dx$.

Maple:

```
with(student): int(log(x),x);  f1:=x->log(x);
I1:=Int(f1(x),x); I2:=value(intparts(I1,log(x)));
f2:=x->cos(2*x)*exp(-2*x); I1:=Int(f2(x),x);
R1:=intparts(I1,exp(-2*x)); R2:=intparts(R1,exp(-2*x));
for i from 1 to nops(R2) do if type(op(i,R2),function)=true
 then I2:=op(i,R2); fi; od; I2; I3:=combine(I1-I2);
I4:=R2-I2; C:=op(1,I3); CC:=remove(has,C,x); I5:=I4/CC;
```

Mathematica:

```
f1[x_]:=Log[x]; g1[x_]:=1;
g2[x_]:=Exp[-2*x]; f2[x_]:=Cos[2*x]; {f1[x], g1[x],
 i1=Integrate[f1[x],x], i2=f1[x]*Integrate[g1[x],x]-
 Integrate[D[f1[x],x]*Integrate[g1[x],x],x], i1===i2,
 i3=Integrate[Cos[2*x]*Exp[-2*x],x], i4=f2[x]*Integrate[g2[x],x]-
 Integrate[D[f2[x],x]*Integrate[g2[x],x],x]//Simplify, i3===i4}
```
□

Problem 7.24 Approximate $\displaystyle\int_0^{\sqrt{\pi}} e^{-x} \cos(x)\, dx$ and evaluate various definite and improper integrals.

Maple:

```
f1:=x->exp(-x)*cos(x); I1:=factor(expand(int(f1(x),
 x=0..sqrt(Pi)))); evalf(I1); f2:=x->sqrt(1+x^2);
int(f2(x),x=0..1); int(1/(x^2+1),x=0..infinity);
int(x^(2*n),x=1..infinity) assuming n<-1;
int(cos(x)/(2*x),x=-2..5,'CauchyPrincipalValue');
evalf(int(cos(4*x)*exp(-4*x^2),x=-2*Pi..2*Pi),30);
int(1/(x-1),x=a..2,'AllSolutions'); g:=x->Int(f1(t),t=1..x);
plot(g(x),x=1..3); evalf(diff(g(x),x)); evalf(g(2));
combine(int(1/(x-8)/sqrt(x+1),x=0..3)); with(student):
f3:=x->1/(x-8)/sqrt(x+1); I1:=Int(f3(x),x=0..3);
Cv:=changevar(x+1=t^2,I1); I2:=combine(value(Cv));
```

Mathematica:

```
f1[x_]:=Exp[-x]*Cos[x];
f2[x_]:=Sqrt[1+x^2]; f3[x_]:=1/(x-8)/Sqrt[x+1];
{i1=Integrate[f1[x],{x,0,Sqrt[Pi]}], i2=i1//N,
 TrigToExp[Integrate[f2[x],{x,0,1}]], Integrate[1/(x^2+1),
 {x,0,Infinity}],Integrate[x^(2*n),{x,1,Infinity},
 Assumptions->{n<-1}], Integrate[Cos[x]/(2*x),{x,-2,5},
 PrincipalValue->True], NIntegrate[Cos[4*x]*Exp[-4*x^2],
 {x,-2*Pi,2*Pi},WorkingPrecision->30], Integrate[1/(x-1),
 {x,a,2},Assumptions->{a>1 && a<2}]}
g[x_]:=Integrate[f1[t],{t,1,x}]; Plot[g[x],{x,1,3}]
 {Simplify[g'[x]],g[2]//N}
{f3[x], Integrate[f3[x],{x,0,3}], i1=Integrate[
 f3[x]*Dt[x]/.{x->t^2-1}/.{Dt[t]->1},{t,1,Sqrt[4]}]}
```
□

Problem 7.25 Let $f(x) = \cos(\cos(\cos(x^2))) + \pi$. Graph the function $f(x)$ on the interval $[\pi/2, \pi]$, approximate $\int_{\pi/2}^{\pi} f(x)dx$ by Simpson's and trapezoidal rules ($n = 10$).

Maple:

```
with(Student[Calculus1]): n:=10; a:=Pi/2; b:=Pi;
f:=x->(cos@@3)(x^2)+Pi; plot(f(x),x=a..b,scaling=constrained,
 color=blue,thickness=5);
App_S:=ApproximateInt(f(x),x=a..b,method=simpson,partition=n);
App_T:=ApproximateInt(f(x),x=a..b,method=trapezoid,partition=n);
evalf(App_S); evalf(App_T); evalf(int(f(x),x=a..b));
```

Mathematica:

```
nD=10; f1[x_]:=Nest[Cos,x^2,3]+Pi;
Plot[f1[x],{x,Pi/2,Pi},PlotRange->All,
 PlotStyle->{Hue[0.7],Thickness[0.01]}]
a=Pi/2; b=Pi; n=10; N[Integrate[f1[x],{x,a,b}],nD]
dxTr=(b-a)/n; xTr[k_]:=a+k*dxTr;
apTr=N[1/2*dxTr*(f1[a]+2*Sum[f1[xTr[i]],{i,1,n-1}]+f1[b]),nD]
dxSim=(b-a)/(2*n); xSim[k_]:=a+k*dxSim;
apSim=N[dxSim/3*(f1[a]+2*Sum[f1[xSim[2*i]],{i,1,n-1}]+
 4*Sum[f1[xSim[2*i-1]],{i,1,n}]+f1[b]),nD]
```

☐

Problem 7.26 Approximate the area on the interval $[-4, 4]$ between the curves $p(x) = x^4 + x^3 - x^2 + x + 1$, $q(x) = -0.1x^3 + 20x^2 - 10x - 20$.

Maple:

```
p:=x^4+x^3-x^2+x+1; q:=-0.1*x^3+20*x^2-10*x-20;
plot([p,q],x=-4..4,color=[blue,green],thickness=3);
I_L:=[fsolve(p=q,x=-4..4)];
I1:=Int(p-q,x=I_L[1]..I_L[2]);
I2:=Int(q-p,x=I_L[2]..I_L[3]); evalf(I1+I2);
```

Mathematica:

```
p=x^4+x^3-x^2+x+1; q=-1/10*x^3+20*x^2-10*x-20;
Plot[{p,q},{x,-4,4},PlotStyle->{{Hue[0.6],Thickness[0.01]},
 {Hue[0.9],Thickness[0.015]}},Filling->{1->{2}}]
CountRoots[p-q,{x,-4,4}]
intervals=RootIntervals[p-q][[1]]
sols=Table[FindRoot[p==q,{x,intervals[[i,2]]}],{i,2,4}]
i1=Integrate[p-q,{x,sols[[1,1,2]],sols[[2,1,2]]}];
i2=Integrate[q-p,{x,sols[[2,1,2]],sols[[3,1,2]]}]; i1+i2//N
```

☐

Problem 7.27 Let $f(x) = \sin(\cos((x + \pi))$. Approximate the arc length of the graph of $f(x)$ from $(\pi, f(\pi))$ to $(5\pi/3, f(5\pi/3))$.

Maple:

```
f:=x->sin(cos(x+Pi)); plot(f(x),x=Pi..5*Pi/3,
 color=blue,thickness=3); df:=D(f)(x);
L_A:=evalf(Int(sqrt(1+df^2),x=Pi..5*Pi/3));
```

Mathematica:

```
nD=10; f[x_]:=Sin[Cos[x+Pi]];
Plot[f[x],{x,Pi,5*Pi/3},PlotStyle->{Hue[0.6],Thickness[0.01]}]
arcL=N[Integrate[Sqrt[1+f'[x]^2],{x,Pi,5*Pi/3}],nD]
```

□

Problem 7.28 Let $f(x) = \cos^2 x$ and D the domain bounded by the graphs of $y = f(x)$, $x = 0$, $x = \pi$. Find the volumes of the solids obtained by revolving the bounded region D and the x-axis about the y-axis.

Maple:

```
f:=x->cos(x)^2; setoptions(axes=boxed);
plot(f(x),x=0..Pi,color=blue,thickness=3,scaling=constrained);
plot3d([r*cos(t),r*sin(t),f(r)],r=0..Pi,t=0..2*Pi,grid=[40,40]);
plot3d([r,f(r)*cos(t),f(r)*sin(t)],r=0..Pi,t=0..2*Pi,
 grid=[40,40]); I_xy:=int(2*Pi*x*f(x),x=0..Pi);
```

Mathematica:

```
f[x_]:=Cos[x]^2; Plot[f[x],{x,0,Pi},
 PlotStyle->{Hue[0.9],Thickness[0.01]}]
GraphicsRow[{ParametricPlot3D[{r*Cos[t],r*Sin[t],f[r]},
 {r,0,Pi},{t,0,2*Pi},BoxRatios->{1,1,1}],
 ParametricPlot3D[{r,f[r]*Cos[t],f[r]*Sin[t]},{r,0,Pi},
 {t,0,2*Pi}]},ImageSize->500]
ixy=Integrate[2*Pi*x*f[x],{x,0,Pi}]
```

□

Problem 7.29 Analyze the convergence of the integral $\displaystyle\int_a^\infty \frac{dx}{x^\lambda}$, $a > 0$.

Maple:

```
interface(showassumed=0): assume(a>0);
int(x^(-lambda),x=a..infinity);
F:=(x,lambda)->1/(1-lambda)*x^(1-lambda);
L1:=limit(F(x,lambda),x=infinity) assuming lambda>1;
L2:=limit(F(x,lambda),x=infinity) assuming lambda<=1;
simplify(Int(x^(-lambda),x=a..infinity)=
 L1-F(a,lambda) assuming lambda>1);
```

Mathematica:

```
Integrate[x^(-lambda),{x,a,Infinity}]
F[x_,lambda_]:=1/(1-lambda)*x^(1-lambda); F[x,lambda]
L1=Assuming[lambda>1,Limit[F[x,lambda],x->Infinity]]
L2=Assuming[lambda<=0&&x!=0,Limit[F[x,lambda],x->Infinity]]
Assuming[lambda>1&&a>0,Simplify[HoldForm[Integrate[
 x^(-lambda), {x, a, Infinity}]]]==L1-F[a,lambda]]
```

$\square$

Problem 7.30 Prove that the improper integral $\displaystyle\int_1^\infty \frac{\sin x}{x^2}\, dx$ is absolute convergent.

Maple:

```
int(sin(x)/x^2,x=1..infinity);
abs(sin(x)/x^2)<=1/x^2; int(1/x^2,x=1..infinity);
```

Mathematica:

```
{Integrate[Sin[x]/x^2,{x,1,Infinity}],
Abs[Sin[x]/x^2]<=1/x^2, Integrate[1/x^2,{x,1,Infinity}]}
```

$\square$

Problem 7.31 Prove that the improper integral $\displaystyle\int_a^\infty \frac{\sin x}{x^\lambda}\, dx$ is convergent for $a > 0$ and $\lambda > 0$.

Maple:

```
int(sin(x)/(x^lambda),x=a..infinity); f:=x->sin(x);
g:=(x,lambda)->x^(-lambda);
(abs(int(sin(x),x=a..A)) assuming a>0)<=2;
limit(g(x,lambda),x=infinity) assuming lambda>0;
```

Mathematica:

```
f[x_]:=Sin[x]; g[x_,lambda_]:=x^(-lambda);
{Integrate[Sin[x]/(x^lambda),{x,a,Infinity}],f[x],g[x,lambda]}
Assuming[a>0 ,Abs[Integrate[Sin[x],{x,a,A}]]<=2]
Assuming[lambda>0, Limit[g[x,lambda],x->Infinity]]
```

$\square$

Problem 7.32 Evaluate the improper integral $\int_0^\infty \frac{\sin(ax)}{x}\,dx$ and the
Frullani integral $\int_0^\infty \frac{e^{-ax} - e^{-bx}}{x}\,dx$, where $a > 0$ and $b > 0$.

Maple:

```
int(sin(a*x)/x,x=0..infinity) assuming a>0;
with(inttrans): Intd:=laplace(sin(a*x),x,p)
 assuming a>0; int(Intd,p=0..infinity) assuming a>0;
int((exp(-a*x)-exp(-b*x))/x,x=0..infinity)
 assuming a>0 and b>0; L:=unapply((laplace(exp(-a*x),x,p)
 assuming a>0 and b>0),a,p); factor(int(L(a,p)-L(b,p),
 p=0..infinity) assuming a>0 and b>0);
```

Mathematica:

```
{Assuming[a>0, Integrate[Sin[a*x]/x,
 {x,0,Infinity}]], intd=Assuming[a>0,
 LaplaceTransform[Sin[a*x],x,p]], Assuming[a>0,
 Integrate[intd,{p,0,Infinity}]], Assuming[a>0 && b>0,
 Integrate[(Exp[-a*x]-Exp[-b*x])/x,{x,0,Infinity}]],
 lt=Assuming[a>0 && b>0, LaplaceTransform[Exp[-a*x],x,p]]}
lt1[A_,P_]:=lt/.{a->A, p->P}; lt1[a,p]
{i1=Factor[Assuming[a>0 && b>0,Integrate[lt1[a,p]-lt1[b,p],
 {p,0,Infinity}]]], Assuming[{a>0&&b>0},FunctionExpand[i1]]}
```

□

Problem 7.33 Let $\varphi(x)$ be an arbitrary continuous function such that
$\varphi(1) = 0$. Prove that the improper integral $\int_0^1 \frac{dx}{\varphi^2(x) + \sqrt{1-x}}$ is convergent and does not exceed 2.

Maple:

```
w:=curry(varphi); intd:=1/(w(x)^2+sqrt(1-x));
intd1:=subs(w(x)^2=0,intd); intd<intd1;
I1:=int(intd,x=0..1); I2:=int(intd1,x=0..1); I1<I2;
```

Mathematica:

```
{intd=1/(varphi[x]^2+Sqrt[1-x]),intd1=intd/.varphi[x]^2->0,
 intd<intd1, i1=Integrate[intd,{x,0,1}],
 i2=Integrate[intd1,{x,0,1}], i1<i2}
```

□

7.6   Series

Numerical, Functional, and Power Series

Maple:

```
sum(a[i],i=i1..infinity);            sum(fi(x),i=i1..infinity);
sum(a[i]*x^i,i=i1..infinity);    sum(a[i],i=i1..in); Order:=val;
add(a[i],i=i1..in); series(expr,x=x0,n);  convert(ser,polynom);
taylor(expr,x=x0,n);taylor(expr,x=0);        coeftayl(expr,eqn,k);
mtaylor(expr,[vars],n,ops);            poisson(expr,[vars],n,ops);
with(powseries): inverse(s);  reversion(s);    multiply(s1,s2);
```

sum, add, summation of finite and infinite symbolic and numeric series, summation of finite numeric series,

series, Order, generalized series expansion and the environment variable,

taylor, mtaylor, poisson, Taylor, Maclaurin, multivariate, and Poisson series expansion,

convert,polynom, a polynomial representation of series,

coeftayl, a coefficient in the (multivariate) Taylor series,

inverse, multiply (of the powerseries package), multiplicative inverse of a formal power series and multiplication of power series, etc.

Mathematica:

```
Sum[a[i],{i,i1,Infinity]          Sum[fi[x],{i,i1,Infinity}]
Sum[a[i]*x^i,{i,i1,Infinity]          Sum[a[i],{i,i1,in,iStep}]
Series[expr,{x,x0,n}]     Normal[ser]       Series[expr,{x,0,n}]
Series[f[x,y],{x,x0,n},{y,y0,m}]        SeriesCoefficient[ser,n]
SeriesCoefficient[f[x,y],{x,x0,n1},{y,y0,m1},...]    O[x-x0]^n
Series[f[x],{x,x0,n}]//InputForm  SeriesData[x,x0,{ai},n1,nm,d]
Series[f[x],{x,x0,n}]//TraditionalForm          InverseSeries[ser]
```

Sum, summation of finite and infinite symbolic and numeric series,

Series, generalized (multivariate) series expansion,

Normal, a polynomial representation of series,

InverseSeries, computing a series for the inverse of the function represented by ser,

SeriesCoefficient, finding series coefficients.

Problem 7.34 Evaluate the series sums

$$\sum_{i=1}^{5} i^2, \quad \sum_{k=1}^{n} \sin k, \quad \sum_{k=1}^{\infty} \frac{1}{k^2}, \quad \sum_{n=1}^{\infty} \frac{1}{2n^2+9n+10}, \quad \sum_{n=1}^{\infty} x^{kn}, \quad \sum_{n=1}^{10000} \frac{\cos n}{n}.$$

Maple:

```
sum(i^2,i=1..5); add(i^2,i=1..5); sum(sin(k),k=1..n);
sum(1/k^2,k=1..infinity); sum(1/(2*n^2+9*n+10),n=1..infinity);
sum(x^(k*n),n=1..infinity); Ser:=n->cos(n)/n;
Points:=[seq([i,Ser(i)],i=7000..10000)]:
plot(Points,style=POINT,color=blue,symbol=circle,
  symbolsize=10);  evalf(sum(Ser(i),i=1..10000));
```

Mathematica:

```
ser[n_]:=Cos[n]/n; {Sum[i^2,{i,1,5}],
 Sum[Sin[k],{k,1,n}], Sum[1/k^2,{k,1,Infinity}],
 Sum[1/(2*n^2+9*n+10),{n,1,Infinity}]//Simplify,
 Sum[x^(k*n),{n,1,Infinity}]]}
points=Table[{i,ser[i]},{i,7000,10000}]//N;
ListPlot[points,PlotStyle->{PointSize[0.002],Blue}]
Sum[ser[i],{i,1,10000}]//N
```

□

Problem 7.35 Let $\displaystyle\sum_{n=1}^{\infty} \frac{n^2}{2^n}, \sum_{n=1}^{\infty} \frac{2^n}{n!}$ be numerical series. Analyze the series convergence.

Maple:

```
S1:=n->n^2/(2^n); S2:=n->2^n/n!; I1:=expand(int(S1(i),i=1..n));
limit(I1,n=infinity); int(S1(i),i=1..infinity);
evalf(sum(S1(i),i=1..1000)); sum(S1(i),i=1..infinity);
Points:=[seq([i,S2(i)],i=1..50)]: plot(Points,style=POINT,
  color=blue,symbol=circle,symbolsize=20);
Fr:=simplify(S2(i+1)/S2(i)); limit(Fr,i=infinity);
sum(S2(i),i=1..infinity);
```

Mathematica:

```
nD=10; ser1[n_]:=n^2/(2^n);
ser2[n_]:=2^n/n!; {i1=Integrate[ser1[i],{i,1,n}]//ExpandAll,
 Limit[i1,n->Infinity]//ExpandAll, Integrate[ser1[i],
 {i,1,Infinity}]//ExpandAll, N[Sum[ser1[i],{i,1,1000}],nD],
 Sum[ser1[i],{i,1,Infinity}]}
points=Table[{i,ser2[i]},{i,1,50}]//N; ListPlot[points,
 PlotStyle->{PointSize[0.02],Blue},PlotRange->All]
{fr=FullSimplify[ser2[i+1]/ser2[i]], Limit[fr,i->Infinity],
 Sum[ser2[i],{i,1,Infinity}]}
```

$\square$

Problem 7.36 Find the interval of convergence of $\displaystyle\sum_{n=0}^{\infty} \frac{2^{2n}}{n^2+2}(x^2-2)^n$.

Maple:

```
Ser:=n->2^(2*n)/(n^2+2)*(x^2-2)^n; C1:=simplify(Ser(n+1)/Ser(n));
C2:=limit(C1,n=infinity); C3:=[solve(abs(C2)<1,x)]; C3[1]; C3[2];
```

Mathematica:

```
ser[n_]:=2^(2*n)/(n^2+2)*(x^2-2)^n;
{c1=FullSimplify[ser[n+1]/ser[n]], c2=Limit[c1,n->Infinity],
 c3=Reduce[Abs[c2]<1 && x\[Element]Reals,x],c3[[1]],c3[[2]]}
```

$\square$

Note. The condition x\[Element]Reals is equivalent to $x \in$ Reals.

Problem 7.37 Find the first n terms of the Taylor and Maclaurin series for $f(x)$ about $x = a$. Find the Maclaurin polynomial for the function $g(x) = \sin(x^2)\cos(x^2)/x^2$. Graph $g(x)$ along with its polynomial approximation on the interval $[-\pi, \pi]$.

Maple:

```
n:=10; series(f(x),x=a,n); series(f(x),x=0,n);
g:=x->sin(x^2)*cos(x^2)/x^2; Serg:=series(g(x),x=0,10);
g9:=unapply(convert(Serg,polynom),x); plot([g(x),g9(x)],
 x=-Pi..Pi,-1..1,color=[blue,green],thickness=[5,3]);
```

Mathematica:

```
n=10; g[x_]:=Sin[x^2]*Cos[x^2]/x^2;
{Series[f[x],{x,a,n}], Series[f[x],{x,0,n}],
 gs=Series[g[x],{x,0,7}]}
g7[t_]:=Normal[gs]/.{x->t};
Plot[Evaluate[{g[x],g7[x]}],{x,-Pi,Pi},PlotRange->
 {{-Pi,Pi},{-1,1}},AspectRatio->1,PlotStyle->
 {{Hue[0.3],Thickness[0.01]},{Hue[0.6],Thickness[0.01]}}]
```

□

Problem 7.38 Let $f(x) = \dfrac{x^2}{x^2+2}$. Approximate $f(x)$ on the interval

$[0,1]$ using the n-th degree Maclaurin polynomial and find the upper

bound of the error.

Maple:

```
n:=6; a:=0; b:=1; f:=x->x^2/(x^2+2);
PM:=proc(k) convert(series(f(x),x=0,k),polynom); end:
df:=proc(k) simplify(eval(diff(f(x),x$(k+1)),x=y)); end:
R:=k->df(k)*x^(k+1)/(k+1)!; PM(n); df(n); R(n);
plot(df(n),y=a..b); plot([f(x),PM(n)],x=a..b,
 color=[blue,green],thickness=[5,3]);
Sol:=[evalf(solve(diff(df(n),y)=0,y))];
ConstS:=select(type,evala(fnormal(Sol)),positive);
cMax:=max(seq(evalf(subs(y=ConstS[i],df(n))),
 i=1..nops(ConstS))); errS:=cMax*b^(n+1)/(n+1)!;
```

Mathematica:

```
nD=10; f[x_]:=x^2/(x^2+2);
pMaclaurin[n_]:=Module[{},Series[f[x],{x,0,n-2}]//Normal];
df[n_]:=Module[{},Simplify[D[f[x],{x,n+1}]/.{x->y}]];
r[n_]:=df[n]*x^(n+1)/(n+1)!;
{n=6, a=0, b=1, pM=pMaclaurin[n], df[n], r[n]}
Plot[df[n],{y,a,b},PlotStyle->{Red,Thickness[0.01]}]
Plot[{f[x],pM},{x,a,b},PlotStyle->
 {{Blue,Thickness[0.009]},{Green,Thickness[0.01]}}]
{sol=N[Solve[D[df[n],y]==0,y],nD], constS=Flatten[Select[
 sol,#[[1,2]]\[Element]Reals&&#[[1,2]]>0&]],
 cMax=Max[Table[N[df[n]/.constS[[i]]],nD],
 {i,1,Length[constS]}]], errS=cMax*b^(n+1)/(n+1)!}
```

□

Problem 7.39 Let $f(x) = \sin x$. Construct the Taylor polynomials of degrees $i = 1, \ldots, 10$ about $a = 1$ and compute their value at $x = \pi/12$ and the corresponding approximation error.

Maple:

```
f:=x->sin(x); exval:=evalf(f(Pi/12)); appval:=n->evalf(
 subs(x=Pi/12,convert(series(f(x),x=1,n),polynom)));
printf("    T(Pi/12)         Error");
for i from 2 to 11 do p1:=appval(i): p2:=evalf(abs(p1-exval)):
 printf(" %12.11f, %12.11f\n",p1,p2); od:
```

Mathematica:

```
nD=10; f[x_]:=Sin[x]; exval=f[Pi/12];
appval[n_]:=Normal[Series[f[x],{x,1,n}]]/.x->Pi/12;
points=N[Table[{appval[i],Abs[appval[i]-exval]},{i,1,10}],nD]
PaddedForm[TableForm[points,TableHeadings->
 {Automatic,{"T(Pi/12)","Error"}}],{nD,9}]
```

                                                                              □

Fourier Series

In Maple, there is no a single function for finding an expansion of a function in terms of a set of complete functions, or one of the simplest class of the Fourier expansions, an expansion in terms of the trigonometric functions

$$1, \quad \cos(x), \quad \sin(x), \quad \cos(2x), \quad \sin(2x), \quad \ldots,$$

or their complex equivalents $\phi_n = e^{-inx}$ $(n = 0, \pm 1, \pm 2, \ldots)$, which are complete and orthogonal on the interval $(-\pi, \pi)$ (or any interval of length 2π).

In Mathematica, with the package `FourierSeries` the Fourier exponential and trigonometric series, and the corresponding Fourier coefficients can be studied:

```
<<FourierSeries`
tsF[t_,n_]:=FourierTrigSeries[Sin[t],t,n]; tsF[t,10]
Plot[Evaluate[tsF[t,10]],{t,-Pi,Pi},PlotStyle->Blue]
```

Problem 7.40 For the set of variables $x, 2x, 3x, \ldots$ create the set of functions of the Fourier series.

Maple:

```
N:=20; var:=x;  L_N:=['i*x'$ 'i'=1..N]; Fourier_sin:=convert(
 map(sin,L_N),set); Fourier_cos:=convert(map(cos,L_N),set);
Fourier_F:=Fourier_sin union Fourier_cos;
```

Mathematica:

```
{n=20,var=x,nL=Table[i*x,{i,1,n}],sinFourier=Map[Sin,nL],
 cosFourier=Map[Cos,nL], fF=Union[sinFourier,cosFourier]}
```
□

Problem 7.41 Find the Fourier coefficients of $f = \left(a_0 + \sum_{i=1}^{4} a_i \sin(ix) \right)^4$.

Maple:

```
f:=(a[0]+add(a[i]*sin(i*x),i=1..4))^4; f:=combine(f,trig);
for i from 1 to 5 do  R1:=collect(f,sin(i*x)):
 R2:=coeff(R1,sin(i*x)): print(i,"the coefficient is",R2); od:
```

Mathematica:

```
f=TrigReduce[(Subscript[a,0]+Sum[Subscript[a,i]*Sin[i*x],
 {i,1,4}])^4//Expand]; For[i=1,i<=5,i++,r1=Collect[f,Sin[i*x]];
r2=Coefficient[r1,Sin[i*x]];Print[i," the coefficient is ",r2]]
```
□

Problem 7.42 Consider $f(x) = \begin{cases} 1, & 0 \le x \le 1 \\ -x, & -1 \le x < 0 \end{cases}$ with period 2.

Approximate the function $f(x)$ by the Fourier series. Compute and graph the first 20 partial sums of the Fourier series.

Maple:

```
f:=proc(x) if x>=0 and x<=1 then 1 elif  x<0 and x>=-1 then -x
 elif x<-1 then f(x+2) end if: end proc; f1:=x->-x; f2:=x->1:
N:=20; L:=1; a[0]:=evalf(1/(2*L)*(int(f1(x),x=-1..0)+
 int(f2(x),x=0..1))); for i from 1 to N do
 a[i]:=evalf(1/L*(int(f1(x)*cos(i*Pi*x/L),x=-1..0)+int(
 f2(x)*cos(i*Pi*x/L),x=0..1))); b[i]:=evalf(1/L*(int(f1(x)*
 sin(i*Pi*x/L),x=-1..0)+int(f2(x)*sin(i*Pi*x/L),x=0..1))); od:
Term_n:=n->a[n]*cos(n*Pi*x/L)+b[n]*sin(n*Pi*x/L);
Appr_f:=n->a[0]+add(Term_n(i),i=1..n);
plot([' f(x)',Appr_f(10)],x=-10..1,color=[green,blue],
 discont=true,thickness=[2,3],numpoints=200);
```

Mathematica:

```
f[x_]:=-x/;-1<=x<0; f[x_]:=1/;0<=x<=1; f[x_]:=f[x+2]/;x<-1;
f1[x_]:=-x; f2[x_]:=1; n=20; L=1;
a[0]=1/(2*L)*(Integrate[f1[x],{x,-1,0}]+
 Integrate[f2[x],{x,0,1}])//N; For[i=1,i<=n,i++,
 a[i]=1/L*(Integrate[f1[x]*Cos[i*Pi*x/L],{x,-1,0}]+
  Integrate[f2[x]*Cos[i*Pi*x/L],{x,0,1}])//N;
 b[i]=1/L*(Integrate[f1[x]*Sin[i*Pi*x/L],{x,-1,0}]+
  Integrate[f2[x]*Sin[i*Pi*x/L],{x,0,1}])//N];
termN[n_]:=a[n]*Cos[n*Pi*x/L]+b[n]*Sin[n*Pi*x/L];
apprF[n_]:=a[0]+Sum[termN[i],{i,1,n}];
Plot[{f[x],apprF[10]},{x,-10,1},PlotStyle->{Blue,Green}]
```

□

Problem 7.43 Approximate the square wave waveform of period 2π.

Maple:

```
with(plots): setoptions(plot,thickness=3); a:=0; b:=2*Pi;
h:=Heaviside(t-Pi); A:=i->1/Pi*int(h*cos(i*t),t=a..b);
B:=i->1/Pi*int(h*sin(i*t),t=a..b); A0:=1/(2*Pi)*int(h,t=a..b);
F:=n->evalf(A0+sum(B(i)*sin(i*t),i=1..n)); G1:=plot(h,t=a..b,
 scaling=constrained,discont=true): G2:=plot(F(30),t=a..b,
 scaling=constrained,color=blue): display({G1,G2});
```

Mathematica:

```
h[t_]:=UnitStep[t-Pi];
A[i_]:=1/Pi*Integrate[h[t]*Cos[i*t],{t,0,2*Pi}];
B[i_]:=1/Pi*Integrate[h[t]*Sin[i*t],{t,0,2*Pi}];
A0=1/(2*Pi)*Integrate[h[t],{t,0,2*Pi}];
F[n_]:=A0+Sum[B[i]*Sin[i*t],{i,1,n}]//N;
g1=Plot[h[t],{t,0,2*Pi},PlotStyle->{Hue[0.9],
 Thickness[0.01]}]; g2=Plot[Evaluate[F[30]],{t,0,2*Pi},
 PlotStyle->Hue[0.7]]; Show[g1,g2]
```

□

7.7 Multivariate and Vector Calculus

In Maple, a variety of functions for performing multivariate and vector
calculus are provided in several packages, Student, LinearAlgebra,
VectorCalculus, and subpackages, LinearAlgebra, VectorCalculus,
MultivariateCalculus (of the Student package).

In Mathematica, a variety of functions for performing calculus in various three-dimensional coordinate systems are provided in the package VectorAnalysis.

Maple:

```
with(LinearAlgebra);with(VectorCalculus); f:=(x,y)->expr; D(f);
D[i](f);  D[i,j](f); D[i](D[j,i](f));  diff(f(x1,...,xn),x1$n);
diff(f(x1,...,xn),x1,...,xn);  int(...int(int(f,x1),x2)...,xn);
with(Student[MultivariateCalculus]); MultiInt(f,xdom,ydom,ops);
extrema(f,cnstr,vars);  LagrangeMultipliers(f,cnstr,vars,ops);
SetCoordinates(v,coordsys);      VF:=VectorField(v,coordsys);
Curl(VF); Gradient(f,vars); Nabla(f,vars);    Divergence(VF);
Laplacian(f,vars);    SurfaceInt(f,dom);    Jacobian(f,vars);
DotProduct(v1,v2);    v1.v2;   v1 &x v2;   CrossProduct(v1,v2);
```

diff, Diff, D, partial differentiation, differential operator,

Doubleint, Tripleint, int, Int, multiple and iterative integrals,

extrema, maximum/minimum, Maximize/Minimize (of the Optimization package), finding relative extrema, maximum and minimum of the expression,

the VectorCalculus package with the vector calculus functions, Gradient, Nabla, Laplacian, VectorField, Divergence, DotProduct, ., Curl, CrossProduct, &x, Jacobian, etc.

Mathematica:

```
<<VectorAnalysis`        f[x_,...,xn_]:=expr;      D[f,x1,...,xn]
D[f,{x1,k1},...{xn,kn}] Derivative[k1,...,kn][f]       Dt[f[x,y]]
Dt[f[x,y],x]            Integrate[f[x,y],{x,x1,x2},{y,y1,y2},ops]
Integrate[f[x1,...,xn], {x1,xMin,xMax},...,{xn,xMin,xMax}, ops]
NIntegrate[f[x1,...,xn],{x1,xMin,xMax},...,{xn,xMin,xMax}, ops]
SetCoordinates[Coord[vars]]    Grad[f[vars]] Laplacian[f[vars]]
Curl[f[vars]]  Div[f[vars]]    v1.v2          DotProduct[v1,v2]
CrossProduct[v1,v2,coordsys]       JacobianMatrix[point,coordsys]
```

D, Derivative, Dt, partial derivatives and the total differential and derivative,

Integrate, NIntegrate, analytical integration (indefinite, definite, multiple) and numerical approximations of integrals,

VectorAnalysis, the package with the vector calculus functions.

Problem 7.44 Let $f(x,y) = x^2y^2/(x^2y^2 + (x-y)^2)$. Graph $f(x,y)$ and find the second order partial derivatives.

Maple:

```
f:=(x,y)->x^2*y^2/(x^2*y^2+(x-y)^2); plot3d(f(x,y),x=-1..1,
 y=-1..1,grid=[40,40],axes=boxed,shading=Z,orientation=[-40,55]);
simplify([diff(f(x,y),y,x), D[2,1](f)(x,y), diff(f(x,y),x,y),
 D[1,2](f)(x,y), diff(f(x,y),x$2),D[1,1](f)(x,y),
 diff(f(x,y),y$2), D[2,2](f)(x,y)]);
```

Mathematica:

```
f[x_,y_]:=x^2*y^2/(x^2*y^2+(x-y)^2);
Plot3D[f[x,y],{x,-1,1},{y,-1,1},BoxRatios->{1,1,1}]
{D[f[x,y],y,x], D[f[x,y],x,y], D[f[x,y],{x,2}],
 D[f[x,y],{y,2}]}//Simplify
f1=Derivative[1,1][f]; f2=Derivative[2,0][f];
f3=Derivative[0,2][f]; Map[Simplify,{f1[x,y],f2[x,y],f3[x,y]}]
```
                                                                        □

Problem 7.45 Let $f(x,y) = x^2\cos(y^2)$. Evaluate f_{xy} at the point $(-\pi, \pi)$, find the total differential, and total derivatives with respect to x and y.

Maple:

```
f:=(x,y)->x^2*cos(y^2); subs({x=-Pi,y=Pi},diff(f(x,y),x,y));
convert(D[1,2](f)(-Pi,Pi),diff);
PDEtools[declare](x(t),y(t),Dt=t);
tDif:=simplify(diff(f(x(t),y(t)),t));
TotDerX:=factor(Dt(y,x)*diff(f(x,y),y)+diff(f(x,y),x));
TotDerY:=factor(Dt(x,y)*diff(f(x,y),x)+diff(f(x,y),y));
```

Mathematica:

```
f[x_,y_]:=x^2*Cos[y^2];
{D[f[x,y],x,y]/.{x->-Pi,y->Pi},Derivative[1,1][f][-Pi,Pi],
 Dt[f[x,y]], Dt[f[x,y],x], Dt[f[x,y],y]}// Simplify
```
                                                                        □

Problem 7.46 For $z = x^2 y^2$, where $y = x^2$, evaluate $\dfrac{dz}{dx}$, for x^{2k} evaluate the total differential.

Maple:

```
with(difforms): z:=x^2*y^2; z1:=subs(y=x^2,z);
eval(subs(d(x)=1,d(z1))); defform(k=scalar); z2:=(x)^(2*k);
z3:=simplify(subs(d(k)=0,d(z2)));
```

Mathematica:

```
z=x^2*y^2; {Dt[z,x]/.{y->x^2}, Dt[x^{2*k},Constants->{k}]}
```

☐

Problem 7.47 Find and classify the critical points of the function $f(x, y) = x^4 + 10x^3 + 2x^2 y^2 + xy^2$.

Maple:

```
f:=(x,y)->x^4+10*x^3+2*x^2*y^2+x*y^2; df_x:=diff(f(x,y),x);
df_y:=diff(f(x,y),y); sols:=allvalues({solve({df_x=0,df_y=0},
 {x,y})}); P_cr:=evalf(convert(map(`union`,{op(sols[1]),
 op(sols[2])}),list)); N:=nops(P_cr); df_xx:=diff(f(x,y),x$2);
df_yy :=diff(f(x,y),y$2); df_xy:=diff(f(x,y),x,y);
d:=simplify(df_xx*df_yy-(df_xy)^2); L:=[x,y,df_xx,d];
Classification:=array([seq(subs(P_cr[i],L),i=1..N)]);
```

Mathematica:

```
nD=10; f[x_,y_]:=x^4+10*x^3+2*x^2*y^2+x*y^2;
{dfx=D[f[x,y],x], dfy=D[f[x,y],y], sols=N[Solve[{dfx==0,dfy==0},
 {x,y}],nD], pcr=Union[sols,{}], n=Length[pcr]}
d=D[f[x,y],{x,2}]*D[f[x,y],{y,2}]-(D[f[x,y],x,y])^2//Simplify
crlist={x,y,D[f[x,y],{x,2}],d}
classification=PaddedForm[Table[crlist/.pcr[[i]],
 {i,1,n}]//TableForm,{12,5}]
```

☐

Problem 7.48 Let $f(x, y) = \exp(-2(x^2 + y^2))$. Find an equation of the tangent plane to the graph of $f(x, y)$ at $(-1, 1)$.

Maple:

```
with(plots): f:=(x,y)->exp(-2*(x^2+y^2)); d_x:=D[1](f)(-1,1);
d_y:=D[2](f)(-1,1); EqTP:=evalf(d_x*(x+1)+d_y*(y-1)+f(-1,1));
setoptions3d(grid=[40,40],axes=boxed,orientation=[68,72]);
xR:=-5..5: yR:=-5..5: G1:=plot3d(f(x,y),x=xR,y=yR,shading=Z):
G2:=plot3d(EqTP,x=xR,y=yR,shading=zhue):with(VectorCalculus):
SetCoordinates('cartesian'[x,y,z]);
tp:=TangentPlane(f(x,y),x=-1,y=1); G3:=plot3d(tp,x=xR,y=yR,
 shading=zhue): display({G1,G2}); display({G1,G3});
```

Mathematica:

```
nD=10; f[x_,y_]:=Exp[-2*(x^2+y^2)];
dx=Derivative[1,0][f][-1,1]; dy=Derivative[0,1][f][-1,1];
Eqtp=N[dx*(x+1)+dy*(y-1)+f[-1,1],nD]//Simplify
SetOptions[Plot3D,BoxRatios->{1,1,1},PlotRange->All];
g1=Plot3D[f[x,y],{x,-5,5},{y,-5,5}]; g2=Plot3D[Eqtp,
 {x,-5,5},{y,-5,5}]; Show[{g1,g2}]
```

□

Problem 7.49 Use the method of Lagrange multipliers to find the maximum and minimum values of $f(x,y) = 2x^2 - 3y^3 + 4(xy)^2$ subject to the constraint $x^2 + 2y^2 = 1$.

Maple:

```
f:=(x,y)->2*x^2-3*y^3+4*x^2*y^2; Cnstr:=(x,y)->x^2+y^2-1;
with(plots): with(Student[MultivariateCalculus]):
Ex:=evalf(extrema(f(x,y),Cnstr(x,y),{x,y},'Sol'));
LSol:=convert(allvalues(Sol),list); N:=nops(LSol);
L:=[x,y,f(x,y)]; array([seq(subs(evalf(LSol[i]),L),i=1..N)]);
C1:=spacecurve([cos(t),sin(t),0],t=0..2*Pi,color=blue,
 thickness=3): C2:=spacecurve([cos(t),sin(t),f(cos(t),sin(t))],
 t=0..2*Pi,color=magenta,thickness=5,orientation=[15,66]):
display3d([C1,C2],axes=boxed);
LMs:=LagrangeMultipliers(f(x,y),[Cnstr(x,y)],[x,y]);
LMNum:=LMs[1..4],evalf(allvalues(LMs[5])); evalf(LSol);
with(Optimization); Minimize(f(x,y),{Cnstr(x,y)=0});
Maximize(f(x,y),{Cnstr(x,y)=0});
```

Mathematica:

```
nD=10; f[x_,y_]:=2*x^2-3*y^3+4*x^2*y^2;
g[x_,y_]:=x^2+y^2-1; conds=Eliminate[
 {D[f[x,y],x]==lambda*D[g[x,y],x],D[f[x,y],y]==lambda*
  D[g[x,y],y],g[x,y]==0},lambda]
{points=Solve[conds]//Simplify, fVal=f[x,y]/.points}
{N[Max[fVal],nD], N[Min[fVal],nD], n=Length[points]}
l1={x,y,f[x,y]}; l2=PaddedForm[Table[l1/.points[[i]],
 {i,1,n}]//N//TableForm,{12,5}]
c1=ParametricPlot3D[{Cos[t],Sin[t],0},{t,0,2*Pi},PlotRange->All,
 PlotStyle->{Blue,Thickness[0.01]}]; c2=ParametricPlot3D[{Cos[t],
 Sin[t],f[Cos[t],Sin[t]]},{t,0,2*Pi},PlotRange->All,PlotStyle->
 {Magenta,Thickness[0.01]}]; Show[{c1,c2},BoxRatios->1]
```
□

Problem 7.50 Evaluate the integrals:

$$\int_0^1 \int_0^1 \sin(\cos xy)dxdy, \qquad \int_0^\pi \int_0^\pi \int_0^1 e^{2xz}\cos(x^2-y^2)dzdxdy.$$

Maple:

```
Ri1:=0..1; Ri2:=0..Pi; evalf(int(int(sin(cos(x*y)),
 x=Ri1),y=Ri1)); evalf(Int(Int(Int(exp(2*x*z)*cos(x^2-y^2),
 z=Ri1),x=Ri2),y=Ri2));
```

Mathematica:

```
Integrate[Sin[Cos[x*y]],{y,0,1},{x,0,1}]//N
Integrate[Exp[2*x*z]*Cos[x^2-y^2],{y,0,Pi},{x,0,Pi},{z,0,1}]//N
```
□

Problem 7.51 Evaluate the integral

$$I = \iint_D \frac{xy\,dx\,dy}{(ax+by)^3}, \qquad D = \{0 \le x \le 1,\ 1 \le y \le 2\},\ a > 0,\ b > 0.$$

Maple:

```
with(Student[MultivariateCalculus]):
xR:=0..1; yR:=1..2; f:=(x,y)->x*y/(a*x+b*y)^3;
MultiInt(f(x,y),x=xR,y=yR,output=integral)=combine(MultiInt(
 x*y/(a*x+b*y)^3,x=xR,y=yR) assuming a>0 and b>0);
with(student): Doubleint(f(x,y),x,y,D)=combine(value(
Doubleint(x*y/(a*x+b*y)^3,x=xR,y=yR)) assuming a>0 and b>0);
```

Mathematica:

```
Assuming[a>0 && b>0, Integrate[Integrate[
  (x*y)/(a*x+b*y)^3,{x,0,1}],{y,1,2}]]//Factor
```

□

Problem 7.52 Let $D(t)$ be a deformable domain plane bounded by an ellipse $x^2/a^2(t) + y^2/b^2(t) = 1$ and let $f(x, y, t) = 1$. Evaluate the derivative of the double integral with respect to t:

$$\frac{d}{dt} \iint_{D(t)} f(x, y, t)\, dx\, dy = \iint_{D(t)} f_t(x, y, t)\, dx\, dy + \iint_{L(t)} (\mathbf{n} \cdot \mathbf{v}) f(x, y, t)\, dl,$$

where $L(t)$ is the boundary of the domain $D(t)$, $\mathbf{n}$ is the outer unit normal to $L(t)$, and $\mathbf{v}$ is the velocity of motion of the points of $L(t)$.

Maple:

```
X:=(lambda,t)->a(t)*cos(lambda); Y:=(lambda,t)->b(t)*sin(lambda);
I1:=Int((N.V)*f(x,y,t),l=L(t)..NULL)=int(f(X(lambda,t),
  Y(lambda,t),t)*(diff(Y(lambda,t),lambda)*diff(X(lambda,t),t)-
  diff(X(lambda,t),lambda)*diff(Y(lambda,t),t)),lambda=0..2*Pi);
I2:=subs(f(X(lambda,t),Y(lambda,t),t)=1,op(2,I1));
with(student): Diff(Doubleint(f(x,y,t),x,y,`D(t)`),t)=
  factor(value(I2));
```

Mathematica:

```
X[lambda_,t_]:=a[t]*Cos[lambda];
Y[lambda_,t_]:=b[t]*Sin[lambda]; {X[lambda,x],Y[lambda,x]}
i1=Integrate[f[X[lambda,t],Y[lambda,t],t]*(D[Y[lambda,t],
  lambda]*D[X[lambda,t],t]-D[X[lambda,t],lambda]*
  D[Y[lambda,t],t]),{lambda,0,2*Pi}]
i2=Factor[i1/.{f[X[lambda,t],Y[lambda,t],t]->1}]
```

□

Problem 7.53 Let $x^2/a^2 + y^2/b^2 = 1$, $0 \le z \le h$, be a bounded homogeneous elliptic cylinder. Determine the moment of inertia about the z-axis, $I_z = \gamma \iiint_D (x^2 + y^2)\, dx\, dy\, dz$.

Maple:

```
with(student): I1:=gamma*Tripleint((x^2+y^2),x,y,z,U);
I2:=factor(changevar({x=a*rho*cos(phi),y=b*rho*sin(phi),z=z},
  Tripleint(x^2+y^2,x,y,z),[rho,phi,z]) assuming a>0 and b>0
  and rho>=0 and rho<=1 and phi>=0 and phi<=2*Pi and z>=0 and
  z<=h and h>0); Intd:=op(1,op(1,op(1,I2)));
I3:=gamma*Tripleint((x^2+y^2),x,y,z,D)=factor(gamma*
  int(int(int(Intd,rho=0..1),phi=0..2*Pi),z=0..h));
```

Mathematica:

```
i1=Assuming[a>0&&b>0&&\[Rho]>=0&&\[Rho]<=1&&\[Phi]>0&&
  \[Phi]<=2*Pi&&z>=0&&z<=h&&h>0, \[Gamma]*Integrate[
  (x^2+y^2)*Dt[y]*Dt[x]/.{x->a*\[Rho]*Cos[\[Phi]],
  y->b*\[Rho]*Sin[\[Phi]]}/.{Dt[\[Rho]]->1,Dt[\[Phi]]->1,
  Dt[a]->0,Dt[a]->0},{z,0,h},{\[Phi],0,2*Pi},{\[Rho],0,1}]]
id=-a*b*\[Rho]^3*(-a^2-b^2*Sin[\[Phi]]^2+a^2*Sin[\[Phi]]^2)
HoldForm[\[Gamma]*Integrate[id,{z,0,h},{\[Phi],0,2*Pi},
  {\[Rho],0,1}]]==Factor[\[Gamma]*Integrate[id,{z,0,h},
  {\[Phi],0,2*Pi},{\[Rho],0,1}]]
```

□

Problem 7.54 Evaluate the integral $\int_{AB} xy\,dl$, where AB is a quarter of an ellipse with semiaxes a and b.

Maple:

```
x:=t->a*cos(t); y:=t->b*sin(t); I1:=Int(f(x,y),l=AB..NULL)=
  Int(x(t)*y(t)*sqrt(diff(x(t),t)^2+diff(y(t),t)^2),t=0..Pi/2);
I2:=Int(f(x,y),l=AB..NULL)=value(op(2,I1)) assuming a>0 and b>0;
```

Mathematica:

```
x[t_]:=a*Cos[t]; y[t_]:=b*Sin[t]; {x[t],y[t]}
i1=Assuming[a>0 && b>0, HoldForm[Integrate[f[x,y],
  {l,AB," "}]]==Factor[Integrate[x[t]*y[t]*Sqrt[D[x[t],t]^2+
  D[y[t],t]^2],{t,0,Pi/2}]]]
```

□

Problem 7.55 Find the volume of the region bounded by the graphs of the surfaces $f(x,y) = x + y$ and $g(x,y) = 10 - 2x^2 - 2y^2$ on the rectangle $[-4,4] \times [-4,4]$.

Maple:

```
with(plots): f:=(x,y)->x+y;  g:=(x,y)->10-2*x^2-2*y^2;
setoptions3d(axes=boxed,grid=[40,40],orientation=[-21,53]);
Gf:=plot3d(f(x,y),x=-4..4,y=-4..4,color=blue):
Gg:=plot3d(g(x,y),x=-4..4,y=-4..4,color=green):
display3d({Gf,Gg}); y0:=[solve(f(x,y)=g(x,y),y)];
x0:=[solve(op(2,y0[1])=0,x)];
Vol:=evalf(Int(Int(g(x,y)-f(x,y),y=y0[2]..y0[1]),
 x=x0[1]..x0[2]));
```

Mathematica:

```
nD=10; f[x_,y_]:=x+y; g[x_,y_]:=10-2*x^2-2*y^2;
SetOptions[Plot3D,PlotRange->All,BoxRatios->{1,1,1}];
gf=Plot3D[f[x,y],{x,-4,4},{y,-4,4}];
gg=Plot3D[g[x,y],{x,-4,4},{y,-4,4}]; Show[gf,gg]
{y0=Solve[f[x,y]==g[x,y],y], t=y0[[1,1,2,2,2]],
 x0=Solve[t==0,x]}
Vol=N[Integrate[g[x,y]-f[x,y],{x,x0[[1,1,2]],x0[[2,1,2]]},
 {y,y0[[1,1,2]],y0[[2,1,2]]}],nD]
```

$\square$

Problem 7.56 Let $f(x,y,z) = x\sin^2(xyz)$. Find ∇f, $\nabla^2 f$, div (∇f).

Maple:

```
f:=(x,y,z)->x*sin(x*y*z)^2;  vars:=[x,y,z];
with(VectorCalculus): with(linalg):
grad1:=grad(f(x,y,z),vars); grad2:=Gradient(f(x,y,z),vars);
Lap:=Laplacian(f(x,y,z),vars); DivGr:=Divergence(grad2,vars);
```

Mathematica:

```
<<VectorAnalysis`
f[x_,y_,z_]:=x*Sin[x*y*z]^2;
gradf={D[f[x,y,z],x],D[f[x,y,z],y],D[f[x,y,z],z]}
SetCoordinates[Cartesian[x,y,z]]
{Grad[f[x,y,z]],Laplacian[f[x,y,z]],Div[Grad[f[x,y,z]]]}
```

$\square$

Problem 7.57 Let $f(x,y,z) = -(xy)^2\,\mathbf{i} + \cos^2(xyz)\,\mathbf{j} + \sin^2 z\,\mathbf{k}$. Find curl f, div f, $\Delta(\text{div}\,f)$, grad $(\Delta(\text{div}\,f))$.

Maple:

```
with(VectorCalculus): SetCoordinates('cartesian'[x,y,z]);
f:=VectorField(<-(x*y)^2,cos(x*y*z)^2,sin(z)^2>);
curl_f:=map(factor,Curl(f)); div_f:=combine(Divergence(f));
lap_div_f:=combine(Laplacian(div_f));
grad_lap_div_f:=map(simplify,Gradient(lap_div_f));
```

Mathematica:

```
<<VectorAnalysis`
f={-(x*y)^2,Cos[x*y*z]^2,Sin[z]^2};
SetCoordinates[Cartesian[x,y,z]]; {Curl[f],Div[f],
Laplacian[Div[f]],Grad[Laplacian[Div[f]]]}//TrigReduce
```

$\square$

Problem 7.58 Let $f(x,y) = \sin^2(xy) + 2\cos(xy)^3$. Find the formula of a unit normal vector to the graph of $f(x,y)$ at $(x,y,f(x,y))$.

Maple:

```
f:=(x,y)->sin(x*y)^2+2*cos(x*y)^3;
vars:=[x,y,z]; fz:=(x,y,z)->z-f(x,y);
with(VectorCalculus): SetCoordinates('cartesian'[x,y,z]);
grad_f:=expand(Gradient(fz(x,y,z),vars));
norm_grad_f:=expand(sqrt((grad_f).(grad_f)));
norm_f:=map(expand,grad_f/norm_grad_f);
plot3d(f(x,y),x=-5..5,y=-5..5,axes=boxed,grid=[40,40],
  shading=zhue,scaling=constrained,orientation=[-13,36]);
```

Mathematica:

```
<<VectorAnalysis`
f[x_,y_]:=Sin[x*y]^2+2*Cos[x*y]^3; fz[x_,y_,z_]:=z-f[x,y];
Map[ExpandAll,{gradf=Grad[fz[x,y,z],Cartesian[x,y,z]],
  normgradf=Sqrt[gradf.gradf],normf=gradf/normgradf}]
Plot3D[f[x,y],{x,-5,5},{y,-5,5},BoxRatios->{2,2,1}]
```

$\square$

Problem 7.59 Evaluate the integral $\oint_C (3y - e^{\sin x})dx + (7x + \sqrt{y^4+1})dy$, where C is $x^2 + y^2 = 9$.

Maple:

```
with(plots):setoptions(axes=boxed,thickness=5,numpoints=400);
P:=(x,y)->3*y-exp(sin(x)); Q:=(x,y)->7*x+sqrt(y^4+1);
P_y:=diff(P(x,y),y); Q_x:=diff(Q(x,y), x); Rt:=0..2*Pi;
Fp:=[3*cos(t)]; I1:=int(int((Q_x-P_y)*r,r=0..3),theta=0..2*Pi);
G1:=polarplot(Fp,t=Rt,coords=polar,color=blue):
G2:=polarplot(Fp,t=Rt,coords=polar,color=grey,filled=true):
display([G1,G2]);
```

Mathematica:

```
p[x_,y_]:=3*y-Exp[Sin[x]];
q[x_,y_]:=7*x+Sqrt[y^4+1]; py=D[p[x,y],y]; qx=D[q[x,y],x];
i1=Integrate[(qx-py)*r,{r,0,3},{theta,0,2*Pi}]
RegionPlot[x^2+y^2<=9,{x,-3,3},{y,-3,3},BoundaryStyle->
 {Blue,Thickness[0.02]},PlotStyle->{Hue[0.8],Thickness[0.04]}]
g1=PolarPlot[{3*Cos[t]},{t,0,2*Pi},
 PlotStyle->{Thickness[0.04]}]; g2=Graphics[{Hue[0.8],
 Disk[{1.5,0},1.5]}]; Show[g1,g2,Frame->True,Axes->False]
```

$\square$

Problem 7.60 Calculate the external flow of the vector field $F(x,y,z) = (xy + x^2yz)\,\mathbf{i} + (yz + xy^2z)\,\mathbf{j} + (xz + xyz^2)\,\mathbf{k}$ through the surface of the first octant bounded by the planes $x = 2$, $y = 2$, $z = 2$.

Maple:

```
F:=(x,y,z)-><x*y+x^2*y*z,y*z+x*y^2*z,x*z+x*y*z^2>;
with(VectorCalculus): SetCoordinates('cartesian'[x,y,z]);
f:=VectorField(F(x,y,z)); div_F:=Divergence(f);
int(int(int(div_F,z=0..2),y=0..2),x=0..2);
```

Mathematica:

```
<<VectorAnalysis`
SetCoordinates[Cartesian[x,y,z]];
F[x_,y_,z_]:={x*y+x^2*y*z,y*z+x*y^2*z,x*z+x*y*z^2};
{divF=Div[F[x,y,z]], Integrate[divF,{z,0,2},{y,0,2},{x,0,2}]}
```

$\square$

Chapter 8

Complex Functions

8.1 Complex Algebra

Maple and Mathematica perform complex arithmetic automatically, all operations are performed by assuming that the basic number system is the *complex field* $\mathbb{C}$. In both systems, the imaginary unit i of the complex number x+y*I is denoted by I.

Complex numbers and variables

Maple:

```
abs(z);    Re(z); Im(z);  conjugate(z); argument(z);  evalc(z);
signum(z); csgn(z); polar(z); polar(r,theta); convert(z,polar);
with(RandomTools):      Generate(complex(integer(range=a..b)));
```

abs, Re,Im, the absolute value and the real and imaginary parts,

conjugate, argument, evalc, the complex conjugate, the complex argument, and complex evaluation function,

signum, csgn, the sign of a real or complex number and the sign function for complex expressions,

polar, convert,polar, polar representation of complex numbers and rewriting an expression in polar form,

Generate,complex (of the RandomTools package), generating pseudorandom complex numbers.

I.K. Shingareva, C. Lizárraga-Celaya, *Maple and Mathematica*, 2nd ed.,
DOI 10.1007/978-3-211-99432-0_8, © Springer-Verlag Vienna 2009

```
z1:=5+I*7; abs(z1);  Re(z1);  Im(z1); conjugate(z1),
argument(z1); signum(z1); csgn(z1); polar(z1);
z2:=x+I*y; evalc(abs(z2));  evalc(Re(z2));
assume(x,real); assume(y,real); about(x,y); Re(z2); Im(z2);
unassign('x','y'); about(x,y); evalc(signum(z2));
evalc(polar(r,theta)); map(evalc,convert(z2, polar));
expand((1-I)^4); evalc(sqrt(-9)); expand((3+I)/(4-I));
with(RandomTools): Generate(list(complex(float(range=1..9)),9));
interface(showassumed=0): assume(k,complex);
f:=x->exp(-(Re(k)+I*Im(k))*(a+I*c*t)); f(x);
simplify(evalc(convert(f(x),polar)));
```

Mathematica:

```
Abs[z]    Re[z]    Im[z]              Conjugate[z]  Arg[z] Sign[z]
ComplexExpand[expr]                   ComplexExpand[expr,{z1,...,zn}]
Element[z,Complexes]  z\[Element]Complexes      ComplexInfinity
RandomComplex[] RandomComplex[{z1,z2}] RandomComplex[{z1,z2},n]
```

Abs, Re, Im, the absolute value and the real and imaginary parts,

Conjugate, Arg, the complex conjugate and the complex argument,

Sign, the sign of a real or complex number,

ComplexExpand, complex expansion function, the trigonometric form of complex numbers,

Complexes, the domain of complex numbers,

RandomComplex, generating pseudorandom complex numbers.

```
z1=5+I*7; z2=x+I*y; f[x_]:=Exp[-k*(a+I*c*t)];
{Abs[z1],Re[z1],Im[z1],Conjugate[z1],Arg[z1],Sign[z1]}
ComplexExpand[z,{z},TargetFunctions->{Arg,Abs}]
{Abs[z2],Re[z2],Im[z2]}//ComplexExpand
{(1-I)^4,Sqrt[-9],(3+I)/(4-I)}//ComplexExpand
{ComplexExpand[f[x],k], ComplexExpand[f[x],k,
 TargetFunctions->{Abs,Arg}], ComplexExpand[f[x],k,
 TargetFunctions->{Abs,Arg}]/.{Abs[k]->\[Alpha],Arg[k]->\[Phi]}}
{RandomComplex[{1,9},9],RandomComplex[{1,9},{3,3}]//MatrixForm}
```

Problem 8.1 Let $z = a + ib$ be a complex number. Rewrite z in polar, exponential, and trigonometric forms.

Maple:

```
z:=3+4*I; zPolar:=polar(z); rho1:=op(1,zPolar);
phi1:=op(2,zPolar); zExp:=rho*exp(I*theta);
zTrig:=convert(zExp,trig); subs({rho=rho1,theta=phi1},zTrig);
```

Mathematica:

```
{z1=3+4*I, zPolar={Abs[z1],Arg[z1]}, rho1=zPolar[[1]],
 theta1=zPolar[[2]],zExp=rho*Exp[I*theta],zTrig=ExpToTrig[zExp],
 zTrig/.{rho->rho1}/.{theta->InputForm[theta1]},
 ComplexExpand[z,{z},TargetFunctions->{Abs,Arg}]}
```
                                                                                      □

Problem 8.2   Prove that $\cos(3\theta) = \cos(\theta)^3 - 3\cos(\theta)\sin(\theta)^2$ (use De Moivre's theorem).

Maple:

```
interface(showassumed=0): assume(theta, real);
z1:=cos(3*theta)+I*sin(3*theta);
z2:=evalc((cos(theta)+I*sin(theta))^3);
R1:=Re(z1); R2:=Re(z2); R1=R2;
```

Mathematica:

```
{z1=Cos[3*theta]+I*Sin[3*theta], z2=(Cos[theta]+
 I*Sin[theta])^3//ComplexExpand, r1=ComplexExpand[Re[z1]],
 r2=ComplexExpand[Re[z2]], r1==r2}
```
                                                                                      □

Problem 8.3   Find and graph all the solutions of the equation $z^9 = 1$.

Maple:

```
Ops:=style=point,symbol=circle,color=blue,symbolsize=30;
Sols:=map(allvalues,{solve(z^9=1,z)});
Points:=map(u->[Re(u),Im(u)],Sols); plot(Points,Ops);
```

Mathematica:

```
{sol=Solve[z^9==1,z],n=Length[sol]}
{points=sol[[Table[i,{i,1,n}],1,2]]//N,
 points=points/.{x_Real->{x,0},Complex[x_,y_]->{x,y}}}
ListPlot[points,PlotStyle->{PointSize[0.03],Hue[0.7]},
 AspectRatio->1]
```
                                                                                      □

8.2   Complex Functions and Derivatives

Analytic and harmonic functions, multivalued "functions"

Maple:

```
f:=z->expr; diff(f(z),z); F:=evalc(f(x+I*y)); limit(f(z),z=z0);
u:=(x,y)->evalc(Re(F));v:=(x,y)->evalc(Im(F));   w:=(x,y)->expr;
w1:=unapply(u(x,y)+I*v(x,y),x,y);      limit(w(x,y),{x=x0,y=y0});
```

Mathematica:

```
f[z_]:=expr;      w:=(x,y)->expr;       F=ComplexExpand[f[x+I*y]];
u[x_,y_]:=ComplexExpand[Re[F]]; v[x_,y_]:=ComplexExpand[Im[F]];
Limit[Limit[w[x,y],x->x0],y->y0,ops]       Limit[f[z],z->z0,ops]
D[f[z],z] u[x,y]   v[x,y] D[u[x,y],x] Reduce[eqs,vars,Complexes]
```

Problem 8.4 Find the derivatives of the functions $f_1(z) = z^2 + i5z - 1$ and $f_2(z) = (f_1(z))^5$.

Maple:

```
f1:=z->z^2+I*5*z-1; f2:=z->z^5; f3:=z->f2(f1(z));
diff(f1(z),z); diff(f3(z),z);
```

Mathematica:

```
f1[z_]:=z^2+I*5*z-1; f2[z_]:=z^5;
f3[z_]:=f2[f1[z]]; {D[f1[z],z], D[f3[z],z]}
```

$\square$

Problem 8.5 Let $u(x,y) = 2\sin x \cosh y - x$ be a real part of the analytic function $w = f(z)$ and let $f(0) = 0$. Applying the Cauchy–Riemann equations, determine the analytic function $w = f(z)$.

Maple:

```
u:=(x,y)->2*sin(x)*cosh(y)-x;
Eq1:=int(diff(v(x,y),y),y)=int(diff(u(x,y),x),y)+g(x);
Eq2:=diff(rhs(Eq1),x)=-diff(u(x,y),y);
Eq3:=diff(g(x),x)=solve(Eq2,diff(g(x),x));
```

```
Eq4:=dsolve(Eq3,g(x)); Eq5:=subs(Eq4,Eq1);
v:=unapply(rhs(Eq5),x,y); v(x,y); sxy:={x=(z+evalc(
  conjugate(z)))/2,y=(z-evalc(conjugate(z)))/(2*I)};
f1:=unapply(simplify(subs(sxy,u(x,y)+I*v(x,y))),z);
f1(z); c1:=solve(f1(0)=0,_C1);
f:=unapply(subs(_C1=c1,f1(z)),z); f(z);
```

Mathematica:

```
u[x_,y_]:=2*Sin[x]*Cosh[y]-x;
{eq1=Integrate[D[v[x,y],y],y]==Integrate[
  D[u[x,y],x],y]+g[x], eq2=D[eq1[[2]],x]==-D[u[x,y],y],
  eq3=g'[x]==Solve[eq2,g'[x]][[1,1,2]], eq4=DSolve[eq3,
  g[x],x][[1,1]], eq5=eq1/.eq4}
v1[x1_,y1_]:=eq5[[2]]/.{x->x1,y->y1}; v1[x,y]
sxy={x->(z+ComplexExpand[Conjugate[z]])/2,
  y->(z-ComplexExpand[Conjugate[z]])/(2*I)};
f1[z1_]:=Simplify[u[x,y]+I*v1[x,y]/.sxy]/.{z->z1}; f1[z]
c1=Solve[f1[0]==0,C[1]]
f[z1_]:=(f1[z]/.c1[[1]])/.{z->z1}; f[z]
```

□

Problem 8.6 Show that $u(x,y) = x^2 - y^2$, $v(x,y) = 2xy$ are harmonic functions.

Maple:

```
f:=z->z^2: F:= evalc(f(x+I*y));
u:=(x,y)->evalc(Re(F)); v:=(x,y)->evalc(Im(F));
u(x,y); v(x,y); evalb(0=diff(u(x,y),x$2)+diff(u(x,y),y$2));
evalb(0=diff(v(x,y),x$2)+diff(v(x,y),y$2));
```

Mathematica:

```
f[z_]:=z^2; f1=ComplexExpand[f[x+I*y]];
u[x_,y_]:=ComplexExpand[Re[f1]]; v[x_,y_]:=ComplexExpand[Im[f1]];
{u[x,y], v[x,y], D[u[x,y],{x,2}]+D[u[x,y],{y,2}]===0,
  D[v[x,y],{x,2}]+D[v[x,y],{y,2}]===0}
```

□

Problem 8.7 Determine the harmonic conjugate $v(x,y)$ for harmonic function $u(x,y) = xy^3 - x^3y$.

Maple:

```
VConj:=proc(u) local v1,v2,v3,v4;
 v1:=int(diff(u,x),y); v2:=-diff(u,y)-diff(v1,x);
 v3:=int(v2,x); v4:=v1+v3; RETURN(v4): end;
u:=(x,y)->x*y^3-x^3*y;
evalb(0=diff(u(x,y),x$2)+diff(u(x,y),y$2));
v:=(x,y)->VConj(u(x,y)); v(x,y); v(2*x,2*y);
evalb(0=diff(v(x,y),x$2)+diff(v(x,y),y$2));
```

Mathematica:

```
vConj[u_]:=Module[{v1,v2,v3,v4},
 v1=Integrate[D[u,x],y]; v2=-D[u,y]-D[v1,x];
 v3=Integrate[v2,x]; v4=v1+v3]; u[x_,y_]:=x*y^3-x^3*y;
D[u[x,y],{x,2}]+D[u[x,y],{y,2}]===0
v[x_,y_]:=vConj[u[x,y]]; {v[x,y], v[2*x,2*y],
 D[v[x,y],{x,2}]+D[v[x,y],{y,2}]===0}
```
□

Problem 8.8 Evaluate the values: $\mathrm{Log}(z_1)$, $\log(z_2)$, z_2^c, where $z_1 = 1+2i$, $z_2 = 2$, $c = (1-i)/30$.

Maple:

```
z1:=1+2*I; w1:=evalc(log(z1)); w2:=evalc(log(z1))+I*2*Pi*n;
z2:=2; c:=(1-I)/30;  w3:=evalc(exp(c*log(z2)));
z3:=n->evalf(exp(c*(log(z2)+2*Pi*I*n))); k:=9;
for n from -k to k do  print(z3(n)) od:
Points:=[[Re(z3(i)),Im(z3(i))] $ i=-k..k];
plot(Points,scaling=constrained,style=point,axes=boxed,
 symbol=circle,color=blue);
```

Mathematica:

```
{z1=1+2*I, w1=ComplexExpand[Log[z1]],
 w2=ComplexExpand[Log[z1]]+I*2*Pi*n, z2=2, c=(1-I)/30,
 w3=ComplexExpand[Exp[c*Log[z2]]]}
z3[n_]:=Exp[c*(Log[z2]+2*Pi*I*n)]//N; k=9;
Do[Print[z3[n]],{n,-k,k}];
points=Table[{Re[z3[i]],Im[z3[i]]},{i,-k,k}]
ListPlot[points,PlotStyle->{PointSize[0.03],Hue[0.7]},
 AspectRatio->1]
```
□

Problem 8.9 Evaluate $\lim\limits_{z \to 1+i} \dfrac{z^2 - 2i}{z^2 - 2z + 2}$.

Maple:

```
f:=z->(z^2-2*I)/(z^2-2*z+2); F:=factor(f(Z));
L1:=subs(Z=1+I,F);  L2:=limit(f(z),z=1+I);
```

Mathematica:

```
f[z_]:=(z^2-2*I)/(z^2-2*z+2); f1=Factor[f[z1]]
{l1=f1/.{z1->1+I}, l2=Limit[f[z],z->1+I]}
```
□

Problem 8.10 Show that the function $f(z) = \dfrac{\sin(z^2)}{z^3 - z^2\pi/4}$ has a removable discontinuity at $z = 0$.

Maple:

```
with(plots): k:=0; f:=z->sin(z^2)/(z^3-Pi/4*z^2);
F:=factor(f(z)); L:=limit(f(z),z=k);
h:=z->proc(z) if z=k then evalf(L) else evalf(f(z)) fi: end;
conformal('h'(z),-2-2*I..2+2*I,axes=boxed,
  color=blue,grid=[200,200]);
```

Mathematica:

```
f[z_]:=Sin[z^2]/(z^3-Pi/4*z^2);
{k=0, f1=Factor[f[z]], l=Limit[f[z],z->k]}
h[z_]:=Piecewise[{{l,z==k}},f[z]]; ParametricPlot[
  Through[{Re,Im}[Evaluate[f[x+I*y]]]],{x,-3,3},{y,-3,3},
  PlotPoints->200,ColorFunction->Function[{x,y},Hue[y]],
  AspectRatio->1,Frame->True,Axes->False]
```
□

Geometric interpretations of complex numbers and functions

Maple:

```
with(plots):          plot(f(theta),theta=theta1..theta2,ops);
pointplot(list,ops);  polarplot(f(theta),theta=theta1..theta2);
       polarplot(f(theta),theta=theta1..theta2,style=point);
              implicitplot(f(x+I*y),x=a..b,y=c..d,ops);
              densityplot(f(x+I*y),x=x1..x2,y=y1..y2,ops);
with(plottools);           z0:=<x0,y0>;     arrow(z0,z1,ops);
```

Mathematica:

```
                              Plot[f[theta],{theta,theta1,theta2}]
PolarPlot[f[theta],{theta,theta1,theta2}]     ListPlot[list,ops]
ListPolarPlot[list,ops]    RegionPlot[Ineqs,{x,x1,x2},{y,y1,y2}]
              DensityPlot[f[x+I*y],{x,x1,x2},{y,y1,y2},ops]
              ContourPlot[f[x+I*y],{x,x1,x2},{y,y1,y2},ops]
z1={x1,y1};   Graphics[{Arrowheads[{spec}],Arrow[{z1,...,zn}]}]
```

Problem 8.11 Graph various sets of complex numbers (e.g, points, lines, curves, and regions) over the complex plane and in the polar coordinate system.

Maple:

```
with(plots): interface(showassumed=0): assume(z,complex);
zPolar:=polar(z); f1:=theta->argument(subs(
 {op(1,zPolar)=5,op(2,zPolar)=theta},zPolar));
plot(f1(theta),theta=-Pi..Pi,thickness=3);
pointplot([seq([i,1/i^2],i=1..50)],color=blue,symbol=circle,
 symbolsize=20); polarplot(2*cos(theta),theta=0..Pi);
polarplot(Pi/theta,theta=1..100,style=point);
implicitplot([argument((x+I*y)+2-I)=-Pi/6,
 argument((x+I*y)+2-I)=Pi/4],x=-2..2,y=-2..2,filledregions,
 coloring=[blue,green]); implicitplot(Re(1/(x+I*y)^2)=1/4,
 x=-2..2,y=-2..2,grid=[100,100],filledregions,
 coloring=[blue,green]); with(plottools):z1:=<2,-2>; z2:=<4,5>;
 A1:=arrow(z1,z2,.1,.4,.1): A2:=arrow(z1,z1+z2,.1,.4,.1):
 display({A1,A2},axes=framed,color=blue);
```

Mathematica:

```
polar=ComplexExpand[z,{z},TargetFunctions->{Arg,Abs}]
f1[theta_]:=Arg[polar/.{Abs[z]->5,Arg[z]->theta}];
Plot[f1[theta],{theta,-Pi,Pi}]
ListPlot[Table[{i,1/i^2},{i,1,50}],PlotStyle->PointSize[0.02]]
PolarPlot[2*Cos[theta],{theta,0,Pi}]
ListPolarPlot[Table[Pi/theta,{theta,1,100}]]
RegionPlot[-Pi/6<=Arg[(x+I*y)+2-I]<=Pi/4,{x,-2,2},{y,-2,2}]
RegionPlot[Re[1/(x+I*y)^2]>1/4,{x,-2,2},{y,-2,2}]
z1={2,-2}; z2={4,5}; g1=Graphics[{Blue,Arrow[{{0,0},z1}]}];
g2=Graphics[{Blue,Arrow[{z1,z2}]}];
```

```
g3=Graphics[{Blue,Arrow[{z2,z1+z2}]}]; Show[g1,g2,g3]
Graphics[{Blue,Arrow[{{0,0},z1,z2,z1+z2}]}]
Graphics[{Arrowheads[{-.1,.1}],Blue,Arrow[{z1,z2}]}]
```

$\square$

Problem 8.12   Graph various complex functions over the complex plane.

Maple:

```
with(plots); ops:=colorstyle=HUE,style=patchnogrid,
 brightness=0.9,contrast=0.7,numpoints=5000,axes=boxed;
f1:=z->sin(z); f2:=z->Re(z); f3:=z->Im(z);
densityplot(evalc(abs(1/f1(x+I*y))),x=-1..1,y=-1..1,ops,
view=[-1/20..1/20,-1/20..1/20]);
densityplot(evalc(f2(sin((x+I*y)^2))),x=-2..2,y=-2..2,ops,
view=[-2..2,-2..2]); implicitplot(evalc(f3(sin((x+I*y)^2))),
 x=-3..3,y=-3..3,color=blue,numpoints=5000,view=[-3..3,-3..3]);
```

Mathematica:

```
f1[z_]:=Sin[z]; f2[z_]:=Re[z]; f3[z_]:=Im[z];
DensityPlot[Abs[1/f1[x+I*y]],{x,-2,2},{y,-2,2},
 ColorFunction->Function[{x,y},Hue[y]]]
DensityPlot[f2[Sin[(x+I*y)^2]],{x,-2,2},{y,-2,2},
 ColorFunction->Function[{x,y},Hue[y]]]
ContourPlot[Evaluate[f3[Sin[(x+I*y)^2]]],{x,-3,3},{y,-3,3}]
```

$\square$

8.3   Complex Integration

Maple:

```
f:=z->expr;       F:=u(x,y)+I*v(x,y); w:=evalc(F);  diff(f(z),z);
U:=evalc(Re(F));   V:=evalc(Im(F));    int(f(z),z,'continuous');
int(f(z),z,'CauchyPrincipalValue'); int(f(z),z,'AllSolutions');
```

Mathematica:

```
f[z_]:=expr; D[f[z],z] F=u[x,y]+I*v[x,y]       w=ComplexExpand[F]
ComplexExpand[Re[F]] ComplexExpand[Im[F]]  Residue[f[z],{z,z0}]
Integrate[f[z],{z,a,b},PrincipalValue->True] NIntegrate[f[z],z]
```

Problem 8.13 The Cauchy's integral theorem: let A be an open subset of B which is simply connected, let $f : A \to B$ be a holomorphic function, and let C be a rectifiable path in A whose starting point is equal to its ending point. Then, $\oint_C f(z)\,dz = 0$. Evaluate this integral for $f(z) = z^2$.

Maple:

```
f:=z->z^2; C:=[r*exp(I*theta),theta=0..2*Pi];
Int(subs(z=C[1], f(z)*Diff(z,theta)),C[2])=
  value(Int(subs(z=C[1], f(z)*Diff(z,theta)),C[2]));
```

Mathematica:

```
f[z_]:=z^2; z[theta_]:=r*Exp[I*theta];
intd=f[z[theta]]*z'[theta]
Integrate[TraditionalForm[intd],{theta,0,2*Pi}]==Integrate[
 f[z[theta]]*z'[theta],{theta,0,2*Pi}]
```

□

Problem 8.14 Applying the Cauchy's integral formula, evaluate the integrals:
$$\int_{|z|=2} \frac{\cosh(iz)}{z^2 + 4z + 3}\,dz, \quad \int_{|z-2|=2} \frac{\cosh(z)}{z^4 - 1}\,dz.$$

Maple:

```
Eq1:=factor(cosh(I*z)/(z^2+4*z+3)); Eq2:=numer(Eq1);
Eq3:=denom(Eq1); f1:=unapply(Eq2/(z+3),z);
z0:=rhs(isolate(Eq3/(z+3),z)); i1:=2*Pi*I*f1(z0);
Eq11:=factor(cosh(z)/(z^4-1)); Eq12:=numer(Eq11);
Eq13:=denom(Eq11); f2:=unapply(Eq12/((z+1)*(z^2+1)),z);
z0:=rhs(isolate(Eq13/((z+1)*(z^2+1)),z));i2:=2*Pi*I*f2(z0);
```

Mathematica:

```
{eq1=Factor[Cosh[I*z]/(z^2+4*z+3)],
 eq2=Numerator[eq1],eq3=Denominator[eq1]}
f1[z1_]:=eq2/(z+3)/.{z->z1};
{z0=Solve[eq3/(z+3)==0,z][[1,1,2]], i1=2*Pi*I*f1[z0]}
{eq11=Factor[Cosh[z]/(z^4-1)], eq12=Numerator[eq11],
 eq13=Denominator[eq11]}
f2[z1_]:=eq12/((z+1)*(z^2+1))/.{z->z1}; {z0=Solve[
 eq13/((z+1)*(z^2+1))==0,z][[1,1,2]],i2=2*Pi*I*f2[z0]}
```

□

Problem 8.15 Evaluate integral around poles:

$$\int_C f(z)\,dz = 2\pi i \sum_{k=1}^{n} \mathrm{Res}(f,a), \quad \text{where } f(z) = \frac{5z-2}{z^2-z}.$$

Maple:

```
f:=(5*z-2)/(z^2-z); R0:=residue(f,z=0); R1:=residue(f,z=1);
2*Pi*I*(R0+R1);
```

Mathematica:

```
f[z_]:=(5*z-2)/(z^2-z); {r0=Residue[f[z],{z,0}],
 r1=Residue[f[z],{z,1}], 2*Pi*I*(r0+r1)}
```

□

8.4   Sequences and Series

Problem 8.16 Evaluate $\lim\limits_{n=\infty} z_n$, where $z_n = (n^{1/4} + i(n^2 + 1))/n^2$.

Maple:

```
z_n:=(n^(1/4)+I*(n^2+1))/n^2; L1:=limit(z_n,n=infinity);
```

Mathematica:

```
{zn=(n^(1/4)+I*(n^2+1))/n^2, l1=Limit[zn,n->Infinity]}
```

□

Problem 8.17 Show that the series $\sum\limits_{n=1}^{\infty}(1 + i(-1)^{n+1}n^2)/n^4$ converges.

Maple:

```
z:=n->(1+I*(-1)^(n+1)*n^2)/n^4; In:=infinity; ns:=1..infinity;
L:=expand(z(n+1)/z(n));limit(limit(subs((-1)^n-K,L),K-In),n=In);
S1:=sum(evalc(Re(z(n))),n=ns)+I*sum(evalc(Im(z(n))),n=ns);
S2:=convert(S1,StandardFunctions);
```

Mathematica:

```
z[n_]:=(1+I*(-1)^(n+1)*n^2)/n^4;
{l=z[n+1]/z[n]//Expand, Limit[l,n->Infinity]}
Sum[ComplexExpand[Re[z[n]]],{n,1,Infinity}]+I*Sum[
 ComplexExpand[Im[z[n]]],{n,1,Infinity}]
```

□

Problem 8.18 Find the radius of convergence of $f(z) = \sum_{n=0}^{\infty} \dfrac{z^{2n}}{2n!}$.

Maple:

```
c:=n->1/(2*n)!; L:=simplify(c(n+1)/c(n)); A:=infinity;
R:=limit(1/L,n=A); sum(c(n)*z^(2*n),n=0..A);
```

Mathematica:

```
c[n_]:=1/(2*n)!; l=c[n+1]/c[n]//Simplify
{r=Limit[1/l,n->Infinity], Sum[c[n]*z^(2*n),{n,0,Infinity}]]}
```

□

The Mandelbrot and Julia sets: let $f:\mathbb{C} \to \mathbb{C}$ be either the complex plane or the Riemann sphere. We consider f as a discrete dynamical system on the phase space $\mathbb{C}$ and study the iteration behavior of f. Let the map $z \mapsto z^2 + c$, where $z = x + iy$, $c \in \mathbb{C}$. We define the sequence $\{z_n\}_{n=0}^{\infty}$ as follows: $z_n = z_{n-1}^2 + c$, $z_0 = 0$.

The Mandelbrot set, M, generated for all values of c, consists of the values of c for which the sequence $\{z_n\}_{n=0}^{\infty}$ does not diverge to infinity under iteration. For the Julia set $J(f)$, we consider a fix point c and generate the sequence $\{z_n\}_{n=0}^{\infty}$ as follows: $z_n = z_{n-1}^2 + c$, $z_0 = z$.

The Julia set, J consists of the values of z_0 for which the sequence $\{z_n\}_{n=0}^{\infty}$ does not diverge to infinity under iteration. The Mandelbrot and Julia sets produce fractals.

Problem 8.19 Construct the *Julia set* in 2D and 3D for the map $z \mapsto z^3 - Cz$, $C = 0.69 + 0.67i$.

Maple:

```
with(plots): d:=3/2; CZ:=0.69+0.67*I; Ops2D:=symbol=circle,
 style=point,color=red,axes=none; Ops3D:=colorstyle=HUE,
 style=patchnogrid,axes=boxed,grid=[200,200];
JuliaSet2D:=proc(C,t,d) local x,y,z,i,L,k,m:
 L:=[]: z:=0: k:=200: m:=3:
 for x from -d*t to d*t do for y from -d*t to d*t do
  z:=x/t+I*y/t: for i from 0 while i<k and evalf(abs(z))<m do
  z:=z^3-C*z: od: if i=k then L:=[op(L),[x,y]]: fi: od: od:
 RETURN(L): end:
JuliaSet3D:=proc(X,Y,C) local Z,m,k,t;
 Z:=X+I*Y; k:=30: m:=3.0:
 for t from 1 while t<k and evalf(abs(Z))<m do
  Z:=Z^3-C*Z;od;-t; end:
L_J:=JuliaSet2D(CZ,30,d): pointplot(L_J,Ops2D);
densityplot('JuliaSet3D'(x,y,CZ),x=-d..d,y=-d..d,Ops3D);
```

Mathematica:

```
{d=3/2,cz=0.69+0.67*I}
Set2DJulia[c_,t_,d_]:=Module[{x,y,z=0,k=200,m=3.,l={},l1,i},
 Do[Do[z=x/t+I*y/t;i=0;While[i<k && N[Abs[z]]< m,z=z^3-c*z;i++];
 If[i==k,l=Join[l,{x,y}]],{y,-d*t,d*t}],{x,-d*t,d*t}];
 l1=Partition[l,2]];
Set3DJulia[x_,y_,c_]:=Module[{z,m=3.,k=30,t=1},z=x+I*y;
 While[t<k && N[Abs[z]]< m,z=z^3-c*z;t++]; Return[-t]];
1J2D=Set2DJulia[cz,30,d];
ListPlot[1J2D,PlotStyle->{PointSize[0.03],Hue[0.7]},
 AspectRatio->1]
pointsJ=Table[Set3DJulia[x,y,cz],{y,-d,d,0.01},{x,-d,d,0.01}];
ListDensityPlot[pointsJ,Mesh->False,ColorFunction->Hue]
```

□

Problem 8.20 Construct the *Mandelbrot (Multibrot) set* in 3D for the map $z \mapsto z^3$.

Maple:

```
with(plots): d:=3/2; k:=200; m:=3; Ops:=colorstyle=HUE,
 style=patchnogrid, axes=boxed,grid=[200,200];
MandelbrotSet:=proc(X,Y) local Z,t; Z:=X+I*Y;
 for t from 1 while t<k and evalf(abs(Z))<m do
  Z:=Z^3+(X+I*Y) od; -t; end:
densityplot('MandelbrotSet'(x,y),x=-d..d,y=-d..d,Ops);
```

Mathematica:

```
{d=3/2,k=200,m=3}
Set3DMandelbrot[x_,y_]:=Module[{z,t=1},z=x+I*y;
 While[t<k && N[Abs[z]]< m,z=z^3+(x+I*y);t++]; Return[-t]];
pointsM=Table[Set3DMandelbrot[x,y],{y,-d,d,0.01},
 {x,-d,d,0.01}]; ListDensityPlot[pointsM,
Mesh->False,ColorFunction->Hue]
```

□

8.5   Singularities and Residue Theory

Singularities, series expansions about singularities, Laurent series

Maple:

```
f:=z->(z^2+a^2)^(-1/2); singular(f(z),z);
series(sqrt(z)+z^(-1/2),z=0,9); series(cos(sqrt(z)),z=0,9);
series((1/sin(z))^2,z=0,9);
with(numapprox): laurent(1/(x^3*sin(x^3)),x=0,19);
laurent(1/((x-1)*(x-3)),x=1,11); laurent(1/((x-1)*(x-3)),x=3,11);
```

Mathematica:

```
f[z_]:=(z^2+a^2)^(-1/2); Solve[1/f[z]==0,z]
{Series[Sqrt[z]+z^(-1/2),{z,0,9}],
 Series[Cos[Sqrt[z]],{z,0,9}], Series[(1/Sin[z])^2,{z,0,9}]}
{Series[1/(x^3*Sin[x^3]),{x,0,9}], Series[1/((x-1)*(x-3)),
 {x,1,9}], Series[1/((x-1)*(x-3)),{x,3,9}]}
```

Limits at singularities

Maple:

```
f0:=(x, y)->evalc(conjugate(x+I*y));
limit(f0(x, y),{x=x0,y=y0}); f1:=z->z^2; f2:=z->z^(1/2);
limit(f1(z),z=0,complex); limit(f2(z),z=0,complex);
f3:=z->z^(-2); f4:=z->z^(-1/2); f5:=z->exp(1/z);
limit(f3(z),z=0,complex); limit(f4(z),z=0,complex);
limit(f5(z),z=0,right);
```

Mathematica:

```
f0[x_,y_]:=ComplexExpand[Conjugate[x+I*y]];
Limit[Limit[f0[x,y],x->x0],y->y0]
f1[z_]:=z^2; f2[z_]:=Sqrt[z];
{Limit[f1[z],z->0], Limit[f2[z],z->0]}
f3[z_]:=z^(-2); f4[z_]:=z^(-1/2); f5[z_]:=Exp[1/z];
{Limit[f3[z],z->0], Limit[f4[z],z->0], Limit[f5[z],z->0]}
```

8.6   Transformations and Mappings

Problem 8.21 The transformation $w = z^2$ maps lines onto lines or parabolas. Find the image of the vertical line $x = 5$.

Maple:

```
z:=(x+I*y)^2; Eq1:={U=evalc(Re(z)),V=evalc(Im(z))};
Eq2:=subs(x=5,Eq1); Eq3:=eliminate(Eq2,y);
Sols:=[solve(Eq3[2][1],U)]; U:=V->expand(Sols[1]);
U(V); plot(U(V), V=-10..10);
```

Mathematica:

```
{z=(x+I*y)^2, eq1={u==ComplexExpand[Re[z]],
 v==ComplexExpand[Im[z]]}, eq2=eq1/.{x->5},
 eq3=Eliminate[eq2,y]}
u[v_]:=Flatten[Solve[eq3,u]//Simplify][[1,2]]; u[v]
Plot[u[v],{v,-10,10},PlotStyle->{Blue,Thickness[0.01]}]
```

□

Conformal mapping $z = f(Z)$ on a rectangular region and the Riemann sphere

Maple:

```
with(plots): f:=z->(5*z-1)/(5*z+1);
conformal(f(z),z=-1-2*I..1+2*I,-5-5*I..5+5*I,
 grid=[25,25],numxy=[140,140]);
conformal(cos(z)-sin(z),z=0-2*I..2*Pi+2*I,color=gold);
conformal3d(cos(z)-sin(z),z=0-2*I..2*Pi+2*I,color=white,
 grid=[20,20],orientation=[17,111]);
```

Mathematica:

```
f1[z_]:=(5*z-1)/(5*z+1); f2[z_]:=Cos[z]-Sin[z];
ParametricPlot[Through[{Re,Im}[f1[x+I*y]]],{x,-5,5},
 {y,-5,5},Mesh->30,PlotPoints->80,PlotStyle->Hue[0.7],
 AspectRatio->1,PlotRange->{{0,2},{-1,1}}]
ParametricPlot[Through[{Re,Im}[f2[r*Exp[I*t]]]],{r,0,1},
 {t,0,2*Pi},PlotStyle->Hue[0.9],AspectRatio->1,
 PlotRange->{{0.6,1.3},{-0.5,0.5}}]
ParametricPlot[Through[{Re,Im}[f2[x+y*I]]],{x,-2,2+2*Pi},
 {y,-2,2},PlotStyle->Hue[0.9],AspectRatio->1,PlotRange->All]
ParametricPlot3D[Evaluate[{ComplexExpand[Re[f2[x+I*y]]]/(1+
 Abs[f2[x+I*y]]^2),ComplexExpand[Im[f2[x+I*y]]]/(1+
 Abs[f2[x+I*y]]^2),Abs[f2[x+I*y]]^2/(1+Abs[f2[x+I*y]]^2)}],
 {x,0,2*Pi},{y,-Pi,Pi},ColorFunction->Function[{x,y},
 Hue[0.1*(x+y)]],BoxRatios->{1,1,1},Mesh->30]
```

Riemann surfaces

We construct the graph of the Riemann surface for the logarithm function $w = \ln z$ considering the map $z = \exp(w) = \exp(u + iv)$.

Maple:

```
with(plots): w:=u+I*v; z:=evalc(exp(w)); x:=evalc(Re(z));
y:=evalc(Im(z)); Eq1:=z=X+I*Y; Eq2:=evalc(Re(Eq1)); Eq3:=
 evalc(Im(Eq1)); eliminate({map(x->x^2,Eq2)+map(x->x^2,Eq3)},u);
plot3d([x,y,v],u=-1..1,v=-4*Pi..4*Pi,grid=[100,100],
 orientation=[-27,67],style=PATCHNOGRID,axes=none,colour=v);
```

Mathematica:

```
{w=u+I*v, z=ComplexExpand[Exp[w]], x=ComplexExpand[Re[z]],
 y=ComplexExpand[Im[z]]}
{eq1=z==X+I*Y, eq2=ComplexExpand[Thread[Re[eq1],Equal]],
 eq3=ComplexExpand[Thread[Im[eq1], Equal]],
 eq31=Thread[eq2^2,Equal], eq32=Thread[eq3^2,Equal],
 Reduce[Simplify[Thread[eq31+eq32,Equal]],{u},Reals]}
ParametricPlot3D[{x,y,v},{u,-1,1},{v,-4*Pi,4*Pi},PlotPoints->70,
 Mesh->None,Axes->None,BoxRatios->{1,1,1},ViewPoint->{-27,77,39},
 ColorFunction->Function[{u,v},Hue[u*v]]]
```

Chapter 9

Special Functions and Orthogonal Polynomials

Special functions is a set of some classes of particular functions that have
attractive or useful properties arising from solutions of theoretical
and applied problems in different areas of mathematics.

Special functions can be defined by means of power and trigonometric
series, series of orthogonal functions, infinite products, generating and distribution functions, integral representations, sequential
differentiation, trascendental, differential, difference, integral, and
functional equations.

Maple includes over 200 special functions. We will consider some the
most important special functions (see ?inifcn, ?index[package],
?index[function], ?FunctionAdvisor).

Mathematica includes all the common special functions of mathematical
physics found in standard handbooks. It should be noted that
the definitions (including normalizations and special values) of any
particular special function can be different in handbooks and also
in *Maple* and *Mathematica* (e.g., the Mathieu functions). We will
discuss some the most important special functions.

9.1 Functions Defined by Integrals

Gamma, Beta, digamma, and polygamma functions

Maple: GAMMA, Beta, Psi(x), Psi(n,x).

```
plot(GAMMA(x),x=-5..5,-20..20,numpoints=500,color=blue);
GAMMA(z+2); convert(GAMMA(z+2), factorial);
```

I.K. Shingareva, C. Lizárraga-Celaya, *Maple and Mathematica*, 2nd ed.,
DOI 10.1007/978-3-211-99432-0_9, © Springer-Verlag Vienna 2009

```
convert(z!, GAMMA); simplify(z*GAMMA(z)); GAMMA(1/2);
GAMMA(-1/2); GAMMA(5/2); Beta(2,9); expand(Psi(73));
evalf(expand(Psi(2,73))); evalf(Psi(2,73));
```

Mathematica: Gamma, Beta, PolyGamma[x], PolyGamma[n,x].

```
Plot[Gamma[x],{x,-5,5},PlotRange->{-20,20},PlotStyle->Hue[0.9]]
{z*Gamma[z]//FunctionExpand, Gamma[1-2*I]//N,
 D[Gamma[z],z], Gamma[1/2], Gamma[-1/2], Gamma[5/2]}
{Beta[2,9], N[{Erf[1],Erfc[1]}]}
{PolyGamma[73], PolyGamma[2,73]//FunctionExpand,
 PolyGamma[2,73]//N}
```

Exponential, logarithmic, polylogarithmic, and trigonometric integrals

Maple: Ei, Li, dilog, polylog, Si, Ci.

```
expand(Ei(5,x)); evalf(Si(5)); convert(Li(x), Ei);
convert(Ci(x),Ei); dilog(-1/2); evalf(polylog(2,-1/2));
evalf(convert(series(Li(x),x=2,9),polynom));
diff(Ci(x),x); limit(int(Si(x),x),x=0);
```

Mathematica: ExpIntegralE, LogIntegral, PolyLog,
SinIntegral, CosIntegral.

```
{ExpIntegralE[5,x]//FunctionExpand, SinIntegral[5]//N}
 Series[LogIntegral[x],{x,2,9}]//Normal//N
{PolyLog[2,-1/2]//N, D[CosIntegral[x],x],
 Limit[Integrate[SinIntegral[x],x],x->0]}
```

Error functions, Fresnel integrals

Maple: erf, erfc, FresnelC, FresnelS.

```
evalf(erf(1)); evalf(erfc(1));
plot([FresnelC(z),FresnelS(z),z=-10..10],-1..1,-1..1,
    scaling=constrained,color=blue,axes=boxed);
```

Mathematica: Erf, Erfc, FresnelC, FresnelS.

```
{Erf[1]//N, Erfc[1]//N}
ParametricPlot[{FresnelC[z],FresnelS[z]},{z,-10,10},
  PlotRange->{{-1,1},{-1,1}},AspectRatio->1,
  PlotStyle->Hue[0.7]]
```

Elliptic integrals and functions

Maple: EllipticF, EllipticK, EllipticE, EllipticPi.

```
EllipticF(0.1,0.2); EllipticK(0.1);   EllipticE(0.1,0.2);
EllipticE(0.1);EllipticPi(0.1,0.1,0.2);EllipticPi(0.1,0.2);
z:=1/2*sqrt(10+25*I); k:=1/2*sqrt(2)-1/2*I*sqrt(2);
evalf(EllipticE(z, k)); z1:=evalf(z); k1:=evalf(k);
int(sqrt(evalc(1-k1^2*t^2))/sqrt(evalc(1-t^2)),t=0..z1);
```

Mathematica: EllipticF, EllipticK, EllipticE, EllipticPi.

```
{EllipticF[0.1,0.2], EllipticK[0.1], EllipticE[0.1,0.2],
  EllipticE[0.1],EllipticPi[0.1,0.1,0.2],EllipticPi[0.1,0.2]}
{z=1/2*Sqrt[10+25*I], k=1/2*Sqrt[2]-1/2*I*Sqrt[2],
  EllipticE[z,k]//N, z1=N[z], k1=N[k],
  NIntegrate[Sqrt[1-k1^2*t^2]/Sqrt[1-t^2],{t,0,z1}]}
```

9.2   Orthogonal Polynomials

Gegenbauer, Hermite, Jacobi, Laguerre, Legendre, Chebyshev polynomials

Maple:

```
with(orthopoly): k:=7; m:=x=1;
P_Gegenbauer:=[G(n,2,x)$n=0..k];subs(m,P_Gegenbauer);
P_Hermite:=[H(n,x) $ n=0..k]; subs(m,P_Hermite);
P_Jacobi:=[P(n,2,4,x) $ n=0..k]; subs(m,P_Jacobi);
P_Laguerre:=[L(n,x) $ n=0..k]; subs(m,P_Laguerre);
P_Legendre:=[P(n,x)$n=0..k]; subs(m,P_Legendre);
P_ChebyshevI:=[T(n,x)$n=0..k];subs(m,P_ChebyshevI);
plot({T(n,x)$n=0..k},x=-1..1,color=blue,axes=boxed);
P_ChebyshevII:=[U(n,x)$n=0..k];subs(m,P_ChebyshevII);
F:=HermiteH(4,x); simplify(F,'HermiteH');
```

Mathematica:

```
k=7; m={x->1}; {pGegenbauer=Table[GegenbauerC[n,2,x],
 {n,0,k}], pHermiteH=Table[HermiteH[n,x],{n,0,k}],
 pJacobiP=Table[JacobiP[n,2,4,x],{n,0,k}]//Simplify,
 pLaguerreL=Table[LaguerreL[n,x],{n,0,k}],
 pLegendreP=Table[LegendreP[n,x],{n,0,k}],
 pChebyshevT=Table[ChebyshevT[n,x],{n,0,k}]}
Plot[Evaluate[pChebyshevT],{x,-1,1},
 PlotStyle->Blue,AspectRatio->1]
pChebyshevU=Table[ChebyshevU[n,x],{n,0,k}]
{pGegenbauer/.m,pHermiteH/.m,pJacobiP/.m,pLaguerreL/.m,
 pLegendreP/.m,pChebyshevT/.m,pChebyshevU/.m}
```

9.3   Functions Defined by Trascendental Equations

The Lambert W function

Maple: `LambertW`

```
with(plots):  alias(W=LambertW): solve(y*exp(y)-z,y);
diff(W(z),z); series(W(z),z,9); W(-1.); A:=thickness=2;
G1:=plot(W(x)^(-1),x=-0.3678..0,-5..1,A,linestyle=2):
G2:=plot(W(x),x=-0.3678..1,-5..1,A): display([G1,G2]);
```

Mathematica: `ProductLog`

```
{Reduce[y*Exp[y]-z==0,y], D[ProductLog[z],z],
 Series[ProductLog[z],{z,0,9}], ProductLog[-1.]}
g1=Plot[ProductLog[x]^(-1),{x,-0.3678,0},PlotRange->
 {{-0.3678,0},{-5.,1.}},PlotStyle->{Blue,Dashing[{0.01}]}];
g2=Plot[ProductLog[x],{x,-0.3678,1},PlotRange->
 {{-0.3678,1},{-5.,1.}},PlotStyle->{Blue,Thickness[0.02]}];
Show[{g1,g2},AspectRatio->1,Axes->False]
```

9.4   Functions Defined by Differential Equations

The Bessel functions, the modified Bessel functions, the Hankel functions

Maple:

```
with(plots): C:=color=[green,blue,magenta];
S:=scaling=constrained; B:=axes=boxed;
plot([BesselJ(n,x) $ n=0..2],x=0..10,C,B);
plot([BesselY(n,x) $ n=0..2],x=0..10,-1..1,C,B);
plot([BesselI(n,x) $ n=1..3],x=-Pi..Pi,C,B);
plot([BesselK(n,x) $ n=0..2],x=0..Pi,-10..10,C,B);
conformal(HankelH1(0,x),x=1+I..-1+2*I,color=blue,S,B);
conformal(HankelH2(0,x),x=1-I..-1-2*I,color=blue,S,B);
expand(asympt(BesselJ(0,x),x,1));convert(HankelH2(n,x),Bessel);
evalf(BesselJZeros(1,1..9));evalf(BesselYZeros(1,1..9));
```

Mathematica:

```
nD=10; SetOptions[Plot,PlotStyle->{Green,Blue,Magenta},
  AspectRatio->1]; SetOptions[ParametricPlot,PlotPoints->20,
  ColorFunction->Function[{x,y},Hue[y]],AspectRatio->1,
  ImageSize->300];
Plot[Evaluate[Table[BesselJ[n,x],{n,0,2}]],{x,0,10}]
Plot[Evaluate[Table[BesselY[n,x],{n,0,2}]],{x,0,10},
    PlotRange->{{0,10},{-1,1}}]
Plot[Evaluate[Table[BesselI[n,x],{n,1,3}]],{x,-Pi,Pi},
    PlotRange->{{-Pi,Pi},{-4,4}}]
Plot[Evaluate[Table[BesselK[n,x],{n,0,2}]],{x,0,Pi},
    PlotRange->{{0,Pi},{-10,10}}]
GraphicsRow[{ParametricPlot[Through[{Re,Im}[Evaluate[
  HankelH1[0,x+I*y]]]],{x,-1,1},{y,2,1}], ParametricPlot[Through[
  {Re,Im}[Evaluate[HankelH2[0,x+I*y]]]],{x,-1,1},{y,-2,-1}]}]
asympt=FullSimplify[Series[BesselJ[0,x],{x,Infinity,1}]]
{N[BesselJZero[1,Range[9]],nD], N[BesselYZero[1,Range[9,17]],
  nD], N[BesselYZero[1,Range[9]],30]}
```

Note. Several functions of the **NumericalMath`BesselZeros`** package (for *ver* < 6) are available at the *Wolfram Website*.

The Mathieu functions

Maple:

```
Ops:=scaling=constrained,axes=boxed;
plot(MathieuCE(0,10,x),x=-Pi..Pi,Ops);
```

```
plot(MathieuSE(1,10,x),x=-Pi..Pi,Ops);
plot(MathieuCE(10,20,x),x=-Pi..Pi,Ops);
MathieuCE(n,0,x); MathieuSE(n,0,x);
s1:=series(MathieuA(2,q),q,10); evalf(s1);
evalf(MathieuB(2,20)); MathieuFloquet(a,0,x);
MathieuExponent(a,0); MathieuC(a,0,x); MathieuS(a,0,x);
MathieuCEPrime(n,0,x);
```

Mathematica:

```
SetOptions[Plot,PlotStyle->Blue,AspectRatio->1,
  ImageSize->300]; SetOptions[ListPlot,AspectRatio->1];
GraphicsRow[{Plot[Re[MathieuC[MathieuCharacteristicA[0,20],
  20,x]],{x,-Pi,Pi}], Plot[Re[MathieuS[
  MathieuCharacteristicB[1,10],10,x]],{x,-Pi,Pi}], Plot[
  Re[MathieuC[MathieuCharacteristicA[10,20],20,x]],{x,-Pi,Pi}]}]
MathieuC[MathieuCharacteristicA[n,0],0,x]//FunctionExpand
Assuming[n>0,MathieuS[MathieuCharacteristicB[n,0],0,x]//
  FunctionExpand//Simplify]
N[MathieuCharacteristicB[2,20],10]
MathieuCharacteristicExponent[a,0]
Normal[N[Series[MathieuCharacteristicA[2,x],{x,0,4}]]]//Expand
{MathieuC[a,0,x], MathieuS[a,0,x], MathieuCPrime[n,0,x]}
{q=20,a1=0,an=50}
g1=ListPlot[Table[{a,MathieuCharacteristicA[a,q]},{a,a1,an}],
  PlotStyle->{Hue[0.7],PointSize[0.02]}]; g2=ListPlot[Table[
  MathieuCharacteristicB[a,q],{a,a1+3,an}],
  PlotStyle->{Hue[0.9],PointSize[0.02]}]; Show[{g1,g2}]
```

The Legendre functions and the associated Legendre functions

Maple:

```
EqLeg:=v->(1-x^2)*Diff(v,x$2)-2*x*Diff(v,x)+n*(n-1);
EqLeg(y(x)); Sol:=dsolve(EqLeg(y(x))=0,y(x)); op(2,Sol);
test1:=simplify(subs(y(x)=op(2,Sol),EqLeg(v)));
evalb(test1=simplify(EqLeg(v)));
F_LP :=LegendreP(2,x);   simplify(F_LP,'LegendreP');
F_LQ :=LegendreQ(2,x);   simplify(F_LQ,'LegendreQ');
F_LPA:=LegendreP(2,2,x); simplify(F_LPA,'LegendreP');
F_LQA:=LegendreQ(2,2,x); simplify(F_LQA,'LegendreQ');
convert(LegendreP(2,a,x),hypergeom);
```

Mathematica:

```
eqLeg=(1-x^2)*y''[x]-2*x*y'[x]+n*(n-1);
{sol=DSolve[eqLeg==0,y[x],x],
 test1=Simplify[eqLeg/.sol][[1]], test1===Simplify[eqLeg]}
Map[FunctionExpand, {LegendreP[2,x],LegendreQ[2,x],
    LegendreP[2,2,x],LegendreQ[2,2,x],LegendreQ[2,2,2,x]}]
```

9.5   Functions Defined by Infinite Series

The Riemann Zeta function, hypergeometric functions

Maple:

```
sum(1/i^9,i=1..infinity)=Zeta(9.);
plot(Zeta(x),x=-1..2,-10..10); map(simplify,
 [z*hypergeom([1,1],[2],-z), hypergeom([-3],[],x),
  hypergeom([-n,beta],[beta],-z), x*hypergeom([2,2],[1],x),
  hypergeom([1/2],[3/2],-x^2), hypergeom([],[1],-(x/2)^2)]);
convert(hypergeom([1,2],[3/2],z),StandardFunctions);
```

Mathematica:

```
{ex1=Sum[1/i^9,{i,1,Infinity}], N[ex1],ex1==Zeta[9]}
Plot[Zeta[x],{x,-1,2},PlotRange->{{-1,2},{-10,10}},
 PlotStyle->Blue,AspectRatio->1]
{z*HypergeometricPFQ[{1,1},{2},-z],HypergeometricPFQ[{-3},{},x],
 HypergeometricPFQ[{-n,beta},{beta},-z]}
{x*HypergeometricPFQ[{2,2},{1},x], HypergeometricPFQ[
 {1/2},{3/2},-x^2], HypergeometricPFQ[{},{1},-(x/2)^2]}
Series[Hypergeometric0F1[a,x],{x,0,10}]
Integrate[Hypergeometric1F1[a,b,x],x]
{Hypergeometric2F1[2,2,2,z],D[HypergeometricU[a,b,x],{x,2}]}
```

9.6   Generalized Functions or Distributions

The Dirac δ-function, the Heaviside step function

Maple: Dirac, Heaviside

```
with(plots): setoptions(thickness=3); delta:=t->Dirac(t);
H:=t->Heaviside(t); Int(delta(t), t)=int(delta(t),t);
Int(H(t),t)=int(H(t),t); Diff(H(t),t)=diff(H(t),t);
Diff(delta(t),t)=diff(delta(t),t);
Diff(delta(t),t$2)=diff(delta(t),t$2);
Int(delta(t-a)*f(t),t=-infinity..infinity)
 =int(delta(t-a)*f(t), t=-infinity..infinity);
convert(Heaviside(t-1)+Heaviside(t-2),piecewise);
assume(b<a): simplify(convert(piecewise(t>=0 and t<b,0,
 t>=b and t<a,1,t>=a,f(t)),Heaviside));
plot(H(cos(x)),x=-9..9,color=blue);
plot([seq(sqrt(i/Pi)*exp(-i*x^2),i=10..50,40)],
 x=-Pi..Pi,color=[blue,green]);
```

Mathematica: DiracDelta, HeavisideTheta (the generalized function),
UnitStep (the piecewise constant function)

```
delta[t1_]:=DiracDelta[t]/.{t->t1};
H[t1_]:=HeavisideTheta[t]/.{t->t1};
SetOptions[Plot,PlotStyle->{{Blue,Thickness[0.01]},
{Purple,Thickness[0.01]}},ImageSize->300];
 {HoldForm[Integrate[delta[t],t]]==TraditionalForm[
 Integrate[delta[t],t]],HoldForm[Integrate[H[t],t]]==
 TraditionalForm[Integrate[H[t],t]]}
{HoldForm[D[H[t],t]]==TraditionalForm[H'[t]],
 HoldForm[D[delta[t],t]]==TraditionalForm[delta'[t]],
 HoldForm[D[delta[t],{t,2}]]==TraditionalForm[delta''[t]]}
HoldForm[Integrate[delta[t-a]*f[t],
 {t,-Infinity,Infinity}]]==Assuming[a\[Element]Reals,
 Integrate[delta[t-a]*f[t],{t,-Infinity,Infinity}]]
{f1=H[(t-1)*(t-2)]//FunctionExpand, PiecewiseExpand[f1],
 TraditionalForm[f1]}
GraphicsRow[{Plot[H[Cos[x]],{x,-9,9}],Plot[UnitStep[Cos[x]],
{x,-9,9},Exclusions->None],Plot[Evaluate[Sqrt[i/Pi]*
 Exp[-i*x^2]/.{i->{10,50}}],{x,-Pi,Pi},PlotRange->All]}]
```

Chapter 10

Integral and Discrete Transforms

10.1 Laplace Transforms

In Maple, the integral transforms (e.g., Fourier, Hilbert, Laplace, Mellin integral transforms) can be studied with the aid of the `inttrans` package.

In Mathematica, the Laplace integral transforms are defined by the two functions `LaplaceTransform`, `InverseLaplaceTransform`.

Problem 10.1 Find the Laplace integral transforms for different functions.

Maple:

```
with(inttrans); with(plots); assume(a>0);
f1:=t->piecewise(t>=0 and t<=1,-1,t>1,1);
f2:=t->2*Heaviside(t-1)-Heaviside(t);
G:=array(1..2); G[1]:=plot(f1(t),t=0..10,color=blue,
 discont=true): G[2]:=plot(f2(t),t=0..10,color=magenta):
display(G,scaling=constrained);
map(simplify,[laplace(f1(t),t,s),laplace(f2(t),t,s)]);
laplace(diff(u(x),x$2),x,s); laplace(Dirac(t-a),t,s);
simplify(laplace(exp(a*t)*erf(sqrt(a*t)),t,s));
simplify(laplace(F(t-a)*Heaviside(t-a),t,s));
laplace(t^2*cos(2*t),t,s);
```

Mathematica:

```
f1[t_]:=Piecewise[{{-1,0<=t<=1},{1,t>1}}];
f2[t_]:=2*HeavisideTheta[t-1]-HeavisideTheta[t]; {f1[t],f2[t]}
```

I.K. Shingareva, C. Lizárraga-Celaya, *Maple and Mathematica*, 2nd ed.,
DOI 10.1007/978-3-211-99432-0_10, © Springer-Verlag Vienna 2009

```
g1=Plot[f1[t],{t,0,10},PlotStyle->{Hue[0.7],
 Thickness[0.01]}]; g2=Plot[f2[t],{t,0,10},PlotStyle->
 {Hue[0.9],Thickness[0.01]}]; GraphicsRow[{g1,g2}]
LaplaceTransform[f[t],t,s]//TraditionalForm
Map[FullSimplify, {LaplaceTransform[f1[t],t,s],
 FullSimplify[LaplaceTransform[f1[x],x,s]]===
 FullSimplify[LaplaceTransform[f2[x],x,s]],
 LaplaceTransform[D[u[x],{x,2}],x,s],Assuming[a>0,
 LaplaceTransform[DiracDelta[t-a],t,s]],LaplaceTransform[
 DiracDelta[t-a],t,s,GenerateConditions->True],
 LaplaceTransform[Exp[a*t]*Erf[Sqrt[a*t]],t,s,
 Assumptions->a>0],LaplaceTransform[t^2*Cos[2*t],t,s]}]
```
□

Problem 10.2 Find the inverse Laplace transforms for different functions.

Maple:

```
interface(showassumed=0): with(inttrans);
f1:=t->2*Heaviside(t-1)-1; F1:=unapply(laplace(f1(s),s,t),t);
T1:=invlaplace(F1(s),s,t); evalb(T1=f1(t));
invlaplace(1/s^3,s,t); invlaplace(1/(s-a),s,t);
assume(a>0,x>0); simplify(invlaplace(exp(-s*x),s,t));
simplify(invlaplace((s^2+a^2)^(-1/2),s,t));
```

Mathematica:

```
f1[t_]:=2*HeavisideTheta[t-1]-1; f1[t]
F1[T_]:=LaplaceTransform[f1[t],t,s]/.{t->T};
InverseLaplaceTransform[F1[s],s,t]===(f1[t]//FullSimplify)
{InverseLaplaceTransform[1/s^3,s,t],
 InverseLaplaceTransform[1/(s-a),s,t], Assuming[{x>0},
 InverseLaplaceTransform[Exp[-x*s],s,t]//FullSimplify],
 Assuming[a>0,InverseLaplaceTransform[
 1/Sqrt[s^2+a^2],s,t]//Simplify]}
```
□

Applications to ordinary differential equations

Problem 10.3 Obtain the solution of the initial value problem

$$y' + ay = e^{-at}, \quad y(0) = 1.$$

Maple:

```
with(inttrans); ODE:=diff(y(t),t)+a*y(t)=exp(-a*t);
Eq1:=laplace(ODE,t,p); Eq2:=subs(y(0)=1,Eq1);
Eq3:=solve(Eq2,laplace(y(t),t,p));
Sol:=invlaplace(Eq3,p,t);
dsolve({ODE,y(0)=1},y(t)); Sol1:=subs(a=7,Sol);
plot(Sol1,t=0..1,color=blue);
```

Mathematica:

```
ode={y'[t]+a*y[t]==Exp[-a*t]}
eq1=LaplaceTransform[ode,t,s]/.{y[0]->1}
eq2=Solve[eq1,LaplaceTransform[y[t],t,s]]
sol=Map[InverseLaplaceTransform[#,s,t]&,eq2,{3}]/.{a->7}
DSolve[{ode,y[0]==1},y[t],t]
Plot[y[t]/.sol,{t,0,1},AspectRatio->1,
 PlotStyle->{Hue[0.7],Thickness[0.01]}]
```

$\square$

Applications to partial differential equations

Problem 10.4 Obtain the solution of the initial boundary-value problem for the wave equation describing the transverse vibration of a semi-infinite string,

$$u_{tt} = c^2 u_{xx}, \quad 0 \le x < \infty, \quad t > 0,$$

with the initial and boundary conditions

$$u(x,0) = 0, \quad u_t(x,0) = 0, \quad u(0,t) = Af(t), \quad u(x,t) \to 0, \text{ as } x \to 0.$$

Maple:

```
interface(showassumed=0): with(inttrans): with(PDEtools):
assume(c>0,s>0); declare(u(x,t),U(x)); ON;
Eq1:=diff(u(x,t),t$2)-c^2*diff(u(x,t),x$2);
Eq2:=laplace(Eq1,t,s); Eq3:=subs({laplace(u(x,t),t,s)=U(x),
 laplace[x,x]=diff(u1(x),x$2)},Eq2);
IC1:={u(x,0)=0,D[2](u)(x,0)=0}, IC2:=laplace(IC1,t,s);
BC1:=u(0,t)=A*f(t); BC2:=laplace(BC1,t,s);
BC3:=subs({laplace(u(0,t),t,s)=U(0),laplace(f(t),t,s)=F(s)},
 BC2); Eq4:=subs(IC1,Eq3); Sol:=dsolve(Eq4,U(x));
```

```
Sol1:=_C1*exp(-s*x/c)+_C2*exp(s*x/c);
l1:=simplify(limit(op(1,Sol1),x=infinity));
l2:=simplify(limit(op(2,Sol1),x=infinity));
Sol2:=subs(_C2=0,Sol1);
Sol3:=subs(_C1=op(2,BC3),Sol2);
U1:=subs(F(s)=laplace(f(t),t,s),Sol3);
Sol_F:=simplify(invlaplace(U1,s,t)) assuming t>0 and x>0;
convert(Sol_F,piecewise,t);
```

Mathematica:

```
eq1=D[u[x,t],{t,2}]==c^2*D[u[x,t],{x,2}]
eq2=LaplaceTransform[eq1,t,s]/.
 {LaplaceTransform[u[x,t],t,s]->u1[x],
  LaplaceTransform[D[u[x,t],{x,2}],t,s]->u1''[x]}
ic1={u[x,0]->0,(D[u[x,t],t]/.{t->0})->0}
ic2=Map[LaplaceTransform[#,t,s]&,ic1,{2}]
eq3=(eq2/.ic1)
bc1=u[0,t]==a*f[t]
bc2=LaplaceTransform[bc1,t,s]/.
 {LaplaceTransform[u[0,t],t,s]->u1[0],
  LaplaceTransform[f[t],t,s]->f1[s]}
{sol=DSolve[eq3,u1[x],x], sol1=u1[x]/.sol}
l1=Limit[sol1[[1,1]],x->Infinity,
    Assumptions->{c>0,s>0}]//Simplify
l2=Limit[sol1[[1,2]],x->Infinity,
    Assumptions->{c>0,s>0}]//Simplify
{sol2=sol1/.{C[1]->0}, sol3=sol2/.{C[2]->bc2[[2]]}}
u2=sol3/.f1[s]->LaplaceTransform[f[t],t,s]
solFin=InverseLaplaceTransform[u2,s,t]//PiecewiseExpand
```

□

Distributions and Laplace transform

Maple:

```
interface(showassumed=0): with(inttrans); assume(a>0,b>0);
simplify(laplace(a*(t-b)^2*Heaviside(t-b),t,s));
factor(invlaplace(exp(-a*s)/s^2,s,t));
```

Mathematica:

```
Assuming[{a>0,b>0},FullSimplify[LaplaceTransform[
  a*(t-b)^2*HeavisideTheta[t-b],t,s]]]
Assuming[a>0,InverseLaplaceTransform[Exp[-a*s]/s^2,s,t]]
```

Problem 10.5 Find the solution of the inhomogeneous Cauchy problem

$$Ay'' + By = f(t), \quad y(0) = 0, \quad y'(0) = 0,$$

where $f(t)$ is a given function representing a source term

$$f(t) = H(t) - H(t - \pi).$$

Maple:

```
with(inttrans); f:=t->Heaviside(t)-Heaviside(t-Pi);
ODE:=A*diff(y(t),t$2)+B*y(t)=f(t);
Eq1:=laplace(ODE,t,s);
Eq2:=subs({y(0)=0,D(y)(0)=0},Eq1);
Eq3:=simplify(solve(Eq2,laplace(y(t),t,s)));
Sol:=invlaplace(Eq3,s,t);
combine(dsolve({ODE,y(0)=0,D(y)(0)=0},y(t)));
convert(convert(Sol,piecewise),cos);
Sol1:=eval(Sol,{A=5,B=10}); plot(Sol1,t=0..10*Pi);
combine(convert(Sol1,cos));
```

Mathematica:

```
f[t_]:=HeavisideTheta[t]-HeavisideTheta[t-Pi];
ode={a*y''[t]+b*y[t]==f[t]}
eq1=LaplaceTransform[ode,t,s]/.{y[0]->0,y'[0]->0}
eq2=Solve[eq1,LaplaceTransform[y[t],t,s]]
eq3=eq2[[1,1,2]]
sol=InverseLaplaceTransform[eq3,s,t]
DSolve[{ode,y[0]==0,y'[0]==0},y[t],t]//Simplify
{sol//FullSimplify, f=sol/.{a->5,b->10}}
Plot[f,{t,0,10*Pi},PlotStyle->{Hue[0.8],Thickness[0.01]}]
```

□

Initial value problems for ODE systems with the Laplace transform

Problem 10.6 Applying the Laplace transform, solve the initial value problem

$$x' - 2y = t, \quad 4x + y' = 0, \quad x(0) = 1, \quad y(0) = 0,$$

and graph the solution.

Maple:

```
with(inttrans): ICs:={x(0)=1,y(0)=0};
OdeSys:={D(x)(t)-2*y(t)=t,4*x(t)+D(y)(t)=0};
Sol1:=dsolve(OdeSys union ICs,{x(t),y(t)});
Sol2:=dsolve(OdeSys union ICs,{x(t),y(t)},method=laplace);
odetest(Sol1,OdeSys union ICs);
Eq1:=laplace(OdeSys,t,s); Eq2:=subs(ICs,Eq1);
Eq3:=solve(Eq2,{laplace(x(t),t,s),laplace(y(t),t,s)});
Sol3:=invlaplace(Eq3,s,t); assign(Sol3):
plot([x(t),y(t),t=0..Pi],color=blue,thickness=3);
```

Mathematica:

```
odeSys={x'[t]-2*y[t]==t,4*x[t]+y'[t]==0}
eq1=LaplaceTransform[odeSys,t,s]
eq2=Solve[eq1,{LaplaceTransform[x[t],t,s],
  LaplaceTransform[y[t],t,s]}]
sol1=Map[InverseLaplaceTransform[#,s,t]&,eq2,{3}]/.
  {x[0]->1,y[0]->0}
sol2=DSolve[{odeSys,x[0]==1,y[0]==0},
  {x[t],y[t]},t]//Simplify
ParametricPlot[Evaluate[{x[t],y[t]}/.sol1],{t,0,Pi},
  PlotStyle->{Hue[0.5],Thickness[0.01]},AspectRatio->1]
```

$\square$

10.2 Integral Fourier Transforms

In Maple, integral Fourier transforms are defined with the functions
fourier, invfourier, fouriercos, fouriersin. It should be noted
that in *Maple* the product of normalization factors is equal to
$1/(2\pi)$.

In Mathematica, integral Fourier transforms are defined with a family of functions FourierTransform, InverseFourierTransform, etc. (for more details, see ?*Fourier*). Note that there is the option FourierParameters allowing one to choose different conventions used for defining Fourier transforms.

Problem 10.7 Find the Fourier transform for $f(t)=e^{-t^2}$, $f(t)=e^{-a(t-b)^2}$.

Maple:

```
interface(showassumed=0): with(inttrans);
f1:=t->exp(-t^2); f1(t); F1:=s->fourier(f1(t),t,s); F1(s);
plot({f1(x),F1(x)}, x=-Pi..Pi, color=[blue, green]);
assume(a>0,b>0); f2:=t->exp(-a*(t-b)^2);
F2:=s->fourier(f2(t),t,s); f2(t); expand(F2(s));
```

Mathematica:

```
f1[t_]:=Exp[-t^2]; f2[t_]:=Exp[-a*(t-b)^2];
g[s_]:=FourierTransform[f1[t],t,s]
g1[s_]:=FourierTransform[f1[t],t,s,
 FourierParameters->{1,-1}]; {f1[t], g[s], g1[s]}
Plot[Evaluate[{f1[x],g1[x]}],{x,-10,10},PlotRange->{{-10,10},
 {0,2}},PlotStyle->{Hue[0.5],Hue[0.7]},AspectRatio->1]
g2[s_]:=Assuming[a>0&&b>0,FourierTransform[f2[t],t,s,
 FourierParameters->{1,-1}]]; {f2[t],Evaluate[g2[s]]}
```

□

Problem 10.8 Find the Fourier transform for $f(t) = \begin{cases} 1, & |t| < a \\ 0, & |t| \geq a. \end{cases}$

Maple:

```
interface(showassumed=0): with(inttrans); assume(a>0);
C:=color=[red,blue]; f1:=piecewise(abs(t)<a,1,0);
f2:=unapply(convert(f1,Heaviside),t);
F1:=s->fourier(f2(t),t,s); f2(t); F1(s); limit(F1(s),k=0);
L:=[f2(x),F1(x)]; L1:=subs(a=1,L[1..2]);
plot(L1,x=-3*Pi..3*Pi,thickness=[3,2],C);
```

Mathematica:

```
f1[t_]:=Piecewise[{{1,Abs[t]<a},{0,Abs[t]>=a}}];
g1[s_]:=Assuming[a>0,FourierTransform[f1[t],t,s,
 FourierParameters->{1,-1}]]; {f1[t]//TraditionalForm, g1[s],
 Limit[g1[s],s->0,Assumptions->{a>0}]}
{l={PiecewiseExpand[f1[x]],g1[x]}, l1=l/.{a->1}}
Plot[Evaluate[l1],{x,-3*Pi,3*Pi},Exclusions->True,
 PlotStyle->{{Red,Thickness[0.01]},{Blue,Thickness[0.01]}}]
```

□

Problem 10.9 Find the inverse Fourier transform for $1/p^2$, the Fourier Sine transform for the exponential integral Ei(x), and the Fourier Cosine transform for the Bessel function BesselY(0,x).

Maple:

```
with(inttrans); t1:=invfourier(1/(1+ p^2),p,t);
convert(t1,piecewise); fouriersin(Ei(x),x,p);
fouriercos(BesselY(0,x),x,p);
```

Mathematica:

```
{InverseFourierTransform[1/(1+p^2),p,t,FourierParameters->
  {1,-1}], FourierSinTransform[ExpIntegralEi[x],x,p,
 FourierParameters->{0,1}]//Expand,
 FourierCosTransform[BesselY[0,x],x,p]}
```

□

Applications to partial differential equations

Problem 10.10 Obtain the d'Alembert solution of the initial-value problem for the wave equation

$$u_{tt} = c^2 u_{xx}, \quad -\infty < x < \infty, \quad t > 0,$$

with the initial conditions $u(x,0) = f(x)$, $u_t(x,0) = g(x)$.

Maple:

```
with(inttrans): with(PDEtools): declare(u(x,t),U(t)); ON;
Eq1:=diff(u(x,t),t$2)-c^2*diff(u(x,t),x$2);
Eq2:=subs(fourier(u(x,t),x,k)=U(t),fourier(Eq1,x,k));
IC1:={u(x,0)=f(x),D[2](u)(x,0)=g(x)};
IC2:=fourier(IC1,x,k);
IC3:=subs({fourier(D[2](u)(x,0),x,k)=D(U)(0),
 fourier(u(x,0),x,k)=U(0),fourier(f(x),x,k)=F(k),
 fourier(g(x),x,k)=G(k)}, IC2);
sys:={Eq2} union IC3; Sol:=dsolve(sys,U(t));
U1:=convert(rhs(Sol),exp); Sol2:=invfourier(U1,k,x);
Sol3:=factor(combine(convert(Sol2*(2*Pi),int)));
N1:=factor(subs([G(k)=0],op(1,Sol3)));
N2:=subsop(2=1, 3=f(x+c*t)+f(x-c*t), N1);
N3:=factor(subs([F(k)=0],op(1,Sol3)));
N4:=simplify(N3/G(k)/int(exp(I*k*xi),xi=x-c*t..x+c*t));
u_dAlembert:=N2+N4*int(g(xi),xi=x-c*t..x+c*t);
```

Mathematica:

```
eq1=D[u[x,t],{t,2}]==c^2*D[u[x,t],{x,2}]
eq2=Map[FourierTransform[#,x,k]&,eq1]/.
 {FourierTransform[u[x,t],x,k]->U[t],
  FourierTransform[D[u[x,t],{t,2}],x,k]->U''[t]}
ic1={u[x,0]==f[x],(D[u[x,t],t]/.{t->0})==g[x]}
ic2=Map[FourierTransform[#,x,k]&,ic1,{2}]
ic3=ic2/.{FourierTransform[(D[u[x,t],t]/.{t->0}),x,k]->U'[0],
 FourierTransform[u[x,0],x,k]->U[0], FourierTransform[
 f[x],x,k]->f1[k], FourierTransform[g[x],x,k]->g1[k]}
{sys=Union[{eq2},ic3], sol=DSolve[sys,U[t],t],
 sol1=U[t]/.sol//Simplify,
 u1=sol1/.{Sin[a_]:>-I/2*(Exp[I*a]-Exp[-I*a]),
  Cos[a_]:>1/2*(Exp[I*a]+Exp[-I*a])}//Expand,
 u2=Integrate[u1*Exp[x*k*I],{k,-Infinity,Infinity}]//Simplify}
{n1=u2[[1,1]]/.g1[k]->0//ExpandAll, n11=Collect[n1,f1[k]/2],
 n2=n11/.{f1[k]->1,n11[[2]]->f[x+c*t]+f[x-c*t]},
 n3=u2[[1,1]]/.f1[k]->0//ExpandAll,
 n4=(n3/g1[k])/Integrate[Exp[I*k*\[Xi]],
  {\[Xi],x-c*t,x+c*t}]//Simplify}
udAlembert=n2+n4*Integrate[g1[\[Xi]],{\[Xi],x-c*t,x+c*t}]
```

□

10.3 Discrete Fourier Transforms

In Maple, with the new `DiscreteTransforms` package (*Maple* ≥ 9), the
fast Fourier transform and inverse transform of a single or multidimensional data can be calculated. It should be noted that these
functions work in hardware precision (`Digits:=15`), but very fast
(in compiled code). Also *fast Fourier transforms* can be calculated
with the `FFT` and `iFFT` functions.

In Mathematica, the *discrete Fourier transforms* are defined with the
`Fourier` and `InverseFourier` functions. It should be noted that
there is the option `FourierParameters` allowing one to choose different conventions used for defining discrete Fourier transforms. In
the algorithm known as the Fast Fourier Transform (FFT), it is
recommended that the length of the original list be a power of 2.
This can reduce the computation time of discrete Fourier transform. But *Mathematica* can work well with lists whose lengths are
different from a power of 2.

Problem 10.11 Generate a signal, construct the corresponding frequency spectrum, and recover the original signal.

Maple:

```
with(DiscreteTransforms): with(plots):
with(LinearAlgebra): setoptions(axes=boxed):
N:=64; f:=0.15; Freq:=500; N_band:=N/2;
Freq_max:=Freq/2; Step_band:=evalf(Freq_max/N_band);
A:=array(1..2); Sig:=Vector(N,n->sin(n*2*Pi*f)):
A[1]:=listplot(Sig,color=magenta,title="Original Signal"):
freq:=FourierTransform(Sig, N);
Points_DFT:=[seq([i,evalc(Re(freq[i]))],i=1..N)]:
pointplot(Points_DFT,color=blue,thickness=3,title="DFT");
Points_Sp:=[seq([i*Step_band,abs(freq[i])],i=1..N_band)]:
plot(Points_Sp,title="Frequency Spectrum",color=green);
Sig_inv:=InverseFourierTransform(freq,N);
Points_inv:=[seq([i,Re(Sig_inv[i])],i=1..N)]:
A[2]:=plot(Points_inv,color=blue,title="Recovered Signal"):
display(A,tickmarks=[2,8]);
```

Mathematica:

```
SetOptions[ListPlot,ImageSize->300,Frame->True,Axes->None,
  AspectRatio->1]; {n=64,f=0.15,freq=500,nBand=n/2,
  freqMax=freq/2,stepBand=freqMax/nBand//N}
sig=Table[Sin[i*2*Pi*f],{i,1,n}]
g1=ListPlot[sig,PlotStyle->Magenta,Joined->True,
 PlotLabel->"Original Signal"]; f1=Fourier[sig]//Chop
pointsFFT=Table[{i,ComplexExpand[Re[f1[[i]]]]},{i,1,n}]
ListPlot[pointsFFT,PlotStyle->{PointSize[0.018],Blue},
 PlotLabel->"FFT",AspectRatio->1]
pointsSp=Table[{i*stepBand,Abs[f1[[i]]]},{i,1,nBand}]
ListPlot[pointsSp,PlotStyle->Green,Joined->True,
 PlotLabel->"Frequency Spectrum",PlotRange->All]
sigInv=InverseFourier[f1]
pointsInv=Table[{i,ComplexExpand[Re[sigInv[[i]]]]},{i,1,n}]
g2=ListPlot[pointsInv,PlotStyle->Blue,Joined->True,
 PlotLabel->"Recovered Signal"]; GraphicsRow[{g1,g2}]
```

☐

Problem 10.12 Generate a signal, obtain the fast Fourier transform, and recover the original signal.

Maple:

```
with(plots): setoptions(axes=boxed): f:=0.15: n:=6; N:=2^n;
A:=array(1..2); Sig_Re:=Vector(N,[sin(2*Pi*f*i)$i=1..N]):
Sig_Im:=Vector(N,0): A[1]:=listplot(Sig_Re,color=blue,
 title="Original Signal"): FFT(n, Sig_Re, Sig_Im);
G1:=listplot(Sig_Re,color=magenta,title="FFT"):
iFFT(n, Sig_Re, Sig_Im);
A[2]:=listplot(Sig_Re,color=red,title="Recovered Signal"):
display(A,tickmarks=[2,8]); display(G1);
```

Mathematica:

```
SetOptions[ListPlot,AspectRatio->1,ImageSize->300,Frame->True];
{f=0.15, n=6, n1=2^n, sigRe=Table[Sin[2*Pi*f*i],{i,1,n1}]}
{f1=Fourier[sigRe]//Chop, f2=InverseFourier[f1]}
ListPlot[ComplexExpand[Re[f1]],PlotStyle->Magenta,
 Joined->True,PlotLabel->"FFT"]
g1=ListPlot[sigRe,PlotStyle->Blue,Joined->True,
 PlotLabel->"Original Signal"]; g2=ListPlot[f2,PlotStyle->Red,
 Joined->True,PlotLabel->"Recovered Signal"];
GraphicsRow[{g1,g2}]
```

☐

Problem 10.13 Let $f(x) = \exp(-(t-\pi))$ for $t \in [0, 2\pi]$. Find the Fourier polynomial of degree n.

Maple:

```
with(DiscreteTransforms): with(plots):
Digits:=15: N:=9; M:=2^N; S:=NULL; L:=Pi; K:=20;
DataRe:=Vector(M,0); cA:=Vector(K+1,0);
cB:=Vector(K+1,0); f:=t->exp(-(t-Pi));
for i from 0 to M-1 do
 x:=evalf(2*Pi*i/M); S:=S,[x,evalf(f(x))]; od:
for i from 1 to M do DataRe[i]:=S[i][2]; od:
DataIm:=Vector(M,0): op(DataRe);
Sol:=evalf(2/sqrt(M)*FourierTransform(DataRe)): op(Sol);
for i from 1 to K+1 do
 cA[i]:=Re(Sol[i]); cB[i]:=-Im(Sol[i]); od:
F:=unapply(cA[1]/2+add(cA[i+1]*cos(X*i*Pi/L)
 +cB[i+1]*sin(X*i*Pi/L),i=1..K),X);
G1:=plot(f(t),t=0..2*L,color=blue):
G2:=plot(F(X),X=0..2*L,view=[0..2*L,0..23],color=magenta,
 thickness=3): G:=array(1..2); G[1]:=G1: G[2]:=G2:
display(G,axes=boxed,scaling=constrained);
display([G1,G2],axes=boxed);
```

Mathematica:

```
SetOptions[Plot,AspectRatio->1,ImageSize->250,Frame->True,
 PlotRange->All]; nD=15; f[t_]:=Exp[-(t-Pi)];
{n=9,m=2^n,s={},L=Pi}
For[i=0,i<=m-1,i++, x=N[2*Pi*i/m,nD];
s=Append[s,{x,N[f[x],nD]}]]; data=Table[s[[i,2]],{i,1,m}]
sol=2/Sqrt[m]*Chop[Fourier[data]]
{k=20, cA=Re[Take[sol,k+1]], cB=Im[Take[sol,k+1]]}
F[x_]:=cA[[1]]/2+Sum[cA[[i+1]]*Cos[x*i*Pi/L]
 +cB[[i+1]]*Sin[x*i*Pi/L],{i,1,k}];
g1=Plot[f[t],{t,0,2*L},PlotStyle->Hue[0.7]];
g2=Plot[F[x],{x,0,2*L},PlotStyle->Hue[0.9]];
GraphicsRow[{g1,g2,Show[{g1,g2}]}]
```

□

In Mathematica, with the `FourierSeries`˙ package, the numerical approximations to the Fourier transforms can be calculated. This package also includes functions for the Fourier series, Fourier coefficients, discrete-time Fourier transforms.

In this package the options of the function NIntegrate and the option FourierParameters are valid. Here we discuss the most important functions:

```
<<FourierSeries`                       NFourierTransform[f,t,w]
DTFourierTransform[f,n,w]              NDTFourierTransform[f,n,w]
NInverseFourierTransform[f,w,t] InverseDTFourierTransform[f,w,n]
                              NInverseDTFourierTransform[f,w,n]
```

```
<<FourierSeries`
f=Exp[-t^4];
FourierTransform[f,t,s]//Timing
NFourierTransform[f,t,-10]//Timing
points=Table[NFourierTransform[f,t,i],{i,-20,20}]
ListPlot[Abs[points],Joined->True,PlotStyle->
 {Thickness[0.01],Blue},PlotRange->All,Frame->True]
Plot[f,{t,-Pi,Pi}]
```

10.4   Hankel Transforms

The Hankel transform is defined formally by

$$\mathcal{H}_n\{f(t)\} = F_n(s) = \int_0^\infty t J_n(st) f(t)\, dt, \quad 0 < s < \infty,$$

where $J_n(st)$ is the Bessel function of the first kind of order n (see Chapter 9). The Hankel transforms of order zero $(n = 0)$ and of order one $(n = 1)$ are useful for solving initial-value and boundary-value problems involving the Laplace or Helmholtz equations in an axisymmetric cylindrical geometry.

In Maple, the Hankel transforms can be obtained with the function

hankel according to the formula $F(s) = \int_0^\infty f(t) J_n(st) \sqrt{st}\, dt$ and the inttrans package.

In Mathematica, there is no built-in function for constructing Hankel transforms, so we define the function HankelTransform.

Problem 10.14   Find the Hankel transforms for the functions $f(r)$, $\delta(a - r)/r$, and $H(a - r)$, where $\delta(r)$ and $H(r)$ are, respectively, the Dirac δ-function and the Heaviside step function (see Chapter 9).

Maple:

```
interface(showassumed=0): with(inttrans); assume(a>0);
convert(hankel(f(r),r,k,0),int);
hankel(Dirac(a-r)/r,r,k,0);
t1:=hankel(Heaviside(a-r),r,k,0); convert(t1,Hankel);
factor(int(Heaviside(a-t)*BesselJ(0,t*s)*t,t=0..infinity))
 assuming s>0 and t>0;
```

Mathematica:

```
HankelTransform[f_,t_,s_,n_,assump_List:{}]:=
 Module[{},FullSimplify[Evaluate[Integrate[
 FunctionExpand[f[t]*BesselJ[n,s*t]*t],{t,0,Infinity},
 Assumptions->{t>0,s>0}],assump]]];
f2[r_]:=DiracDelta[a-r]/r; f3[r_]:=HeavisideTheta[a-r];
{HankelTransform[f1,r,k,0,{a>0,k>0}],
 HankelTransform[f2,r,k,0,{a>0,r\[Element]Reals,r>0,k>0}],
 HankelTransform[f3,r,k,0,{a>0,r\[Element]Reals,r>0,k>0}]}//
 FullSimplify
```

□

Problem 10.15   Find the Hankel transforms for the functions $\exp(-ar)$ and $\exp(-ar)/r$.

Maple:

```
interface(showassumed=0): with(inttrans); assume(a>0);
hankel(exp(-a*r),r,k,0);
addtable(hankel,exp(-a*r),k*(a^2+k^2)^(-3/2),r,k,
 hankel=n::Range(-infinity,infinity));
hankel(exp(-a*r),r,k,0);
factor(int(1/t*exp(-a*t)*BesselJ(0,t*s)*t,t=0..infinity))
 assuming s>0 and t>0;
```

Mathematica:

```
HankelTransform[f_,t_,s_,n_,assump_List:{}]:=Module[{},
 FullSimplify[Evaluate[Integrate[FunctionExpand[
 f[t]*BesselJ[n,s*t]*t],{t,0,Infinity},
 Assumptions->{t>0,s>0}],assump]]];
f1[r_]:=Exp[-a*r]; f2[r_]:=1/r*Exp[-a*r];
HankelTransform[f1,r,k,0,
 {a>0,r\[Element]Reals,r>0,k>0}]//FullSimplify
HankelTransform[f2,r,k,0,
 {a>0,r\[Element]Reals,r>0,k>0}]//FullSimplify
```

The Hankel transforms of unknown functions (for *Maple*) can be added with
the addtable function. □

Problem 10.16 The Hankel transform is self-inverting for $n > -1/2$.
Find the forward and inverse Hankel transforms for $1/\sqrt{r}$.

Maple:

```
with(inttrans); F:=hankel(r^(-1/2),r,k,0); hankel(F,k,r,0);
```

Mathematica:

```
f[r_]:=1/Sqrt[r];
HankelTransform[f_,t_,s_,n_,assump_List:{}]:=Module[{},
 FullSimplify[Evaluate[Integrate[FullSimplify[f[t]*
 BesselJ[n,s*t]*Sqrt[s*t]],{t,0,Infinity},
 Assumptions->{s>0,t>0}],assump]]];
HankelTransform[f,r,k,0,{r\[Element]Reals,r>0,k>0}]
HankelTransform[f,k,r,0,{r\[Element]Reals,r>0,k>0}]
```
□

Applications to partial differential equations

Problem 10.17 Obtain the solution of the free vibration of a large
circular membrane described by the initial-value problem:

$$u_{tt} = c^2(u_{rr} + u_r/r), \quad 0 \le r < \infty, \quad t > 0,$$
$$u(r,0) = f(r), \quad u_t(r,0) = g(r), \quad 0 \le r \le \infty,$$

where $f(r)$ and $g(r)$ are arbitrary functions (use the *Maple* function
hankel).

Maple:

```
with(inttrans): with(PDEtools):
declare(u(r,t),U(t)); ON;
Eq1:=c^2*(diff(u(r,t),r$2)+1/r*diff(u(r,t),r))=
 diff(u(r,t),t$2);
Eq2:=hankel(Eq1,r,k,0);
Eq3:=subs({hankel(u(r,t),r,k,0)[t,t]=diff(U(t),t$2),
 hankel(u(r,t),r,k,0)=U(t)},Eq2);
IC1:={u(r,0)=f(r),D[2](u)(r,0)=g(r)};
IC2:=hankel(IC1,r,k,0);
IC3:=subs({hankel(D[2](u)(r,0),r,k,0)=D(U)(0),
 hankel(u(r,0),r,k,0)=U(0),hankel(f(r),r,k,0)=F(k),
 hankel(g(r),r,k,0)=G(k)},IC2);
sys:={Eq3} union IC3; Sol:=dsolve(sys,U(t));
Sol_Fin:=expand(convert(hankel(rhs(Sol),k,r,0),int));
```

Mathematica:

```
f1[r_]:=c^2*(D[u[r,t],{r,2}]+1/r*D[u[r,t],r]);
HankelTransform[f_,t_,s_,n_,assump_List:{}]:=
 Module[{},FullSimplify[Evaluate[Integrate[FunctionExpand[
 f[t]*BesselJ[n,s*t]*Sqrt[s*t]],{t,0,Infinity},
 Assumptions->{s>0,t>0}],assump]]];
eq21=HankelTransform[f1,r,k,0,{r\[Element]Reals,r>0,k>0}]
eq3=-c^2*k^2*U[t]==D[U[t],{t,2}]
ic3={U[0]==F[k],(D[U[t],t]/.{t->0})==G[k]}
{sys={eq3,ic3}, sol=DSolve[sys,U[t],t]//Expand}
f2[k1_]:=(sol[[1,1,2]]//InputForm)/.{k->k1};
solFin=HankelTransform[f2,k,r,0,{r\[Element]Reals,r>0,k>0}]
```

□

Chapter 11

Mathematical Equations

11.1 Algebraic and Trascendental Equations

Exact solutions of algebraic equations and systems of equations. Exact representations of roots of trascendental equations

Maple:

```
solve(Eq,var);                 solve({Eq1,...,Eqn},{var1,...varn});
RootOf(expr,x);       allvalues(expr,ops);       isolate(Eq,expr);
```

solve, RootOf, representing for roots of equations,

allvalues, computing all possible values of expressions that contain RootOfs,

isolate, isolating a subexpression to left side of an equation.

Mathematica:

```
Solve[eq,var]                    Solve[{eq1,...,eqn},{var1,...,varn}]
Solve[eq,var,VerifySolutions->False]    Reduce[eqs,vars,domain]
Root[eq,var]      Roots[eq,var]          CountRoots[f,{x,x1,x2}]
RootIntervals[f]RootIntervals[f,Complexes] IsolatingInterval[f]
```

Solve, Root, Roots, representing for roots of equations,

Reduce, finding all possible solutions or verifying identities,

CountRoots, RootIntervals, IsolatingInterval, counting and isolating roots of polynomials.

Note. The roots obtained by Solve are expressed as a list of transformation rules x->x1. This notation indicates that the solution x=x1, but x is not replaced by

I.K. Shingareva, C. Lizárraga-Celaya, *Maple and Mathematica*, 2nd ed.,
DOI 10.1007/978-3-211-99432-0_11, © Springer-Verlag Vienna 2009

the value x1. For accessing the values of the solutions, we can use the replacement operator (/.) or list operation functions Part or [[]].

In *Mathematica 7*, a new approach for solving trascendental equations has been developed. This approach allows us to obtain *exact representations* of trascendental equation roots in a symbolic form and *numerical approximations* of transcendental roots with arbitrary precision. For more details, see the functions Root, Reduce, Maximize, Minimize.

Problem 11.1 Find the exact solutions of the equations: $3x + 11 = 5$, $x^3 - 5x^2 + 2x - 1 = 0$, $\cos^2 x - \cos x - 1 = 0$, $(x^2 - 1)/(x^2 + 5) + 1 = 0$. Find the inverse function of $f(x) = (2x - 3)/(1 - 5x)$. Solve the equation $V = \pi r^2 / h$ for h.

Maple:

```
solve(3*x+11=5,x); solve((x^2-1)/(x^2+5)+1=0,x);
sol1:=solve(x^3-5*x^2+2*x-1=0,x); evala(product(op(i,[sol1]),
 i=1..3)); solve(cos(x)^2-cos(x)-1=0,x);
F_inv:=unapply(solve(y=(2*x-3)/(1-5*x),x),y); F_inv(x);
solve(v=Pi*r^2/h, h); isolate(v=Pi*r^2/h, h);
solve(sqrt(x)+x=1,x); evalb(factor(x^2-1)=(x-1)*(x+1));
evalf(solve([x^5-x^3+2*x-10=0,x>0],{x}));
```

Mathematica:

```
{Solve[3*x+11==5], Solve[Cos[x]^2-Cos[x]-1==0]}
sol1=Solve[x^3-5*x^2+2*x-1==0]//Simplify
Product[sol1[[i,1,2]],{i,1,3}]//Simplify
fInv[t_]:=Solve[y==(2*x-3)/(1-5*x),x]/.y->t; fInv[x]
{Solve[v==Pi*r^2/h, h], Reduce[v==Pi*r^2/h, h]}
Solve[Sqrt[x]+x==1,x,VerifySolutions->False]
Reduce[x^2-1==(x-1)*(x+1),x]
Reduce[x^5-x^3+2*x-10==0 && x\[Element]Reals, x]//N
```

$\square$

Problem 11.2 Solve the systems of equations:

$$\begin{cases} x^2 + y^2 = 4, \\ y = 2x, \end{cases} \qquad \begin{cases} 2x - 3y + 4z = 2, \\ 3x - 2y + z = 0, \\ x + y + z = 1. \end{cases}$$

Find the values of the expression $\sin^2(x) + \cos^2(y)$ at the solution points.

Maple:

```
Sys1:={x^2+y^2=4,y=2*x}; Var1:={x,y};
Sol1:=evalf(allvalues(solve(Sys1,Var1)));
Sys2:={2*x-3*y+4*z=2,3*x-2*y+z=0,x+y+z=1};
Var2:={x,y,z}; Sol2:=solve(Sys2, Var2);
Expr:=sin(x)^2+cos(y)^2; [[evalf(subs(Sol1[1],Expr)),
  evalf(subs(Sol1[2],Expr))],eval(subs(Sol2,Expr))];
```

Mathematica:

```
nD=10; sys1={x^2+y^2==4,y==2*x}; var1={x,y};
sol1=N[Solve[sys1,var1],nD]
sys2={2*x-3*y+4*z==2,3*x-2*y+z==0,x+y+z==1}; var2={x,y,z};
sol2=Solve[sys2,var2]
Sin[x]^2+Cos[y]^2/.{sol1,sol2}
```

□

Numerical solutions of algebraic and trascendental equations

For a more detail discussion over numerical solutions of algebraic and trascendental equations, see Chapter 12 and Sect. 12.1).

Maple:

```
Digits:=n;            evalf(solve(Eq,var));      evalf(root(x,n));
fsolve(Eq,var,ops);                       fsolve(Eq,var=a..b,ops);
fsolve(Eq,var,complex);                   fsolve(f,var=a..b,ops);
with(RootFinding);    Isolate(f);         NextZero(f(x),x0,ops);
Isolate([f1,f2],[x1,x2]);                 Homotopy([f1,f2],ops);
                      use RealDomain in solve(eq,var); end use;
```

Mathematica:

```
NSolve[eqs,vars,ops]  NSolve[eqs,vars,n]    N[Solve[eqs,vars],n]
FindRoot[eq,{x,x0,a,b}]                     FindRoot[eq,{x,{x0,x1}}]
FindRoot[eqs,{x1,x10},{x2,x20},...]   FindRoot[eq,{var,I},ops]
CountRoots[f,x]          NRoots[eq,var]              N[Root[f,x,k]
RootIntervals[f,domain]     FindInstance[eq1&&eq2,vars,domain,n]
<<FunctionApproximations`        InterpolateRoot[eq,{x,x0,x1}]
```

Here x0, x1 are the initial values of the variable x and $[a, b]$ is the interval outside of which the iteration process stops.
The function FindRoot solves equations using iterative methods (Newton's method, the secant method) and has many options (see Options[FindRoot]).

Problem 11.3 Approximate the values of x that satisfy the equations $x^5 - 2x^2 = 1 - x$, $1 - x^2 = x^3$.

Maple:

```
eq1:=x^5-2*x^2=1-x; map(evalf,[solve(eq1,x)]);
map(evalf,[solve(eq1,x)],20); map(evalf,[solve(1-x^2=x^3,x)]);
```

Mathematica:

```
{NSolve[x^5-2*x^2==1-x,x,20], N[Solve[x^5-2*x^2==1-x,x],20],
 NSolve[1-x^2==x^3,x],          N[Solve[1-x^2==x^3,x]]}
```

□

Problem 11.4 Find the numerical solutions of the equation $5x^5 - 4x^4 - 3x^3 + 2x^2 - x - 1 = 0$. Approximate the roots of $\cos(\sin x) - x^2 = 0$. Find a numerical approximation to the solution of the equation $\sin x = x/2$ for $x \in (0, \pi]$.

Maple:

```
Eq1:=5*x^5-4*x^4-3*x^3+2*x^2-x-1=0;
fsolve(Eq1,x); fsolve(Eq1,x,complex);
f:=x->cos(sin(x))-x^2; plot(f(x),x=-1..1);
S1:=fsolve(f(x),x=-1..0); S2:=fsolve(f(x),x=0..1);
fsolve(sin(x)=x/2,x=0..Pi,avoid={x=0});
```

Mathematica:

```
nD=10; eq1=5*x^5-4*x^4-3*x^3+2*x^2-x-1==0
{sol1=Reduce[eq1&&x\[Element]Complexes,x],N[sol1,nD]}
{sol2=Reduce[eq1&&x\[Element]Reals,x],N[sol2,nD]}
{FindRoot[eq1,{x,0},MaxIterations->30],FindRoot[eq1,{x,I}],
 FindRoot[eq1,{x,-I}],FindRoot[eq1,{x,1}],FindRoot[eq1,{x,-0.5}]}
f[x_]:=Cos[Sin[x]]-x^2; Plot[f[x],{x,-1,1}]
sol1=FindRoot[f[x],{x,-0.9,-1,0},WorkingPrecision->30]
sol2=FindRoot[f[x],{x,0.1,0,1},DampingFactor->2]
FindRoot[Sin[x]==x/2,{x,0.0001,Pi}]
```

□

Problem 11.5 Find exactly and approximately the intersection points of the graphs $f_1(x) = x^2 + x + 10$ and $f_2(x) = 9x^2$.

Maple:

```
f1:=x->x^2+x+10; f2:=x->9*x^2;
plot({f1(x),f2(x)},x=-3..3); sols:=[solve(f1(x)=f2(x),x)];
for i from 1 to nops(sols) do
 evalf(sols[i]), evalf(f1(sols[i])); evalf(f2(sols[i])) od;
```

Mathematica:

```
f1[x_]:=x^2+x+10; f2[x_]:=9*x^2;
Plot[{f1[x],f2[x]},{x,-3,3}]
sols=Solve[f1[x]==f2[x],x]; {x,f1[x],f2[x]}/.sols//Simplify//N
```

□

Problem 11.6 Find the numerical solutions of $\begin{cases} \sin x - \cos y = 1, \\ e^x - e^{-y} = 1. \end{cases}$

Maple:

```
with(plots): sys:={sin(x)-cos(y)=1,exp(x)-exp(-y)=1};
implicitplot(sys,x=-Pi..Pi,y=-Pi..Pi);
fsolve(sys,{x=0,y=3}); fsolve(sys,{x=0,y=-3});
fsolve(sys,{x=2,y=-3});
```

Mathematica:

```
sys={Sin[x]-Cos[y]==1,Exp[x]-Exp[-y]==1}
ContourPlot[{Sin[x]-Cos[y]==1,Exp[x]-Exp[-y]==1},
 {x,-Pi,Pi},{y,-Pi,Pi},AspectRatio->1]
{FindRoot[sys,{x,0},{y,3}],FindRoot[sys,{x,0},{y,-3}],
 FindRoot[sys,{x,2},{y,-3}]}
```

□

Approximate analytical solutions of algebraic equations

In Maple and Mathematica, there is no single function for finding approximate analytical solutions to equations, so various methods can be applied or developed.

Problem 11.7 Find approximate analytical solutions to the algebraic equation $F(x; \varepsilon) = 0$, where F is a real function of x and a small parameter ε.

According to the regular perturbation theory, the equation $F(x; 0) = 0$ has a solution x_0 and the solution of the perturbed equation is near x_0 and can be represented as a power series of ε.

For example, we solve the equation $x^2 - ax + \varepsilon = 0$, $|\varepsilon| \ll 1$. If $\varepsilon = 0$, the roots are $x_0 = 0, a$. If $\varepsilon \to 0$, we have $X_1(\varepsilon) \to 0$, $X_2(\varepsilon) \to a$. If $|\varepsilon| \ll 1$, the roots are: $X_i = x_0 + \varepsilon x_1 + \ldots + \varepsilon^k x_k + \ldots$, $i = 1, 2$, $k = 1, 2, \ldots$, where x_k are the unknown coefficients to be determined. Substituting the X_i series into the original equation and matching the coefficients of like powers of ε, we arrive to a system of algebraic equations for the i-th approximation, which can be solved for x_k.

Maple:

```
RegPertPoly:=proc(Expr,var,param) local i,j,Expr1,X,y,k,m,Sers;
 y:=var[0]; Expr1:=collect(Expr(y),param);
 X[0]:=[solve(coeff(Expr1,param,0),var[0])]; k:=nops(X[0]);
 for j from 1 to k do  y[j]:=X[0][j];
 for i from 1 to n do  y[j]:=y[j]+param^i*var[i];
 Expr1:=Expr(y[j]); X[i]:=solve(coeff(Expr1,param,i),var[i]);
 y[j]:=X[0][j]+add(X[m]*param^m,m=1..i); od:
 Sers:=sort(convert(y,list)): od: RETURN(Sers): end:
n:=5; Eq:=x->x^2-a*x+epsilon; s1:=RegPertPoly(Eq,x,epsilon);
```

Mathematica:

```
regPertPoly[f_,var_,param_]:=Module[
 {y,expr,X0,k,m,i,j,X},y=var[0];expr=Collect[f[var[0]],param];
  X0=Solve[Coefficient[expr,param,0]==0,y]; k=Length[X0];
  For[j=1,j<=k,j++, y[j]=X0[[j,1,2]]];
  For[i=1,i<=n,i++, y[j]=y[j]+param^i*var[i]; expr=f[y[j]];
  X[i]=Solve[Coefficient[expr,param,i]==0,var[i]];
  y[j]=X0[[j,1,2]]+Sum[X[m][[1,1,2]]*param^m,{m,1,i}];];
  serSol=Table[y[j],{j,1,k}];]; Return[serSol];];
n=5; eq[x_]:=x^2-a*x+\[Epsilon]; s1=regPertPoly[eq,x,\[Epsilon]]
```

□

11.2 Ordinary Differential Equations

In Maple, there exists a large set of functions to solve (analytically, numerically, graphically) ordinary and partial differential equations or the systems of differential equations. We discuss the most important functions and packages.

Exact solutions to ordinary differential equations

Maple:

```
dsolve(ODE,y(x),ops);           dsolve[interactive](ODEs,ops);
dsolve(ODEsys,funcs);           dsolve({ODEs,ICs},funcs,ops);
        dsolve(ODE,y(x),'formal_series','coeffs'=coeff_type);
dsolve(ODEs,funcs,method=transform);           dsolve(ODE,Lie);
dsolve(ODEs,funcs,'series',ops);with(DEtools);liesol(ODE,y(x));
```

dsolve, finding closed form solutions for a single ODE or a system of ODEs (see
   ?dsolve),

dsolve,ICs, solving ODEs or a system of them with given initial or boundary
   conditions,

dsolve,formal_series, finding formal power series solutions to a homogeneous
   linear ODE with polynomial coefficients,

dsolve,series, finding series solutions to ODEs problems,

dsolve,method, finding solutions using integral transforms,

dsolve[interactive], interactive symbolic and numeric solving of ODEs,

dsolve,Lie, solving ODEs using the Lie method of symmetries,

liesol (of the DEtools package), finding solutions of a first order Lie ODE, and
   a variety of other functions.

Mathematica:

```
DSolve[ODE,y[x],x] DSolve[{ODEs,ICs},y[x],x]      DSolve[ODE,y,x]
      DSolve[ODE,y[x],x,GeneratedParameters->(Subscript[c,#]&)]
DSolve[{eq1,...},{y1,...},x]                 DSolve[{ODEs,ICs},y,x]
```

DSolve, find the general solution y[x] or y (expressed as a "pure" function) for a
   single ODE or a system of ODEs,

DSolve,ICs, solve ODEs or a system of them with given initial or boundary conditions.

Explicit and implicit forms of exact solutions, graphs of solutions

Maple:

```
with(plots): Sol_Ex:=dsolve(diff(y(t),t)+t^2/y(t)=0,y(t));
Sol_Imp:=dsolve(diff(y(t),t)+t^2/y(t)=0, y(t),implicit);
G:=subs({y(t)=y},lhs(Sol_Imp));Gs:=seq(subs(_C1=i,G),i=-5..5);
contourplot({Gs},t=-5..5,y=-10..10,color=blue);
d1:=dsolve(diff(y(t),t$2)+y(t)=0,y(t));
subs({_C1=A[1],_C2=A[2]},d1);
```

Mathematica:

```
DSolve[y'[t]+t^2/y[t]==0,y[t],t]
{sol=DSolve[y'[t]+t^2/y[t]==0,y,t],y'''[t]+y[t]/.sol}
f:=sol[[1,1,2]]; {f[x],f'[x],f''[x]}
ContourPlot[Evaluate[Table[y^2==(f[t])^2/.{C[1]->i},{i,-5,5}]],
 {t,-5,5},{y,-10,10},PlotRange->{{-5,5},{-5,5}}]
DSolve[D[y[t],{t,2}]+y[t]==0,y[t],t,GeneratedParameters->A]
```

ODE classification and solution methods suggestion

In Maple, there is a variety of functions for solving various classes of ODEs. The `odeadvisor` function of the `DEtools` package allows us to classify a given ODE according to standard reference books [34] and suggests methods for solving it (for more details see `?DEtools`, `?odeadvisor`).

In Mathematica, the unique function `DSolve` allows us to obtain analytical solutions for almost all classes of ODEs whose solutions are given in standard reference books.

Maple:

```
with(DEtools): ODE1:=diff(y(t),t)=y(t)*sin(t)^2/(1-y(t));
ODE2:=(t^2-y(t)*t)*diff(y(t),t)+y(t)^2=0;
S_expl:=dsolve(ODE1,y(t)); S_impl:=dsolve(ODE1,y(t),implicit);
odeadvisor(ODE1); Sol_sep:=separablesol(ODE1,y(t));
odeadvisor(ODE2); Sol1:=dsolve(ODE2,y(t));
Sol2:=genhomosol(ODE2,y(t)); Sol3:=dsolve(ODE1,y(t),'series');
```

Mathematica:

```
Off[InverseFunction::ifun]; Off[General::stop];
Off[Solve::ifun]; ODE1=D[y[t],t]==y[t]*Sin[t]^2/(1-y[t]);
ODE2=(t^2-y[t]*t)*D[y[t],t]+y[t]^2==0;
{DSolve[ODE1,y,t],DSolve[ODE1,y[t],t],DSolve[ODE2,y,t],
 DSolve[ODE2,y[t],t]}
```

Solving these ODEs, *Mathematica* generates a warning message (the `Solve` function obtains the solution using inverse functions). This warning message can be ignored or suppressed with the `Off` function.

Higher-order ODE: exact and numeric solutions, and their graphs

Maple:

```
with(plots): setoptions(axes=boxed,scaling=constrained,
 numpoints=200); ODE:=diff(x(t),t$2)-diff(x(t),t)+(t-1)*x(t)=0;
ICs:=D(x)(0)=0,x(0)=1; Sol_ex:=dsolve({ODE,ICs},x(t));
Sol_num:=dsolve({ODE,ICs},x(t),numeric); G:=Array(1..3);
G[1]:=odeplot(Sol_num,[t,x(t)],0..10,color=blue):
G[2]:=plot(rhs(Sol_ex),t=0..10,color=red): G[3]:=odeplot(
 Sol_num,[x(t),diff(x(t),t)],0..10,color=magenta): display(G);
```

Mathematica:

```
SetOptions[Plot,ImageSize->300,ColorFunction->
 Function[{x,t},Hue[x]],PlotPoints->50,PlotStyle->
 {Thickness[0.01]},PlotRange->All,AspectRatio->1];
ODE={x''[t]-x'[t]+(t-1)*x[t]==0}; ICs={x'[0]==0,x[0]==1};
{eq1=DSolve[{ODE,ICs},x[t],t], eq2=DSolve[{ODE,ICs},x,t],
 solEx=eq2[[1,1,2]]]}
g1=Plot[x[t]/.eq1,{t,0,10}]; g11=Plot[solEx[t],{t,0,10}];
eq3=NDSolve[{ODE,ICs},x,{t,0,10}]; solN=eq3[[1,1,2]]
Table[{t,solN[t]},{t,0,10}]//TableForm
g2=Plot[solN[t],{t,0,10}]; g3=Plot[solN'[t],{t,0,10}];
GraphicsGrid[{{g1,g11},{g2,g3}}]
ParametricPlot[{solN[t],solN'[t]},{t,0,10},AspectRatio->1]
```

ODE systems: exact solutions, the integral Laplace transforms, graphs of solutions

For a more detail discussion over integral transforms, see Chapter 10 and Sect. 10.1).

Maple:

```
with(DEtools): with(inttrans):
ODE_sys1:={D(x)(t)=-2*x(t)+5*y(t),D(y)(t)=4*x(t)-3*y(t)};
Sol_sys1:=dsolve(ODE_sys1,{x(t),y(t)});
A1:=array([[-2,5],[4,-3]]); matrixDE(A1,t);
ODE_sys2:={diff(x(t),t)=-y(t)+cos(2*t),
 diff(y(t),t)=5*x(t)+2*sin(2*t)};
Eq1:=laplace(ODE_sys2,t,p); Eq2:=subs({x(0)=2,y(0)=0},Eq1);
Eq3:=solve(Eq2,{laplace(x(t),t,p),laplace(y(t),t,p)});
Sol_sys2:=invlaplace(Eq3,p,t); assign(Sol_sys2):
plot([x(t),y(t),t=-3..3],color=blue,grid=[90,90],axes=boxed);
```

Mathematica:

```
sys1={x'[t]==-2*x[t]+5*y[t],y'[t]==4*x[t]-3*y[t],x[0]==1,y[0]==0}
solsys1=DSolve[sys1,{x[t],y[t]},t]
sys2={x'[t]==-y[t]+Cos[2*t],y'[t]==5*x[t]+2*Sin[2*t]}
eq1=LaplaceTransform[sys2,t,p]/.{x[0]->2,y[0]->0}
eq2=Solve[eq1,{LaplaceTransform[x[t],t,p],
 LaplaceTransform[y[t],t,p]}]
xsol=InverseLaplaceTransform[eq2[[1,1,2]],p,t]
ysol=InverseLaplaceTransform[eq2[[1,2,2]],p,t]
ParametricPlot[{xsol,ysol},{t,-3,3},PlotStyle->Hue[0.9],
 AspectRatio->1]
```

Numerical and graphical solutions to ordinary differential equations

For a more detail discussion over numerical approximations of differential equations see Chapter 12 and Sect. 12.5).

Maple:

```
dsolve(ODEs,numeric,vars,ops);      dsolve(numeric,proc_ops,ops);
dsolve(ODEs,numeric,method=m);      dsolve(ODEs,numeric,output=n);
dsolve[interactive](ODEs,ops);      NS:=dsolve(ODEs,numeric,vars);
with(plots):                                  odeplot(NS,vars,tR,ops);
with(DETools);                   DEplot(ODEs,vars,tR,ICs,xR,yR,ops);
                                DEplot3d(ODEs,vars,tR,xR,yR,ICs,ops);
dfieldplot(ODEs,vars,tR,xR,yR);phaseportrait(ODEs,vars,tR,ICs);
```

Note. Here tR, xR, yR is, respectively, the range of the independent and dependent variables, t=t1..t2, x=x1..x2, y=y1..y2.

dsolve,numeric, finding numerical solutions to ODEs problems,

odeplot, constructing graphs or animations of 2D and 3D solution curves obtained from the numerical solution,

phaseportrait, constructing phase portraits for a system of first order differential equations or a single higher order differential equation with initial conditions,

DEplot, DEplot3d, constructing graphs or animations of 2D and 3D solutions to a system of differential equations using numerical methods, etc.

Mathematica:

```
s=NDSolve[{ODEs,ICs},y,{t,t1,t2},ops]        s1=Evaluate[y[t]/.s]
NDSolve[{ODEs,ICs},y[t],{t,t1,t2},ops]           Plot[s1,{t,t1,t2}]
                    NDSolve[{ODEs,ICs},y,{t,t1,t2},Method->m]
         NDSolve[{ODEs,ICs},y,{t,t1,t2},Method->{m1,Method->m2}]
<<VectorFieldPlots`
               VectorFieldPlot[{ft,fy},{t,t1,t2},{y,y1,y2},ops]
```

NDSolve, finding numerical solutions to ODEs problems,

VectorFieldPlot of the VectorFieldPlots package, constructing vector fields (see Sect. 6.10 and Sect. 12.5).

Note. In *Mathematica* 7, the new functions VectorPlot, VectorPlot3D have been introduced and incorporated into the *Mathematica* kernel.

Initial value problems, graphs of solutions

Maple:

```
with(plots): N:=7; Sols:=Vector(N,0): Gr:=NULL:
IVP:={D(y)(t)=3*t+2*y(t),y(0)=n}; Sol:=dsolve(IVP,y(t));
for i from 1 to N do Sols[i]:=subs(n=-3+(i-1),Sol); od:
op(Sols); SList:=['rhs(Sols[i])' $ 'i'=1..N];
for i from 1 to N do
 G||i:=plot(SList[i],t=0..2.5,color=blue): Gr:= Gr,G||i: od:
VField:=fieldplot([1,3*t+2*y],t=0..2.5,y=-600..600,grid=[30,30],
 arrows=slim, color=t): display({Gr,VField},axes=boxed);
```

Mathematica:

```
<<VectorFieldPlots`
IVP={y'[t]==3*t+2*y[t],y[0]==n}
sol=DSolve[IVP,y[t],t]//Expand
sols=Table[DSolve[IVP,y[t],t],{n,-3,3}]
slist=Table[sols[[i,1,1,2]],{i,1,7}]
Do[g[i]=Plot[slist[[i]],{t,0,2.5},PlotRange->All,
 PlotStyle->Hue[0.7]],{i,1,7}]; gr=Table[g[i],{i,1,7}];
vField=VectorFieldPlot[{1,3*t+2*y},{t,0,2.5},{y,-500,500},
 ColorFunction->Function[{t},Hue[t]]];
Show[gr,vField,Frame->True,AspectRatio->1]
```

Boundary value problems, graphs of solutions

Maple:

```
BVP:={diff(y(t),t$2)+2*y(t)=0,y(0)=1,y(Pi)=0};
Sol:=dsolve(BVP,y(t));
plot(rhs(Sol),t=0..Pi,axes=boxed,color=blue,thickness=3);
```

Mathematica:

```
BVP={y''[t]+2*y[t]==0,y[0]==1,y[Pi]==0}
sol=DSolve[BVP,y[t],t]
Plot[sol[[1,1,2]],{t,0,Pi},Frame->True,Axes->False,
 PlotStyle->{Hue[0.7],Thickness[0.01]},AspectRatio->1]
```

Problem 11.8 Solve the linear system of ordinary differential equations with constant coefficients $x' = x + 3y$, $y' = -2x + y$, and graph the solutions together with the direction field associated with the system.

Maple:

```
with(LinearAlgebra): with(DEtools): with(plots):
A:=Matrix([[1,3],[-2,1]]): Eigenvectors(A); matrixDE(A,t);
ODE_sys:=Equate(Matrix([[diff(x(t),t)],[diff(y(t),t)]]),
 A.Matrix([[x(t)],[y(t)]]));
Sol_sys:=dsolve(ODE_sys,{x(t),y(t)});
assign(Sol_sys); Curves:={seq(seq(subs({_C1=i,_C2=j},
 [x(t),y(t),t=-3..3]),i=-3..3),j=-3..3)}:
```

```
G1:=plot(Curves,view=[-10..10,-10..10],color=blue,
  axes=boxed): unassign('x','y'); G2:=DEplot(ODE_sys,
  [x(t),y(t)],t=-3..3,x=-10..10,y=-10..10,color=red):
display({G1,G2});
```

Mathematica:

```
<<VectorFieldPlots`
sys3={x'[t]==x[t]+3*y[t],y'[t]==-2*x[t]+y[t]};
sol=Table[DSolve[sys3,{x[t],y[t]},t]/.{C[1]->i,C[2]->j},
  {i,1,5},{j,1,5}];
g1=Table[ParametricPlot[Evaluate[{sol[[i,j,1,1,2]],
  sol[[i,j,1,2,2]]},{t,-3,3}],PlotStyle->Hue[0.9],
  Frame->True,Background->LightPurple],{i,1,5},{j,1,5}];
g2=VectorFieldPlot[{x+3*y,-2*x+y},{x,-10,10},{y,-10,10},
  ColorFunction->Function[{t},Hue[t]]]; Show[g1,g2,
  AspectRatio->1,Frame->True,PlotRange->{{-10,10},{-10,10}}]
```
□

Problem 11.9 Solve numerically the initial value problem $x' = xy$, $y' = x + y$, $x(0) = 1$, $y(0) = 1$, and graph the solutions.

Maple:

```
with(plots): setoptions(scaling=constrained);
A:=Array(1..3); IC:={x(0)=1,y(0)=1};
ODE:={D(x)(t)=x(t)*y(t),D(y)(t)=x(t)+y(t)};
Sol:=dsolve(ODE union IC,numeric,output=operator);
A[1]:=plot(rhs(Sol[2](t)),t=0..1):
A[2]:=plot(rhs(Sol[3](t)),t=0..1):
A[3]:=plot({rhs(Sol[2](t)),rhs(Sol[3](t))},t=0..1): display(A);
```

Mathematica:

```
SetOptions[Plot,ImageSize->300,AspectRatio->1,Frame->True,
  Axes->False]; {ODE={x'[t]==x[t]*y[t],y'[t]==x[t]+y[t]},
  ICs={x[0]==1,y[0]==1}}
eq1=NDSolve[{ODE,ICs},{x,y},{t,0,1}];
{solNX=eq1[[1,1,2]],solNY=eq1[[1,2,2]]}
Table[{t,solNX[t]},{t,0,1,0.1}]//TableForm
Table[{t,solNY[t]},{t,0,1,0.1}]//TableForm
g1=Plot[solNX[t],{t,0,1},PlotStyle->{Hue[0.8],Thickness[0.01]}];
g2=Plot[solNY[t],{t,0,1},PlotStyle->{Hue[0.9],Thickness[0.01]}];
g12=Show[g1,g2,AspectRatio->1,Frame->True,ImageSize->300]
GraphicsRow[{g1,g2,g12}]
```
□

Problem 11.10 Solve the nonlinear initial value problem

$$y' = -e^{yt}\cos(t^2), \quad y(0) = p.$$

Graph the solutions $y(t)$ for various values of the parameter p on the interval $[0, \pi]$. Graph the solutions for various initial conditions $p = 0.1i$ $(i = 1, 2, \ldots, 5)$.

Maple:

```
with(plots): NGSol:=proc(IC) local Eq,Eq_IC,L1,Sol_N,N_IC,i;
 Eq:=D(y)(t)=-exp(y(t)*t)*cos(t^2); L1:=NULL; N_IC:=nops(IC);
 for i from 1 to N_IC do Eq_IC:=evalf(y(0)=IC[i]);
  Sol_N:=dsolve({Eq,Eq_IC},y(t),type=numeric,range=0..evalf(Pi));
  L1:=L1,odeplot(Sol_N,[t,y(t)],0..evalf(Pi),numpoints=100,
   color=blue,thickness=2,axes=boxed): od; display([L1]); end:
List1:=[seq(0.1*i,i=1..5)]; NGSol(List1);
```

Mathematica:

```
eq=y'[t]==-Exp[y[t]*t]*Cos[t^2];
Do[{ICs={y[0]==0.1*i}; sN=NDSolve[{eq,ICs},y[t],{t,0,Pi},
 MaxSteps->1000]; solN[t_]:=sN[[1,1,2]]; g[i]=Plot[
 solN[t],{t,0,Pi},PlotStyle->Hue[0.5+i*0.07]];},{i,1,10}]
Show[Table[g[i],{i,1,5}],Axes->False,PlotRange->All]
```

□

Problem 11.11 Construct a phase portrait of the dynamical system that describes the evolution of the amplitude and the slow phase of a fluid under the subharmonic resonance

$$\frac{dv}{dt} = -\nu v + \varepsilon u \left[\delta + \frac{1}{4} - \frac{1}{2}\phi_2(u^2 + v^2) + \frac{1}{4}\phi_4(u^2 + v^2)^2\right],$$

$$\frac{du}{dt} = -\nu u + \varepsilon v \left[-\delta + \frac{1}{4} + \frac{1}{2}\phi_2(u^2 + v^2) - \frac{1}{4}\phi_4(u^2 + v^2)^2\right].$$

This system has been obtained in [40] by averaging transformations with *Maple* . Here ν is the fluid viscosity, ε is the small parameter, ϕ_2, ϕ_4 are the second and the fourth corrections to the nonlinear wave frequency, δ is the off resonance detuning. Choosing the corresponding parameter values (for the six regions where the solution exists), we can obtain a phase portrait.

Maple:

```
with(plots): with(DEtools): Ops:=arrows=medium,dirgrid=[20,20],
  stepsize=0.1,thickness=2,linecolour=blue,color=green;
delta:=-1/2: phi_2:=1; phi_4:=1; nu:=0.005; epsilon:=0.1;
Eq1:=D(v)(t)=-nu*v(t)+epsilon*u(t)*(delta+1/4-phi_2/2*
  (u(t)^2+v(t)^2)+phi_4/4*(u(t)^2+v(t)^2)^2);
Eq2:=D(u)(t)=-nu*u(t)+epsilon*v(t)*(-delta+1/4+phi_2/2*
  (u(t)^2+v(t)^2)-phi_4/4*(u(t)^2+v(t)^2)^2); Eqs:=[Eq1,Eq2];
vars:=[v(t),u(t)]; IC:=[[u(0)=0,v(0)=1.1033],[u(0)=0,
  v(0)=-1.1033],[u(0)=1.1055, v(0)=0],[u(0)=-1.1055, v(0)=0],
  [u(0)=0,v(0)=1.613],[u(0)=0,v(0)=-1.613]];
phaseportrait(Eqs,vars,t=-48..400,IC,Ops);
```

Mathematica:

```
<<VectorFieldPlots`
delta=-1/2; phi2=1; phi4=1; nu=0.005; epsilon=0.1;
eq1=-nu*v[t]+epsilon*u[t]*(delta+1/4-phi2/2*
  (u[t]^2+v[t]^2)+phi4/4*(u[t]^2+v[t]^2)^2);
eq2=-nu*u[t]+epsilon*v[t]*(-delta+1/4+phi2/2*
  (u[t]^2+v[t]^2)-phi4/4*(u[t]^2+v[t]^2)^2);
ICs={{0,1.1033},{0,-1.1033},{1.1055, 0},{-1.1055,0},
  {0,1.613},{0,-1.613}}; n=Length[ICs];
Do[{sys[i]={v'[t]==eq1,u'[t]==eq2,v[0]==ICs[[i,1]],
  u[0]==ICs[[i,2]]}; sols=NDSolve[sys[i],{v,u},{t,-48,400}];
  cv=v/.sols[[1]]; cu=u/.sols[[1]]; c[i]=ParametricPlot[
  Evaluate[{cv[t],cu[t]}],{t,-48,400},PlotStyle->
  {Hue[0.1*i+0.2],Thickness[.001]}];},{i,1,n}]
fv=eq1/.{v[t]->v,u[t]->u}; fu=eq2/.{v[t]->v,u[t]->u};
fd=VectorFieldPlot[{fv,fu},{v,-2.2,2.2},{u,-2.2,2.2},
  Frame->True,ColorFunction->Function[{u},Hue[u]]];
Show[fd,Table[c[i],{i,1,6}]]
```

□

Problem 11.12 Animate the phase portrait associated with the linear ODE system $x' = \pi x - 2y$, $y' = 4x - y$, $x(0) = 1/2C$, $y(0) = 1/2C$, where $C \in [-1, 1]$ (use the *Maple* functions animate and phaseportrait).

Maple:

```
with(plots): with(DEtools): Vars:=[x(t),y(t)];
Eqs:=[(D(x))(t)=Pi*x(t)-2*y(t),(D(y))(t)=4*x(t)+y(t)];
```

```
ICs:=[[x(0)=-1/2*C,y(0)=1/2*C]];
animate(phaseportrait,[Eqs,Vars,t=0..Pi,ICs,
 x=-40..40,y=-40..40,stepsize=0.1,scaling=constrained,
 linecolor=blue],C=-1..0);
```

Mathematica:

```
<<VectorFieldPlots`
g={}; n=50; subs={x[t]->x,y[t]->y};
eqs={x'[t]==Pi*x[t]-2*y[t],y'[t]==4*x[t]+y[t]}
ICs={x[0]==-1/2*C1,y[0]==1/2*C1}
{fx=eqs[[1,2]]/.subs, fy=eqs[[2,2]]/.subs}
vf=VectorFieldPlot[{fx,fy},{x,-40,40},{y,-40,40},Frame->True,
 ColorFunction->Function[{x},Hue[x]],PlotRange->{{-40,40},
 {-40,40}}]; sys=Table[Flatten[{eqs,ICs}],{C1,-1,0,1/n}];
sols=Table[NDSolve[sys[[i]],{x[t],y[t]},{t,0,Pi}],{i,1,n}];
ll=Table[{cx=x[t]/.sols[[i,1]],cy=y[t]/.sols[[i,1]]},{i,1,n}];
Do[g=Append[g,Show[vf,Evaluate[ParametricPlot[ll[[i]],{t,0,Pi},
 AspectRatio->1,PlotStyle->{Blue,Thickness[0.01]},
 PlotRange->{{-40,40},{-40,40}}]]]],{i,1,n}]; ListAnimate[g]
```

                                                                        □

11.3   Partial Differential Equations

Analytical solutions to partial differential equations

Maple:

```
sol:=pdsolve(PDE);      pdsolve(PDEsys);      pdetest(sol,PDEs);
        pdsolve(PDE,funcs,HINT=val,INTEGRATE,build,singsol=val);
with(PDEtools);    declare(funcs); ON;      dchange(rules,PDE);
diff_table(funcs); Laplace(PDE,func); separability(PDE,func);
casesplit(PDEs);  Infinitesimals(PDE);  ReducedForm(sys1,sys2);
```

where HINT=val are some hints, with build can be constructed an explicit expression for the indeterminate function func.

pdsolve, finding analytical solutions for a given partial differential equation PDE and systems of PDEs,

PDEtools, a collection of functions for finding analytical solutions for PDEs, for example, declare, declaring functions and derivatives on the screen for a simple, compact display,

casesplit, split into cases and sequentially decouple a system of differential equations,

separability, determine under what conditions it is possible to obtain a complete solution through separation of variables, etc.

Mathematica:

```
DSolve[PDE,u[x1,...,xn],{x1,...,xn}]    DSolve[PDE,u,{x1,..,xn}]
```

DSolve, finding analytical solutions for a partial differential equation PDE and systems of PDEs.

Problem 11.13 Find a general solution to the wave equation $u_{tt} = c^2 u_{xx}$.

Maple:

```
with(PDEtools); declare(u(x,t)); ON;
pde:=diff(u(x,t),t$2)=c^2*diff(u(x,t),x$2); casesplit(pde);
separability(pde,u(x,t)); pdsolve(pde,build);
```

Mathematica:

```
pde1=D[u[x,t],{t,2}]-c^2*D[u[x,t],{x,2}]==0
Assuming[c\[Element]Reals && c>0,
 DSolve[pde1,u[x,t],{x,t}]//Simplify]
pde2=(D[#,{x2,2}]-c^2*D[#,{x1,2}])&[y[x1,x2]]==0
Assuming[c\[Element]Reals && c>0,
 DSolve[pde2,y[x1,x2],{x1,x2}]//Simplify]
```

□

Problem 11.14 Solve the initial-value problem $u_t + t^2 u_x = 9$, $u(x,0) = x$ using the *method of characteristics*. Graph the characteristics.

Maple:

```
with(plots); Rt:=0..4; Rx:=-40..40; ODE:=diff(U(t),t)=9;
Sol_Ch:=dsolve({ODE,U(0)=X[0]}); Eq_Ch:=diff(x(t),t)=t^2*U(t);
Eq_Ch:=subs(Sol_Ch,Eq_Ch); Cur_Ch:=dsolve({Eq_Ch, x(0)=X[0]});
display([seq(plot([subs(X[0]=x,eval(x(t),Cur_Ch)),t,t=Rt],
 color=blue,thickness=2),x=Rx)],view=[Rx,Rt]);
u:=unapply(subs(X[0]=solve(subs(x(t)=x,Cur_Ch),X[0]),
 eval(U(t),Sol_Ch)),x,t);
```

Mathematica:

```
solCh=DSolve[{U'[t]==9,U[0]==X[0]},U[t],t]
eqCh=x'[t]==t^2*U[t]/.solCh[[1]]
curCh=DSolve[{eqCh,x[0]==X[0]},x[t],t]//FullSimplify
g=Table[Plot[{curCh[[1,1,2]]/.X[0]->x},{t,0,4},
 PlotStyle->Hue[0.7],PlotRange->{-40,40},
 AspectRatio->1],{x,-40,40}]; Show[g]
uu=solCh[[1]]/.Solve[curCh[[1,1,2]]==x,X[0]]
u[X_,T_]:=uu[[1,1,2]]/.{x->X,t->T}; u[X,T]
```

□

Numerical and graphical solutions to PDEs

For a more detail discussion over numerical solutions of partial differential equations, see Chapter 12 and Sect. 12.5).

Maple:

```
                   Sol:=pdsolve(PDEs,ICsBCs,numeric,funcs,ops);
Num_vals:=Sol:-value();           Sol:-plot3d(func,t=t0..t1,ops);
Num_vals(num1,num2);   Sol:-animate(func,t=t0..t1,x=x0..x1,ops);
```

pdsolve,numeric, finding numerical solutions to a partial differential equation PDE or a system of PDEs.

Note. The solution obtained is represented as a module (similar to a procedure, the operator `:-` is used, see Sect. 1.3.4) which can be used for obtaining visualizations (`plot`, `plot3d`, `animate`, `animate3d`) and numerical values (`value`), in more detail, see `?pdsolve[numeric]`.

Mathematica:

```
          NDSolve[PDE,u,{x,x1,x2},{t,t1,t2},...,ops]
NDSolve[PDE,{u1,...,un},{x,x1,x2},{t,t1,t2},...,ops]
```

NDSolve, finding numerical solutions to **PDE** or a system of **PDEs**.

Problem 11.15 Find numerical and graphical solutions to the boundary value problem for the wave equation,

$$u_{tt} = 0.01\,u_{xx}, \quad u(0,t) = 0, \quad u(1,t) = 0,$$
$$u(x,0) = 0, \quad u_t(x,0) = \sin(2\pi x),$$

in the domain $\mathcal{D} = \{0 < x < 1,\ 0 < t < \infty\}$.

Maple:

```
with(VectorCalculus): with(plots): with(PDEtools): C:=0.01:
Ops:=spacestep=1/100, timestep=1/100;
PDE:=diff(u(x,t),t$2)-C*Laplacian(u(x,t),'cartesian'[x])=0;
Ics:={D[2](u)(x,0)=sin(2*Pi*x),u(x,0)=0};
Bcs:={u(0,t)=0,u(1,t)=0};
Sol:=pdsolve(PDE,Ics union Bcs,numeric,u(x,t),Ops);
Num_vals:=Sol:-value(); Num_vals(1/2,Pi); Sol:-animate(u(x,t),
  t=0..5*Pi,x=0..1,frames=30,numpoints=100,thickness=3);
Sol:-plot3d(u(x,t),t=0..5*Pi,shading=zhue,axes=boxed);
```

Mathematica:

```
pde={D[u[x,t],{t,2}]-0.01*D[u[x,t],{x,2}]==0};
ICs={(D[u[x,t],t]/.{t->0})==Sin[2*Pi*x],u[x,0]==0}
BCs={u[0,t]==0,u[1,t]==0}
sol=NDSolve[{pde,ICs,BCs},u,{x,0,1},{t,0,5*Pi}];
f=u/.sol[[1]]; Plot3D[f[x,t],{x,0,1},{t,0,5*Pi},BoxRatios->1]
PaddedForm[Table[{x,f[x,Pi]},{x,0,1,0.1}]//TableForm,{12,5}]
Animate[Plot[f[x,t],{t,0,5*Pi},AspectRatio->1,PlotStyle->
  Thickness[0.02],PlotRange->{{0,5*Pi},{-Pi,Pi}}],{x,0,1}]
```

                                                                            □

Application to nonlinear standing waves in a fluid using the Lagrangian formulation

Problem 11.17 Construct approximate analytical solutions describing nonlinear standing waves on the free surface of a fluid.

Statement of the problem. The classical two-dimensional standing waves problem consists of solving the Euler equations for a one- or two-layer fluid with boundary conditions. The assumption is made that the flow is irrotational. The boundary value problem needs to be solved in a flow domain, $\mathcal{D} = \{0 \leq x \leq L, -h \leq y \leq \eta(x,t)\}$ for the surface elevation $\eta(x,t)$ and the velocity potential $\phi(x,y,t)$. The fluid depth h, the surface tension constant T, and the horizontal size of the domain L are given. We study periodic solutions (in x and t) of a standing waves problem.

In general, there are two ways for representing the fluid motion: the Eulerian approach, in which the coordinates are fixed in the reference frame of the observer, and the Lagrangian approach, in which the coordinates are fixed in the reference frame of the moving fluid.

Approximate solutions in Lagrangian variables are constructed analytically. The analytic Lagrangian approach proposed by [39] is followed for constructing approximate solutions for the nonlinear waves. We generalize the solution method to allows us to solve a set of problems, for example, the problems of infinite- and finite-depth surface standing waves and infinite- and finite-depth internal standing waves. This method can be useful for extending a series solutions to higher order, solving a problem that is not solvable in Eulerian formulation, or to solve other set of problems. We develop computer algebra procedures to aid in the construction of higher-order approximate analytical solutions.

Most of the approximate analytic solutions have been obtained using the Eulerian formulation, this section deals with the alternative formulation, which deserves to be considered. Therefore, we compare the analytic frequency-amplitude dependences obtained in Lagrangian variables with the corresponding ones known in Eulerian variables. The analysis has shown that the analytic frequency-amplitude dependences are in complete agreement with previous results obtained by [36], [31], [6], [47], [30] in Eulerian variables, and by [40], [41], [43] in Lagrangian variables.

The analysis of solutions has shown that the use of the Lagrangian approach to solve standing waves problems presents some advantages with respect to the Eulerian formulation, particulary because it allows us to simplify the boundary conditions (the unknown free boundary is a line), the radius of convergence of an expansion parameter is larger than in the Eulerian variables (this allows us to observe steep standing waves).

Asymptotic solution. Let us consider two-dimensional nonlinear wave motions in the fluid domain $\mathcal{D} = \{0 \leq x \leq L, -\infty \leq y \leq \eta(x,t)\}$. On the free surface the pressure is constant and equal to zero. We consider this as a basic model and other models can be derived from this one.

The rectangular system of coordinates xOy in the plane of motion is chosen so that the x-axis coincides with the horizontal level of fluid at rest and the y-axis is directed vertically upwards so that the unperturbed free surface has coordinates $y = 0$ and $x \in [0, L]$.

The transformation of variables from Eulerian (x, y) to Lagrangian (a, b) are carried out adding the following requirements:

the Jacobian $J = \partial(x, y)/\partial(a, b) = 1$,

the free surface $y = \eta(x, t)$ is equivalent to the parametric curve $\{x(a, 0, t), y(a, 0, t)\}$,

at $t = 0$ the free surface is $\{x(a, 0, 0), y(a, 0, 0)\}$,

at the vertical lines $a = 0$ and $a = L$ the horizontal velocity $x_t = 0$,

the infinite depth is $b = -\infty$.

The Lagrange equations for wave motions in fluid, the continuity equation and the boundary conditions are then given as:

$$x_{tt}x_a + (y_{tt} + g)y_a + \frac{p_a}{\rho} = 0, \quad x_{tt}x_b + (y_{tt} + g)y_b + \frac{p_b}{\rho} = 0,$$

$$\frac{\partial(x, y)}{\partial(a, b)} = 1,$$

$$x(0, b, t) = 0, \quad x(L, b, t) = L, \quad y(a, -\infty, t) = -\infty, \quad p(a, 0, t) = p_T,$$

where $x(a, b, t)$ and $y(a, b, t)$ are the coordinates of an individual fluid particle in motion, $p(a, 0, t)$ is the pressure on the free surface due to surface tension, ρ is the fluid density. We study a weak capillarity and include surface tension similar to the parametric expression (2.7) written by Schultz et al. [44]. In Lagrangian coordinates p_T can be computed by means of

$$p_T = -T\left[\frac{y_{aa}x_a - y_a x_{aa}}{(x_a^2 + y_a^2)^{3/2}}\right], \tag{11.1}$$

where T is the surface tension constant. The dispersion relation for linear waves is $\omega_{(0)}^2 = g\kappa(1 + T_z)$, where the dimensionless surface tension is $T_z = \kappa^2 T/(\rho g)$.

Note that we follow the notation of Concus [12], $\delta = \dfrac{T_z}{1 + T_z}$, for more compact presentation of the result.

Let us consider weakly nonlinear standing waves or waves of small amplitude and steepness, for which the amplitude and the ratio of wave height to wavelength is assumed to be of order ε, where ε is a small parameter.

We introduce the dimensionless amplitude ε, the wave phase ψ, Lagrangian variables α, β (instead of a, b), and space coordinates and pressure ξ, η, and σ (instead of x, y, and p):

$$\kappa A = \varepsilon, \quad \psi = \omega t, \quad \alpha = a\kappa, \quad \beta = b\kappa,$$
$$\kappa x = \alpha + \varepsilon\xi, \quad \kappa y = \beta + \varepsilon\eta, \quad \kappa^2 p = -\kappa(\rho g)\kappa y + \varepsilon\rho\omega_{(0)}^2\sigma,$$

where $\kappa = \pi n/L$ is the wave number (n is the number of nodes of the wave), ω is the nonlinear frequency, $\omega_{(0)}^2 = g\kappa$ is the dispersion relation for linear periodic waves, and g is the acceleration due to gravity.

In terms of the dimensionless variables, the equations of motion and the boundary conditions can be rewritten in the form

$$\mathcal{L}^1(\xi, \sigma) = -\varepsilon(\xi_{\psi\psi}\xi_\alpha + \eta_{\psi\psi}\eta_\alpha), \quad \mathcal{L}^2(\eta, \sigma) = -\varepsilon(\xi_{\psi\psi}\xi_\beta + \eta_{\psi\psi}\eta_\beta),$$
$$\mathcal{L}^3(\xi, \eta) = -\varepsilon\frac{\partial(\xi, \eta)}{\partial(\alpha, \beta)},$$
$$\xi(0, \beta, \psi) = 0, \quad \xi(\pi n, \beta, \psi) = 0, \quad \eta(\alpha, -\infty, \psi) = 0,$$
$$\sigma(\alpha, 0, \psi) - \eta(\alpha, 0, \psi) = -T_z\varepsilon\kappa\left[\frac{\eta_{\alpha\alpha} + \varepsilon(\eta_{\alpha\alpha}\xi_\alpha - \eta_\alpha\xi_{\alpha\alpha})}{[1 + 2\varepsilon\xi_\alpha + \varepsilon^2(\xi_\alpha^2 + \eta_\alpha^2)]^{3/2}}\right],$$

where linear differential operators $\mathcal{L}^i$ ($i = 1, 3$) are

$$\mathcal{L}^1(\xi, \sigma) = \xi_{\psi\psi} + \sigma_\alpha, \quad \mathcal{L}^2(\eta, \sigma) = \eta_{\psi\psi} + \sigma_\beta, \quad \mathcal{L}^3(\xi, \eta) = \xi_\alpha + \eta_\beta.$$

The construction of asymptotic solutions is based on perturbation theory. Defining the formal power series in the amplitude parameter ε

$$u = \mathcal{F}^{111}(u) + \sum_{i=2}^{N} \varepsilon^{i-1}u^{(i)} + O(\varepsilon^N), \quad u = \xi, \eta, \sigma,$$

where $\xi^{(i)}, \eta^{(i)}$, and $\sigma^{(i)}$ ($i = 2, \ldots, N$) are unknown 2π-periodic in ψ functions of variables α, β, and ψ, and the linear terms $\mathcal{F}^{111}(u)$, $u = \xi, \eta, \sigma$, are defined by the following expressions:

$$\mathcal{F}^{111}(\xi) = -\sin(\alpha)e^\beta\cos(\psi), \quad \mathcal{F}^{111}(\eta) = \mathcal{F}^{111}(\sigma) = \cos(\alpha)e^\beta\cos(\psi).$$

Considering weakly nonlinear standing waves, we can assume that the nonlinear wave frequency ω is close to the linear wave frequency $\omega_{(0)}$:

$$\omega(\varepsilon) \equiv \psi_t = \omega_{(0)} + \sum_{i=1}^{N-1} \varepsilon^i\omega_{(i)} + O(\varepsilon^N),$$

where $\omega_{(i)}$ are new unknown corrections to the nonlinear wave frequency.

Substituting these expansions into the equations of motion and the boundary conditions and matching the coefficients of like powers of ε, we arrive at the following linear inhomogeneous system of partial differential equations and boundary conditions for the i-th approximation:

$$\mathcal{L}^1(\xi^{(i)}, \sigma^{(i)}) = \mathcal{S}^1(\xi^{(i)}), \qquad \mathcal{L}^2(\eta^{(i)}, \sigma^{(i)}) = \mathcal{S}^2(\eta^{(i)}),$$
$$\mathcal{L}^3(\xi^{(i)}, \eta^{(i)}) = \mathcal{S}^3(\sigma^{(i)}),$$

$$\text{(11.2)}$$

$$\xi^{(i)}(0, \beta, \psi) = 0, \quad \xi^{(i)}(\pi n, \beta, \psi) = 0, \qquad \eta^{(i)}(\alpha, -\infty, \psi) = 0,$$
$$\sigma^{(i)}(\alpha, 0, \psi) - \eta^{(i)}(\alpha, 0, \psi) = -T_z \varepsilon \kappa \mathcal{E}^{(i)},$$

where

$$\mathcal{E}^{(i)} = \left[\frac{\eta_{\alpha\alpha}^{(i)} + \varepsilon(\eta_{\alpha\alpha}^{(i)} \xi_\alpha^{(i)} - \eta_\alpha^{(i)} \xi_{\alpha\alpha}^{(i)})}{[1 + 2\varepsilon \xi_\alpha^{(i)} + \varepsilon^2((\xi_\alpha^{(i)})^2 + (\eta_\alpha^{(i)})^2)]^{3/2}} \right],$$

$$\mathcal{S}^m(u^{(i)}) = \sum_{j,k,l=0}^{i} [F_{mi}^{jkl} \mathcal{F}^{jkl}(u^{(i)}) + G_{mi}^{jkl} \mathcal{F}_\psi^{jkl}(u^{(i)})], \quad u^{(i)} = \xi^{(i)}, \eta^{(i)}, \sigma^{(i)},$$

$$\mathcal{F}^{ijk}(\xi^{(i)}) = \sin(i\alpha) e^{(j\beta)} \cos(k\psi),$$
$$\mathcal{F}^{ijk}(\eta^{(i)}) = \mathcal{F}^{ijk}(\sigma^{(i)}) = \cos(i\alpha) e^{(j\beta)} \cos(k\psi).$$

We omit the expressions for all coefficients $F_{mi}^{jkl}, G_{mi}^{jkl}$ $(m=1,2,3, j,k,l= 0, \ldots, i, i = 2, \ldots N)$ because of their great complexity.

We look for a solution, 2π-periodic in ψ, to this system of equations and the boundary conditions in the form:

$$v^{(i)} = \sum_{j,k,l=0}^{i} V_i^{jkl} \mathcal{F}^{jkl}(v), \quad v = \xi, \eta, \sigma, \quad V = \Xi, H, \Sigma,$$

where Ξ_i^{jkl}, H_i^{jkl}, and Σ_i^{jkl} are unknown constants.

By using formulas described above, we obtain the asymptotic solution of the order of $O(\varepsilon^N)$ and the unknown corrections to the nonlinear wave frequency $\omega_{(i)}$. Setting $\beta = 0$ in the parametric equations for κx and κy, we can obtain the profiles of surface standing waves in Lagrangian variables for the i-th approximation $(i = 1, \ldots, N)$. Changes in the amplitude ε influence the surface configuration by changing both the shape of the surface and the amplitude of motion.

Approximate analytical solution with Maple. We present the *Maple* solution for every stage of the method and up to the second approximation, $N = 2$ (NA=2).

(1) We find the second derivative with respect to dimensionless time ψ for the variables $\xi(\alpha, \beta, t)$ and $\eta(\alpha, \beta, t)$:

```
NA:=2; NP:=NA-1; NN:=NA+1; TZ:=-delta/(delta-1): lambda:=1+TZ;
psiT:=omega+add(epsilon^i*omega||i,i=1..NP);
setsub:={diff(psi(t),t)=psiT}; subt:={psi(t)=psi};
xi:=(x,y,z)->-sin(x)*exp(y)*cos(z); eta:=(x,y,z)->
 cos(x)*exp(y)*cos(z); sigma:=(x,y,z)->cos(x)*exp(y)*cos(z);
Fxi:=(x-> xi(alpha,beta,psi(x)));
Feta:=(x->eta(alpha,beta,psi(x)));  dxi:=diff(Fxi(t),t);
deta:=diff(Feta(t),t);              xi1T:=subs(setsub,dxi);
xi2T:=subs(setsub,diff(xi1T,t));    eta1T:=subs(setsub,deta);
eta2T:=subs(setsub,diff(eta1T,t));  Xi:=xi(alpha,beta,psi);
Eta:=eta(alpha,beta,psi);           Sigma:=sigma(alpha,beta,psi);
```

It should be noted that, for convenience, we exclude the symbolic expressions of linear operators $\mathcal{L}^1$, $\mathcal{L}^2$, and $\mathcal{L}^3$ in the time derivatives and the governing equations. We are working with the right-hand sides (trigonometric parts) of the expressions in (11.2). For example, in the second time derivative

$$\frac{\partial^2 \eta^{(2)}}{\partial t^2} = -\omega^2 \cos \alpha^2 \exp \beta \cos \psi$$

$$+\varepsilon\left(\omega^2 \frac{\partial^2 \eta^{(2)}}{\partial \psi^2} - 2\omega \cos \alpha^2 \exp \beta \cos \psi\right) + O(\varepsilon^2),$$

and in $\mathcal{L}^2$, the second equation in (11.2), we omit the symbolic expression $\eta^{(2)}_{\psi\psi}$ (the second approximation, NA=2).

(2) The equations of motion and the continuity equation:

```
F1:=-omega^2*diff(Sigma,alpha)-epsilon*(xi2T*diff(Xi,alpha)
 +eta2T*diff(Eta, alpha))-xi2T;
F2:=-omega^2*diff(Sigma,beta)-epsilon*(xi2T*diff(Xi,beta)
 +eta2T*diff(Eta, beta))-eta2T;
F3:=-diff(Xi,alpha)-diff(Eta,beta)+epsilon*(diff(Xi,beta)
 *diff(Eta,alpha)-diff(Eta,beta)*diff(Xi,alpha));
for i from 1 to 3 do
 F||i||S:=coeff(subs(subt,F||i),epsilon,NP)*epsilon^NP: od;
```

(3) The boundary conditions:

```
Bc1:=xi=evala(subs(alpha=0,coeff(xi(alpha,beta,psi),
  epsilon,NP)));
Bc2:=xi=evala(subs(alpha=Pi*n,coeff(xi(alpha,beta,psi),
  epsilon,NP)));
Bc3:=eta=evala(subs(beta=-infinity,
  coeff(eta(alpha,beta,psi),epsilon,NP)));
Bc4:=(x,y)->lambda*x-y+TZ*diff(y,alpha$2)+
  TZ*(S-diff(y,alpha$2));
Bc4_1:=Bc4(sigma(alpha,beta,psi),eta(alpha,beta,psi));
S_numer:=((1+epsilon*diff(xi(alpha,beta,psi),alpha))*
  diff(eta(alpha,beta,psi),alpha$2)-
  epsilon*diff(xi(alpha,beta,psi),alpha$2)*
  diff(eta(alpha,beta,psi),alpha));
S_denom:=(1+epsilon*diff(xi(alpha,beta,psi),alpha))^2+
  (epsilon*diff(eta(alpha,beta,psi),alpha))^2;
S:=convert(series(S_numer/S_denom^(3/2),epsilon,NA),polynom);
Bc4_2:=evala(subs(beta=0,Bc4_1));
F4S:=coeff(Bc4_2,epsilon,NP)*epsilon^NP;
```

(4) We rewrite the governing equations describing the standing wave motion and the fourth boundary condition at the free surface in the form of *Maple* functions:

```
Eq1:=(N,M,K,L)->-K^2*(Xi||N||M||K||L)-N*(Sigma||N||M||K||L)
  =combine(EqC(F1S,N,M,K)/epsilon^NP);
Eq2:=(N,M,K,L)->-K^2*(Eta||N||M||K||L)+M*(Sigma||N||M||K||L)
  =combine(EqC(F2S,N,M,K)/epsilon^NP);
Eq3:=(N,M,K,L)->N*(Xi||N||M||K||L)+M*(Eta||N||M||K||L)
  =combine(EqC(F3S,N,M,K)/epsilon^NP);
Bc4:=(N,M,K,L)->lambda*(Sigma||N||M||K||L)-(Eta||N||M||K||L)
  -N^2*TZ*(Eta||N||M||K||L);
```

(5) The coefficients F_{mi}^{jkl} and G_{mi}^{jkl} ($m = 1, 2, 3$, $j, k, l = 0, \ldots, i$) can be calculated according to the orthogonality conditions. On the basis of the orthogonality property of eigenfunctions, we create the procedures S1_NMK, S2_NMK, and Bc4_NK, which are the functions in the procedures EqC and BcC:

```
S1_NMK:=proc(x,N,M,K,OM2)
 local i,I_N,I_M,I_K,T,TI,CTI,Z3,NZ3,TII,CTII,TT,I_T;
 I_N:=factor(combine(1/Pi*int(combine(x,exp)*sin(N*alpha),
  alpha=0..2*Pi),exp,trig));
 if K=0 then I_K:=1/(2*Pi)*int(I_N*cos(K*psi),psi=0..2*Pi);
  else I_K:=1/Pi*int(I_N*cos(K*psi),psi=0..2*Pi); fi:
 I_M:=0: Z3:=collect(I_K,[exp,delta]):
 if Z3=0 then I_T:=0: else if type(Z3,`+`)=true then
 NZ3:=nops(Z3):
  for i from 1 to NZ3 do TI[i]:=select(has,op(i,Z3),beta);
   CTI[i]:=remove(has,op(i,Z3),beta);
   if TI[i]=0 then T[i]:=0: else
   T[i]:=evala(subs(beta=0,diff(TI[i],beta))); fi:
   if T[i]=0 or T[i]<>M then I_M:=I_M+0; else
    I_M:=I_M+CTI[i]/OM2; fi: od: I_T:=I_M:
 else TII:=select(has,Z3,beta); CTII:=remove(has,Z3,beta);
 if TII=0 then TT:=0: else
  TT:=evala(subs(beta=0,diff(TII,beta))); fi:
 if TT=0 or TT<>M then I_M:=I_M+0; else I_M:=I_M+CTII; fi:
 I_T:=I_M/OM2: fi: fi: RETURN(I_T); end;


S2_NMK:=proc(x,N,M,K,OM2)
 local i,I_N,I_M,I_K,T,TI,CTI,Z3,NZ3,TII,CTII,TT,I_T;
 if N=0 then I_N:=1/(2*Pi)*int(x*cos(N*alpha),alpha=0..2*Pi);
  else I_N:=1/Pi*int(x*cos(N*alpha),alpha=0..2*Pi); fi:
 if K=0 then I_K:=1/(2*Pi)*int(I_N*cos(K*psi),psi=0..2*Pi);
  else I_K:=1/Pi*int(I_N*cos(K*psi),psi=0..2*Pi); fi:
 I_M:=0: Z3:=expand(I_K,exp,sin,cos):
 if Z3=0 then I_T:=0: else if type(Z3,`+`)=true then
 NZ3:=nops(Z3):
  for i from 1 to NZ3 do TI[i]:=select(has,op(i,Z3),beta);
   CTI[i]:=remove(has,op(i,Z3),beta);
   if TI[i]=0 then T[i]:=0: else
   T[i]:=evala(subs(beta=0,diff(TI[i],beta))); fi:
   if T[i]=0 or T[i]<>M  then I_M:=I_M+0; else
   I_M:=I_M+CTI[i]/OM2; fi:
  od: I_T:=I_M:
 else TII:=select(has,Z3,beta); CTII:=remove(has,Z3,beta);
 if TII=0 then TT:=0: else
  TT:=evala(subs(beta=0,diff(TII,beta))); fi:
 if TT=0 or TT<>M then I_M:=I_M+0; else I_M:=I_M+CTII; fi:
 I_T:=I_M/OM2: fi: fi: RETURN(I_T); end;
```

```
Bc4_NK:=proc(x,N,K) local I_T;
 I_T:=1/Pi^2*int(int(x*cos(N*alpha)*cos(K*psi),alpha=0..2*Pi),
 psi=0..2*Pi); if N=0 then I_T:=I_T/2 fi;
 if K=0 then I_T:=I_T/2 fi; RETURN(I_T); end;
```

(6) To find the coefficients in the governing equations and the fourth
boundary condition, we create the procedures EqC and BcC:

```
EqC:=proc(Eq,N,M,K) local SS; SS:=0:
 if Eq=F1S then SS:=S1_NMK(Eq,N,M,K,omega^2):
 elif Eq=F2S then SS:=S2_NMK(Eq,N,M,K,omega^2):
 elif Eq=F3S then SS:=S2_NMK(Eq,N,M,K,1): fi: RETURN(SS) end;
BcC:=proc(Eq,N,K) local SS,NEq,i; SS:=0:
 if type(Eq,`+`)=false then SS:=SS+Bc4_NK(Eq,N,K):
 else NEq:=nops(Eq); for i from 1 to NEq do
 SS:=SS+Bc4_NK(op(i,Eq),N,K): od: fi: RETURN(SS) end;
```

(7) We solve these systems and obtain the asymptotic solution of the
order of $O(\varepsilon^N)$, using the equations described above:

```
system1:={Eq1(2,0,0,2),Eq2(2,0,0,2),Eq3(2,0,0,2),
          Eq1(2,2,0,2),Eq2(2,2,0,2),Eq3(2,2,0,2),
Bc4(2,0,0,2)+Bc4(2,2,0,2)=simplify(BcC(F4S,2,0)/epsilon^NP)};
system1:=system1 union {Eta200||NA=0, Eta200||NA=0};
   var1:={Xi200||NA,Eta200||NA,Sigma200||NA,
          Xi220||NA,Eta220||NA,Sigma220||NA};
   DF1:=solve(system1,var1);
system2:={Eq1(2,0,2,2),Eq2(2,0,2,2),Eq3(2,0,2,2),
          Eq1(2,2,2,2),Eq2(2,2,2,2),Eq3(2,2,2,2),
Bc4(2,0,2,2)+Bc4(2,2,2,2)=simplify(BcC(F4S,2,2)/epsilon^NP)};
system2:=system2 union {Eta202||NA=0,Eta202||NA=0}:
   var2:={Xi202||NA,Eta202||NA,Sigma202||NA,
          Xi222||NA,Eta222||NA,Sigma222||NA};
    DF2:=solve(system2,var2);
system3:={Eq1(0,2,0,2),Eq2(0,2,0,2),Eq3(0,2,0,2),
          Eq1(0,0,0,2),Eq2(0,0,0,2),Eq3(0,0,0,2),
Bc4(0,2,0,2)+Bc4(0,0,0,2)=simplify(BcC(F4S,0,0)/epsilon^NP)};
system3:=system3 union
          {Xi020||NA=0,Xi000||NA=0,Sigma000||NA=0}:
   var3:={Xi020||NA,Eta020||NA,Sigma020||NA,
          Xi000||NA,Eta000||NA,Sigma000||NA};
   DF3:=solve(system3,var3);
```

```
system4:={Eq1(0,2,2,2),Eq2(0,2,2,2),Eq3(0,2,2,2),
          Eq1(0,0,2,2),Eq2(0,0,2,2),Eq3(0,0,2,2),
Bc4(0,2,2,2)+Bc4(0,0,2,2)=simplify(BcC(F4S,0,2)/epsilon^NP)};
system4:=system4 union {Eta002||NA=0}:
   var4:={Xi022||NA,Eta022||NA,Sigma022||NA,
          Xi002||NA,Eta002||NA,Sigma002||NA};
    DF4:=solve(system4,var4);
system5:={Eq1(1,1,1,2),Eq2(1,1,1,2),Eq3(1,1,1,2),
     Bc4(1,1,1,2)=simplify(BcC(F4S,1,1)/epsilon^NP)};
system5:=system5 union {Sigma111||NA=0};
   var5:={Xi111||NA,Eta111||NA,Sigma111||NA,omega||NP};
   DF5S:=solve(system5,var5);
for i from 1 to nops(DF5S) do if has(op(i,DF5S),omega||NP) then
   ZZ:=i: oomega1:=op(i,DF5S): fi: od:
   DF5:=subsop(ZZ=NULL,DF5S);
```

(8) We generate the approximate analytical solution for the second
approximation:

```
FF:=proc(x::list) local Xi_S,Eta_S,Sigma_S,i,Nx,a,b,c,s,N,M,K;
Nx:=nops(x); Xi_S:={}: Eta_S:={}: Sigma_S:={}:
for i from 1 to Nx do a:=lhs(x[i]); for N from 0 to NA do
for M from 0 to NA do for K from 0 to NA do
if a=Xi||N||M||K||NA then CXi||N||M||K||NA:=rhs(x[i])*
 sin(N*alpha)*exp(M*beta)*cos(K*psi):
Xi_S:=Xi_S union {CXi||N||M||K||NA};
 elif a=Eta||N||M||K||NA then CEta||N||M||K||NA:=rhs(x[i])*
  cos(N*alpha)*exp(M*beta)*cos(K*psi):
Eta_S:=Eta_S union {CEta||N||M||K||NA};
 elif a=Sigma||N||M||K||NA then CSigma||N||M||K||NA:=rhs(x[i])*
  cos(N*alpha)*exp(M*beta)*cos(K*psi):
Sigma_S:=Sigma_S union {CSigma||N||M||K||NA}; fi:
od: od: od: od: RETURN(Xi_S, Eta_S, Sigma_S); end;
ASet||NA:={'op(DF||i)'$'i'=1..5}; ASetf||NA:={};
BSet||NA:=convert(ASet||NA,list); NS:=nops(BSet||NA);
for i from 1 to NS do V||i:=op(i,BSet||NA):
 if rhs(V||i)=0 then
 ASetf||NA:=ASetf||NA union {subs(rhs(V||i)=0,V||i)}: else
 ASetf||NA:=ASetf||NA union {subs(rhs(V||i)=lhs(V||i),V||i)}:
fi: od:
BSetf||NA:=convert(ASetf||NA,list);
Xi||NA:=convert(FF(BSetf||NA)[1],`+`);
```

```
Eta||NA:=convert(FF(BSetf||NA)[2],`+`);
Sigma||NA:=convert(FF(BSetf||NA)[3],`+`);
Xi_Ser:=xi(alpha,beta,psi)+epsilon^NN*Xi||NA;
Eta_Ser:=eta(alpha,beta,psi)+epsilon^NN*Eta||NA;
Sigma_Ser:=sigma(alpha,beta,psi)+epsilon^NN*Sigma||NA;
```

The case $T=0$. If surface tension is neglected, then the free boundary problem is well-posed in a finite interval of time if the sign condition $-\nabla p \cdot \mathbf{n} > 0$ is satisfied [55]. Here p is the pressure and $\mathbf{n}$ is the outward unit normal to the free boundary. Note that when the forces of surface tension are taken into account, then the free boundary problem is always well-posed irrespective of the sign condition (see [4], [55]). This condition is somewhat similar to the fact that $p > 0$ inside the water region. In our case, according to the solution obtained, $\eta^{(i)}, \sigma^{(i)}$, the pressure takes the form:

$$p(\alpha, \beta, t) = -\frac{1}{12\kappa}\rho g(12\beta + \varepsilon^2[-6e^{2\beta}\cos(2\psi) + 6\cos(2\psi)]$$
$$+\varepsilon^3[-3\cos(\alpha)e^{3\beta}\cos(\psi) + 3\cos(\alpha)e^{\beta}\cos(\psi)$$
$$-5\cos(\alpha)e^{3\beta}\cos(3\psi) + 5\cos(\alpha)e^{\beta}\cos(3\psi)]) + O(\varepsilon^4).$$

The parametrization of the free surface by Lagrangian coordinates (α, β), that is, $\mathcal{B} = \big(x(\alpha, \beta, t), y(\alpha, \beta, t)\big)$, has the following form:

$$x(\alpha, \beta, t) = -\frac{1}{96\kappa}(-96\alpha + 96\varepsilon\sin(\alpha)e^{\beta}\cos(\psi)$$
$$+\varepsilon^3[-5\sin(\alpha)e^{\beta}\cos(3\psi) + 24\sin(\alpha)e^{\beta}\cos(\psi)]) + O(\varepsilon^4),$$
$$y(\alpha, \beta, t) = \frac{1}{96\kappa}(96\beta + 96\varepsilon\cos(\alpha)e^{\beta}\cos(\psi)$$
$$+\varepsilon^2[24e^{2\beta} + 24e^{2\beta}\cos(2\psi)]$$
$$+\varepsilon^3[24\cos(\alpha)e^{3\beta}\cos(\psi) + 24\cos(\alpha)e^{\beta}\cos(\psi)$$
$$+8\cos(\alpha)e^{3\beta}\cos(3\psi) - 5\cos(\alpha)e^{\beta}\cos(3\psi)]) + O(\varepsilon^4).$$

It is easy to verify that the solution obtained satisfies the sign condition and the condition $p > 0$ inside the water region.

Frequency-amplitude dependence. Applying the above method, we obtain the high-order asymptotic solution to the problem of capillary-gravity waves in an infinite-depth fluid. The analytic frequency-amplitude dependence is

$$\frac{\omega}{\omega_{(0)}} = 1 + \varepsilon^2 \left[\frac{81\delta^3 + 36\delta^2 + 27\delta - 8}{64\,(3\delta+1)(1-3\delta)} \right] - \varepsilon^4 \left[\frac{P^\infty}{16384\,Q^\infty} \right] + O(\varepsilon^5),$$

where polynomials P^∞ and Q^∞ are:

$$
\begin{aligned}
P^\infty &= -16691184\delta^9 + 13314456\delta^8 - 876987\delta^7 - 726327\delta^6 + 3458214\delta^5 \\
&\quad - 3099546\delta^4 + 554373\delta^3 + 65721\delta^2 - 1216\delta - 1472, \\
Q^\infty &= (1+12\delta)(4\delta-1)(-1+3\delta)^3(3\delta+1)^2.
\end{aligned}
$$

This dependence coincides with the analytic result in Eulerian coordinates obtained by Concus [12] up to the 5-th order.

If surface tension is neglected, $T = 0$, applying the above method, we obtain the asymptotic solution to the problem of gravity waves in an infinite-depth fluid and we write out the frequency-amplitude dependence:

$$\frac{\omega}{\omega_{(0)}} = 1 - \frac{1}{8}\varepsilon^2 - \frac{23}{256}\varepsilon^4 + O(\varepsilon^5).$$

This dependence is equal to the previous results obtained by [40], [43] in Lagrangian variables and equal to the analytic solution obtained by [36] in Eulerian variables, where $\omega_1 = 0$ and $\omega_2 = -\frac{1}{8}A^2\omega_0$. This expression coincides with the results obtained by [31] and [6] in Eulerian variables, where $\omega_3 = 0$ and $\omega_4 = -\frac{15}{256}A^4\omega_0$.

The coincidence of the results follows from determining the maximum amplitudes $A_{\max}$ that corresponds to the maximum wave profiles $y_{\max}$ in both types of variables and calculating the relation between them. In our case, after the passage from Lagrangian variables back to Eulerian, we have

$$\kappa y(x,t) = \left(\varepsilon + \frac{5}{32}\varepsilon^3\right)\cos\kappa x \cos t; \quad \kappa y_{\max} = \left(\varepsilon + \frac{5}{32}\varepsilon^3\right)\cos\kappa x, \quad (11.3)$$

where $A_{\max} = \varepsilon + \frac{5}{32}\varepsilon^3$. In the papers by [31] and [6] the maximum wave profile and the maximum amplitude are

$$y(x,t) = \left(A + \frac{1}{32}A^3\right)\cos x \sin \sigma t; \quad y_{\max} = \left(A + \frac{1}{32}A^3\right)\cos x, \quad (11.4)$$

where $A_{\max} = A + \frac{1}{32}A^3$. Equating the maximum amplitudes in both cases, we obtain the relation $A = \varepsilon + \frac{1}{8}\varepsilon^3$. Substituting the value of A into the frequency-amplitude dependence obtained by [31] and [6], we finally obtain

$$\frac{\omega}{\omega_{(0)}} = 1 - \frac{1}{8}A^2 - \frac{15}{256}A^4 + O(A^5) = 1 - \frac{1}{8}\varepsilon^2 - \frac{23}{256}\varepsilon^4 + O(\varepsilon^5), \quad (11.5)$$

that is equal to the frequency-amplitude dependence obtained.

Surface profiles. We construct the standing wave profiles, using the approximate analytical solutions obtained up to the third approximation. According to the Lagrangian formulation, the free surface $y = \eta(x, t)$ is defined by the parametric curve $\{x(a, 0, t), y(a, 0, t)\}$, $b = 0$. Setting $b = 0$ in the solutions obtained for x and y, we can observe the standing wave motion:

```
with(plots): Digits:=30; Omega1:=0; Omega3:=0;
Omega2:=(81*delta^3+36*delta^2+27*delta-8)/
  (64*(3*delta+1)*(1-3*delta));
Param:=[T=72,g=981.7,n=2,L=50,rho=1];
kappa :=evalf(subs(Param,(Pi*n)/L));
TTZ:=evalf(subs(Param,T*kappa^2/(rho*g)));
lambda:=subs(Param,subs(delta=TTZ,1-delta/(delta-1)));
A:=7; Epsilon:=A*kappa;
omega0:=evalf(subs(Param,sqrt(lambda*g*kappa)));
xiT_2:=evala(subs(Param,subs(BSet2,delta=TTZ,beta=0,
  alpha=a*kappa,Xi_Ser)));
etaT_2:=evala(subs(Param,subs(BSet2,delta=TTZ,
  beta=0, alpha=a*kappa,Eta_Ser)));
OM_2:=subs(delta=TTZ, epsilon=Epsilon,
  omega0+Omega1*epsilon+Omega2*epsilon^2);
Y_2:=unapply(evalf(subs(epsilon=Epsilon,psi=OM_2*t,
  epsilon*etaT_2/kappa)),[a,t]);
X_2:=unapply(evalf(subs(epsilon=Epsilon,psi=OM_2*t,
  a+epsilon*xiT_2/kappa)),[a,t]);
animate([X_2(a,t),Y_2(a,t),a =0..50],t=0..5,
  color=blue,thickness=4,scaling=constrained,frames=300);
```

$\square$

11.4   Integral Equations

Integral equations arise in various areas of science and numerous applications, for example, mathematical physics, fluid mechanics, theory of elasticity, biomechanics, economics, medicine, control theory, etc.

In Maple and Mathematica, there are no built-in functions to find exact, approximate analytical and numerical solutions of integral equations. Following the analytical approach, [24], [32], [33], we show how to solve the most important integral equations with *Maple* and *Mathematica*.

Integral equations can be divided into two main classes: linear and nonlinear integral equations. In general, linear integral equations can be written as follows:

$$\beta f(x) + \int_D K(x,t) f(t)\, dt = g(x), \quad x \in D, \qquad (11.6)$$

where $f(x)$ is the unknown function; β (the coefficient), $K(x,t)$ (the kernel), $g(x)$ (the free term or the right-hand side of the integral equation) are given functions of the integral equation; D is a bounded or unbounded domain in a finite-dimensional Euclidean space, x and t are points of this space, and dt is the volume element. It is required to determine $f(x)$ such that Eq.(11.6) holds for all (or almost all, if the integral is taken in the sense of Lebesgue) $x \in D$. If $g(x) \equiv 0$, then the integral equation is said to be homogeneous, otherwise it is called inhomogeneous.

There are three distinct types of linear integral equations, depending on the coefficient β:

(i) $\beta = 0$ for all $x \in D$, then Eq. (11.6) is called an equation of the first kind;

(ii) $\beta \neq 0$ for all $x \in D$, an equation of the second kind;

(iii) $\beta = 0$ on some non-empty subset of $S \subset D$, an equation of the third kind.

We consider the most important linear integral equations of the first and second kind in the one-dimensional case with variable integration limit and constant limits of integration, respectively:

$$\beta f(x) - \lambda \int_a^x K(x,t) f(t)\, dt = g(x), \quad K(x,t) \equiv 0, \ t > x, \quad (11.7)$$

$$\beta f(x) - \lambda \int_a^b K(x,t) f(t)\, dt = g(x), \qquad\qquad\qquad (11.8)$$

where $x \in [a, b]$, λ is the parameter of integral equations.

If the kernels and the right-hand sides of integral equations satisfy special conditions, then the integral equations (11.7) and (11.8) are called, respectively, the Volterra and Fredholm equations of the first/second kind. Usually, these special conditions are: the kernel $K(x,t)$ is continuous or square-integrable in $\Omega = \{a \le x \le b, a \le t \le b\}$, and $g(x)$ is continuous or square-integrable on $[a, b]$.

If a linear integral equation is not of the form (11.8), then it is called a singular equation. In these equations one or both of the limits of integration are infinity or the integral is to be understood as a Cauchy principal value. The equations of the second kind arise more frequently in problems of mathematical physics.

The most important nonlinear integral equations can be written in a general form as follows:

$$\beta f(x) - \lambda \int_a^x K\big(x,t,f(t)\big)\, dt = g(x), \quad K\big(x,t,f(t)\big) \equiv 0, \ t > x, \quad (11.9)$$

$$\beta f(x) - \lambda \int_a^b K\big(x,t,f(t)\big)\, dt = g(x), \qquad\qquad (11.10)$$

where $x \in [a, b]$.

Linear integral equations of the first kind with variable integration limit

$$\int_a^x K(x,t) f(t)\, dt = g(x).$$

Problem 11.18 Reducing the Volterra integral equation of the first kind with $g(x) = x$, $K(x,t) = \exp(x)$, to the Volterra integral equation of the second kind and find the exact solution of the integral equation. Show that the constructed function $f(x)$ is the integral equation solution.

Maple:

```
with(inttrans): g:=x->x; K:=(x,t)->exp(x); a:=0;
Eq1:=Int(K(x,t)*f(t),t=a..x)=g(x);
Eq2:=expand(isolate(diff(Eq1,x),f(x)));
laplace(g(x),x,p); laplace(K(x,t),x,p); laplace(f(x),x,p);
Eq2_L:=laplace(Eq2,x,p);
Sol:=factor(solve(subs(laplace(f(x),x,p)=F(p),Eq2_L),F(p)));
f:=x->invlaplace(Sol,p,x); f(x);
simplify(value(rhs(Eq2)-lhs(Eq2)));
simplify(value(rhs(Eq1)-lhs(Eq1)));
```

Mathematica:

```
g[x_]:=x; K[x_,t_]:=Exp[x]; a=0;
h[u_*v_]:=u*Integrate[v,{t,a,x}]/;FreeQ[u,v]
{Eq1=h[K[x,t]*f[t]]==g[x],Eq2=Solve[D[Eq1,x],f[x]]//FullSimplify}
{LaplaceTransform[g[x],x,p],LaplaceTransform[K[x,t],x,p],
 LaplaceTransform[f[x],x,p]}
{Eq2Lap=LaplaceTransform[Eq2,x,p]/.
 {LaplaceTransform[f[x],x,p]->F[p]},
 Eq3=Eq2Lap[[1,1,1]]==Eq2Lap[[1,1,2]],Sol=Solve[Eq3,F[p]],
 InvLap=Map[InverseLaplaceTransform[#,p,x]&,Sol,{3}]}
f[t_]:=InvLap[[1,1,2]]/.{x->t}; {f[t],Eq1}
```
                                                                            □

Problem 11.19 Applying the Laplace transform, solve the Volterra integral equation of the first kind of convolution type $\int_0^x K(x-t)f(t)\,dt = g(x)$, $g(x) = x^n$, $K(x-t) = \exp(x-t)$. Show that the constructed function $f(x)$ is the integral equation solution.

Maple:

```
with(inttrans): n:=2; g:=x->x^n; K:=(x,t)->exp(x-t); a:=0;
Eq1:=Int(K(x,t)*f(t),t=a..x)=g(x);
laplace(g(x),x,p); laplace(K(x,t),x,p); laplace(f(x),x,p);
Eq1_L:=laplace(Eq1,x,p);
Sol:=factor(solve(subs(laplace(f(x),x,p)=F(p),Eq1_L),F(p)));
f:=x->invlaplace(Sol,p,x); f(x); Eq1;
simplify(value(Eq1));
```

Mathematica:

```
n=2; g[x_]:=x^n; K[x_,t_]:=Exp[x-t]; a=0;
Eq1=Integrate[K[x,t]*f[t],{t,a,x}]==g[x]
{LaplaceTransform[g[x],x,p], LaplaceTransform[K[x,t],x,p],
 LaplaceTransform[f[x],x,p]}
Eq1Lap=LaplaceTransform[Eq1,x,p]/.
 {LaplaceTransform[f[x],x,p]->F[p]}
Sol=Solve[Eq1Lap,F[p]]
InvLap=Map[InverseLaplaceTransform[#,p,x]&,Sol,{3}]
f[t_]:=InvLap[[1,1,2]]/.{x->t}; {Expand[f[t]],Eq1}
```

□

Linear integral equations of the second kind with variable integration limit

$$f(x) - \lambda \int_a^x K(x,t) f(t)\, dt = g(x).$$

Problem 11.20 Show that the function $f(x) = xe^x$ is the solution of the Volterra integral equation of the second kind with $g(x) = \sin(x)$, $K(x,t) = \cos(x-t)$, $\lambda = 2$.

Maple:

```
lambda:=2: a:=0; f:=x->x*exp(x); g:=x->sin(x);
K:=(x,t)->cos(x-t);
Eq1:=f(x)-lambda*Int(K(x,t)*f(t),t=a..x)=g(x); value(Eq1);
intsolve(F(x)-lambda*Int(K(x,t)*F(t),t=a..x)=g(x),F(x));
```

Mathematica:

```
f[x_]:=x*Exp[x]; g[x_]:=Sin[x];
K[x_,t_]:=Cos[x-t]; {lambda=2, a=0,
 Eq1=f[x]-lambda*Integrate[K[x,t]*f[t],{t,a,x}]==g[x]}
```

□

Problem 11.21 Construct the Volterra integral equation of the second kind corresponding to the linear ordinary differential equation $y''_{xx} + xy'_x + y = 0$.

Maple:

```
a:=0; DifEq:=diff(y(x),x$2)+x*diff(y(x),x)+y=0;
IC:=[1,0]; Eq1:=diff(y(x),x$2)=f(x);
Eq11:=subs(_C1=IC[2],diff(y(x),x)=int(f(t),t=a..x)+_C1);
Eq2:=dsolve(Eq1, y(x)); EqSub:=Int(Int(f(x),x),x)=
 subs(n=2,1/(n-1)!*int((x-t)^(n-1)*f(t),t=a..x));
Eq3:=subs(op(1,rhs(Eq2))=rhs(EqSub), Eq2);
Eq4:=y=rhs(subs(_C1=IC[2],_C2=IC[1], Eq3));
Eq5:=subs({Eq1,Eq11,Eq4},DifEq);
IntEq:=factor(combine(isolate(Eq5,f(x))));
```

Mathematica:

```
{a=0,DifEq=y''[x]+x*y'[x]+y[x]==0,IC={1,0}}
fun[x_]:=f[x]; Eq1=y''[x]->fun[x]
Eq11=y'[x]->Integrate[fun[t],{t,a,x}]+C[2]/.{C[2]->IC[[2]]}
Eq2=DSolve[{y''[x]==fun[x]},y[x],x];
Eq21=Eq2/.{Eq2[[1,1,2,3]]->1/(n-1)!*Integrate[(x-t)^(n-1)
 *fun[t],{t,a,x}]/.{n->2}}
{Eq3=Eq21[[1,1]]/.{C[1]->IC[[1]],C[2]->IC[[2]]},
 Eq4=DifEq/.{Eq1,Eq11,Eq3}, IntEq=Solve[Eq4,f[x]],
 IntEq[[1,1,1]]==IntEq[[1,1,2]]}
```

$\square$

Problem 11.22 Applying the resolvent kernel method to the Volterra integral equation of the second kind with $\lambda = 1$, $a = 0$, $g(x) = \exp(x^2)$, $K(x, t) = \exp(x^2 - t^2)$, construct the resolvent kernel $R(x, t)$ and the exact solution. Show that the constructed function $f(x)$ is the integral equation solution.

Maple:

```
K:=(x,t)->exp(x^2-t^2); K1:=(x,t)->K(x,t); g:=x->exp(x^2);
lambda:=1; a:=0; k:=10; for i from 2 to k do
 K||i:=unapply(factor(value(int(K(x,s)*K||(i-1)(s,t),
  s=t..x))),x,t): print(K||i(x,t)); od:
R:=unapply(sum('exp(x^2-t^2)*(x-t)^n/n!','n'=0..infinity),x,t);
Sol:=unapply(g(x)+lambda*int(R(x,t)*g(t),t=a..x),x);
Eq1:=value(Sol(x)-lambda*Int(K(x,t)*Sol(t),t=a..x)=g(x));
```

Mathematica:

```
K[x_,t_]:=Exp[x^2-t^2]; KP[x_,t_]:=K[x,t];
g[x_]:=Exp[x^2]; {lambda=1, a=0, k=10, L=Array[x,k,0],
 M=Array[x,k,0], L[[1]]=KP[x,t], M[[1]]=KP[s,t]}
Do[KP[X_,T_]:=Integrate[K[x,s]*M[[i-1]],{s,t,x}]/.{x->X,t->T};
 L[[i]]=KP[x,t]; M[[i]]=KP[s,t],{i,2,k}];
T1=Table[L[[i]],{i,1,k}]//FullSimplify
R[X_,T_]:=Sum[Exp[x^2-t^2]*(x-t)^n/n!,
 {n,0,Infinity}]/.{x->X,t->T}; Sol[X_]:=g[x]+lambda*
 Integrate[R[x,t]*g[t],{t,a,x}]/.{x->X}; Factor[Sol[x]]
Eq1=Sol[x]-lambda*Integrate[K[x,t]*Sol[t],{t,a,x}]==g[x]
```

$\square$

Problem 11.23 Representing the kernel $K(x,t) = (x-t)$ of the Volterra integral equation of the second kind (with $\lambda = 1$, $g(x) = \sin(x)$) as a polynomial of the order $n-1$ and solving the corresponding initial value problem for ODE, find the resolvent kernel $R(x,t)$ and construct the exact solution.

Maple:

```
n:=10; K:=(x,t)->x-t; L:=NULL; lambda:=1: a:=0;
g:=x->sin(x); K_P:=a0+add(a||i*(x-t)^i,i=1..n-1);
for i from 0 to n-1 do
 if i=1 then a||i:=1 else a||i:=0 fi: L:=L,a||i: od: L:=[L];
H:=unapply(convert(rhs(simplify(dsolve(
 {diff(h(x),x$2)-h(x)=0,h(t)=0,D(h)(t)=1},h(x)))),trig),x);
R:=unapply(1/lambda*diff(H(x),x$2),x,t);
Sol:=unapply(g(x)+lambda*int(R(x,t)*g(t),t=a..x),x);
simplify(Sol(x)); combine(convert(Sol(x),trig));
Eq1:=value(Sol(x)-lambda*Int(K(x,t)*Sol(t),t=a..x)=g(x));
```

Mathematica:

```
K[x_,t_]:=x-t; g[x_]:=Sin[x]; x[i_]:=0;
{n=10,L=Array[x,n,0],ck=Array[x,n,0],lambda=1,a=0}
KP=A[0]+Sum[A[i]*(x-t)^i,{i,1,n-1}]==K[x,t]
Do[ck[[i]]=Coefficient[KP[[1]],A[i-1]]/KP[[2]];
 If[ck[[i]]\[Element] Reals,L[[i]]-ck[[i]],L[[i]]=0],{i,1,n}];
T1=Table[L[[i]],{i,1,n}]
solDif=DSolve[{h''[x]-h[x]==0,h[t]==0,h'[t]==1},h[x],x]
H[X_]:=FullSimplify[ExpToTrig[solDif[[1,1,2]]]]/.{x->X};
```

```
R[X_,T_]:=1/lambda*H''[x]/.{x->X,t ->T};   {H[X],R[x,t]}
Sol[X_]:=g[x]+lambda*Integrate[R[x,t]*g[t],
 {t,a,x}]/.{x->X}; Sol[x]//FullSimplify
Eq1=Sol[x]-lambda*Integrate[K[x,t]*Sol[t],
 {t,a,x}]==g[x]//FullSimplify
```

$\square$

Problem 11.24 Applying the Laplace transform to the Volterra integral equation of the second kind of convolution type, $K(x-t)=e^{-(x-t)}\sin(x-t)$, $\lambda=1$, $g(x)=\cos x$, and using the Convolution Theorem for the Laplace transform, find the resolvent kernel of this equation and construct the exact solution. Show that the constructed function $f(x)$ is the integral equation solution.

Maple:

```
with(inttrans): lambda :=1; g:=x->cos(x);
K:=(x,t)->exp(-(x-t))*sin(x-t); a:=0;
K_L:=laplace(exp(-X)*sin(X),X,p); R_L:=factor(K_L/(1-K_L));
R:=unapply(invlaplace(R_L,p,X),X); R(x-t);
Sol:=unapply(g(x)+lambda*int(R(x-t)*g(t),t=a..x),x);
Eq1:=combine(value(Sol(x)-lambda*Int(K(x,t)*Sol(t),
     t=a..x)=g(x)));
```

Mathematica:

```
lambda=1; g[x_]:=Cos[x];
K[x_,t_]:=Exp[-(x-t)]*Sin[x-t]; a=0;
{KL=LaplaceTransform[Exp[-X]*Sin[X],X,p],RL=Factor[KL/(1-KL)]}
R[z_]:=InverseLaplaceTransform[RL,p,X]/.{X->z}; R[x-t]
Sol[z_]:=g[x]+lambda*Integrate[
 R[x-t]*g[t],{t,a,x}]/.{x->z}; Sol[x]
Eq1=FullSimplify[Sol[x]-lambda*Integrate[K[x,t]*Sol[t],
 {t,a,x}]==g[x]]
```

$\square$

Problem 11.25 Applying the Laplace transform find the exact solution of the Volterra integral equation of the second kind of convolution type, $g(x)=\sinh(x)$, $K(x-t)=\cosh(x-t)$, $\lambda=1$. Show that the constructed function $f(x)$ is the integral equation solution.

Maple:

```
with(inttrans): lambda:=1; a:=0; g:=x->sinh(x);
K:=(x,t)->cosh(x-t); laplace(g(x),x,p);
laplace(K(x,t),x,p); laplace(f(x),x,p);
Eq1:=f(x)-lambda*int(K(x,t)*f(t),t=a..x)=g(x);
Eq1_L:=laplace(Eq1,x,p);
Sol:=factor(solve(subs(laplace(f(x),x,p)=F(p),Eq1_L),F(p)));
f:=x->invlaplace(Sol,p,x); f(x);
B1:=convert(Eq1,exp); simplify(rhs(B1)-lhs(B1));
```

Mathematica:

```
g[x_]:=Sinh[x]; K[x_,t_]:=Cosh[x-t];
lambda=1; a=0; {LaplaceTransform[g[x],x,p],
 LaplaceTransform[K[x,t],x,p],LaplaceTransform[f[x],x,p]}
Eq1=f[x]-lambda*Integrate[K[x,t]*f[t],{t,a,x}]==g[x]
Eq1Lap=LaplaceTransform[Eq1,x,p]/.
 {LaplaceTransform[f[x],x,p]->F[p]}
Sol:=Solve[Eq1Lap,F[p]]
InvLap=Map[InverseLaplaceTransform[#,p,x]&,Sol,{3}]
f[t_]:=FullSimplify[ExpToTrig[InvLap[[1,1,2]]]/.{x->t}]];
{f[x], Eq1}
```

□

Linear integral equations of the first kind with constant limits of integration

$$\int_a^b K(x,t)f(t)\,dt = g(x).$$

Ill-posed problems: if $K(x,t)$ is a square integrable function in Ω, $g(x) \in L^2(a,b)$, $f(x) \in L^2(a,b)$, the problem of finding solutions of linear integral equations of the first kind with constant limits of integration belongs to the class of ill-posed problems, i.e. this problem is unstable with respect to small changes of the right-hand side of the integral equation.

The most important methods for studying linear integral equations of the first kind are the methods for constructing approximate solutions of ill-posed problems. Here we consider the successive approximations method and the regularization method.

Problem 11.26 Applying the successive approximations method solve
the Fredholm integral equation of the first kind with $f_0(x) = 0$, $a = 0$,
$b = 1$, $g(x) = 1$, $K(x, t) = 1$. Show that the constructed function $f(x)$ is
the integral equation solution.

Maple:

```
k:=10; f0:=x->0; g:=x->1; a:=0; b:=1; lambda:=1; K:=(x,t)->1;
for i from 1 to k do  f||i:=unapply(f||(i-1)(x)+
 lambda*(g(x)-int(K(x,t)*f||(i-1)(t),t=a..b)),x); od:
for i from 0 to k-1 do simplify(f||i(x)); od;
Sol:=x->1; Eq1:=value(lambda*Int(K(x,t)*Sol(t),t=a..b)=g(x));
```

Mathematica:

```
fP[x_]:=0; g[x_]:=1; K[x_,t_]:=1;
{k=10,a=0,b=1,lambda=1,L=Array[x,k,0],M=Array[x,k,0],
 L[[1]]=fP[x],M[[1]]=fP[t]}
Do[fP[X_]:=L[[i-1]]+lambda*(g[x]-Integrate[K[x,t]*M[[i-1]],
 {t,a,b}])/.{x->X}; L[[i]]=fP[x]; M[[i]]=fP[t],{i,2,k}];
T1=Table[L[[i]],{i,1,k}]//FullSimplify
Sol[x_]:=1; Eq1=lambda*Integrate[K[x,t]*Sol[t],{t,a,b}]==g[x]
```

$\square$

Problem 11.27 Applying the Tikhonov regularization method solve the
Fredholm integral equation of the first kind with $a = 0$, $b = 1$, $g(x) = 1$,
$K(x, t) = 1$, $\lambda = 1$.

Maple:

```
K:=(x,t)->1; K1:=(x,t)->K(x,t); g:=x->1; lambda:=1; a:=0; b:=1;
for i from 2 to 10 do  K||i:=unapply(factor(value(int(
 K(x,s)*K||(i-1)(s,t),s=a..b))),x,t): print(K||i(x,t)); od:
R:=unapply(sum('1*(lambda/epsilon)^(n-1)','n'=1..infinity),x,t);
Sol:=unapply(g(x)/epsilon+lambda/epsilon*int(R(x,t)*g(t),
 t=a..b),x); Eq1:=factor(value(epsilon*Sol(x)-
 lambda*Int(K(x,t)*Sol(t),t=a..b)=g(x)));
Eq2:=factor(value(lambda*Int(K(x,t)*Sol(t),t=a..b)));
B:=sqrt(int((g(x)-Eq2)^2,x=a..b)); subs(epsilon=0.382,B);
plot(B,epsilon=0.38..0.4); evalf(subs(epsilon=382/1000,Sol(x)));
```

Mathematica:

```
K[x_,t_]:=1; KP[x_,t_]:=K[x,t]; g[x_]:=1;
{lambda=1,a=0,b=1,k=10,L=Array[x,k,0],M=Array[x,k,0]}
{L[[1]]=KP[x,t], M[[1]]=KP[s,t]}
Do[KP[X_,T_]:=Integrate[K[x,s]*M[[i-1]],{s,a,b}]/.{x->X,t->T};
  L[[i]]=KP[x,t]; M[[i]]=KP[s,t],{i,2,k}];
T1=Table[L[[i]],{i,1,k}]//FullSimplify
R[X_,T_]:=Sum[1*(lambda/\[Epsilon])^(n-1),{n,1,Infinity}]/.
  {x->X,t->T}; Sol[X_]:=g[x]/\[Epsilon]+lambda/\[Epsilon]*
  Integrate[R[x,t]*g[t],{t,a,b}]/.{x->X}; Factor[Sol[x]]
{Eq1=\[Epsilon]*Sol[x]-lambda*Integrate[K[x,t]*Sol[t],
  {t,a,b}]==g[x],Eq2=lambda*Integrate[K[x,t]*Sol[t],{t,a,b}]}
{B=Sqrt[Integrate[(g[x]-Eq2)^2,{x,a,b}]],B/.{\[Epsilon]->0.382}}
Plot[B,{\[Epsilon],0.38,0.4},PlotRange->All,Frame->True]
N[Sol[x]/.{\[Epsilon]->382/1000},10]
```
□

Linear integral equations of the second kind with constant limits of integration:

$$f(x) - \lambda \int_a^b K(x,t) f(t)\, dt = g(x).$$

Problem 11.28   Show that the function $f(x) = \cos(2x)$ is the solution of the Fredholm integral equation of the second kind with

$$g(x) = \cos(x),\ \ \lambda = 3,\ \ a = 0,\ \ b = \pi,\ \ K(x,t) = \begin{cases} \sin x \cos t & a \le x \le t, \\ \sin t \cos x & t \le x \le b. \end{cases}$$

Maple:

```
a:=0; b:=Pi; f:=x->cos(2*x); g:=x->cos(x); lambda:=3;
K:=(x,t)->piecewise(x>=a and x <=t,sin(x)*cos(t),
  x>=t and x<=b,sin(t)*cos(x));
Eq1:=f(x)-lambda*Int(K(x,t)*f(t),t=a..b,'AllSolutions')=g(x);
B1:=combine(value(Eq1)) assuming x>a and x<b;
```

Mathematica:

```
f[x_]:=Cos[2*x]; g[x_]:=Cos[x]; {a=0,b=Pi,lambda=3}
K[x_,t_]:=Piecewise[{{Sin[x]*Cos[t],a<=x<=t},
  {Sin[t]*Cos[x],t<=x<=b}}];
Eq1=f[x]-lambda*Integrate[K[x,t]*f[t],{t,a,b}]==g[x]
B1=FullSimplify[Assuming[x>a && x<b,PiecewiseExpand[Eq1]]]
```
□

Problem 11.29 Applying the Fredholm determinant method, construct the Fredholm resolvent kernel $R(x, y; \lambda)$ of the Fredholm integral equation of the second kind with $a = 0$, $b = 1$, $g(x) = \exp(-x)$, $K(x,t) = x \exp(t)$. Show that the constructed function $f(x)$ is the integral equation solution.

Maple:

```
with(LinearAlgebra): a:=0; b:=1; g:=x->exp(-x);
K:=(x,t)->x*exp(t); lambda:=2; B0:=K(x,t); C0:=1;
Eq1:=f(x)-lambda*Int(K(x,t)*f(t),t=a..b)=g(x);
for k from 2 to 5 do
 DF||k:=Matrix(1..k,1..k,[]); DF||k[1,1]:=K(x,t);
 i:=1; for j from 2 to k do DF||k[i,j]:=K(x,t||(j-1)) od:
 j:=1; for i from 2 to k do DF||k[i,j]:=K(t||(i-1),t) od:
 for i from 2 to k do
  for j from 2 to k do
   DF||k[i,j]:=K(t||(i-1),t||(j-1))
od: od: od:
DF2; DF3; DF4; DF5;
B1:=value(Int(Determinant(DF2),t1=a..b));
B2:=value(Int(Int(Determinant(DF3),t1=a..b),t2=a..b));
B3:=value(Int(Int(Int(Determinant(DF4),t1=a..b),t2=a..b),
 t3=a..b));
B4:=value(Int(Int(Int(Int(Determinant(DF5),t1=a..b),
 t2=a..b),t3=a..b),t4=a..b));
for k from 1 to 4 do
 DDF||k :=Matrix(1..k,1..k,[]);
 for i from 1 to k do
  for j from 1 to k do
   DDF||k[i,j]:=K(t||(i),t||(j))
od: od: od:
DDF1; DDF2; DDF3; DDF4;
C1:=value(Int(Determinant(DDF1),t1=a..b));
C2:=value(Int(Int(Determinant(DDF2),t1=a..b),t2=a..b));
C3:=value(Int(Int(Int(Determinant(DDF3),t1=a..b),
 t2=a..b),t3=a..b));
C4:=value(Int(Int(Int(Int(Determinant(DDF4),t1=a..b),
 t2=a..b),t3=a..b),t4=a..b));
DN:=K(x,t)+add((-1)^n/n!*B||n*lambda^n,n=1..3);
DD:=1+add((-1)^n/n!*C||n*lambda^n,n=1..3);R:=unapply(DN/DD,x,t);
Sol:=unapply(value(g(x)+lambda*Int(R(x,t)*g(t),t=a..b)),x);
Eq1:=value(Sol(x)-lambda*Int(K(x,t)*Sol(t),t=a..b)=g(x));
```

Mathematica:

```
g[x_]:=Exp[-x]; K[x_,t_]:=x*Exp[t];
{a=0, b=1, n=5, lambda=2, BA=Array[x1,n,0], CA=Array[x2,n,0],
 tt=Array[t,n,1], BA[[0]]=K[x,t], CA[[0]]=1}
Eq1=f[x]-lambda*Integrate[K[x,t]*f[t],{t,a,b}]==g[x]
DFF[z_]:=Array[x4,{z,z},{1,1}];
DA[k_]:=Module[{},DF=Array[x5,{k,k},{1,1}];DF[[1,1]]=K[x,t];
 i=1; Do[DF[[i,j]]=K[x,tt[[j-1]]],{j,2,k}];
 j=1; Do[DF[[i,j]]=K[tt[[i-1]],t],{i,2,k}];
 Do[Do[DF[[i,j]]=K[tt[[i-1]],tt[[j-1]]],{i,2,k}],{j,2,k}];DF];
Map[MatrixForm,{DA[2],DA[3],DA[4],DA[5]}]
BA[[1]]=Integrate[Det[DA[2]],{t[1],a,b}]
BA[[2]]=Integrate[Integrate[Det[DA[3]],{t[1],a,b}],{t[2],a,b}]
BA[[3]]=Integrate[Integrate[Integrate[Det[DA[4]],
 {t[1],a,b}],{t[2],a,b}],{t[3],a,b}]
BA[[4]]=Integrate[Integrate[Integrate[Integrate[Det[DA[5]],
 {t[1],a,b}],{t[2],a,b}],{t[3],a,b}],{t[4],a,b}]
DDA[k_]:=Module[{},DDF=Array[x6,{k,k},{1,1}];
 Do[Do[DDF[[i,j]]=K[tt[[i]],tt[[j]]],{i,1,k}],{j,1,k}];DDF];
Map[MatrixForm,{DDA[1],DDA[2],DDA[3],DDA[3]}]
CA[[1]]=Integrate[Det[DDA[1]],{t[1],a,b}]
CA[[2]]=Integrate[Integrate[Det[DDA[2]],{t[1],a,b}],{t[2],a,b}]
CA[[3]]=Integrate[Integrate[Integrate[Det[DDA[3]],
 {t[1],a,b}],{t[2],a,b}],{t[3],a,b}]
CA[[4]]=Integrate[Integrate[Integrate[Integrate[
 Det[DDA[4]],{t[1],a, b}],{t[2],a,b}],{t[3],a,b}],{t[4],a,b}]
DN=BA[[0]]+Sum[(-1)^n/n!*BA[[n]]*lambda^n,{n,1,3}]
DD=CA[[0]]+Sum[(-1)^n/n!*CA[[n]]*lambda^n,{n,1,3}]
R[X_,T_]:=(DN/DD)/.{x->X,t->T};
Sol[X_]:=g[x]+lambda*Integrate[R[x,t]*g[t],{t,a,b}]/.{x->X};
{R[x,t], Sol[x]}
Eq1=Sol[x]-lambda*Integrate[K[x,t]*Sol[t],{t,a,b}]==g[x]
```

□

Problem 11.30 Applying the Fredholm determinant method and the recurrence relations, construct the Fredholm resolvent kernel $R(x, y; \lambda)$ of the Fredholm integral equation of the second kind with $a = 0$, $b = 1$, $g(x) = x$, $K(x,t) = x - 2t$.

Show that the constructed function $f(x)$ is the integral equation solution.

Maple:

```
a:=0; b:=1; g:=x->x; K:=(x,t)->x-2*t; lambda:=1; C0:=1;
Eq1:=f(x)-lambda*Int(K(x,t)*f(t),t=a..b)=g(x);
B0:=(x,t)->K(x,t);
for k from 1 to 9 do
 C||k:=value(Int(B||(k-1)(s,s),s=a..b));
 B||k:=unapply(value(C||k*K(x,t)-
  k*Int(K(x,s)*B||(k-1)(s,t),s=a..b)),x,t); od;
DN:=K(x,t)+add((-1)^n/n!*B||n(x,t)*lambda^n,n=1..9);
DD:=1+add((-1)^n/n!*C||n*lambda^n,n=1..9);
R:=unapply(DN/DD,x,t); collect(simplify(R(x,t)),[x,t]);
Sol:=unapply(value(g(x)+lambda*Int(R(x,t)*g(t),t=a..b)),x);
Eq1:=value(Sol(x)-lambda*Int(K(x,t)*Sol(t),t=a..b)=g(x));
```

Mathematica:

```
g[x_]:=x; K[x_,t_]:=x-2*t;
KP[x_,t_]:=K[x, t]; {a=0, b=1, lambda=1, n=9,
 BL=Array[x1,n,0], BM=Array[x2,n,0],BN=Array[x3,n,0],
 CA=Array[x4,n,0], BL[[0]]=KP[s,t], BM[[0]]=KP[s,s],
 BN[[0]]=KP[x,t], CA[[0]]=1}
Eq1=f[x]-lambda*Integrate[K[x,t]*f[t],{t,a,b}]==g[x]
Do[CA[[k]]=Integrate[BM[[k-1]]],{s,a,b}];
 KP[X_,T_]:=CA[[k]]*K[x,t]-k*Integrate[K[x,s]*BL[[k-1]]],
 {s,a,b}]/.{x->X,t->T}; BL[[k]]=KP[s,t]; BM[[k]]=KP[s,s];
 BN[[k]]=KP[x,t],{k,1,n}];
T1=Table[BN[[i]],{i,1,n}]//FullSimplify
T2=Table[CA[[i]],{i,1,n}]//FullSimplify
DN=BN[[0]]+Sum[(-1)^n/n!*BN[[n]]*lambda^n,{n,1,9}]
DD=CA[[0]]+Sum[(-1)^n/n!*CA[[n]]*lambda^n,{n,1,9}]
R[X_,T_]:=(DN/DD)/.{x->X,t->T}; Sol[X_]:=g[x]+lambda*
 Integrate[R[x,t]*g[t],{t,a,b}]/.{x->X}; {Factor[R[x,t]],Sol[x]}
Eq1=Sol[x]-lambda*Integrate[K[x,t]*Sol[t],{t,a,b}]==g[x]
```

□

Problem 11.31 Applying the resolvent kernel method to the Fredholm integral equation of the second kind with $g(x) = x$, $K(x,t) = xt$, $a = 0$, $b = 1$, construct the resolvent kernel $R(x,t;\lambda)$ and the exact solution. Show that the constructed function $f(x)$ is the integral equation solution.

Maple:

```
K:=(x,t)->x*t; K1:=(x,t)->K(x,t); g:=x->x; a:=0; b:=1;
for i from 2 to 10 do K||i:=unapply(factor(value(int(
 K(x,s)*K||(i-1)(s,t),s=a..b))),x,t): print(K||i(x,t)); od:
R:=unapply(sum('(x*t)/3^(n-1)*lambda^(n-1)','n'=1..infinity),
 x,t,lambda);
Sol:=unapply(g(x)+lambda*int(R(x,t,lambda)*g(t),t=a..b),x);
Eq1:=simplify(value(Sol(x)-lambda*Int(K(x,t)*Sol(t),
 t=a..b)=g(x)));
B:=sqrt(int(int(K(x,t)^2,x=a..b),t=a..b)); abs(lambda)<1/B;
```

Mathematica:

```
K[x_,t_]:=x*t; KP[x_,t_]:=K[x,t];
g[x_]:=x; {a=0, b=1, n=10, L=Array[x1,n,1], M=Array[x2,n,1],
 L[[1]]=KP[s,t], M[[1]]=KP[x,t]}
Do[KP[X_,T_]:=Integrate[K[x,s]*L[[i-1]],{s,a,b}]/.{x->X,t->T};
 L[[i]]=KP[s,t]; M[[i]]=KP[x,t], {i,2,n}];
T1=Table[M[[i]],{i,1,n}]//FullSimplify
R[X_,T_,Lambda_]:=Sum[(x*t)/3^(n-1)*lambda^(n-1),
 {n,1,Infinity}]/.{x->X,t->T,lambda->Lambda};
Sol[X_]:=g[x]+lambda*Integrate[R[x,t,lambda]*g[t],
 {t,a,b}]/.{x->X}; {R[x,t,lambda], Sol[x]}
{Eq1=Sol[x]-lambda*Integrate[K[x,t]*Sol[t],{t,a,b}]==g[x],
 B=Sqrt[Integrate[Integrate[K[x,t]^2,{x,a,b}],{t,a,b}]],
 Abs[lambda]<1/B//TraditionalForm}
```

□

Problem 11.32 Let $a = -1$, $b = 1$. Verify that the kernels $KA(x,t) = xt$ and $KB(x,t) = x^2t^2$ are orthogonal kernels in $[a,b]$.

Applying the resolvent kernel method to the Fredholm integral equation of the second kind with $g(x) = x$, $K(x,t) = KA(x,t) + KB(x,t)$, construct the resolvent kernel $R(x,t;\lambda)$ and the exact solution.

Show that the constructed function $f(x)$ is the integral equation solution.

Maple:

```
KA:=(x,t)->x*t; KA1:=(x,t)>KA(x,t);
KB:=(x,t)->x^2*t^2; KB1:=(x,t)->KB(x,t);
g:=x->x; a:=-1; b:=1;
int(KA(x,s)*KB(s,t),s=a..b); int(KB(x,s)*KA(s,t),s=a..b);
```

```
for i from 2 to 10 do
 KA||i:=unapply(factor(value(int(KA(x,s)*KA||(i-1)(s,t),
 s=a..b))),x,t): print(KA||i(x,t));
 KB||i:=unapply(factor(value(int(KB(x,s)*KB||(i-1)(s,t),
 s=a..b))),x,t): print(KB||i(x,t)); od:
RA:=unapply(sum('(x*t)*2^(n-1)/3^(n-1)*lambda^(n-1)',
 'n'=1..infinity),x,t,lambda);
RB:=unapply(sum('(x^2*t^2)*2^(n-1)/5^(n-1)*lambda^(n-1)',
 'n'=1..infinity),x,t,lambda);
Sol:=unapply(g(x)+lambda*int((RA(x,t,lambda)+
 RB(x,t,lambda))*g(t),t=a..b),x);
Eq1:=simplify(value(Sol(x)-lambda*Int((KA(x,t)+
 KB(x,t))*Sol(t),t=a..b)=g(x)));
B:=sqrt(int(int((KA(x,t))^2,x=a..b),t=a..b)); abs(lambda)<1/B;
```

Mathematica:

```
KA[x_,t_]:=x*t; KPA[x_,t_]:=KA[x,t];
KB[x_,t_]:=x^2*t^2; KPB[x_,t_]:=KB[x,t];
g[x_]:=x; {a=-1, b=1, n=10, LA=Array[x1,n,1], MA=Array[x2,n,1],
 LB=Array[x3,n,1], MB=Array[x4,n,1], LA[[1]]=KPA[s,t],
 MA[[1]]=KPA[x,t], LB[[1]]=KPB[s,t], MB[[1]]=KPB[x,t]}
Integrate[KA[x,s]*KB[s,t],{s,a,b}]
Integrate[KB[x,s]*KA[s,t],{s,a,b}]
Do[KPA[X_,T_]:=Integrate[KA[x,s]*LA[[i-1]],{s,a,b}]/.{x->X,t->T};
 LA[[i]]=KPA[s,t]; MA[[i]]=KPA[x,t]; KPB[X_,T_]:=Integrate[
 KB[x,s]*LB[[i-1]],{s,a,b}]/.{x->X,t->T}; LB[[i]]=KPB[s,t];
 MB[[i]]=KPB[x,t], {i,2,n}];
T1=Table[MA[[i]],{i,1,n}]//FullSimplify
T2=Table[MB[[i]],{i,1,n}]//FullSimplify
RA[X_,T_,Lambda_]:=Sum[(x*t)*2^(n-1)/3^(n-1)*lambda^(n-1),
 {n,1,Infinity}]/.{x->X,t->T,lambda->Lambda};
RB[X_,T_,Lambda_]:=Sum[(x^2*t^2)*2^(n-1)/5^(n-1)*lambda^(n-1),
 {n,1,Infinity}]/.{x->X,t->T,lambda->Lambda};
Sol[X_]:=g[x]+lambda*Integrate[(RA[x,t,lambda]+
 RB[x,t,lambda])*g[t],{t,a,b}]/.{x->X};
{RA[x,t,lambda], RB[x,t,lambda], Sol[x]}
{Eq1=FullSimplify[Sol[x]-lambda*Integrate[(KA[x,t]+KB[x,t])*
 Sol[t],{t,a,b}]==g[x]], B=Sqrt[Integrate[Integrate[KA[x,t]^2,
 {x,a,b}],{t,a,b}]], Abs[lambda]<1/B//TraditionalForm}
```

□

Problem 11.33 Let $a = -\pi$, $b = \pi$. Solve the Fredholm integral equation of the second kind with the degenerate kernel $K(x,t) = x\cos t + t^2\sin x + \cos x \sin t$, and $g(x) = x$. Show that the constructed function $f(x)$ is the integral equation solution.

Maple:

```
with(LinearAlgebra): g:=x->x; a:=-Pi; b:=Pi;
K:=(x,t)->x*cos(t)+t^2*sin(x)+cos(x)*sin(t);
k:=nops(K(x,t)); var:=[C1,C2,C3];
Eq1:=f(x)-lambda*Int(K(x,t)*f(t),t=a..b)=g(x);
Eq2:=f(x)-add(lambda*Int(op(i,K(x,t))*f(t),t=a..b),
 i=1..k)=g(x); Eq3:=isolate(Eq2,f(x));
CC1:=expand(op(2,rhs(Eq3))/x/lambda);
CC2:=expand(op(3,rhs(Eq3))/sin(x)/lambda);
CC3:=expand(op(4,rhs(Eq3))/cos(x)/lambda);
F:=lambda*x*C1+lambda*sin(x)*C2+lambda*cos(x)*C3+g(x);
Eq4:=subsop(2=F,Eq3); f:=unapply(rhs(Eq4),x); sys:={};
for i from 1 to 3 do  CC||i;
 sys:=sys union {value(expand(C||i=CC||i))}; od;
sys:=convert(sys,list);
M1:=GenerateMatrix(sys,var,augmented=true);
(M2,b2):=GenerateMatrix(sys,var); Determinant(M2);
b1:=LinearSolve(M1); subC:={};
for i from 1 to k do subC:=subC union {C||i=b1[i]}; od;
Sol:=unapply(subs(subC,F),x); Eq11:=simplify(value(Sol(x)
 -lambda*Int(K(x,t)*Sol(t),t=a..b)=g(x)));
```

Mathematica:

```
a=-Pi; b=Pi; f1={x,Cos[x],Sin[x]};
f2={Cos[t],Sin[t],t^2}; g[x_]:=x; {K12=f1.f2, CA={c1,c2,c3}}
k=Length[K12]; Eq1=f[x]-lambda*Sum[f1[[i]]*
 Integrate[f2[[i]]*f[t],{t,a,b}],{i,1,k}]==g[x]//ExpandAll
Eq2=Solve[Eq1,f[x]]
CC={Eq2[[1,1,2,2]]/x/lambda,Eq2[[1,1,2,3]]/Cos[x]/lambda,
    Eq2[[1,1,2,4]]/Sin[x]/lambda}//Cancel
F[x1_]:=(lambda*(f1.CA)+g[x]//Expand)/.{x->x1}; F[x]
CCC=Table[Integrate[f2[[i]]*F[t],{t,a,b}],{i,1,k}]
{sys=Table[CA[[i]]==CCC[[i]],{i,1,k}], S=Solve[sys,CA]}
Sol[X_]:=Cancel[(F[x]/.S[[1]])/.{x->X}]; Sol[x]
Eq12=FullSimplify[Sol[x]-
 lambda*Integrate[K12*Sol[t],{t,a,b}]]==g[x]
```

□

Problem 11.34 Let $a=0$, $b=\pi$. Find the eigenvalues and eigenfunctions of the Fredholm integral equation of the second kind with the degenerate kernel $K(x,t) = \cos^2 x \cos 2t + \cos 3x \cos^3 t$, and $g(x) = x$.

Construct the general solution $f(x)$ and verify the solution constructed.

Maple:

```
with(LinearAlgebra): g:=x->0; a:=0; b:=Pi;
K:=(x,t)->cos(x)^2*cos(2*t)+cos(3*x)*cos(t)^3;
k:=nops(K(x,t)); var:=[C1,C2];
Eq1:=f(x)-lambda*Int(K(x,t)*f(t),t=a..b)=g(x);
Eq2:=f(x)-add(lambda*Int(op(i,K(x,t))*f(t),
 t=a..b),i=1..k)=g(x);
Eq3:=isolate(Eq2,f(x));
CC1:=combine(op(1,rhs(Eq3))/cos(x)^2/lambda);
CC2:=combine(op(2,rhs(Eq3))/cos(3*x)/lambda);
F:=lambda*cos(x)^2*C1+lambda*cos(3*x)*C2+g(x);
Eq4:=subsop(2=F,Eq3); f:=unapply(rhs(Eq4),x); sys:=NULL;
for i from 1 to k do
 CC||i; sys:=sys,value(combine(C||i=CC||i)); od;
sys:=[sys];
(M1,b1):=GenerateMatrix(sys,var);
Eq5:=Determinant(M1)=0;
EigenVal:=[solve(Eq5,lambda)];
M2:=subs(lambda=EigenVal[1],M1);
M3:=subs(lambda=EigenVal[2],M1);
sys1:=M1.Vector(var)=0;
Eq6:=subs(lambda=EigenVal[1],sys1);
Eq7:=subs(lambda=EigenVal[2],sys1);
P1:={C2=0,C1=C1}; P2:={C1=0,C2=C2};
EigenFun1:=unapply(subs(P1,lambda=EigenVal[1],
 C1=1/EigenVal[1],F),x);
EigenFun2:=unapply(subs(P2,lambda=EigenVal[2],
 C2=1/EigenVal[2],F),x);
f:=(lambda,x)->piecewise(lambda=EigenVal[1],
 C*EigenFun1(x),lambda=EigenVal[2],C*EigenFun2(x),0);
combine(value(f(EigenVal[1],x)-EigenVal[1]*
 Int(K(x,t)*f(EigenVal[1],t),t=a..b)=g(x)));
combine(value(f(EigenVal[2],x)-EigenVal[2]*
 Int(K(x,t)*f(EigenVal[2],t),t=a..b)=g(x)));
value(f(Pi,x)-Pi*Int(K(x,t)*f(Pi,t),t=a..b)=g(x));
```

Mathematica:

```
{a=0, b=Pi}
f1={Cos[x]^2,Cos[3*x]}; f2={Cos[2*t],Cos[t]^3}; g[x_]:=0;
{K12=f1.f2, k=Length[K12], CA=Array[c,k,1], CC=Array[x1,k,1]}
{Eq1=f[x]-lambda*Sum[f1[[i]]*Integrate[f2[[i]]*f[t],{t,a,b}],
 {i,1,k}]==g[x]//ExpandAll, Eq2=Solve[Eq1,f[x]]}
CC={Eq2[[1,1,2,1]]/(Cos[x]^2)/lambda,
    Eq2[[1,1,2,2]]/Cos[3*x]/lambda}//Cancel
F[x1_]:=(lambda*(f1.CA)+g[x]//Expand)/.{x->x1}; F[x]
CCC=Table[Integrate[f2[[i]]*F[t],{t,a,b}],{i,1,k}]
sys0=Table[CA[[i]]==CCC[[i]],{i,1,k}]
genMat=Normal[CoefficientArrays[sys0,CA]]
{Mat1=genMat[[2]], b1=genMat[[1]], Eq5=Det[Mat1]==0}
{EigenVal=Solve[Eq5,lambda], Mat2=Mat1/.EigenVal[[1]],
 Mat3=Mat1/.EigenVal[[2]], sys1=Mat1.CA==0}
{Eq6=sys1/.EigenVal[[1]], Eq7=sys1/.EigenVal[[2]]}
P1={c[2]->0,c[1]->1/EigenVal[[1,1,2]],EigenVal[[1,1]]}
P2={c[1]->0,c[2]->1/EigenVal[[2,1,2]],EigenVal[[2,1]]}
EigenFun1[X_]:=(F[x]/.P1)/.{x->X}; EigenFun1[x]
EigenFun2[X_]:=(F[x]/.P2)/.{x->X}; EigenFun2[x]
F1[Lambda_,X_]:=Piecewise[{{C1*EigenFun1[x],
 lambda==EigenVal[[1,1,2]]},{C1*EigenFun2[x],
 lambda==EigenVal[[2,1,2]]}}]/.{lambda->Lambda,x->X};
F1[EigenVal[[1,1,2]],x]-EigenVal[[1,1,2]]*
 Integrate[K12*F1[EigenVal[[1,1,2]],t],{t,a,b}]==g[x]
F1[EigenVal[[2,1,2]],x]-EigenVal[[2,1,2]]*
 Integrate[K12*F1[EigenVal[[2,1,2]],t],{t,a,b}]==g[x]
F1[Pi,x]-Pi*Integrate[K12*F1[Pi,t],{t,a,b}]==g[x]
```

$\qquad\qquad\qquad\qquad\qquad\qquad\qquad\qquad\qquad\qquad\qquad\square$

Nonlinear integral equations with variable integration limit

$$\beta f(x) - \lambda \int_a^x K\big(x, t, f(t)\big)\, dt = g(x).$$

Problem 11.35 Applying the successive approximations method solve the nonlinear Volterra integral equation of the second kind with

$$f_0(x) = x, \quad \lambda = 1, \quad a = 0, \quad g(x) = 0, \quad K\big(x, t, f(t)\big) = \frac{1 + f(t)^2}{1 + t^2}.$$

Show that the constructed function $f(x) = x$ is the integral equation solution.

Maple:

```
k:=10; lambda:=1; f0:=x->x; g:=x->0; a:=0;
for i from 1 to k do
 F||(i-1):=unapply((1+f||(i-1)(t)^2)/(1+t^2),x,t,f||(i-1));
 f||i:=unapply(g(x)+lambda*int(F||(i-1)(x,t,f||(i-1)),
 t=a..x),x); od;
f:=x->x; F:=(x,t,f)->(1+f(t)^2)/(1+t^2);
Eq1:=f(x)-lambda*Int(F(x,t,f),t=a..x)=g(x); simplify(value(Eq1));
```

Mathematica:

```
fP[x_]:=x; g[x_]:=0;
{k=10, lambda=1, a=0, Lf=Array[x1,k,0], LF=Array[x2,k,0],
 Mf=Array[x3,k,0], Lf[[0]]=fP[t], Mf[[0]]=fP[x]}
Do[F[X_,T_]:=((1+Lf[[i-1]]^2)/(1+t^2))/.{x->X,t->T};
 LF[[i-1]]=F[x,t];
 ff[X_]:=g[x]+lambda*Integrate[LF[[i-1]],{t,a,x}]/.{x->X};
 Lf[[i]]=ff[t]; Mf[[i]]=ff[x],{i,1,k}];
T1=Table[Mf[[i]],{i,0,k-1}]//FullSimplify
T2=Table[LF[[i]],{i,0,k-1}]//FullSimplify
f[x_]:=x; F[x_,t_,f1_]:=(1+f1[t]^2)/(1+t^2); F[x,t,f1]
Eq1=f[x]-lambda*Integrate[F[x,t,f],{t,a,x}]==g[x]
```

$\square$

Problem 11.36 Applying the successive approximations method solve the nonlinear Volterra integral equation of the second kind with

$$f_0(x) = 0, \quad \lambda = 1, \quad a = 0, \quad g(x) = 0, \quad K\big(x,t,f(t)\big) = \frac{t^2 f(t)}{1+t^2+f(t)}.$$

Show that the $f(x) \equiv 0$ is the integral equation solution.

Maple:

```
k:=10; lambda:=1; f0:=x->0; g:=x->0; a:=0;
for i from 1 to k do
 F||(i-1):=unapply((t^2*f||(i-1)(t))/
          (1+t^2+f||(i-1)(t)),x,t,f||(i-1));
 f||i:=unapply(g(x)+lambda*
       int(F||(i-1)(x,t,f||(i-1)),t=a..x),x); od;
f:=x->0; F:=(x,t,f)->(t^2*f(t))/(1+t^2+f(t));
Eq1:=f(x)-lambda*Int(F(x,t,f),t=a..x)=g(x);
simplify(value(Eq1));
```

Mathematica:

```
fP[x_]:=0; g[x_]:=0;
{k=10, lambda=1, a=0, Lf=Array[x1,k,0], LF=Array[x2,k,0],
 Mf=Array[x3,k,0], Lf[[0]]=fP[t], Mf[[0]]=fP[x]}
Do[F[X_,T_]:=((t^2*Lf[[i-1]])/(1+t^2+Lf[[i-1]]))/.{x->X,t->T};
 LF[[i-1]]=F[x,t];
 ff[X_]:=g[x]+lambda*Integrate[LF[[i-1]],{t,a,x}]/.{x->X};
 Lf[[i]]=ff[t]; Mf[[i]]=ff[x],{i,1,k}];
T1=Table[Mf[[i]],{i,0,k-1}]//FullSimplify
T2=Table[LF[[i]],{i,0,k-1}]//FullSimplify
f[x_]:=0; F[x_,t_,f1_]:=(t^2*f1[t])/(1+t^2+f1[t]); F[x,t,f1]
Eq1=f[x]-lambda*Integrate[F[x,t,f],{t,a,x}]==g[x]
```

□

Problem 11.37 Applying the Laplace transform, solve the Volterra non-linear homogeneous integral equation of the second kind of convolution type

$$f(x) - \lambda \int_a^x f(t)f(x-t)\, dt = g(x), \quad g(x) = -x^9, \quad \lambda = \frac{1}{2}, \quad a = 0.$$

Show that the constructed function $f(x)$ is the integral equation solution.

Maple:

```
with(inttrans): lambda:=1/2; g:=x->-x^9; a:=0;
Eq1:=0-lambda*Int(f(t)*f(x-t),t=a..x)=g(x);
laplace(g(x),x,p); laplace(f(x),x,p);
Eq1_L:=laplace(Eq1,x,p);
Sol:=solve(subs(laplace(f(x),x,p)=F(p),Eq1_L),F(p));
f1:=x->invlaplace(expand(Sol[1]),p,x); f1(x);
f2:=x->invlaplace(expand(Sol[2]),p,x); f2(x);
E1:=0-lambda*Int(f1(t)*f1(x-t),t=a..x)=g(x);
E2:=0-lambda*Int(f2(t)*f2(x-t),t=a..x)=g(x);
simplify(value((E1))); simplify(value((E2)));
```

Mathematica:

```
g[x_]:=-x^9; {lambda=1/2, a=0, beta=0}
Eq1=beta*f[t]-lambda*Integrate[f[t]*f[x-t],{t,a,x}]==g[x]
{LaplaceTransform[g[x],x,p],LaplaceTransform[f[x],x,p]}
```

```
Eq1Lap=LaplaceTransform[Eq1,x,p]/.
    {LaplaceTransform[f[x],x,p]->F[p]}
Sol=Solve[Eq1Lap,F[p]]
InvLap1=InverseLaplaceTransform[Sol[[1]],p,x]
f1[t_]:=InvLap1[[1,2]]/.{x->t}; f1[t]
InvLap2=InverseLaplaceTransform[Sol[[2]],p,x]
f2[t_]:=InvLap2[[1,2]]/.{x->t}; f2[t]
E1=beta*f1[t]-lambda*Integrate[f1[t]*f1[x-t],{t,a,x}]==g[x]
E2=beta*f2[t]-lambda*Integrate[f2[t]*f2[x-t],{t,a,x}]==g[x]
```

$\square$

Nonlinear integral equations with constant limits of integration

$$\beta f(x) - \lambda \int_a^b K\big(x,t,f(t)\big)\, dt = g(x).$$

Problem 11.38 Let $a = 0$, $b = 1$. Solve the nonlinear integral equation of the second kind

$$f(x) - \lambda \int_a^b K(x,t)f^2(t)\, dt = 0, \quad K(x,t) = xt + x^2 t^2,$$

is the degenerate kernel and $g(x){=}0$. Show that the constructed function $f(x)$ is the integral equation solution.

Maple:

```
lambda:=1: g:=x->0; a:=-1; b:=1;
K:=(x,t)->x*t+x^2*t^2; k:=nops(K(x,t)); var:={C1,C2};
Eq1:=f(x)-lambda*Int(K(x,t)*(f(t))^2,t=a..b)=g(x);
Eq2:=f(x)-add(lambda*Int(op(i,K(x,t))*(f(t))^2,
 t=a..b),i=1..k)=g(x);
Eq3:=isolate(Eq2,f(x)); CC1:=expand(op(1,rhs(Eq3))/x/lambda);
CC2:=expand(op(2,rhs(Eq3))/x^2/lambda);
F:=lambda*x*C1+lambda*x^2*C2+g(x);
Eq4:=subsop(2=F,Eq3); f:=unapply(rhs(Eq4),x); sys:={};
for i from 1 to k do
 CC||i; sys:=sys union {value(expand(C||i=CC||i))}; od;
S1:=allvalues({solve(sys, var)});
S2:=S1[1] union S1[2]; m:=nops(S2);
for i from 1 to m do f||i:=unapply(subs(S2[i],F),x); od;
for i from 1 to m do simplify(value(f||i(x)-lambda*
 Int(K(x,t)*(f||i(t))^2,t=a..b)=g(x))); od;
```

Mathematica:

```
g[x_]:=0; K[x_,y_]:=x*y+x^2*y^2
h[u_*v_]:=u*Integrate[v,{y,a,b}]/;FreeQ[u,v]
{lambda=1, a=-1, b=1, k=Length[K[x,y]],
 CA=Array[c,k,1], CC=Array[x1,k,1]}
{Eq1=f[x]-Sum[lambda*h[K[x,y][[i]]*(f[y])^2],{i,1,k}]==g[x],
 Eq2=Solve[Eq1,f[x]]}
CC[[1]]=Eq2[[1,1,2,1]]/x/lambda
CC[[2]]=Eq2[[1,1,2,2]]/x^2/lambda
sys={}; F=lambda*x*CA[[1]]+lambda*x^2*CA[[2]]+g[x]
f[X_]:=F/.{x->X}; f[x]
Do[CC[[i]]; sys=Union[sys,{CA[[i]]==CC[[i]]}],{i,1,k}]; sys
{S=Solve[sys,CA], m=Length[S], fA=Array[x1,m,1],
 fB=Array[x2,m,1]}
Do[f[X_]:=(F/.S[[i]])/.{x->X}; fA[[i]]=f[x];
   fB[[i]]=f[y],{i,1,m}]; {fA,fB}
Do[Eq1=fA[[i]]-lambda*Integrate[K[x,y]*(fB[[i]])^2,
   {y,a,b}]==g[x]; Print[FullSimplify[Eq1]],{i,1,m}];
```
□

Singular integral equations $f(x) - \lambda \int_a^\infty K(x,t)f(t)\,dt = g(x)$

Problem 11.39 Let $a = 0$, $b = \infty$. Solve the singular integral equation with $K(x,t) = \cos(xt)$ and $g(x) = 0$. Show that the constructed functions are the integral equation solutions.

Maple:

```
interface(showassumed=0): with(inttrans): g:=x->0;
a:=0; b:=infinity; K:=(x,t)->cos(x*t); assume(A>0,t>0,x>0);
F1:=x->exp(-A*x);    F2:=unapply(fouriercos(F1(t),t,x),x);
f1:=x->F1(x)+F2(x); f2:=x->F1(x)-F2(x); f1(x); f2(x);
for i from 1 to 2 do
 In||i:=value(map(Int,expand(f||i(t)*K(x,t)),t=a..b));
 Inn||i:=convert(map(simplify,In||i),exp);
 Eq||i:=f||i(x)-lambda*Inn||i=g(x);
 EigenVal||i:=(-1)^(i-1)*sqrt(2/Pi);
 simplify(subs(lambda=EigenVal||i,Eq||i));
 EigenFun||i:=f||i(x);
EqFin||i:=convert(simplify(f||i(x)-EigenVal||i*
 value(map(Int,expand(K(x,t)*f||i(t)),t=a..b))=g(x)),exp)
 assuming A>0 and x>0; od;
```

Mathematica:

```
{a=0, b=Infinity}
g[x_]:=0; K[x_,t_]:=Cos[x*t]; F1[x_]:=Exp[-A*x];
F2[X_]:=FourierCosTransform[F1[t],t,x]/.{x->X}; {F1[t],F2[t]}
f[x_,m_]:=Module[{},F1[x]+(-1)^(m-1)*F2[x]]; {f[t,1],f[t,2]}
Sol[m_]:=Module[{}, Int1=Assuming[A>0&&x>0,
 Integrate[f[t,m]*K[x,t],{t,a,b}]]; Print[Int1];
 Eq1=f[x,m]-lambda*Int1==g[x];          Print[Eq1];
 EigenVal=(-1)^(m-1)*Sqrt[2/Pi];        Print[EigenVal];
 Eq1/.{lambda->EigenVal};               Print[Eq1];
 EigenFun=FullSimplify[f[x,m]-EigenVal*Assuming[A>0&&x>0,
  Integrate[K[x,t]*f[t,m],{t,a,b}]]==g[x]]; {Sol[1],Sol[2]}
```

Problem 11.40 Let $a = 0$, $b = \infty$. Applying the Laplace transform and the Efros theorem on the generalized convolution, solve the singular integral equation with $K(x,t) = \exp(-t^2/(4x))$ and $g(x) = \cos x$. Show that the constructed function is the integral equation solution.

Maple:

```
with(inttrans): g:=x->cos(x); a:=0; b:=infinity;
K:=(x,t)->exp(-t^2/(4*x)); assume(x>0,u>0);
Eq1:=1/(sqrt(Pi*x))*Int(K(x,t)*f(t),t=a..b)=g(x);
Eq2:=F(sqrt(p))/sqrt(p)=laplace(g(x),x,p);
Eq3:=solve(Eq2,F(sqrt(p)));
Eq4:=simplify(subs(p=u^2,Eq3));
Sol:=unapply(combine(invlaplace(Eq4,u,x)),x);
Eq11:=simplify(value(1/(sqrt(Pi*x))*Int(K(x,t)*Sol(t),
 t=a..b)=g(x)));
```

Mathematica:

```
g[x_]:=Cos[x]; K[x_,t_]:=Exp[-t^2/(4*x)];
{a=0, b=Infinity}
Eq1=1/(Sqrt[Pi*x])*Integrate[K[x,t]*f[t],{t,a,b}]==g[x]
Eq2=F[Sqrt[p]]/Sqrt[p]==LaplaceTransform[g[x],x,p]
Eq3=Solve[Eq2,F[Sqrt[p]]]/.{p->u^2}
Eq4=Assuming[u>0, FullSimplify[Eq3]]
Sol[X_]:=FullSimplify[InverseLaplaceTransform[
 Eq4,u,x]/.{x->X}]; Sol[x][[1,1,2]]
Eq11=Assuming[x>0,FullSimplify[1/(Sqrt[Pi*x])*
 Integrate[K[x,t]*Factor[TrigToExp[Sol[t][[1,1,2]]]],
 {t,a,b}]]==g[x]]
```

Systems of linear integral equations

Problem 11.41 Applying the Laplace transform find the exact solution of the system of linear integral equations

$$f_1(x) - \lambda_{11} \int_a^x K_{11} f_1(t)\, dt - \lambda_{12} \int_a^x K_{12} f_2(t)\, dt = g_1(x),$$

$$f_2(x) - \lambda_{21} \int_a^x K_{21} f_1(t)\, dt - \lambda_{22} \int_0^x K_{22} f_2(t)\, dt = g_2(x),$$

where

$$a = 0, \quad g_1(x) = \sin(x), \quad g_2(x) = \cos(x),$$
$$K_{11}(x,t) = K_{12}(x,t) = \cos(x-t), \quad K_{21}(x,t) = K_{22}(x,t) = \sin(x-t),$$
$$\lambda_{11} = \lambda_{12} = \lambda_{21} = \lambda_{22} = 1.$$

Show that the functions $f_1(x)$, $f_2(x)$ are the solutions of the system of integral equations.

Maple:

```
with(inttrans): lambda11:=1; lambda12:=1: lambda21:=1;
lambda22:=1; a:=0; g1:=x->sin(x); g2:=x->cos(x);
K11:=(x,t)->cos(x-t);     K12:=(x,t)->cos(x-t);
K21:=(x,t)->sin(x-t);     K22:=(x,t)->sin(x-t);
Eq1:=f1(x)-lambda11*int(K11(x,t)*f1(t),t=a..x)
           -lambda12*int(K12(x,t)*f2(t),t=a..x)=g1(x);
Eq2:=f2(x)-lambda12*int(K12(x,t)*f1(t),t=a..x)
           -lambda22*int(K22(x,t)*f2(t),t=a..x)=g2(x);
sys:={Eq1, Eq2}; sys_L:=laplace(sys,x,p);
Sol:=factor(solve(subs({laplace(f1(x),x,p)=F1(p),
 laplace(f2(x),x,p)=F2(p)},sys_L),{F1(p),F2(p)}));
F11:=rhs(op(select(has,Sol,F1)));
f1:=x->invlaplace(F11,p,x);
F12:=rhs(op(select(has,Sol,F2)));
f2:=x->invlaplace(F12,p,x); f1(x); f2(x);
E1:=f1(x)-lambda11*Int(K11(x,t)*f1(t),t=a..x)
 -lambda12*Int(K12(x,t)*f2(t),t=a..x)=g1(x);
E2:=f2(x)-lambda12*Int(K12(x,t)*f1(t),t=a..x)
 -lambda22*Int(K22(x,t)*f2(t),t=a..x)=g2(x);
B1:=simplify(value(E1)); B2:=simplify(value(E2));
```

Mathematica:

```
g1[x_]:=Sin[x]; g2[x_]:=Cos[x];
K11[x_,t_]:=Cos[x-t]; K12[x_,t_]:=Cos[x-t];
K21[x_,t_]:=Sin[x-t]; K22[x_,t_]:=Sin[x-t];
{lambda11=1, lambda12=1, lambda21=1, lambda22=1, a=0}
Eq1=f1[x]-lambda11*Integrate[K11[x,t]*f1[t],{t,a,x}]-
 lambda12*Integrate[K12[x,t]*f2[t],{t,a,x}]==g1[x]
Eq2=f2[x]-lambda12*Integrate[K12[x,t]*f1[t],{t,a,x}]-
 lambda22*Integrate[K22[x,t]*f2[t],{t,a,x}]==g2[x]
sys={Eq1,Eq2};
sysLap=LaplaceTransform[sys,x,p]/.
 {LaplaceTransform[f1[x],x,p]->F1[p]}/.
 {LaplaceTransform[f2[x],x,p]->F2[p]}
Sol=Solve[sysLap,{F1[p],F2[p]}]
f1[X_]:=InverseLaplaceTransform[Sol[[1,1,2]],p,x]/.{x->X};
f2[X_]:=InverseLaplaceTransform[Sol[[1,2,2]],p,x]/.{x->X};
{f1[x],f2[x]}
 Map[FullSimplify,
 {E1=f1[x]-lambda11*Integrate[K11[x,t]*f1[t],{t,a,x}]
 -lambda12*Integrate[K12[x,t]*f2[t],{t,a,x}]==g1[x],
  E2=f2[x]-lambda12*Integrate[K12[x,t]*f1[t],{t,a,x}]
 -lambda22*Integrate[K22[x,t]*f2[t],{t,a,x}]==g2[x]}]
```

□

Chapter 12

Numerical Analysis and Scientific Computing

We know that *scientific computing* and *numerical analysis* are concerned with constructing and investigating approximate methods to obtain numerical solutions to complicated mathematical, scientific, and engineering problems using computers. Numerical analysis is one of the most important fields of mathematics that has numerous applications in all fields of engineering and sciences.

We will discuss the most important numerical methods for solving various mathematical problems in both computer algebra systems, *Maple* and *Mathematica*. In our solutions we will combine the advantages of these systems for scientific computation:

1) traditional approximate numerical computation,

2) symbolic computation, and

3) arbitrary-precision approximate numerical computation.

In this chapter we will show, solving mathematical problems in both systems, how to construct numerical algorithms and derive numerical methods analytically, to obtain exact and numerical solutions of problems and compare the results, to evaluate computational errors and derive convergence rates, to simplify proofs and visualize the solutions obtained, etc.

As we know, the use of a computer for numerical approximations introduces errors, therefore it is important to control their propagation. The errors (discrepancies) can be of different nature, e.g., the errors between the physical and mathematical models, the errors between the solutions obtained in exact arithmetics and in computer floating-point arithmetics (roundoff errors), the errors between the exact and approximate analytical or numerical solutions (truncation or discretization errors). These *roundoff* and *truncation errors* can be combined

I.K. Shingareva, C. Lizárraga-Celaya, *Maple and Mathematica*, 2nd ed.,
DOI 10.1007/978-3-211-99432-0_12, © Springer-Verlag Vienna 2009

in the unique concept of a *computational error*, E (for more details on floating-point arithmetics and computational errors in both systems, see Sect. 4.2). Recall that there are the two types of computational errors, the absolute and relative errors, $E_{abs} = |X - x|$ and $E_{rel} = |X - x|/|x|$ ($x \neq 0$), where x and X are, respectively, the exact solution of mathematical model and the corresponding approximate solution. Therefore obtaining numerical solutions, it is important to introduce a new parameter $\varepsilon \in \mathbb{R}$, the tolerance, we wish to set for a computational error.

In theory, since the computational errors depend on the unknown exact solution, it is not possible to compute them. In practice, therefore it is necessary to introduce the computable errors (or *error estimators*), for estimating the computational errors. Usually, the computable errors for numerical methods are the difference between two successive iterations[1] $E_{abs}^{(i)} = |X^{(i)} - X^{(i-1)}|$ or $E_{rel}^{(i)} = |X^{(i)} - X^{(i-1)}|/|X^{(i-1)}|$.

If, in the general case, the quantities x and X are not real numbers (e.g., they can also be vectors or matrices), we shall consider normed spaces. If we denote by $||x||$ a vector norm, then the absolute and relative errors for the vector $||X||$ are defined by $||X - x||$ and $||X - x||/||x||$, respectively.

Problem 12.1 *The computational errors.* Compute the absolute and relative errors for a given solution of the mathematical model and the corresponding approximate solution.

Maple:

```
E_abs:=(x,X)->abs(X-x);
E_rel:=(x,X)->if x<>0 then abs((X-x)/x); else 0; fi;
for x from 0 to 0.25 by 0.05 do
 s1:=evalf(sin(x)); s2:=evalf(x-x^3/6);
 printf("%15.9e %15.9e %15.9e %15.9e\n",
 s1,s2,E_abs(s1,s2),E_rel(s1,s2)): od:
```

Mathematica:

```
eAbs[x_,X_]:=Abs[X-x];
eRel[x_,X_]:=If[x!=0.,Abs[(X-x)/x],0.];
Do[s1=N[Sin[x]]; s2=N[x-x^3/6.]; Print[PaddedForm[s1,{15,9}],
 PaddedForm[s2,{15,9}],PaddedForm[eAbs[s1,s2],{15,9}],
 PaddedForm[eRel[s1,s2],{15,9}]],{x,0.,0.25,0.05}];
```

□

---

[1] from "iteratio", the Latin word for "repetition"

In general, any numerical process (or a finite number of approximations of the mathematical model) can be represented as a function $\mathcal{F}(E_i)$ of a computational error (absolute or relative) E_i ($E_i > 0$). If $\lim_{E_i \to 0} \mathcal{F}(E_i) = x$, i.e. the numerical solution coincides with the exact solution of the mathematical model, then the numerical process is *convergent*. Moreover, if $\lim_{i \to \infty} E_i / E_{i-1}^q = C$ (constant $C > 0$), then the numerical method is *convergent of order q*.

There are several important parameters in the analysis of algorithms, e.g., the *computational cost* of an algorithm. Since we are interested in algorithms which involve a finite number of steps, the computational cost is defined by the number of required floating-point operations (flops) to obtain a numerical solution. In *Maple*, we can compute the computational cost of an algorithm with the `cost` function (of the `codegen` package).

However in some cases it is important to determine the order of magnitude of an algorithm as a function of a problem dimension parameter n and therefore we recall here the concept of the *complexity of an algorithm*. For example, a *constant complexity* algorithm requires a number of operations independent of the problem dimension n, i.e. $O(1)$ operations, a *linear complexity* algorithm requires $O(n)$ operations, a *polynomial complexity* algorithm requires $O(n^k)$ operations ($k \in \mathbb{Z}, k > 0$), etc.

Another relevant parameter is an indicator of the performance of an algorithm, i.e., the CPU (central processing unit) time that is required to perform the different sets of code instructions and operations for a specific computational task that may involve other computer processes like read/write into memory and disks or screen displaying. In *Maple* and *Mathematica*, this parameter can be obtained, respectively, with the functions `time` and `Timing` (examples of these functions can be found in Chapters 1 and 2).

It should be noted that all operations in *Maple* and *Mathematica* are performed by assuming that the basic number system is the *complex field* $\mathbb{C}$.

12.1   Nonlinear Equations

In this Section we consider a problem of computing the *zeros*, α, of a function $f : \mathbb{R}^n \to \mathbb{R}^n$ $(n \geq 1)$ or the *roots of the equation*

$$f(x)=0, \ x=(x_1,\ldots,x_n)^{\mathrm{T}}, \ f(x)=(f_1(x_1,\ldots,x_n),\ldots,f_n(x_1,\ldots,x_n))^{\mathrm{T}}.$$

By the fundamental theorem of algebra, an univariate polynomial of degree m has m complex zeros. The various proofs of this result (for $m > 4$) give no constructive procedure for computing these zeros. In contrast with linear equations, there are no explicit solution methods in general for nonlinear equations. So, this problem cannot be solved in a finite number of operations and also the problem is more difficult if $f(x)$ is an arbitrary function. Numerical analysis provides *iterative methods* for computing the zeros of $f(x)$. Starting from one or several initial data, we can construct a sequence of values $x^{(i)} \in \mathbb{R}^n$ that will converge to a zero of $f(x)$. This result is based on the *Banach fixed point theorem*, which is one of the most important tools in numerical analysis and is the fundamental basis for solving linear and nonlinear systems by iterative methods.

Numerical solutions of nonlinear equations and systems

Maple:

```
Digits:=n;            evalf(solve(Eq,var));        evalf(root(x,n));
fsolve(Eq,var,ops);                        fsolve(Eq,var=a..b,ops);
fsolve(Eq,var,complex);                    fsolve(f,var=a..b,ops);
with(RootFinding); Isolate(f);             NextZero(f(x),x0,ops);
Isolate([f1,f2],[x1,x2]);                  Homotopy([f1,f2],ops);
use RealDomain in solve(eq,var); end use;
```

fsolve, solving equations using iterative methods (the Newton methods) with options (see ?fsolve[details]),

fsolve(f,var=a..b,ops), for solving an equation defined as a procedure f,

NextZero, Isolate, Homotopy, functions of the RootFinding package, finding next real zero of a function f, isolating the real roots of a univariate polynomial or polynomial system, finding numerical approximations to roots of systems of polynomial equations,

root, finding n-th root of an algebraic expression,

the functions of the RealDomain package, performing computations under the assumption that the basic number system is the field of real numbers $\mathbb{R}$.

```
Digits:=trunc(evalhf(Digits)); F:=proc(x) sin(x)^2-1 end;
fsolve(F(x)=0,x=0..Pi/2); fsolve(sin(x)^2-1,x=0..Pi/2);
fsolve(x->F(x),0..Pi/2);
evalf(root(5,3));evalf(root[3](5)); evalf(solve(sin(x)^2-1,x));
with(RootFinding); NextZero(x->sin(x),evalf(Pi/10));
Isolate(x^2-2*x+1); Isolate([x^2-2*x+1,y^2+3*y-5],[x,y]);
Homotopy([x^2+x+1,y^2-y-1]); use RealDomain in solve(x^4-1,x);
end use; with(RealDomain): fsolve(x^4-1,x,complex);
```

Mathematica:

NSolve[eqs,vars,ops]	NSolve[eqs,vars,n]	N[Solve[eqs,vars],n]
FindRoot[eq,{x,x0}]	FindRoot[eq,{x,{x0,x1}}]	N[Root[f,x,k]]
FindRoot[eqs,{x1,x10},{x2,x20},...]		FindRoot[eq,{var,I},ops]
CountRoots[f,x]	NRoots[eq,var]	NRoots[eq,var]
RootIntervals[f,domain]	FindInstance[eq1&&eq2,vars,domain,n]	
<<FunctionApproximations`	InterpolateRoot[eq,{x,x0,x1}]	

FindRoot, solving equations using iterative methods (Newton methods or the secant method, depending on the number of initial data, x_0 or x_0, x_1) with options (for more details see ??FindRoot). Here $[a, b]$ is the interval outside of which the iteration process stops,

CountRoots, the number of real roots of the polynomial f in x,

NRoots, NSolve, finding numerical approximations to the roots of a polynomial equation,

Root, finding the k-th root of the polynomial f in x,

FindInstance, finding an instance of variables that makes the statement be True.

```
nD=14; f[x_]:=Sin[x]^2-1; g=x^4-1;
{N[Solve[f[x]==0,x, InverseFunctions->True],nD],
  FindRoot[f[x]==0,{x,N[Pi/2]}],
  NSolve[g==0,x,WorkingPrecision->nD], CountRoots[g,x],
  NRoots[g==0,x],N[Root[g,x,3]],N[5^(1/3)],
  RootIntervals[g,Complexes],FindInstance[g==0,x,Complexes,4]}
<<FunctionApproximations`
InterpolateRoot[g==0,{x,0,0.9}]
```

Problem 12.2 *The bisection method.* Let be a continuous real function of a single real variable in $I = [a, b]$ which satisfies the condition $f(a)f(b) < 0$. Then (according to the intermediate value theorem) $f(x)$ has at least one zero in (a, b). Assume that it is unique (for several zeros, we can isolate an interval which contains only one zero).

1) Compute a zero of $f(x)$ by the bisection method for a given tolerance ε, i.e. by generating a sequence of intervals $I^{(i)}$ whose length is halved at each iteration and a sequence of the midpoints $x^{(i)}$, where $x^{(i)} \to \alpha$, $|I^{(i)}| \to 0$ as $i \to \infty$.

2) Analyze the convergence rate of the method numerically at the i-th iteration evaluating the constant $C = \dfrac{\Delta x^{(i)}}{\Delta x^{(i-1)}} = \dfrac{(x^{(i+1)} - x^{(i)})}{(x^{(i)} - x^{(i-1)})}$.

3) Find the number of iterations needed to solve the problem with a given tolerance, i.e. to guarantee that $|E_{\text{abs}}^{(i)}| = |x^{(i)} - \alpha| < \frac{1}{2}|I^{(i)}| = (\frac{1}{2})^{i+1}(b - a) < \varepsilon$.

Maple:

```
Digits:=30; a:=0: b:=1: epsilon:=10^(-15): N_max:=100:
E_abs:=(x,X)->abs(X-x); E_rel:=(x,X)->if x<>0 then
 abs((X-x)/x); else 0; fi; f:=x->evalf(sin(x)-exp(-x));
Sol:=fsolve(f(x)=0,x=0..1); plot({sin(x),exp(-x)},x=0..7);
FA:=f(a): K:=0:
for i from 1 to N_max do
 alpha:=evalf(a+(b-a)/2): X[i]:=alpha: FP:=evalf(f(alpha)):
 printf("%d %15.9e %15.9e %15.9e %15.9e %15.9e %15.10e\n",
 i,a,b,alpha,f(alpha),E_abs(Sol,alpha),E_rel(Sol,alpha)):
 if FP=0 or abs(f(alpha))<epsilon then
printf("the result is:\n"):
  printf("%d %15.9e %15.9e %15.9e %15.9e %15.9e %15.9e\n",
  i,a,b,alpha,f(alpha),E_abs(Sol,alpha),E_rel(Sol,alpha)):
  K:=1: iIter:=i: break: else if FA*FP>0 then a:=alpha: FA:=FP:
  else b:=alpha: fi: fi: od:
for i from 2 to iIter-1 do C:=abs((X[i+1]-X[i])/(X[i]-X[i-1]));
 print(C); od:
if K=0 then print(`the bisection method stopped without
converging for the given tolerance`,epsilon,`because the maximum
number of iterations`,N_max,`was reached`); fi:
NumIt:=solve(0.5^(N+1)*(1.-0.)<epsilon,N);
```

Mathematica:

```
nD=30; f[x_]:=Sin[x]-Exp[-x]; eAbs[x_,X_]:=Abs[X-x];
eRel[x_,X_]:=If[x!=0,Abs[(X-x)/x],0];
sol=FindRoot[f[x]==0,{x,0},WorkingPrecision->nD][[1,2]]
{a=N[0,nD],b=N[1,nD],epsilon=N[10^(-15),nD],nMax=100,
 FA=N[f[a],nD],K=0,iIter=0,X=N[Table[0,{i,1,nMax}],nD]}
Plot[{Sin[x],Exp[-x]},{x,0,7}]
Do[alpha=N[a+0.5*(b-a),nD]; X[[i]]=alpha; FP=N[f[alpha],nD];
 Print[PaddedForm[i,{3,0}],PaddedForm[a,{15,9}],
 PaddedForm[b,{15,9}],PaddedForm[alpha,{15,9}],
 PaddedForm[f[alpha],{15,9}],
 PaddedForm[eAbs[sol,alpha],{15,9}],
 PaddedForm[eRel[sol,alpha],{15,9}]];
 If[FP==0.||Abs[f[alpha]]<epsilon,{Print["the result is:"],
 Print[PaddedForm[i,{3,0}],PaddedForm[a,{15,9}],
 PaddedForm[b,{15,9}],PaddedForm[alpha,{15,9}],
 PaddedForm[f[alpha],{15,9}],PaddedForm[eAbs[sol,alpha],{15,9}],
 PaddedForm[eRel[sol,alpha],{15,9}]],K=1,iIter=i,Break[]}];
 If[FA*FP>0,{a=alpha,FA=FP},b=alpha],{i,1,nMax}];
Do[c=Abs[(X[[i+1]]-X[[i]])/(X[[i]]-X[[i-1]])];
Print[c],{i,2,iIter-1}]; If[K==0,Print["the bisection method
stopped without converging for the given tolerance", epsilon,
"because the maximum number of iterations",nMax,"was reached"]]
FindRoot[0.5^(x+1.)*(1.-0.)==epsilon,{x,1.}]
```
□

Problem 12.3 *The Newton method* computes a zero α of a continuously differentiable function $f(x)$ calculating the values of $f(x)$ and of its derivative, i.e. for a given approximation $x^{(i)}$ of the root α a better approximation is $x^{(i+1)} = F(x^{(i)}) = x^{(i)} - \dfrac{f(x^{(i)})}{f'(x^{(i)})}$, $f'(x^{(i)}) \neq 0$.

1) Solve the previous problem by the Newton method taking a given initial datum $x^{(0)}$ in a sufficiently small neighborhood of α and the tolerance ε.

2) Compare the number of iterations in the bisection method and the Newton method.

Maple:

```
Digits:=90: f:=x->evalf(sin(x)-exp(-x));
Sol:=fsolve(f(x)=0,x=0..1); X[0]:=0.1; epsilon:=10^(-15):
N_max:=100: fD:=unapply(diff(f(x),x),x);
```

```
for i from 1 to 10 do X[i]:=evalf(
X[i-1]-f(X[i-1])/fD(X[i-1])): print(i,X[i]); od:
for i from 1 to N_max do
X[i]:=evalf(X[i-1]-f(X[i-1])/fD(X[i-1])):
E_rel[i]:=evalf(abs((X[i]-X[i-1])/X[i-1]));
if abs(f(X[i]))>=epsilon then
printf("%d %15.9e %15.9e\n", i,X[i],E_rel[i]): else
lprint(`the result is`); printf("%d %15.9e %15.9e\n",
i,X[i],E_rel[i]): break: fi: od:
```

Mathematica:

```
nD=90; f[x_]:=Sin[x]-Exp[-x]; fD[x1_]:=D[f[x],x]/.{x->x1};
sol=FindRoot[f[x]==0,{x,0},WorkingPrecision->nD]
{epsilon=N[10^(-15),nD],nMax=100,X=N[Table[0,{i,1,nMax}],nD],
  eRel=N[Table[0,{i,1,nMax}],nD],X[[1]]=N[1/10,nD]};
Print[1, "  ", X[[1]]]
Do[X[[i]]=N[X[[i-1]]-f[X[[i-1]]]/fD[X[[i-1]]],nD];
  Print[i," ",X[[i]]],{i,2,10}]; Print[" "]
Do[X[[i]]=N[X[[i-1]]-f[X[[i-1]]]/fD[X[[i-1]]],nD];
  eRel[[i]]=N[Abs[(X[[i]]-X[[i-1]])/X[[i-1]]],nD];
  If[Abs[f[X[[i]]]]>=epsilon,{Print[i-1," ",
  PaddedForm[X[[i]],{15,9}]," ",PaddedForm[eRel[[i]],{15,9}]]},
  {Print["the result is"],Print[i-1," ",PaddedForm[X[[i]],{15,9}],
  " ",PaddedForm[eRel[[i]],{15,9}]],Break[]}],{i,2,nMax}];
```

                                                                    □

Problem 12.4 *The Newton method* is quadratically convergent only if α is a simple root of $f(x)$. If $f(x)$ has a zero of multiplicity m $(m > 1)$, then the convergence is linear.

  1) Prove these results analytically assuming that $f(x)$ is a continuously differentiable function up to its second derivative.

  2) For a given function $f(x)$ that has a zero of multiplicity m $(m > 1)$, solve the previous problem by the *modified Newton method*.

Maple:

```
F1:=x->x-f(x)/D(f)(x); dF1:=x->D(F1)(x); dF1(x);
dF2:=x->D(D(F1))(x); dF2(x); subs(f(alpha)=0,dF1(alpha));
subs(f(alpha)=0,dF2(alpha)); f:=x->(x-alpha)^m*h(x); dF1(x);
C_L:=expand(limit(dF1(x),x=alpha)); subs(m=3,C_L);
Digits:=90: f:=x->x^3-6*x^2+9*x-4; X[0]:=0.1;
epsilon:=10^(-15): N_max:=100: Sol:=fsolve(f(x)=0,x=0..5);
```

```
fD1:=unapply(diff(f(x),x),x); fD2:=unapply(diff(fD1(x),x),x);
for i from 1 to N_max do
 X[i]:=evalf(X[i-1]-f(X[i-1])*fD1(X[i-1])/((fD1(X[i-1]))^2
 -fD2(X[i-1])*f(X[i-1]))):
 if abs(f(X[i]))>=epsilon then print(i,X[i]); else
 lprint(`the result is`); print(i,X[i]); iIter:=i: break: fi:
od:
lprint(`the relative error is`);
for i from 1 to iIter do lprint(abs((X[i]-X[i-1])/X[i-1])); od:
```

Mathematica:

```
F1[x_]:=x-f[x]/D[f[x],x];
dF1[x_]:=D[F1[x],x]; dF2[x_]:=D[F1[x],{x,2}];
{F1[x],dF1[x],dF2[x]}
{dF1[alpha]/.{f[alpha]->0}, dF2[alpha]/.{f[alpha]->0}}
f[x_]:=(x-alpha)^m*h[x]; dF1[x]
{cL=Expand[Limit[dF1[x],x->alpha,Analytic->True]],cL/.{m->3}}
nD=90; f[x_]:=x^3-6*x^2+9*x-4;
{epsilon=N[10^(-15),nD],nMax=100,X=N[Table[0,{i,1,nMax}],nD],
 X[[1]]=N[1/10,nD]}
sol=FindRoot[f[x]==0,{x,0},WorkingPrecision->nD,
 MaxIterations->1000]
fD1[x1_]:=D[f[x],x]/.{x->x1}; fD2[x1_]:=D[fD1[x],x]/.{x->x1};
Do[X[[i]]=N[X[[i-1]]-f[X[[i-1]]]*fD1[X[[i-1]]]/
 ((fD1[X[[i-1]]])^2-fD2[X[[i-1]]]*f[X[[i-1]]]),nD];
If[Abs[f[X[[i]]]]>=epsilon,Print[i-1," ",X[[i]]],
 {Print["the result is"],Print[i-1," ",X[[i]]],
 iIter=i,Break[]}],{i,2,nMax}]; Print["the relative error is"]
Do[Print[Abs[(X[[i]]-X[[i-1]])/X[[i-1]]]],{i,2,iIter}];
```

□

Problem 12.5 *Analytical derivation of the Newton iteration formula.*
Let $f : [a, b] \to \mathbb{R}$ be a real-valued twice continuously differentiable function of a single variable x. Assume that there exists a number $\alpha \in [a, b]$ of $f(x)$ such that $f(\alpha)=0$ and $f'(\alpha)\neq 0$. Then there exist a small number ε such that the sequence $\{X^{(i)}\}_{i=0}^{\infty}$ defined by the *Newton iteration formula* will converge to α for any initial approximation $X^{(0)} \in [\alpha - \varepsilon, \alpha + \varepsilon]$.

1) Perform the analytical derivation of the Newton iteration formula.

2) Determine analytically that the convergence rate of the Newton method is equal to 2 (if α is a simple root of $f(x)$).

Maple:

```
interface(showassumed=0); assume(q,positive); assume(C,positive);
assume(E[0],positive);assume(E[1],positive);
Eq1:=convert(taylor(f(x),x=X0,2),polynom); Eq2:=f(x)=Eq1;
Eq3:=subs(x=X1,Eq2); Eq4:=subs(f(X1)=0,Eq3);
EqNewton:=X1=sort(expand(solve(Eq4,X1)));
F:=unapply(rhs(EqNewton),X0); Eq5:=X[i+1]=F(X[i]);
Eq51:=subs(X[i]=E[i]+alpha,Eq5)-alpha; i:=1;
Eq6:=E[i+1]=convert(taylor(rhs(Eq51),E[i],3),polynom);
Eq7:=subs(f(alpha)=0,Eq6); Eq8:=expand(Eq7/(E[i]^2));
Eq9:=simplify(subs(E[i+1]=C*E[i]^q,E[i]=C*(E[i-1])^q,Eq8));
Eq10:=Eq9/C^(-1+q); Eq11:=solve({op(2,op(1,Eq10))=0,q>0},q);
```

Mathematica:

```
{eq1=Normal[Series[f[x],{x,X0,1}]],eq2=f[x]==eq1,
 eq3=eq2/.{x->X1},eq4=eq3/.{f[X1]->0},
 eqNewton=Expand[Solve[eq4,X1]]}
F[x0_]:=eqNewton[[1,1,2]]/.{X0->x0};
{eq5=X[i+1]==F[X[i]],eq51=(eq5/.{X[i]->e[i]+alpha}),
 eq511=Thread[eq51-alpha,Equal], i=1,
 eq6=e[i+1]==Normal[Series[eq511[[2]],{e[i],0,2}]],
 eq7=eq6/.{f[alpha]->0},eq8=Expand[Thread[eq7/(e[i]^2),Equal]],
 eq9=Thread[Thread[eq8/.{e[i+1]->C*e[i]^q},Equal]/.{e[i]->
 C*(e[i-1])^q},Equal], eq10=Assuming[{q >0,C>0,
 e[0]>0,e[1]},FullSimplify[Thread[eq9/C^(-1+q),Equal]]],
 eq11=Reduce[eq10[[1,2,2]]==0 && q>0,q]}
```

□

Problem 12.6 The geometrical interpretation of the *Newton method* shows that the intersection of the tangent line to $f(x)$ at the point $(x^{(i)}, f(x^{(i)}))$ with the x-axis is the next Newton approximation. Prove this result analytically.

Maple:

```
EqTan:=x->a*x+b; ab:=solve({EqTan(X[i])=f(X[i]),
 D(EqTan)(X[i])=D(f)(X[i])},{a,b});
X[i+1]=expand(subs(ab,solve(EqTan(x)=0,x)));
```

Mathematica:

```
eqTan[x_]:=a*x+b; ab=Solve[{eqTan[X[i]]==f[X[i]],
 (D[eqTan[x],x]/.{x->X[i]})==(D[f[x],x]/.{x->X[i]})},{a,b}]
X[i+1]==Expand[Solve[eqTan[x]==0,x]/.ab][[1,1,1,2]]
```

□

Problem 12.7 *The secant method* has the important advantage that no derivatives are needed. The derivative that appears in the Newton method is approximated by a finite difference.

1) Prove analytically that the convergence rate for the secant method is equal to $q = (1 + \sqrt{5})/2$.

2) For a given function $f(x)$ and the tolerance ε solve the equation $f(x) = 0$ by the secant method.

Maple:

```
interface(showassumed=0); assume(q,positive);
assume(C,positive); assume(E[0],positive);
assume(E[1],positive); F:=(x,y)->x-f(x)*(x-y)/(f(x)-f(y));
Eq1:=X[i+1]=F(X[i],X[i-1]);
a:=solve(E[i+1]=X[i+1]-alpha,X[i+1]); Eq11:=subs(X[i+1]=a,Eq1);
Eq12:=subs({X[i]=E[i]+alpha,X[i-1]=E[i-1]+alpha},Eq11)-alpha;
i:=1; Eq2:=lhs(Eq12)=factor(mtaylor(rhs(Eq12),[E[i-1],E[i]],4));
Eq3:=subs(f(alpha)=0,Eq2); Eq4:=Eq3/(E[i]*E[i-1]);
Eq5:=simplify(subs(E[i+1]=C*E[i]^q,E[i]=C*(E[i-1])^q,Eq4));
Eq6:=Eq5/C^q; Eq7:=solve({op(2,op(1,Eq6))=0,q>0},q);
Digits:=90: N_max:=100; X[0]:=0.1; X[1]:=0.5; epsilon:=10^(-15):
f:=x->evalf(sin(x)-exp(-x)); Sol:=fsolve(f(x)=0,x=0..1);
for i from 2 to N_max do
 X[i]:=F(X[i-1],X[i-2]);
 E_rel[i]:=evalf(abs((X[i]-X[i-1])/X[i-1]));
 if abs(f(X[i]))>=epsilon then
  printf("%d %15.9e %15.9e\n", i,X[i],E_rel[i]):
 else lprint(`the result is`); print(i,X[i]): break: fi: od:
```

Mathematica:

```
F[x_,y_]:=x-f[x]*(x-y)/(f[x]-f[y]);
{eq1=X[i+1]==F[X[i],X[i-1]],
 a=Solve[e[i+1]==X[i+1]-alpha, X[i+1]][[1,1,2]],
 eq11=eq1/.{X[i+1]->a}, eq121=eq11/.{X[i]->e[i]+alpha,
 X[i-1]->e[i-1]+alpha}, eq12=Thread[eq121-alpha,Equal]}
```

```
{i=1,eq2=eq12[[1]]==Normal[FullSimplify[Series[eq12[[2]],
 {e[i-1],0,1},{e[i],0,1}]]], eq3=eq2/.{f[alpha]->0},
 eq4=Thread[eq3/(e[i]*e[i-1]),Equal],
 eq5=Simplify[Thread[Thread[eq4/.{e[i+1]->C*e[i]^q},
  Equal]/.{e[i]->C*(e[i-1])^q},Equal]],
 eq6=Assuming[{q>0,C>0,e[0]>0,e[1]>0},
 Simplify[Thread[eq5/C^q,Equal]]], eq61=Thread[eq6/e[0],Equal],
 eq7=Reduce[eq61[[1,2,2]]==0&&q>0,q]}
nD=90; f[x_]:=Sin[x]-Exp[-x];
sol=FindRoot[f[x]==0,{x,0},WorkingPrecision->nD]
{nMax=100,epsilon=N[10^(-15),nD],X=N[Table[0,{i,1,nMax}],nD],
 eRel=N[Table[0,{i,1,nMax}],nD],X[[1]]=N[1/10,nD],
 X[[2]]=N[1/2,nD]}
Do[X[[i]]=F[X[[i-1]],X[[i-2]]]; eRel[[i]]=N[Abs[
 (X[[i]]-X[[i-1]])/X[[i-1]]],nD]; If[Abs[f[X[[i]]]]>=epsilon,
 Print[i-1," ",PaddedForm[X[[i]],{15,9}]," ",
 PaddedForm[eRel[[i]],{15,9}]],{Print["the result is"],
 Print[i-1," ",X[[i]]],Break[]}],{i,3,nMax}];
```

□

Problem 12.8 *The fixed point iterations.* Let $f(x)$ be a real-valued function of a real single variable (defined in the previous problems).

1) Find the zero of the function $f(x)$ by the *fixed point iterations* (for a given tolerance ε). Let us obtain the *iteration function* $g : [a, b] \to \mathbb{R}$ and a fixed point α ($\alpha \in [a, b]$) of $g(x)$ such that $\alpha = g(\alpha)$ according to the following algorithm: $x^{(i+1)} = g(x^{(i)})$, $i \geq 0$, where $x^{(0)}$ is an initial guess.

2) Verify the conditions of the fixed point theorem, i.e. that $g(x)$ has a unique fixed point $\alpha \in [a, b]$ and the sequence generated by the fixed point iterations converges to α.

3) Verify that the convergence is linear and evaluate the value of the constant C according to the formula of the fixed point theorem:

$$\lim_{i\to\infty} \frac{x^{(i+1)} - \alpha}{x^{(k)} - \alpha} = g'(\alpha) = C.$$

Maple:

```
a:=0.5; b:=0.8; Eq1:=sin(x)=exp(-x); assume(x,positive);
Eq2:=lhs(Eq1)=solve(Eq1,lhs(Eq1)); Eq3:=evala(map(arcsin,Eq2));
g:=unapply(rhs(Eq3),x); g(x); diff({g(x),x},x);
limit(g(x),x=a); limit(g(x),x=b);
plot(abs(diff(g(x),x)),x=a..b); plot({g(x),x},x=a..b,a..b);
```

```
Digits:=30: K:=0: f:=x->evalf(sin(x)-exp(-x)); N_max:=100;
Sol:=fsolve(Eq1,x=0..1); X[0]:=0.9; epsilon:=10^(-15):
for i from 1 to N_max do X[i]:=g(X[i-1]):
 E_rel[i]:=evalf(abs((X[i]-X[i-1])/X[i-1]));
 if E_rel[i]>=epsilon then print(i,X[i],evalf(f(X[i])),E_rel[i]):
 else lprint(`the result is`); print(i,X[i],evalf(f(X[i])),
 E_rel[i]): K:=1: iIter:=i: break: fi: od:
if K=0 then print(`the fixed point iterations stopped without
converging for the given tolerance`,epsilon,`because the
maximum number of iterations`,N_max,`was reached`); fi:
C_L:=abs(evalf(subs(x=X[iIter],diff(g(x),x))));
```

Mathematica:

```
{a=0.5, b=0.8, eq1=Sin[x]==Exp[-x]}
eq2=Assuming[x>0,eq1[[1]]==Solve[eq1,eq1[[1]]][[1,1,2]]]
eq3=Thread[ArcSin[eq2],Equal]
g[x1_]:=eq3[[2]]/.{x->x1}; g[x]
{D[{g[x],x},x],Limit[g[x],x->a],Limit[g[x],x->b]}
Plot[Evaluate[Abs[D[g[x],x]]],{x,a,b}]
Plot[Evaluate[{g[x],x}],{x,a,b},PlotStyle->Hue[0.9]]
nD=30; K=0; f[x_]:=Sin[x]-Exp[-x];
sol=FindRoot[eq1,{x,0},WorkingPrecision->nD]
{epsilon=N[10^(-15),nD],nMax=100,X=N[Table[0,{i,1,nMax}],nD],
 eRel=N[Table[0,{i,1,nMax}],nD],X[[1]]=N[9/10,nD]}
Do[X[[i]]=g[X[[i-1]]];
 eRel[[i]]=N[Abs[(X[[i]]-X[[i-1]])/X[[i-1]]],nD];
 If[eRel[[i]]>=epsilon,Print[i-1," ",X[[i]]," ",
 N[f[X[[i]]],nD]," ",eRel[[i]]],{Print["the result is"],
 Print[i-1," ",X[[i]]," ",N[f[X[[i]]],nD]," ",eRel[[i]]],
 K=1,iIter=i,Break[]}],{i,2,nMax}];
If[K==0,Print["the fixed point iterations stopped without
converging for the given tolerance",epsilon,"because
the maximum number of iterations",nMax,"was reached"]]
cL=Abs[N[D[g[x],x]/.{x->X[[iIter]]},nD]]
```

                                                                       ☐

Problem 12.9 *The system of nonlinear equations.* Let $D \in \mathbb{R}^n$ be an open domain and let $f : D \to \mathbb{R}^n$ be a continuously differentiable function such that the Jacobian matrix $f'(x)$ is nonsingular for all $x \in D$. The *Newton method* can be applied to find zeros of a system of n nonlinear equations. Let $n=2$ and $\{f(x,y)=0, g(x,y)=0\}$ be a system of nonlinear equations.

1) Analyze the functions $f(x,y)$ and $g(x,y)$. Choose the initial data $\{x^{(0)}, y^{(0)}\}$ for which the sequence generated by the Newton method will converge.

2) Find the solution of the system of nonlinear equations according to the *Newton iteration formula* for a given tolerance ε.

Maple:

```
with(plots): Digits:=90: N_max:=100: epsilon:=10^(-15);
f:=(x,y)->x^2+y^2-2: g:=(x,y)->x^2-y^2-1;
G1:=implicitplot(f(x,y)=0,x=-2..2,y=-2..2):
G2:=implicitplot(g(x,y)=0,x=-2..2,y=-2..2): display([G1,G2]);
X[0]:=0.1: Y[0]:=0.1:
f_x:=unapply(diff(f(x,y),x),x,y);
f_y:=unapply(diff(f(x,y),y),x,y);
g_x:=unapply(diff(g(x,y),x),x,y);
g_y:=unapply(diff(g(x,y),y),x,y);
for i from 1 to N_max do
 Eq1:=f_x(X[i-1],Y[i-1])*H[i-1]+
  f_y(X[i-1],Y[i-1])*K[i-1]=-f(X[i-1],Y[i-1]);
 Eq2:=g_x(X[i-1],Y[i-1])*H[i-1]+g_y(X[i-1],Y[i-1])*K[i-1]=
 -g(X[i-1],Y[i-1]);
 Sol:=solve({Eq1,Eq2},{H[i-1],K[i-1]}); assign(Sol);
 X[i]:=evalf(X[i-1]+H[i-1]);
 Y[i]:=evalf(Y[i-1]+K[i-1]);
 E_relX[i]:=abs((X[i]-X[i-1])/X[i-1]);
 E_relY[i]:=abs((Y[i]-Y[i-1])/Y[i-1]);
 if abs(f(X[i],Y[i]))>=epsilon or abs(g(X[i],Y[i]))>=epsilon
 then printf("%d %15.9e %15.9e %15.9e %15.9e\n",i,X[i],Y[i],
 E_relX[i],E_relY[i]): else lprint(`the result is`);
 printf("%d %15.9e %15.9e %15.9e %15.9e\n",i,X[i],Y[i],
 E_relX[i],E_relY[i]): break: fi: od:
```

Mathematica:

```
f[x_,y_]:=x^2+y^2-2;
g[x_,y_]:=x^2-y^2-1; fx[x1_,y1_]:=D[f[x,y],x]/.{x->x1,y->y1};
fy[x1_,y1_]:=D[f[x,y],y]/.{x->x1,y->y1};
gx[x1_,y1_]:=D[g[x,y],x]/.{x->x1,y->y1};
gy[x1_,y1_]:=D[g[x,y],y]/.{x->x1,y->y1}; g1=ContourPlot[
f[x,y]==0,{x,-2,2},{y,-2,2},ContourStyle->Hue[0.9]];
g2=ContourPlot[g[x,y]==0,{x,-2,2},{y,-2,2}]; Show[{g1,g2}]
```

```
{nD=90,nMax=100,epsilon=N[10^(-15),nD],
 X=N[Table[0,{i,1,nMax}],nD],Y=N[Table[0,{i,1,nMax}],nD],
 eRelX=N[Table[0,{i,1,nMax}],nD],
 eRelY=N[Table[0,{i,1,nMax}],nD],
 X[[1]]=N[1/10,nD],Y[[1]]=N[1/10,nD]}
Do[
 eq1=fx[X[[i-1]],Y[[i-1]]]*h[i-1]+fy[X[[i-1]],Y[[i-1]]]*
 k[i-1]==-f[X[[i-1]],Y[[i-1]]];
 eq2=gx[X[[i-1]],Y[[i-1]]]*h[i-1]+gy[X[[i-1]],Y[[i-1]]]*
 k[i-1]==-g[X[[i-1]],Y[[i-1]]];
 sol=Solve[{eq1,eq2},{h[i-1],k[i-1]}];
 X[[i]]=N[X[[i-1]]+sol[[1,1,2]],nD]; Y[[i]]=N[Y[[i-1]]+
 sol[[1,2,2]],nD];
 eRelX[[i]]=Abs[(X[[i]]-X[[i-1]])/X[[i-1]]];
 eRelY[[i]]=Abs[(Y[[i]]-Y[[i-1]])/Y[[i-1]]];
 If[Abs[f[X[[i]],Y[[i]]]]>=epsilon || Abs[g[X[[i]],Y[[i]]]]>=
 epsilon,Print[i-1," ",PaddedForm[X[[i]],{15,9}]," ",
 PaddedForm[Y[[i]],{15,9}]," ",PaddedForm[eRelX[[i]],{15,9}],
 " ",PaddedForm[eRelY[[i]],{15,9}]],{Print["the result is"],
 Print[i-1," ",X[[i]]," ",Y[[i]]," ",eRelX[[i]],
 " ",eRelY[[i]]],Break[]}],{i,2,nMax}];
```

This problem can be solved in a more compact form according to the Newton formula: $x^{(i+1)} = x^{(i)} - J^{-1}f(x^i)$, where $x^{(i)} \in \mathbb{R}^n$. In this case we use the JACOBIAN function (of the codegen package in *Maple*) and we write the JacobianMatrix function (in *Mathematica*) for generating the Jacobian matrix of $f(x)$.

Maple:

```
with(codegen): Digits:=90: N_max:=100: epsilon:=10^(-15);
f:=(x,y)->x^2+y^2-2: g:=(x,y)->x^2-y^2-1; X[0]:=<0.1,0.1>;
J:=JACOBIAN([f,g],result_type=array);
F:=X->X-convert(J(X[1],X[2]),
 Matrix)^(-1).<f(X[1],X[2]),g(X[1],X[2])>;
for i from 1 to N_max do
 X[i]:=F(X[i-1]);
 if abs(f(X[i][1],X[i][2]))>=epsilon or
 abs(g(X[i][1],X[i][2]))>=epsilon then
  print(i,X[i][1],X[i][2]):
 else lprint(`the result is`), print(i,X[i][1],X[i][2]): break:
 fi:
od:
```

Mathematica:

```
JacobianMatrix[f_List?VectorQ,x_List]:=
 Outer[D,f,x]/;Equal@@(Dimensions/@{f,x});
f[x_,y_]:=x^2+y^2-2; g[x_,y_]:=x^2-y^2-1;
JInv[x1_,y1_]:=Inverse[JacobianMatrix[{f[x,y],g[x,y]},
 {x,y}]]/.{x->x1,y->y1};
{nD=90,nMax=100,epsilon=N[10^(-15),nD],X[1]=N[{1/10,1/10},nD]}
F[X_List?VectorQ]:=X-JInv[X[[1]],X[[2]]].{f[X[[1]],X[[2]]],
 g[X[[1]],X[[2]]]};
Do[X[i]=F[X[i-1]]; If[Abs[f[X[i][[1]],X[i][[2]]]]>=epsilon ||
Abs[g[X[i][[1]],X[i][[2]]]]>=epsilon,Print[i-1," ",
X[i][[1]]," ",X[i][[2]]],{Print["the result is"],
Print[i-1," ",X[i][[1]]," ",X[i][[2]]],Break[]}},{i,2,nMax}];
```
                                                                                              □

12.2 Approximation of Functions and Data

In general, the approximation of a set of data $S=\{x_i,y_i\}$ $(i=1,\ldots,n)$ or a function $f(x)$ in $[a,b]$ consists of constructing another function $F(x)$ of more simpler form (or from some predetermined class) that represents S or $f(x)$ with a prescribed accuracy. This strategy is used frequently in numerical integration. In sciences and engineering, the function $f(x)$ is frequently defined as a number of data points (e.g., obtained in experiments), and we have to construct a continuous function which *closely fits* those data points. This strategy is called *curve fitting*. The *regression analysis* is the *least squares curve fitting*. The *interpolation* process is a specific case of curve fitting, in which the continuous function $F(x)$ must *go exactly* through the data points, i.e. $F(x_i)=f(x_i)$ or $F(x_i)=y_i$.

In *Maple*, there are the several packages, CurveFitting, Statistics, and numapprox, in which various functions allow us to perform curve fitting, regression analysis or least square curve fitting, interpolation and approximation of functions and data. The numapprox package contains functions for numerical approximations of functions.

In *Mathematica*, there are various functions for performing interpolation, curve fitting, least square approximations (for example, Interpolation, FunctionInterpolation, InterpolatingPolynomial, Fit, FindFit), the FunctionApproximations package with various functions for interpolation and approximation, and the Splines package.

Polynomial, trigonometric, rational, and spline interpolation of functions and data

Maple:

```
with(CurveFitting);          PolynomialInterpolation(XY,var,ops);
                  PolynomialInterpolation(XY,var,form=Lagrange);
PolynomialInterpolation(XY,var,form=Newton);convert(p,ratpoly);
ArrayInterpolation(XY,xValues,ops);Spline(XY,var,degree=d,ops);
RationalInterpolation(XY,var,ops);    ThieleInterpolation(XY,x);
convert(p,horner);
```

Note. In the CurveFitting package the data can have the form XY or X, Y, i.e., $[[x_1,y_1],\ldots,[x_n,y_n]]$ or $[x_1,\ldots,x_n],[y_1,\ldots,y_n]$.

PolynomialInterpolation, constructing an interpolating polynomial (including the Lagrange, monomial, Newton, or power representations),

ArrayInterpolation, performing n-dimensional data interpolation,

RationalInterpolation, constructing a rational interpolating function,

Spline, constructing a *piecewise polynomial* (of degree d) in variable var that approximates data.

Mathematica:

```
Interpolation[data,ops]    FunctionInterpolation[f,{x,a,b},ops]
InterpolatingPolynomial[data,x,ops]    ListInterpolation[array]
<<FunctionApproximations`                              <<Splines`
RationalInterpolation[f,{x,d1,d2},{xi}]    SplineFit[data,type]
```

Note. The data can have various forms $\{f_1,\ldots,f_n\}$, or $\{\{x_1,f_1\},\ldots,\{x_n,f_n\}\}$, or $\{\{x_1,y_1,z_1,\ldots,f_1\},\{x_2,y_2,z_1,\ldots,f_2\},\ldots\}$.

Interpolation, constructing an interpolation of the function values f_i corresponding to the coordinate values x_i, or an interpolation of multidimensional data,

InterpolatingPolynomial, constructing an interpolant polynomial (univariate or multivariate) in a Horner form,

FunctionInterpolation, evaluating an expression in [a,b] and constructing an interpolating function,

RationalInterpolation, constricting a rational interpolant of degree (d_1,d_2),

SplineFit, constructing a spline function of the specified type (Cubic, Bezier, and CompositeBezier) that approximates the multidimensional data,

ListInterpolation, constructing an approximate function that interpolates the array of values.

Problem 12.10 *Interpolating polynomials.* Let $\{x_i, y_i\}$ be $n + 1$ given data pairs $(i = 0, \ldots, n)$, the nodes x_i being all distinct. Then there exists a unique polynomial of degree less than or equal to n interpolating the set of values y_i at the nodes x_i. There are several kinds of the approximate function or *interpolant* $F(x)$ satisfying the interpolation conditions: $F(x_i) = y_i$, $i = 0, \ldots, n$.

  1) Construct interpolant polynomials in the Lagrange, Newton, and Horner representations to the data given.

  2) Plot the interpolating polynomials and the data.

Maple:

```
with(CurveFitting); with(plots): X:=[1,2,3,4,5,6];
Y:=[25,26,26,25,24,22]; X1:=array(0..5,X); Y1:=array(0..5,Y);
p1:=PolynomialInterpolation(X,Y,x); convert(p1,horner);
p2:=PolynomialInterpolation(X,Y,x,form=Newton);
p3:=PolynomialInterpolation(X,Y,x,form=Lagrange);
L:=proc(x,i,k) local j,P; P:=1: for j from 0 to k do
 if j <> i then P:=P*(x-X1[j])/(X1[i]-X1[j]); fi; od; P; end;
L(x,1,2); n:=nops(X)-1; pLagrange:=add(L(x,i,n)*Y1[i],i=0..n);
subs(x=2,pLagrange); factor(pLagrange-p3);
XY:=[seq([X[i],Y[i]],i=1..nops(X))];
G1:=plot(XY,style=point,color=blue): G2:=plot(pLagrange,x=0..7):
display({G1,G2}); for i from 0 to n do FD[i,0]:=Y1[i]; od;
pNewton:=add(FD[i,i]*product(x-X1[j],j=0..i-1),i=0..n);
for i from 1 to n do for j from 1 to i do
 FD[i,j]:=(FD[i,j-1]-FD[i-1,j-1])/(X1[i]-X1[i-j]); od; od;
pNewton; subs(x=2,pNewton); factor(p2-pNewton);
G3:=plot(pNewton,x=0..7,color=magenta): display({G1,G3});
```

Mathematica:

```
data={{1,25},{2,26},{3,26},{4,25},{5,24},{6,22}}
{X1=Transpose[data][[1]],Y1=Transpose[data][[2]],n=Length[X1]-1}
pHorner=InterpolatingPolynomial[data,x]
L[x_,i_,k_]:=Module[{j,P},P=1; Do[If[j!=i,
 P=P*(x-X1[[j+1]])/(X1[[i+1]]-X1[[j+1]])],{j,0,k}];P]; L[x,1,2]
{pLagrange=Sum[L[x,i,n]*Y1[[i+1]],{i,0,n}],pLagrange/.{x->2},
 Factor[pLagrange-pHorner]}
```

```
g1=ListPlot[data,PlotStyle->{PointSize[.02],Hue[0.9]}];
g2=Plot[pLagrange,{x,0,7}]; Show[{g1,g2}]
Do[FD[i,0]=Y1[[i+1]]; Print[FD[i,0]],{i,0,n}];
pNewton=Sum[FD[i,i]*Product[x-X1[[j+1]],{j,0,i-1}],{i,0,n}]
Do[Do[FD[i,j]=(FD[i,j-1]-FD[i-1,j-1])/(X1[[i+1]]-X1[[(i-j)+1]]),
 {j,1,i}],{i,1,n}]; {pNewton, pNewton/.{x->2},
Factor[pHorner-pNewton]}
g3:=Plot[pNewton,{x,0,7}]; Show[{g1,g3}]
```

□

Problem 12.11 *Hermite interpolating polynomials.* Let $f(x) \in C^1[a, b]$ and $\{x_i\}$ be $n + 1$ distinct nodes $(i = 0, \ldots, n)$, and let $\{f(x_i), f'(x_i)\}$ be $2n + 2$ values of $f(x)$ and $f'(x)$. Then there exists a unique interpolating polynomial $F(x)$ of degree $2n + 1$ with the properties: $F(x_i) = f(x_i)$, $F'(x_i) = f'(x_i)$ $i = 0, \ldots, n$. This *Hermite interpolation polynomial* is defined by the formula: $F(x) = \sum_{i=0}^{n} \left(H_{i,n}^1(x)f(x_i) + H_{i,n}^2 f'(x_i) \right)$, where the Hermite factors are expressed in terms of the Lagrange factors: $H^1(i, n)(x) = \left(1 - 2(x - x_i)L_{i,n}'(x_i) \right) L_{i,n}^2(x)$ and $H^2(i, n)(x) = (x - x_i)L_{i,n}^2(x)$. Construct and plot the Hermite interpolating polynomial.

Maple:

```
with(CurveFitting); with(plots): n:=5; a:=-2; b:=2;
f:=x->x*exp(-x)+exp(x); fD:=unapply(diff(f(x),x),x); f(x); fD(x);
X:=[seq(a+(b-a)/n*i,i=0..n)]; Y:=NULL: YD:=NULL:
for i from 1 to n+1 do Y:=Y,f(X[i]); YD:=YD,fD(X[i]); od:
Y:=evalf([Y]); YD:=evalf([YD]); X1:=array(0..5,X);
Y1:=array(0..5,Y); YD1:=array(0..5,YD);
L:=proc(x,i,k) local j,P; P:=1: for j from 0 to k do
 if j <> i then P:=P*(x-X1[j])/(X1[i]-X1[j]); fi; od; P; end;
H1:=(x,i,n)->(1-2*(x-X1[i])*subs(x=X1[i],diff(L(x,i,n),x)))
    *(L(x,i,n))^2; H2:=(x,i,n)->(x-X1[i])*(L(x,i,n))^2;
FHermite:=add(Y1[i]*H1(x,i,n)+YD1[i]*H2(x,i,n),i=0..n);
subs(x=2,FHermite);  XY:=[seq([X1[i],Y1[i]],i=1..n)];
G1:=plot(XY,style=point,color=blue): G2:=plot(FHermite,x=a..b):
display({G1,G2});
```

Mathematica:

```
f[x_]:=x*Exp[-x]+Exp[x];
fD[x1_]:=D[f[x],x]/.{x->x1}; {f[x],fD[x]}
{n=5,a=-2,b=2,X=Table[a+(b-a)/n*i,{i,0,n}],Y={},YD={}}
Do[Y=Append[Y,f[X[[i]]]];YD=Append[YD,fD[X[[i]]]],{i,1,n+1}];
{Y=N[Y],YD=N[YD]}
L[x_,i_,k_]:=Module[{j,P},P=1;
 Do[If[j!=i,P=P*(x-X[[j+1]])/(X[[i+1]]-X[[j+1]])],{j,0,k}];P];
H1[x_,i_,n_]:=(1-2*(x-X[[i+1]])*(D[L[x,i,n],x]/.{x->X[[i+1]]}))*
 (L[x,i,n])^2; H2[x_,i_,n_]:=(x-X[[i+1]])*(L[x,i,n])^2;
FHermite=Sum[Y[[i+1]]*H1[x,i,n]+YD[[i+1]]*H2[x,i,n],{i,0,n}]
{FHermite/.{x->2}, XY=Table[{X[[i]],Y[[i]]},{i,1,n}]}
g1=ListPlot[XY,PlotStyle->{PointSize[.02],Hue[0.9]}];
g2=Plot[FHermite,{x,a,b}]; Show[{g1,g2}]
```

$\square$

Problem 12.12 *The upper bound of the Lagrange interpolation.* Let $I = [a, b]$ be a bounded interval and let $\{x_i\}$ $(i = 0, \ldots, n)$ be $n+1$ equidistant interpolation nodes, $x^{(i)} = x^{(i-1)} + h$, on I. Let $f(x)$ be a continuously differentiable function up to $n+1$ order in I. Evaluate the upper bounds of the Lagrange interpolation polynomials (increasing the degree n) according to the formula: $\max_{x \in I} |E^{(n)} f(x)| \leq \dfrac{h^{n+1}}{4(n+1)} \max_{x \in I} |f^{(n+1)}(x)|, \quad \forall x \in I.$

Maple:

```
with(plots): n:=8; E:=(N,fM,H)->evalf(fM*H^(N+1)/(4*(N+1)));
f1:=x->cos(x)+sin(x); f1(x); a:=-Pi; b:=Pi; h:=(b-a)/n;
X:=[seq(a+h*i,i=0..n)]; fD:=unapply(abs(diff(f1(x),x$(n+1))),x);
plot(fD(x),x=a..b); xMax1:=fsolve(diff(fD(x),x)=0,x=a..0);
xMax2:=fsolve(diff(fD(x),x)=0,x=0..b); fMax1:=fD(xMax1);
fMax2:=fD(xMax2); E1[n]=E(n,fMax1,h); E2[n]=E(n,fMax2,h);
```

Mathematica:

```
n=8; Er[N1_,fM_,H_]:=N[fM*H^(N1+1)/(4*(N1+1))];
f1[x_]:=Cos[x]+Sin[x]; fD[x1_]:=Abs[D[f1[x],{x,n+1}]]/.{x->x1};
{f1[x],a=N[-Pi],b=N[Pi],h=N[(b-a)/n],X=N[Table[a+h*i,{i,0,n}]]]}
Plot[fD[x],{x,a,b}]
xMax=N[Solve[D[fD[x],x]==0,x,InverseFunctions->True]]
{x1=xMax[[1,1,2]],x2=xMax[[2,1,2]],fMax1=fD[x1],fMax2=fD[x2],
 e1=Er[n,fMax1,h],e2=Er[n,fMax2,h]}
```

Since in the general case we cannot deduce from the above formula that the interpolation error tends to 0 as $n \to \infty$, we have to verify this conclusion for every function we are working. In this case, if we compute the interpolation errors by increasing the parameter n, we can see that the interpolation error tends to 0. But there exist functions $f(x)$ for which this is not true and we consider such functions in the next problem. □

Problem 12.13 *Interpolation errors for equidistant interpolation nodes.* For equidistant interpolation nodes on the interval I, the interpolation error at any point of I may equal to ∞ as the degree n tends to infinity. Let us consider the function $f(x) = 1/(1+25x^2)$ on $I = [-1, 1]$, that was investigated by Runge in 1901.

1) Verify that $\lim\limits_{n\to\infty} \max\limits_{x\in I} |E^{(n)} f(x)| = \infty$.

2) Construct and plot the Lagrange interpolating polynomials (increasing the degree) for Runge's function and verify that the approximations works satisfactorily in the central part of the interval, but if the degree n tends to ∞, the interpolation polynomials diverge for $0.7 \le |x| \le 1$.

Maple:

```
with(plots): n:=9;
E:=(N,fM,H)->evalf(fM*H^(N+1)/(4*(N+1)));
fRunge:=x->1/(1+25*x^2); fRunge(x);
a:=-1; b:=1; h:=(b-a)/n; X:=[seq(a+h*i,i=0..n)];
q:=evalf(0.3/(n+1));
fD:=unapply(simplify(diff(fRunge(x),x$(n+1))),x);
plot(abs(fD(x)),x=a..b);
xMax:=fsolve(diff(fD(x),x)=0,x=-q..q);
fMax:=abs(fD(xMax)); E[n]=E(n,fMax,h);
with(CurveFitting); Y:=NULL:
for i from 1 to n+1 do Y:=Y,fRunge(X[i]) od:
pLagrange:=PolynomialInterpolation(X,[Y],x,form=Lagrange);
XY:=[seq([X[i],Y[i]],i=1..n+1)];
G1:=plot(XY,style=point,color=blue):
G2:=plot(pLagrange,x=a..b):
G3:=plot(fRunge(x),x=a..b,color=magenta):
display({G1,G2,G3});
```

Mathematica:

```
fRunge[x_]:=1/(1+25*x^2); n=9; Er[N1_,fM_,H_]:=N[fM*H^(N1+1)/
 (4*(N1+1))]; fD[x1_]:=Simplify[D[fRunge[x],{x,n+1}]]/.{x->x1};
{fRunge[x],a=-1,b=1,h=(b-a)/n,X=Table[a+h*i,{i,0,n}]}
Plot[Evaluate[Abs[fD[x]]],{x,a,b},PlotRange->All]
xMax=Solve[D[fD[x],x]==0,x,InverseFunctions->True]
MaxT=Table[N[Abs[fD[x]]/.xMax[[i]]],{i,1,Length[xMax]}]
{Max[MaxT], k=Position[MaxT,Max[MaxT]][[1,1]],
 fMax=Abs[fD[xMax[[k]][[1,2]]]], e[n]=Er[n,fMax,h]}
Y={}; Do[Y=Append[Y,fRunge[X[[i]]]],{i,1,n+1}];
L[x_,i_,k_]:=Module[{j,P},P=1; Do[If[j!=i,
 P=P*(x-X[[j+1]])/(X[[i+1]]-X[[j+1]])],{j,0,k}];P];
pLagrange=Sum[L[x,i,n]*Y[[i+1]],{i,0,n}]
XY=Table[{X[[i]],Y[[i]]},{i,1,n+1}]
g1=ListPlot[XY,PlotStyle->{PointSize[.02],Hue[0.7]}];
g2=Plot[pLagrange,{x,a,b},PlotStyle->Hue[0.5]];
g3=Plot[fRunge[x],{x,a,b},PlotStyle->Hue[0.9]]; Show[{g1,g2,g3}]
```

Computing the interpolation errors for Runge's function (varying the parameter n up to 15 within the given accuracy) we can see that the interpolation error tends to ∞. This behavior is known as *Runge's phenomenon*. Also we can observe the lack of convergence by the presence of oscillations (near the endpoints of the interval) in the graph of the interpolating polynomial versus the graph of $f(x)$. $\square$

Problem 12.14 *Chebyshev interpolation.* Let us consider a special distribution of nodes such that Runge's phenomenon can be avoided. There exist some special distributions of nodes x_i, e.g., the *Chebyshev nodes*, for which the convergence property is valid for all continuous functions $f(x)$ on $[a,b]$. Let $\{x_i, f(x_i)\}$ be $n+1$ given data pairs $(i=0,\ldots,n)$, where the nonuniform distribution of nodes, the Chebyshev nodes x_i are given as $x_i = \dfrac{a+b}{2} - \dfrac{b-a}{2} \cos\left(\dfrac{2i+1}{n+1}\dfrac{\pi}{2}\right)$.

1) Construct and plot the interpolating Chebyshev polynomial $F(x)$ that approximates Runge's function of the previous problem.

2) 1) Verify that the amplitude of oscillations near the endpoints decreases as the degree increases and the maximum of the interpolation error decreases as n increases, i.e. $\lim_{n\to\infty} \max_{x\in I} |E^{(n)} f(x)| = 0$.

3) Graph the distribution of the Chebyshev nodes on the interval $[-1,1]$.

Maple:

```
Digits:=30: with(CurveFitting); with(plots): n:=5; a:=-1; b:=1;
fRunge:=x->1/(1+25*x^2); fRunge(x);
XCh:=[seq((a+b)/2-(b-a)/2*cos(((2*i+1)*Pi)/((n+1)*2)),i=0..n)];
YCh:=NULL: for i from 1 to n+1 do YCh:=YCh,fRunge(XCh[i]) od:
pCh:=PolynomialInterpolation(XCh,[YCh],x,form=Lagrange):
XYCh:=[seq([XCh[i],YCh[i]],i=1..n+1)]:
G1:=plot(XYCh,style=point,color=blue):G2:=plot(pCh,x=a..b):
G3:=plot(fRunge(x),x=a..b,color=magenta): display({G1,G2,G3});
fErr:=abs(expand(evalf(pCh-fRunge(x)))); plot(fErr,x=-1..1);
with(Optimization): pMax:=Maximize(fErr,x=-1..1); E[n]=pMax[1];
NodesCh:=evalf([seq([XCh[i],0],i=1..n+1)]);
plot(NodesCh,-1..1,-0.01..0.01,style=point,symbolsize=20,color=
 blue,scaling=constrained); with(numapprox): with(orthopoly):
pCh1:=unapply(chebyshev(fRunge(x),x=-1..1,.001),x);
G4:=plot(evalf(pCh1(x)),x=-1..1,color=green): display({G1,G4});
```

Mathematica:

```
nD=30; fRunge[x_]:=1/(1+25*x^2); {n=5,a=-1,b=1,fRunge[x],
XCh=Table[(a+b)/2-(b-a)/2*Cos[((2*i+1)*Pi)/((n+1)*2)],{i,0,n}]}
YCh={}; Do[YCh=Append[YCh,fRunge[XCh[[i]]]],{i,1,n+1}];
L[x_,i_,k_]:=Module[{j,P},P=1; Do[If[j!=i,
 P=P*(x-XCh[[j+1]])/(XCh[[i+1]]-XCh[[j+1]])],{j,0,k}];P];
pCh=Sum[L[x,i,n]*YCh[[i+1]],{i,0,n}];
XYCh=Table[{XCh[[i]],YCh[[i]]},{i,1,n+1}];
g1=ListPlot[XYCh,PlotStyle->{PointSize[.02],Hue[0.9]}];
g2=Plot[pCh,{x,a,b},PlotStyle->Hue[0.8]];
g3=Plot[fRunge[x],{x,a,b},PlotStyle->Hue[0.7]]; Show[{g1,g2,g3}]
fErr[x1_]:=Abs[Expand[pCh-fRunge[x]]]/.{x->x1}
Plot[fErr[x],{x,-1,1},PlotRange->All]
{pMax=FindMaximum[fErr[x],{x,-0.1},AccuracyGoal->10,
 PrecisionGoal->10,WorkingPrecision->nD,Method->"Newton"],
 e[n]=pMax[[1]], NodesCh=N[Table[{XCh[[i]],0.},{i,1,n+1}],nD]}
ListPlot[NodesCh,DataRange->{-1,1},PlotRange->{-0.01,0.01},
 PlotStyle->{PointSize[.02],Hue[0.7]},AspectRatio->0.05]
```

In *Maple*, in the numapprox package there is the function chebyshev for computing the Chebyshev series expansion of the function $f(x)$ on the interval $[a, b]$ with the accuracy ε. The package orthopoly is loaded for evaluating the i-th Chebyshev polynomial of the first kind $T_k(x)$. □

Problem 12.15 *Trigonometric interpolation.* Let $f : [a, b] \to \mathbb{R}$ be a periodic function, i.e. $f(t) = f(t+T)$ $(t \in \mathbb{R})$, and let $\{t_i, f(t_i)\}$ be $2n+1$ given data pairs $(i = 0, \ldots, 2n)$, where $t_i = 2\pi i / (2n + 1)$. Polynomial interpolation is not appropriate for periodic functions, since algebraic polynomials are not periodic. So let us construct the trigonometric polynomial $F(t)$ which interpolates $f(t)$ at $2n+1$ nodes such that $F(t_i) = f(t_i)$, $i = 0, \ldots, 2n$. This approach was first proposed independently by Clairaut and Lagrange in 1759 and 1762, respectively. Let us assume that $T = 2\pi$ and $I = [a, b] = [0, 2\pi]$.

1) Construct and plot the uniquely determined trigonometric interpolating polynomial of degree n in the *Lagrange representation* according to the formula:

$$F(t) = \sum_{i=0}^{2n} f(t_i) L_{i,n}(t), \quad L_{i,n}(t) = \prod_{j=0, j \neq i}^{2n} \frac{\sin((t - t_j)/2)}{\sin((t_i - t_j)/2)}.$$

Maple:

```
with(plots): n:=20; a:=0; b:=2*Pi; f:=x->x*exp(-x); f(x);
T:=evalf([seq(2*Pi*i/(2*n+1),i=0..2*n)]); Y:=NULL:
for i from 1 to 2*n+1 do Y:=Y,f(T[i]) od: Y:=evalf([Y]);
T1:=array(0..2*n,T); Y1:=array(0..2*n,Y);
L:=proc(t,i,k) local j,P; P:=1:
 for j from 0 to k do if j <> i then
 P:=P*(sin((t-T1[j])/2))/(sin((T1[i]-T1[j])/2)); fi; od; P;
end; FTrig:=add(Y1[i]*L(t,i,2*n),i=0..2*n):
evalf(subs(t=2,FTrig)); TY:=[seq([T1[i],Y1[i]],i=1..2*n)]:
G1:=plot(TY,style=point,color=blue):
G2:=plot(FTrig,t=a..b,color=green): display({G1,G2});
```

Mathematica:

```
f[x_]:=x*Exp[-x]; {nD=10,n=20,a=0,b=2*Pi,f[x]}
T=Table[2*Pi*i/(2*n+1),{i,0,2*n}];
Y={}; Do[Y=Append[Y,f[T[[i]]]],{i,1,2*n+1}];
L[x_,i_,k_]:=Module[{j,P},P=1; Do[If[j!=i,
 P=P*(Sin[(t-T[[j+1]])/2])/(Sin[(T[[i+1]]-T[[j+1]])/2)]),
 {j,0,k}];P]; FTrig=Sum[L[t,i,2*n]*Y[[i+1]],{i,0,2*n}];
{N[T,nD], N[Y,nD], N[FTrig/.{t->2},nD]}
TY=Table[{T[[i]],Y[[i]]},{i,1,2*n}];
g1=ListPlot[TY,PlotStyle->{PointSize[.02],Hue[0.7]}];
g2=Plot[FTrig,{t,a,b},PlotStyle->Hue[0.9]]; Show[{g1,g2}]
```

□

Problem 12.16 *Piecewise linear interpolation* or *composite linear interpolation*. In engineering, ship-building designers developed an elastic ruler, called a *spline*, which can be curved so that it passes through a given set of points $\{x_i, y_i\}$ $(i = 0, \ldots, n)$. Let $a = x_0 < x_1 < \ldots < x_n = b$ be a subdivision of the interval $[a, b]$. A spline function of degree p with the nodes x_i is a function $s(x)$ with the properties:
1) On each subinterval $[x_i, x_{i+1}]$ $(i = 0, \ldots, n - 1)$, $s(x)$ is a polynomial of degree d which interpolates the pairs of values $(x_k, f(x_k))$.
2) $s(x)$ and its first $(d - 1)$ derivatives are continuous on $[a, b]$.

In numerical analysis splines were introduced by Schoenberg in 1946. In the simple case, $d = 1$, $s(x)$ is a piecewise linear function. Let us construct the piecewise linear interpolation polynomial such that on each interval, $[x_i, x_{i+1}]$, it is defined by the line segment joining the two points $(x_i, f(x_i))$ and $(x_{i+1}, f(x_{i+1}))$ according to the formula (of linear interpolation), e.g. $s(x) = f(x_i) + \dfrac{f(x_{i+1}) - f(x_i)}{x_{i+1} - x_i}(x - x_i)$, where $x_i \le x \le x_{i+1}$, $i = 0, \ldots, n - 1$.

1) Construct the piecewise linear interpolating polynomial (or the spline of degree 1) and the Lagrange interpolating polynomial for the given data. Compare the results and plot the graphs.

2) Verify that with this interpolation method we avoid Runge's phenomena when the number of nodes increases.

Maple:

```
with(CurveFitting); with(plots): X:=[1,2,3,4,5,6];
Y:=[25,26,26,25,24,22]; n:=nops(X);
p1:=PolynomialInterpolation(X,Y,x,form=Lagrange);
s1:=Spline(X,Y,x); G1:=plot(p1,x=X[1]..X[n],colour=blue):
G2:=plot(s1,x=X[1]..X[n],colour=green): display({G1,G2});
nR:=9; fRunge:=x->1/(1+25*x^2); fRunge(x); a:=-1; b:=1;
h:=(b-a)/nR; XR:=[seq(a+h*i,i=0..nR)]; YR:=NULL:
for i from 1 to nR+1 do YR:=YR,fRunge(XR[i]) od:
pLagrange:=PolynomialInterpolation(XR,[YR],x,form=Lagrange):
XYR:=evalf([seq([XR[i],YR[i]],i=1..nR+1)]);sR:=Spline(XYR,x);
G1:=plot(XYR,style=point,color=blue): G2:=plot(pLagrange,
  x=a..b,color=red): G3:=plot(fRunge(x),x=a..b,color=green):
G4:=plot(sR,x=a..b,color=magenta): display({G1,G2,G3,G4});
display({G2,G4});
```

Mathematica:

```
<<Splines`
nD=10; data={{1,25},{2,26},{3,26},{4,25},{5,24},{6,22}}
intPHorner=InterpolatingPolynomial[data,x]
{X=Transpose[data][[1]],Y=Transpose[data][[2]],n=Length[X]-1}
L[x_,i_,k_]:=Module[{j,P}, P=1;
 Do[If[j!=i,P=P*(x-X[[j+1]])/(X[[i+1]]-X[[j+1]])],{j,0,k}];
 P];
p1=Sum[L[x,i,n]*Y[[i+1]],{i,0,n}]
s1=SplineFit[data,Cubic]
g1=Plot[p1,{x,X[[1]],X[[n]]},PlotStyle->Hue[0.7]];
g2=ParametricPlot[Evaluate[s1[x]],{x,X[[1]],X[[n]]},
 PlotStyle->Hue[0.8],PlotRange->All]; Show[{g1,g2}]
fRunge[x_]:=1/(1+25*x^2);
{fRunge[x],nR=9,a=-1,b=1,h=(b-a)/nR,XR=Table[a+h*i,{i,0,nR}]}
YR={}; Do[YR=Append[YR,fRunge[XR[[i]]]],{i,1,nR+1}];
L1[x_,i_,k_]:=Module[{j,P},P=1;
 Do[If[j!=i,P=P*(x-XR[[j+1]])/(XR[[i+1]]-XR[[j+1]])],{j,0,k}];
 P];
pLagrange=Sum[L1[x,i,nR]*YR[[i+1]],{i,0,nR}]
{XYR=Table[{XR[[i]],YR[[i]]},{i,1,nR+1}],XYR=N[XYR,nD]}
sR=SplineFit[XYR,Cubic]
g1=ListPlot[XYR,PlotStyle->{PointSize[.02],Hue[0.7]}];
g2=Plot[pLagrange,{x,a,b},PlotStyle->Hue[0.7]];
g3=Plot[fRunge[x],{x,a,b},PlotStyle->Hue[0.3]];
g4=ParametricPlot[Evaluate[sR[x]],{x,0,9},PlotStyle->Hue[0.9]];
Show[{g1,g2,g3,g4},AspectRatio->1.2]
Show[{g2,g4}]
```

We can observe that the Chebyshev interpolating polynomial produces sufficiently accurate approximations of smooth functions $f(x)$. But if $f(x)$ is non smooth or $f(x)$ is defined on a set of points (that are different from the Chebyshev nodes), the piecewise linear interpolation method is more appropriate. $\square$

Problem 12.17 *Spline interpolation.* Let $a = x_0 < x_1 < \ldots < x_n = b$ be a subdivision of the interval $[a, b]$ and let $\{x_i, f(x_i)\}$ be $n + 1$ given data pairs $(i = 0, \ldots, n)$. Let us perform interpolation of the given function $f(x)$ by a piecewise cubic function $s(x)$, i.e. by a spline of degree 3, which is continuous together with its first and second derivatives.

1) Construct and plot the interpolant polynomial (in the Lagrange representation) and the interpolant cubic spline to the function $f(x)$.

2) Varying the parameter n, verify that spline interpolation corrects the deficiency of interpolation by high-degree polynomials through a piecewise polynomial interpolation of low degree.

Maple:

```
Digits:=30: with(CurveFitting); with(plots):
n:=15; f:=x->sin(3*Pi*x)/exp(2*x); f(x);
a:=-Pi/2; b:=Pi/2; h:=(b-a)/n;
X:=[seq(a+h*i,i=0..n)]; Y:=NULL:
for i from 1 to n+1 do Y:=Y,f(X[i]) od:
pLagrange:=PolynomialInterpolation(X,[Y],x,form=Lagrange):
XY:=evalf([seq([X[i],Y[i]],i=1..n+1)]);
s3:=Spline(XY,x,degree=3);
G1:=plot(XY,style=point,color=blue):
G2:=plot(pLagrange,x=a..b,color=red):
G3:=plot(f(x),x=a..b,color=green):
G4:=plot(s3,x=a..b,color=magenta):
display({G1,G3,G2}); display({G3,G4});
```

Mathematica:

```
<<Splines`
nD=30; n=15; f[x_]:=Sin[3*Pi*x]/Exp[2*x];
{f[x],a=-Pi/2,b=Pi/2,h=(b-a)/n,X=Table[a+h*i,{i,0,n}]}
Y={}; Do[Y=Append[Y,f[X[[i]]]],{i,1,n+1}];
L[x_,i_,k_]:=Module[{j,P},P=1;
 Do[If[j!=i,P=P*(x-X[[j+1]])/(X[[i+1]]-X[[j+1]])],
 {j,0,k}];P];
pLagrange=Sum[L[x,i,n]*Y[[i+1]],{i,0,n}]
{XY=Table[{X[[i]],Y[[i]]},{i,1,n+1}], XYN=N[XY,nD]}
s3=SplineFit[XY,Cubic]
g1=ListPlot[XY,PlotStyle->{PointSize[.02],Hue[0.7]},
 PlotRange->All];
g2=Plot[pLagrange,{x,a,b},PlotStyle->Hue[0.8],
 PlotRange->Full];
g3=Plot[f[x],{x,a,b},PlotStyle->Hue[0.3],PlotRange->Full];
g4=ParametricPlot[Evaluate[s3[x]],{x,0.,15.},
 PlotStyle->Hue[0.9],PlotRange->Full]; Show[{g1,g2,g3}]
Show[{g3,g4}]
```

☐

Rational and trigonometric approximations of functions and data, orthogonal polynomials and Chebyshev approximation

Maple:

```
with(numapprox):  with(orthopoly): minimax(f,x=a..b,[d1,d2],w);
pade(f,x=a,[d1,d2]);        convert(series(f,x=a),ratpoly,d1,d2);
hornerform(f,x);  convert(f,horner);          confracform(f,x);
chebpade(f,x=a..b,[d1,d2]);          chebyshev(f,x=a..b,epsilon);
with(Statistics); Fit(TrigModel,X,Y,t);          taylor(f,x=a,n);
with(DiscreteTransforms);FourierTransform(X,Y);series(f,x=a,n);
```

Note. We consider various approximations using a rational function $r(x)$ of degree n ($n = d_1 + d_2$) which has the form $r(x) = p(x)/q(x)$, where $p(x)$ and $q(x)$ are polynomials of degree d1 and d2, respectively.

minimax, computing the best minimax rational approximation of degree [d1,d2] for a real function f(x) on $[a, b]$ with respect to the positive weight function $w(x)$ by Remez algorithm,

hornerform, confracform, converting $f(x)$ to Horner form or converting $r(x)$ to continued-fraction form,

pade, convert,ratpoly, computing the Padé approximation or converting series to a rational polynomial,

chebyshev, chebpade, computing the Chebyshev series expansion of $f(x)$ with epsilon accuracy or the Chebyshev-Padé approximation.

Mathematica:

```
<<FunctionApproximations` PadeApproximant[f,{x,x0,{d1,d2}},ops]
MiniMaxApproximation[f,{x,{a,b},d1,d2},ops]      HornerForm[f,x]
EconomizedRationalApproximation[f,{x,{a,b},d1,d2},ops]
ContinuedFraction[x]                FromContinuedFraction[list]
Rationalize[x]   RationalInterpolation[f,{x,d1,d2},{x,a,b},ops]
```

EconomizedRationalApproximation, constructing the economized rational approximation of degree d1,d2 to a function $f(x)$ on the interval [a,b],

PadeApproximant, computing the Padé approximation to $f(x)$ about the point x0, with numerator and denominator orders d1 and d2, respectively,

MiniMaxApproximation, constructing the rational polynomial function (with numerator and denominator orders d1 and d2) that produces a minimax approximation to $f(x)$ on the interval $[a, b]$,

HornerForm, converting a polynomial in Horner form,

ContinuedFraction, FromContinuedFraction, converting (or reconstructing) a number to (or from) continued fraction form,

Rationalize, converting an approximate number to a nearby rational with small denominator,

RationalInterpolation, constructing the rational interpolant with the interpolation points chosen automatically on the interval $[a, b]$.

Problem 12.18 *Rational approximation and interpolation.* Let $f : \mathbb{R} \to \mathbb{R}$ be a real function of real variable and let $\{x_i, y_y\}$ $(i = 0, \ldots, n)$ be $n + 1$ data pairs satisfying the interpolation conditions. Let us construct rational approximate functions and rational interpolating polynomials using various methods.

1) Compute the best minimax rational approximation, the Padé approximation, the Chebyshev-Padé approximation.

2) Convert the rational polynomials obtained to Horner form and continued-fraction form.

3) Construct rational polynomial interpolants for the given data.

4) Plot the function $f(x)$ and the rational approximations obtained.

Maple:

```
Digits:=14; with(CurveFitting); with(numapprox):
with(orthopoly): with(plots): a:=Pi; b:=2*Pi;
f:=x->x*exp(-x); f(x); r1:=minimax(f(x),x=a..b,[2,3]);
normal(r1);  plot(f(x)-r1,x=a..b,color=green);
G1:=plot(f(x),x=a..b,color=red):
G2:=plot(r1,x=a..b,color=blue):
display({G1,G2});
r2:=pade(f(x),x=a,[2,3]);
r3:=convert(series(f(x),x=0),ratpoly,2,3); evalb(r2=r3);
r4:=hornerform(r2,x); r5:=convert(r2,horner);
r6:=confracform(r2,x); r7:=chebpade(f(x),x=a..b,[2,3]);
evala(r7); X1:=[1,2,3,4,5,6];Y1:=[25,26,26,25,24,22];
r8:=RationalInterpolation(X1,Y1,x,degrees=[2,3]);
X2:=[1,2,4,5,9,11];Y2:=[3,4,5,8,9,10];
r9:=ThieleInterpolation(X2,Y2,x);
```

Mathematica:

```
<<FunctionApproximations`
nD=14; f[x_]:=x*Exp[-x];
{f[x], a=Pi, b=2*Pi, r1=MiniMaxApproximation[f[x],
 {x,{a,b},2,3},WorkingPrecision->nD], r1[[2,1]]}
Plot[1-r1[[2,1]]/f[x],{x,a,b}]
g1=Plot[f[x],{x,a,b},PlotStyle->Hue[0.8]];
g2=Plot[r1[[2,1]],{x,a,b},PlotStyle->Hue[0.9]]; Show[{g1,g2}]
r2=PadeApproximant[f[x],{x,a,{2,3}}]
r3=RationalInterpolation[f[x],{x,2,3},{x,a,b},
   WorkingPrecision->nD]
r4=EconomizedRationalApproximation[f[x],{x,{N[a,nD],N[b,nD]},
   2,3}]//Cancel
```

□

Problem 12.19 *Discrete approximate trigonometric polynomials.* Let $f : [a, b] \to \mathbb{R}$ be a periodic function and let $\{t_i, f(t_i)\}$ be $2n + 1$ given data pairs $(i = 0, \ldots, 2n)$, where $t_i = 2\pi i/(2n + 1)$ and $2n$ is an even number. For approximating the periodic function $f(t)$ we construct the discrete trigonometric polynomial $F(x)$ as a linear combination of sine and cosine functions.

1) Applying the *least squares method*, construct and plot the discrete approximate trigonometric polynomial of degree n according to the formula: $F(t) = \dfrac{A_0}{2} + \sum\limits_{k=1}^{n} (A_i \cos(kt) + B_i \sin(kt))$.

Here $A_i = \dfrac{2}{2n + 1} \sum\limits_{i=0}^{2n} f(t_i) \cos(t_i k)$, $B_i = \dfrac{2}{2n + 1} \sum\limits_{i=0}^{2n} f(t_i) \sin(t_i k)$, and $2n$ is even.

2) Compute the mean square error $\sum\limits_{i=0}^{n} [f(t_i) - F(t_i)]^2$.

Maple:

```
Digits:=20: with(Statistics): with(plots):
n:=20; a:=0; b:=2*Pi; f:=x->x*exp(-x); f(x);
X:=evalf([seq(2*Pi*i/(2*n+1),i=0..2*n)]); Y:=NULL:
for i from 1 to 2*n+1 do Y:=Y,f(X[i]) od: Y:=evalf([Y]);
model:=A[0]/2+add(A[k]*cos(k*t)+B[k]*sin(k*t),k=1..n);
X1:=Vector(X,datatype=float); Y1:=Vector(Y,datatype=float);
FTrig:=Fit(model,X1,Y1,t); FTrig1:=unapply(subs(t=x,FTrig),x);
```

```
FTrig2:=[seq([X[i],FTrig1(X[i])],i=1..2*n+1)];
XY:=[seq([X[i],Y[i]],i=1..2*n+1)]:
G1:=plot(XY,style=point,color=blue):
G2:=plot(FTrig2,style=point,symbolsize=20,color=red):
G3:=plot(f(x),x=a..b,color=magenta): display({G1,G3});
display({G2,G3}); for i from 1 to 2*n+1 do
print(i,X[i],Y[i],evalf(abs(Y[i]-FTrig1(X[i])))); od;
```

Mathematica:

```
nD=20; f[x_]:=x*Exp[-x]; {f[x], n=20, a=0, b=2*Pi,
 X=N[Table[2*Pi*i/(2*n+1),{i,0,2*n}],nD], Y={}}
Do[Y=Append[Y,f[X[[i]]]],{i,1,2*n+1}]; Y=N[Y,nD]
model=A[0]/2+Sum[A[k]*Cos[k*t]+B[k]*Sin[k*t],{k,1,n}]
{Apar=Table[A[k],{k,1,n}],Bpar=Table[B[k],{k,1,n}]}
data=N[Table[{X[[i]],Y[[i]]},{i,1,2*n+1}],nD]
FTrig=FindFit[data,model,Flatten[{A[0],Apar,Bpar}],t,
 WorkingPrecision->nD-1,MaxIterations->1000]
Ftrig10=model/.FTrig
FTrig1[x_]:=Ftrig10/.{t->x};
FTrig2=Table[{X[[i]],FTrig1[X[[i]]]},{i,1,2*n+1}]
XY=Table[{X[[i]],Y[[i]]},{i,1,2*n+1}]
g1=ListPlot[XY,PlotStyle->{PointSize[.02],Hue[0.8]},
 PlotRange->All]; g2=ListPlot[FTrig2,
 PlotStyle->{PointSize[.02],Hue[0.99]},PlotRange->All];
 g3=Plot[f[x],{x,a,b},PlotStyle->Hue[0.9]]; Show[{g1,g3}]
Show[{g2,g3}]
Do[Print[i," ",X[[i]]," ",Y[[i]]," ",
 N[Abs[Y[[i]]-FTrig1[X[[i]]]],nD]],{i,1,2*n+1}];
```

$\qquad\qquad\qquad\qquad\qquad\qquad\qquad\qquad\qquad\qquad\qquad\qquad$ □

Problem 12.20 *Discrete approximate trigonometric polynomials.* Let $f : [a, b] \to \mathbb{R}$ be a periodic function and let $\{t_i, f(t_i)\}$ be $2n + 1$ given data pairs $(i = 0, \ldots, 2n)$, where $t_i = 2\pi i/(2n + 1)$ and $2n$ is an even number. The approximate trigonometric polynomials can be represented as a discrete Fourier transform. The *fast Fourier transform* or FFT, which was described by Cooley and Tukey (in 1965) and which was known to Gauss, is a very efficient algorithm for computing the Fourier coefficients of a trigonometric interpolant satisfying the interpolation conditions $F(t_i) = f(t_i)$, $i = 0, \ldots, 2n$ and admitting the equally *fast inverse Fourier transform.*

1) Applying the FFT transform, construct and plot the discrete trigonometric interpolating polynomial of degree n.

2) Compute the fast inverse Fourier transform.

Maple:

```
with(DiscreteTransforms): with(plots): Digits:=14:
a:=0; b:=2*Pi; f:=x->x*exp(-x); f(x);
p:=6; N:=2^p; m:=N/2;
X:=evalf([seq(a+(b-a)*i/m,i=1..N)]); Y:=NULL:
for i from 1 to N do Y:=Y,f(X[i]) od: Y:=evalf([Y]);
X1:=Vector(N,X): Y1:=Vector(N,Y):
FourierTransform(X1,Y1);
XY:=[seq([X[i],Y[i]],i=1..N)]:
XY1:=[seq([X1[i],Y1[i]],i=1..N)]:
G1:=plot(XY,style=point,color=blue):
G2:=plot(XY1,color=magenta): display({G1,G2});
InverseFourierTransform(X1,Y1);
XYI:=[seq([X1[i],Y1[i]],i=1..N)]:
G3:=plot(XYI,color=green): display({G1,G3});
```

Mathematica:

```
f[x_]:=x*Exp[-x]; {nD=14, a=0, b=2*Pi,f[x],
 p=6, n=2^p, m=n/2, L=b-a, k=40,
 X=N[Table[a+(b-a)*i/m,{i,0,n-1}],nD]}
Y={}; Do[Y=Append[Y,f[X[[i]]]],{i,1,n}]; Y=N[Y,nD]
{f1=Fourier[Y], sol=2/Sqrt[n]*Fourier[Y],
 A=Re[sol], B=Im[sol]}
F[t_]:=A[[1]]/2+Sum[A[[i+1]]*Cos[t*i*Pi/L]
 +B[[i+1]]*Sin[t*i*Pi/L],{i,1,k}];  F[x]
XY:=Table[{X[[i]],Y[[i]]},{i,1,n}];
g1=ListPlot[XY,PlotStyle->{PointSize[.02],Hue[0.8]}];
g2=Plot[Evaluate[F[x]],{x,X[[1]],X[[m]]},
 PlotStyle->Hue[0.9]]; Show[{g1,g2}]
YI=N[InverseFourier[f1],nD];
XYI=Table[{X[[i]],YI[[i]]},{i,1,n}];
g3=ListPlot[XYI,Joined->True,PlotStyle->
 {PointSize[.02],Hue[0.3]}]; Show[{g1,g3}]
```

For more details on the discrete Fourier transforms, see Chapter 10. □

The least squares method, curve fitting, linear and nonlinear regression analysis

Maple:

```
with(Statistics);                      Fit(f,X,Y,var,ops);
ExponentialFit(X,Y,var,ops);           LinearFit([f],X,Y,var,ops);
PowerFit(X,Y,var,ops);                 LogarithmicFit(X,Y,var,ops);
NonlinearFit(f,X,Y,var,ops);           PolynomialFit(d,X,Y,var,ops);
with(CurveFitting);                    LeastSquares(XY,var,ops);
```

Fit, fitting a model function to data,

ExponentialFit, LogarithmicFit, fitting exponential and logarithmic functions to data,

LinearFit, NonlinearFit, fitting linear and nonlinear model functions to data,
    PolynomialFit, PowerFit, fitting polynomial and power functions to data,

LeastSquares (of the CurveFitting package), computing a least squares approximation.

Mathematica:

```
Fit[data,funs,vars]          FindFit[data,{f,cons},pars,vars,ops]
FindFit[data,f,pars,vars,ops]            LeastSquares[m,b]
```

Fit, finding a least squares fit to a list of data as a linear combination of the functions funs of variables vars,

FindFit, finding a best fit to data for a function f of vars with parameters pars (displaying numerical values of pars),

LeastSquares, finding an approximate solution of the linear least squares problem for the matrix equation $MX = B$.

Problem 12.21 *The least squares approximation.* Let $\{x_i, y_i\}$ be $n + 1$ given data pairs $(i = 0, \ldots, n)$. Let us apply the *least squares approximation method*[2] and construct an approximate function or an *approximant* $F(x)$ such that $\sum_{i=0}^{n} [y_i - F(x_i)]^2$ $(i = 0, \ldots, n)$ becomes as small as possible.

---

[2] Gauss was the first to developed the least squares method in 1795.

1) Solve the least squares problem symbolically, obtaining the *normal equations*, i.e. the linear system of equations for determining the coefficients of the polynomial $F(x)$ of degree d. Perform these computations for $1 \leq d < n$.

2) Construct the polynomials $F(x)$ of degree d $(1 \leq d < n)$ that minimize the mean square error for the given data.

3) Plot the approximate polynomials and the data.

4) Evaluate the mean square error for $1 \leq d < n$ and compare the results.

Maple:

```
with(plots): F:=(x,D)->add(a||i*x^i,i=0..D);
LeastSquaresP:=proc(d,n) local x, p1, S, k, Sol;
 p1:=F(X[i],d); S:=sum((Y[i]-p1)^2,i=0..n);
 for k from 1 to d+1 do Eq||k:=diff(S,coeff(F(x,d),x,k-1)); od;
 Sol:=solve({seq(Eq||j,j=1..d+1)},{seq(a||(j-1),j=1..d+1)});
 RETURN(Sol); end;
nData:=5; X:=array(0..nData,[1,2,3,4,5,6]);
Y:=array(0..nData,[25,26,26,25,24,22]);
L:=[[X[i],Y[i]]$i=0..nData]; G0:=plot(L,style=point,color=blue):
LeastSquaresD:=proc(deg,N,X,Y) SolSymb:=LeastSquaresP(deg,n);
 print(SolSymb); Sol||deg:=LeastSquaresP(deg,N);
 F||deg:=unapply(subs(Sol||deg,F(x,deg)),x);
 G||deg:=plot(F||deg(x),x=X[0]..X[N],color=magenta):
 print(display({G0,G||deg}));
 Er||deg:=evalf(add((Y[i]-(F||deg(X[i])))^2,i=0..N));
 RETURN(Er||deg); end;
for k from 1 to nData-2 do
 print(F(x,k)); LeastSquaresD(k,nData,X,Y); od;
```

Mathematica:

```
$HistoryLength=0; F[x_,d_]:=Sum[a[i]*x^i,{i,0,d}];
{X={1,2,3,4,5,6},Y={25,26,26,25,24,22},nData=Length[X]-1}
L=Table[{X[[i]],Y[[i]]},{i,1,nData}]
g0=ListPlot[L,PlotStyle->{PointSize[.02],Hue[0.8]}];
LeastSquaresP[d_,n_]:=Module[{x,p1,S,k,Sol},p1=F[X1[i1],d];
 S=Sum[(Y1[i1]-p1)^2,{i1,0,n}];
 Do[Eq[k]=D[S,Coefficient[F[x,d],x,k-1]],{k,1,d+1}];
 Sol=Solve[Expand[Table[Eq[j]==0,{j,1,d+1}]],
  Table[a[j-1],{j,1,d+1}],VerifySolutions->False,Sort->False]];
```

```
subs=Flatten[{Table[X1[l]->X[[l+1]],{l,0,nData}],
    Table[Y1[l]->Y[[l+1]],{l,0,nData}]}];
Do[SolSymb=LeastSquaresP[deg,n]; Print[SolSymb];
 SolN=LeastSquaresP[deg,nData];
 SolN1=Table[SolN[[1,j]]/.subs,{j,1,deg+1}]; Print[SolN1];
 F1[x1_]:=(F[x,deg]/.SolN1)/.{x->x1};
 g=Plot[F1[x],{x,X[[1]],X[[nData+1]]},PlotStyle->Hue[0.9]];
 Print[Show[{g0,g}]];
 Er=N[Sum[(Y[[i]]-F1[X[[i]]])^2,{i,1,nData+1}]]; Print[Er],
 {deg,1,nData-2}];
```

$\square$

Problem 12.22 *The least squares approximation.* Let $\{x_i, y_i\}$ be $n+1$ given data pairs $(i = 0, \ldots, n)$.

1) Construct the least squares approximate polynomial $F(x)$ of degree $d = n$ that minimizes the mean square error for the given data.

2) Verify that if $d = n$ the least square polynomial $F(x)$ at the nodes x_i $(i = 0, \ldots, n)$ coincides with the Lagrange interpolating polynomial at the same nodes and the mean square error in this case is equal to 0.

Maple:

```
with(plots):with(CurveFitting):
F:=(x,D)->add(a||i*x^i,i=0..D);
LeastSquaresP:=proc(d,n) local x, p1, S, k, Sol;
 p1:=F(X[i],d); S:=sum((Y[i]-p1)^2,i=0..n);
 for k from 1 to d+1 do Eq||k:=diff(S,coeff(F(x,d),x,k-1)); od;
 Sol:=solve({seq(Eq||j,j=1..d+1)},{seq(a||(j-1),j=1..d+1)});
 RETURN(Sol); end;
nData:=5; X:=array(0..nData,[1,2,3,4,5,6]);
Y:=array(0..nData,[25,26,26,25,24,22]);
L:=[[X[i],Y[i]]$i=0..nData];
G0:=plot(L,style=point,color=blue):
deg:=nData; F(x,deg); Sol||deg:=LeastSquaresP(deg,nData);
F||deg:=unapply(subs(Sol||deg,F(x,deg)),x);
G||deg:=plot(F||deg(x),x=X[0]..X[nData],color=magenta):
pLagrange:=expand(PolynomialInterpolation(L,x,form=Lagrange));
evalb(F||deg(x)=pLagrange);
display({G0,G||deg});
Er||deg:=evalf(add((Y[i]-(F||deg(X[i])))^2,i=0..nData));
```

Mathematica:

```
F[x_,d_]:=Sum[a[i]*x^i,{i,0,d}]; {X={1,2,3,4,5,6},
 Y={25,26,26,25,24,22}, nData=Length[X]-1,
 F[x,nData], L=Table[{X[[i]],Y[[i]]},{i,1,nData}]}
g0=ListPlot[L,PlotStyle->{PointSize[.02],Hue[0.8]}];
LeastSquaresP[d_,n_]:=Module[{x,S,k,Sol},
 S=Sum[(Y[[i+1]]-F[X[[i+1]],d])^2,{i,0,n}];
 Do[Eq[k]=D[S,Coefficient[F[x,d],x,k-1]],{k,1,d+1}];
 Sol=Solve[Expand[Table[Eq[j]==0,{j,1,d+1}]],
 Table[a[j-1],{j,1,d+1}],VerifySolutions->False,Sort->False]];
{deg=nData,F[x,deg],SolN=LeastSquaresP[deg,nData]}
SolN1=Table[SolN[[1,j]],{j,1,deg+1}]
F1[x1_]:=(F[x,deg]/.SolN1)/.{x->x1}; g=Plot[F1[x],
 {x,X[[1]],X[[nData+1]]},PlotStyle->Hue[0.9]]; Show[{g0,g}]
Er=N[Sum[(Y[[i]]-F1[X[[i]]])^2,{i,1,nData+1}]]
L1[x_,i_,k_]:=Module[{j,P},P=1; Do[If[j!=i,
 P=P*(x-X[[j+1]])/(X[[i+1]]-X[[j+1]])],{j,0,k}]; P];
{pLagrange=Sum[L1[x,i,nData]*Y[[i+1]],{i,0,nData}]//Expand,
 F1[x], pLagrange===F1[x]}
```

□

Problem 12.23 *The least squares approximation.* Let $\{x_i, y_i\}$ be $n+1$ given data pairs $(i = 0, \ldots, n)$.

1) Perform the least squares fit using various model functions: a linear function, $a+bx$; an exponential function, $y=a\exp(bx)$; a polynomial function, $ax^2 + bx + c$; a power function, ax^b; a logarithmic function, $a + b\ln x$; a nonlinear function, $ax^2 + c$; etc.

2) Plot the approximate polynomials and the data.

Maple:

```
with(CurveFitting); with(Statistics): with(plots):
X:=[1,2,3,4,5,6]; Y:=[25,26,26,25,24,22]; n:=nops(X);
X1:=Vector(X,datatype=float);Y1:=Vector(Y,datatype=float);
p1:=LeastSquares(X,Y,x);
p2:=LeastSquares(X,Y,x,curve=a*x^2+b*x+c);
p3:=Fit(a*x^2+b*x+c,X1,Y1,x);p4:=Fit(a*exp(b*x)+c,X1,Y1,x);
p5:=ExponentialFit(X1,Y1,x); p6:=PowerFit(X1,Y1,x);
p7:=PolynomialFit(4,X1,Y1,x);p8:=LogarithmicFit(X1,Y1,x);
p9:=NonlinearFit(a*x^2+c,X1,Y1,x);
for i from 1 to 9 do G||i:=plot(p||i,x=X[1]..X[n],
 color=COLOR(RGB,rand()/10^12,rand()/10^12,rand()/10^12)): od:
display({seq(G||i,i=1..9)});
```

Mathematica:

```
{X={1,2,3,4,5,6},Y={25,26,26,25,24,22},n=Length[X],nD=20}
data=Table[{X[[i]],Y[[i]]},{i,1,n}]
{p[1]=Fit[data,{1,x},x], Rationalize[p1],
 p[2]=Fit[data,{x^2,x,1},x], Rationalize[p1]}
{mod3=a*x^2+b*x+c, mod4=a*Exp[b*x]+c, mod6=a*x^b,
 mod7=a*x^4+b*x^3+c*x^2+d*x+e, mod8=a+b*Log[x]}
{p3=FindFit[data,mod3,{a,b,c},x],p[3]=mod3/.p3,
 p4=FindFit[data,mod4,{a,b,c},x],p[4]=mod4/.p4,
 p5=FindFit[data,mod4/.c->0,{a,b},x],p[5]=(mod4/.c->0)/.p5,
 p6=FindFit[N[data,nD],mod6,{a,b},x],p[6]=mod6/.p6,
 Fit[data,{x^2},x],
 p7=FindFit[data,mod7,{a,b,c,d,e},x],p[7]=mod7/.p7,
 p8=FindFit[data,mod8,{a,b},x],p[8]=mod8/.p8,
 p9=FindFit[data,mod3/.b->0,{a,c},x],p[9]=(mod3/.b->0)/.p9}
Do[g[i]=Plot[p[i],{x,X[[1]],X[[n]]},PlotStyle->Hue[0.15*i]],
 {i,1,9}]; Show[Table[g[i],{i,1,9}]]
```

$\square$

Problem 12.24 *The least squares approximation.* Let $\{x_i, y_i\}$ be $n+1$ given data pairs $(i = 0, \ldots, n)$.

1) Perform the least squares fit using an exponential function of the form $y = a \exp(bx)$, using the transformed model function $\ln(y) = a_0 + bx$.

2) Obtain the symbolic and numerical solutions of the problem.

3) Plot the approximate polynomial and the data.

4) Compare the results with those obtained using the corresponding predefined functions (ExponentialFit and FindFit).

Maple:

```
with(plots): with(Statistics): Digits:=20; F:=x->a0+a1*x;
FExp:=x->exp(a0)*exp(a1*x);
LeastSquaresP:=proc(n) local x, p1, S,k, Sol,Eq1,Eq2;
 p1:=F(X[i],d); S:=sum((log(Y[i])-p1)^2,i=0..n);
 Eq1:=diff(S,a0); Eq2:=diff(S,a1);
 Sol:=solve({Eq1,Eq2},{a0,a1}); RETURN(Sol); end;
nData:=5; X:=array(0..nData,[1,2,3,4,5,6]);
Y:=array(0..nData,[25,26,26,25,24,22]);
X1:=Vector(X,datatype=float); Y1:=Vector(Y,datatype=float);
L:=[[X[i],Y[i]]$i=0..nData]; G0:=plot(L,style=point,color=blue):
F(x); LeastSquaresP(n); SolExp:=LeastSquaresP(nData);
FExp1:=unapply(subs(SolExp,FExp(x)),x);
```

```
GExp:=plot(FExp1(x),x=X[0]..X[nData],color=magenta):
pExp:=ExponentialFit(X1,Y1,x); display({G0,GExp});
ErExp:=evalf(add((Y[i]-(FExp1(X[i])))^2,i=0..nData));
pExp=evalf(FExp1(x));
```

Mathematica:

```
nD=20; F[x_]:=a0+a1*x; FExp[x_]:=N[Exp[a0],nD]*Exp[N[a1*x,nD]];
{X={1,2,3,4,5,6},Y={25,26,26,25,24,22},nData=Length[X]-1}
L=Table[{X[[i]],Y[[i]]},{i,1,nData+1}]
g0=ListPlot[L,PlotStyle->{PointSize[.02],Hue[0.8]}];
LeastSquaresP[n_]:=Module[{x,p1,S,k,Sol}, p1=F[X1[i1]];
 S=Sum[(Log[Y1[i1]]-p1)^2,{i1,0,n}]; Print[S];
 Eq1=D[S,a0]; Eq2=D[S,a1]; Print[Eq1]; Print[Eq2];
 Sol=Solve[{Eq1==0,Eq2==0},{a0,a1},VerifySolutions->False]];
{F[x],LeastSquaresP[n]}
subs=Flatten[{Table[X1[l]->X[[l+1]],{l,0,nData}],Table[
 Y1[l]->Y[[l+1]],{l,0,nData}]}]; SolExp=LeastSquaresP[nData]
SolExp1=N[Table[SolExp[[1,j]]/.subs,{j,1,2}],nD]
FExp1[x1_]:=(FExp[x]/.SolExp1)/.{x->x1}; FExp1[x]
gExp=Plot[Evaluate[FExp1[x]],{x,X[[1]],X[[nData+1]]},
 PlotStyle->Hue[0.9]];  {modelExp=A*Exp[B*x],
 pExp=FindFit[L,modelExp,{A,B},x,WorkingPrecision->nD],
 pExp1=modelExp/.pExp}
Show[{g0,gExp}]
Er=Sum[(Y[[i]]-FExp1[X[[i]]])^2,{i,1,nData+1}]
```

□

12.3 Numerical Differentiation and Integration

Let us consider methods for numerical approximations of derivatives and integrals of arbitrary functions, since there exist some functions for which either the antiderivative of the integrand cannot be expressed in an explicit form or the functions are defined at a set of discrete points (e.g., from experimental data). Moreover, the quadrature formulas are very important for finding numerical solutions of differential and integral equations. This problem belongs to one of the oldest problems in science, some of these type of evaluations were known in ancient Babylon, Egypt, and Greece. But the fundamental systematic development and analysis were performed by Newton in 1704 (the Newton–Cotes equidistant interpolatory quadratures), Euler in 1738 and Maclaurin in 1737 (the Euler–Maclaurin expansion for numerical integration of periodic functions), Gauss in 1814 (Gaussian nonequidistant interpolatory quadratures), Romberg in 1955 (the Romberg integration using the extrapolation method for increasing the degree of accuracy).

Numerical approximations of derivatives of arbitrary functions

Maple:

```
fdiff(f,x,x=a,ops);        fdiff(f(vars),[vars],[varI=valI],ops);
evalf(D[1,2](f)(a,b));      fdiff(f,[1,2],[a,b])=D[1,2](f)(a,b);
diff(f(x),x); D[1](f)(x);        evalf(eval(diff(f(x),x),x=a));
evalf(eval(diff(f(vars),vars),{varI=valI}));
                 evalf(eval(fracdiff(f,var,nu,ops),var=val));
```

`fdiff`, computing numerical approximations of derivatives of functions,

`fracdiff`, computing the ν-th fractional derivative ($\nu \notin \mathbb{Z}$) of a function with respect to a variable `var`.

Mathematica:

```
<<NumericalCalculus`    ND[f[x],x,a,ops]    ND[f[x],{x,n},a,ops]
ND[f[x],x,a,Method->EulerSum]                D[f[x],x]/.x->val
ND[f,x,a,Method->NIntegrate]        D[f[x,y],x,y]/.{x->a,y->b}
ND[f[x],x,a,Terms->val,WorkingPrecison->val]
```

ND (of the `NumericalCalculus` package), computing numerical approximations to derivatives of functions with respect to a variable at the point a,

ND with the `Method` option, applying Richardson's extrapolation (for non analytic functions in the neighborhood of a), `EulerSum`, or applying Cauchy's integral formula for analytic functions, `NIntegrate`,

ND with the `Scale->val`, computing directional derivatives (the left and right derivatives of nonanalytic functions),

ND with the `Terms->val` and `WorkingPrecison->val` options, increasing the number of terms and a working precision to reduce an error.

Problem 12.25 *Approximation of derivatives by finite differences.* Let $F : [a, b] \to \mathbb{R}$ be an arbitrary function sufficiently many times continuously differentiable.

1) Derive symbolic formulas for higher order approximations to the first derivative (for example, the *forward* and *backward difference approximations* with second order error, and the *centered difference approximation* with fourth order error) using the corresponding interpolation polynomials:

$$F'(x) = \frac{-3F(x) + 4F(x+h) - F(x+2h)}{2h} + O(h^2),$$

$$F'(x) = \frac{3F(x) - 4F(x-h) + F(x-2h)}{2h} + O(h^2),$$

$$F'(x) = \frac{-F(x+2h) + 8F(x+h) - 8F(x-h) + F(x-2h)}{12h} + O(h^4).$$

2) Determine the discretization errors in the approximate formulas using the corresponding series expansions.

Maple:

```
with(CurveFitting): XF:=[seq(x+i*h,i=0..2)]; n:=nops(XF);
YF:=[seq(F(XF[i]),i=1..n)]; pF:=PolynomialInterpolation(XF,YF,t);
DF:=simplify(subs(t=x,diff(pF,t))); ErF:=DF-D(F)(x);
ErF:=convert(series(ErF,h,4),polynom); XB:=[seq(x-i*h,i=0..2)];
n:=nops(XB); YB:=[seq(F(XB[i]),i=1..n)];
pB:=PolynomialInterpolation(XB,YB,t); DB:=factor(subs(t=x,
 diff(pB,t))); ErB:=DB-D(F)(x); ErB:=convert(series(ErB,h,4),
 polynom); XC:=[seq(x+i*h,i=-2..2)]; n:=nops(XC);
YC:=[seq(F(XC[i]),i=1..n)]; pC:=PolynomialInterpolation(XC,YC,t);
DC:=factor(subs(t=x,diff(pC,t)));ErC:=DC-D(F)(x);
ErC:=convert(series(ErC,h,6),polynom);
```

Mathematica:

```
{XF=Table[x+i*h,{i,0,2}],n=Length[XF], YF=Table[F[XF[[i]]],
 {i,1,n}],XYF=Table[{XF[[i]],YF[[i]]},{i,1,n}],
 pF=Collect[InterpolatingPolynomial[XYF,t],h],
 DF=Simplify[D[pF,t]/.t->x], ErF=DF-D[F[x],x],
 ErF=Normal[Series[ErF,{h,0,2}]]}
{XB=Table[x-i*h,{i,0,2}],n=Length[XB], YB=Table[F[XB[[i]]],
 {i,1,n}],XYB=Table[{XB[[i]],YB[[i]]},{i,1,n}],
 pB=Collect[InterpolatingPolynomial[XYB,t],h],
 DB=Factor[D[pB,t]/.{t->x}], ErB=DB-D[F[x],x],
 ErB=Normal[Series[ErB,{h,0,2}]]}
```

```
{XC=Table[x+i*h,{i,-2,2}],n=Length[XC], YC=Table[F[XC[[i]]],
 {i,1,n}], XYC=Table[{XC[[i]],YC[[i]]},{i,1,n}],
 pC=Collect[InterpolatingPolynomial[XYC,t],h],
 DC=Factor[D[pC,t]/.{t->x}], ErC=DC-D[F[x],x],
 ErC=Normal[Series[ErC,{h,0,4}]]}
```

□

Problem 12.26 *Approximation of derivatives by finite differences.* Let
$F : [a, b] \to \mathbb{R}$ be an arbitrary function sufficiently many times continuously differentiable.

1) Derive symbolic formulas for higher order approximations of higher
order derivatives, for example, the second derivative (centered difference
approximations), using the corresponding interpolation polynomials:

$$F''(x) = \frac{F(x+h) - 2F(x) + F(x-h)}{h^2} + O(h^2),$$

$$F''(x) = \frac{-F(x+2h)+16F(x+h)-30F(x)+16F(x-h)-F(x-2h)}{12h^2} + O(h^4).$$

2) Determine the discretization errors in the approximate formulas
using the corresponding series expansions.

Maple:

```
with(CurveFitting): XC2:=[seq(x+i*h,i=-1..1)]; n:=nops(XC2);
YC2:=[seq(F(XC2[i]),i=1..n)];
pC2:=PolynomialInterpolation(XC2,YC2,t);
DC2:=simplify(diff(pC2,t$2));
ErC2:=DC2-D(D(F))(x); ErC2:=convert(series(ErC2,h,6),polynom);
XC4:=[seq(x+i*h,i=-2..2)]; n:=nops(XC4); YC4:=[seq(F(XC4[i]),
 i=1..n)]; pC4:=PolynomialInterpolation(XC4,YC4,t);
DC4:=factor(subs(t=x,(diff(pC4,t$2)))); ErC4:=DC4-D(D(F))(x);
ErC4:=convert(series(ErC4,h,8),polynom);
```

Mathematica:

```
{XC2=Table[x+i*h,{i,-1,1}], n=Length[XC2],
 YC2=Table[F[XC2[[i]]],{i,1,n}],
 XYC2=Table[{XC2[[i]],YC2[[i]]},{i,1,n}],
 pC2=Collect[InterpolatingPolynomial[XYC2,t],h],
 DC2=Simplify[D[pC2,{t,2}]], ErC2=DC2-D[F[x],{x,2}],
 ErC2=Normal[Series[ErC2,{h,0,2}]]}
```

```
{XC4=Table[x+i*h,{i,-2,2}], n=Length[XC4],
 YC4=Table[F[XC4[[i]]],{i,1,n}],
 XYC4=Table[{XC4[[i]],YC4[[i]]},{i,1,n}],
 pC4=Collect[InterpolatingPolynomial[XYC4,t],h],
 DC4=Factor[D[pC4,{t,2}]/.{t->x}], ErC4=DC4-D[F[x],{x,2}],
 ErC4=Normal[Series[ErC4,{h,0,4}]]}
```

☐

Problem 12.27 *Richardson extrapolation.* Let $F : [a, b] \to \mathbb{R}$ be twice continuously differentiable function. Let us perform the numerical differentiation applying the Richardson extrapolation method proposed by Richardson in 1927.

1) Compute $F'(2)$ for $h_0 = 0.5$ and the given tolerance 10^{-25} applying the Richardson extrapolation

2) Determine the symbolic expression of $F'(x)$ and its numerical approximation at $x = 2$. Compare the results obtained.

Maple:

```
Digits:=30; F:=x->x*exp(-x); x:=2.; h:=0.5; Nmax:=100;
epsilon:=10^(-25); R[0,0]:=(F(x+h)-F(x-h))/(2*h);
for i from 1 to Nmax do h:=h/2; R[i,0]:=(F(x+h)-F(x-h))/(2*h);
 for j from 1 to i do
  R[i,j]:=(4^j*R[i,j-1]-R[i-1,j-1])/(4^j-1);
  lprint(N(i,j)=R[i,j]); od:
 if abs(R[i,i]-R[i-1,i-1])<epsilon then break;
 elif i=Nmax then error "Richardson extrapolation stopped
 without converging for the given tolerance"; fi;
od: evalf(subs(t=2.,diff(F(t),t)));
```

Mathematica:

```
nD=30; F[x_]:=N[x*Exp[-x],nD];
{x=N[2,nD], h=N[1/2,nD], nMax=100, epsilon=N[10^(-25),nD]}
R=Table[0,{i,1,nMax},{j,1,nMax}];
R[[1,1]]=(F[x+h]-F[x-h])/(2*h)
Do[h=h/2; R[[i+1,1]]=(F[x+h]-F[x-h])/(2*h);
 Do[R[[i+1,j+1]]=(4^j*R[[i+1,j]]-R[[i,j]])/(4^j-1);
  Print["N","[",i,",",j,"]=",R[[i+1,j+1]]],{j,1,i}];
 If[Abs[R[[i+1,i+1]]-R[[i,i]]]<epsilon,Break[]];
 If[i==nMax,Print["Richardson extrapolation stopped
 without converging for the given tolerance"]],{i,1,nMax}];
N[D[F[t],t]/.{t->2},nD]
```

☐

Numerical approximations of integrals of arbitrary functions

Maple:

```
with(Student[Calculus1]);        RiemannSum(Int(f(x),x=a..b,ops);
RiemannSum(f(x),x=a..b,method=m);  ApproximateInt(f(x),x=a..b);
ApproximateInt(f(x),x=a..b,method=m,ops); I1:=Int(f(x),x=a..b);
ApproximateInt(I1,ops);          evalf(int(f(x),x=a..b,ops));
evalf(value(I1));         with(Student[MultivariateCalculus]):
                            ApproximateInt(f(x,y),x=a..b,y=c..d);
```

RiemannSum with the option **method**, computing the Riemann sum of $f(x)$ on $[a, b]$ using the given method (**left, right, lower, upper, midpoint** Riemann sum),

int, **Int**, **value**, **evalf**, performing closed form integration and obtaining a numerical result applying **evalf**,

ApproximateInt, approximating definite integrals using the given **method**, the Riemann sums (**lower, upper, left, right, midpoint**) and the equidistant interpolatory quadratures (**trapezoid, simpson, simpson[3/8], boole**, and **newtoncotes[N]** of order N). The results can be obtained as numerical values, plots or animations, and approximate sums,

ApproximateInt of the **MultivariateCalculus** subpackage, approximating integrals of bivariate functions $f(x, y)$.

Mathematica:

```
Integrate[f[x],{x,a,b}]//N          NIntegrate[f[x],{x,a,b}]
NIntegrate[f[x],{x,a,b},ops] NIntegrate[f[x],{x,a,b},Method->m]
          NIntegrate[f[x,y,...],{x,a,b},{y,c,d},...,ops]
```

Integrate, **N**, performing closed form integration and obtaining a numerical result applying **N**,

NIntegrate, numerical approximating definite integrals applying various methods of integration.

Problem 12.28 *Riemann sums.* Let $f : [a, b] \to \mathbb{R}$ be an arbitrary continuous function.

1) Illustrate (graphically, numerically, and symbolically) the definition of Riemann Sums for the function $f(x)$ in $[a, b]$ using various methods.

2) Show numerically that a Riemann sum on a given partition is bounded by the lower and the upper Riemann sums. Verify whether $f(x)$ is a *Riemann integrable function*, i.e. if the lower and upper Riemann sums converge to a common value as the partition is refined.

Maple:

```
with(Student[Calculus1]): f:=x->x*exp(-x); a:=0; b:=2*Pi;
RiemannSum(f(x),x=a..b,method=lower);
RiemannSum(f(x),x=a..b,method=upper,output=plot,partition=100);
A1:=RiemannSum(f(x),x=a..b,method=left,output=sum);
expand(value(A1));
RiemannSum(f(x),x=a..b,method=right,output=animation);
for i from 300 to 1000 by 100 do
 i,RiemannSum(f(x),x=a..b,method=lower,partition=i),
   RiemannSum(f(x),x=a..b,method=upper,partition=i); od;
```

Mathematica:

```
f[x_]:=x*Exp[-x]; {a=0,b=2*Pi,nD=11}
RiemannSumLeft[n_,Symb_,Num_]:=Module[{dx,X},dx=(b-a)/n;
 X=Table[a+(i-1)*dx,{i,1,n}]; LSum=Sum[f[X[[i]]]*dx,{i,1,n}];
 If[Symb==1,Print["The Left Riemann Sum\n",LSum]];
 If[Num==1,N[LSum,nD]]];
RiemannSumRight[n_,Symb_,Num_]:=Module[{dx,X,Y,g,g1,L1},
 dx=(b-a)/n; X=Table[a+i*dx,{i,1,n}];
 RSum=Sum[f[X[[i]]]*dx,{i,1,n}]; Y=Table[f[X[[i]]],{i,1,n}];
 g=Plot[f[x],{x,a,b},PlotStyle->Red]; L1={};
Do[L1=Append[L1,g1=Graphics[{Blue,{Line[{{a,Y[[1]]}},
 {X[[1]],Y[[1]]}}],Line[{{X[[i]],Y[[i+1]]},
 {X[[i+1]],Y[[i+1]]}}],Line[{{X[[i]],0},{X[[i]],Y[[i+1]]}}],
 Line[{{X[[i+1]],0},{X[[i+1]],Y[[i+1]]}}]}]],{i,1,n-1}];
 Print[Show[{g,L1}]]; If[Symb==1,Print[
 "The Right Riemann Sum\n",RSum]]; If[Num==1,N[RSum, nD]]];
RiemannSumLower[n_]:=Module[{dx,X,X1,Ymin},dx=(b-a)/n;
 X=N[Table[a+i*dx,{i,0,n}],nD]; XFLw[k_]:=a+k*dx;
 LwSum=0; Do[X1=Select[X,XFLw[i-1]<=#1<=XFLw[i]&];
 Ymin=Min[f[X1]]; LwSum=LwSum+Ymin*dx,{i,1,n}];LwSum];
RiemannSumUpper[n_]:=Module[{dx,X,X1,Ymax},dx=N[(b-a)/n,nD];
 X=N[Table[a+i*dx,{i,0,n}],nD]; XFU[k_]:=a+k*dx; USum=0;
 Do[X1=Select[X,XFU[i-1]<=#1<=XFU[i]&];
 Ymax=Max[f[X1]]; USum=USum+Ymax*dx,{i,1,n}];USum];
```

```
{RiemannSumLeft[10,1,1],RiemannSumRight[10,1,1],
 RiemannSumLower[10],RiemannSumUpper[10],RiemannSumUpper[100]}
Do[Print[i," ",RiemannSumLower[i]," ",RiemannSumUpper[i]],
   {i,300,1000,100}];
```

□

Problem 12.29 *Equidistant interpolatory quadratures.* Let $f : [a, b] \to \mathbb{R}$ be an arbitrary function.

1) Find (graphically, numerically, and symbolically) approximations of the definite integral $\int_a^b f(x)\,dx$ using various equidistant interpolatory quadrature rules and the predefined functions in both systems.

2) Approximate the multiple integral $\int_a^b \int_c^d f(x, y)\,dy\,dx$.

Maple:

```
with(Student[Calculus1]): f:=x->x*exp(-x); a:=0; b:=2*Pi;
I1:=Int(f(x),x=a..b); evalf(ApproximateInt(I1));
ApproximateInt(I1,output=plot);
ApproximateInt(I1,output=animation,partition=100);
ApproximateInt(I1,method=trapezoid);
ApproximateInt(I1,method=trapezoid,output=plot,partition=100);
ApproximateInt(I1,method=trapezoid,output=animation);
ApproximateInt(I1,method=simpson);
ApproximateInt(I1,method=simpson,output=plot,partition=100);
ApproximateInt(I1,method=simpson,output=animation);
ApproximateInt(I1,method=simpson[3/8]);
ApproximateInt(I1,method=simpson[3/8],output=plot,partition=100);
ApproximateInt(I1,method=simpson[3/8],output=animation);
I2:=f(x),x=a..b; for k from 1 to 6 do
  ApproximateInt(I2,method=newtoncotes[k]); od;
f:=(x,y)->x*exp(-y); A:=x=0..1,y=0..1;
with(Student[MultivariateCalculus]):
  ApproximateInt(f(x,y),A,method=midpoint);
```

Mathematica:

```
f[x_]:=x*Exp[-x]; f1[x_,y_]:=x*Exp[-y];
{a=0, b=2*Pi, nD=10, Integrate[f[x],{x,a,b}]}
N[Integrate[f[x],{x,a,b}],nD]
NIntegrate[f[x],{x,a,b},WorkingPrecision->nD]
```

```
NIntegrate[f[x],{x,a,b},Method->"TrapezoidalRule",
 WorkingPrecision->nD]
NIntegrate[f[x],{x,a,b},MaxPoints->100,
 Method->"NewtonCotesRule",WorkingPrecision->nD]
XM[k_]:=a+(k-1/2)*h; XS[k_]:=a+k*h;
{n=100, h=(b-a)/n,MidPoint=N[h*Sum[f[XM[i]],{i,1,n}],nD]}
{m=50,h=(b-a)/(2*m), Simpson=N[h/3*(f[a]+f[b]+4*Sum[
 f[XS[2*i-1]],{i,1,m}]+2*Sum[f[XS[2*i]],{i,1,m-1}]),nD]}
{m=50,h=(b-a)/(3*m), Simpson38=N[3/8*h*Sum[f[XS[3*i-3]]
 +3*f[XS[3*i-2]]+3*f[XS[3*i-1]]+f[XS[3*i]],{i,1,m}],nD]}
NIntegrate[f1[x,y],{x,0,1},{y,0,1}]
```
                                                                    □

Problem 12.30 *The midpoint and trapezoidal quadrature formulas.* Let $f : [a, b] \to \mathbb{R}$ be an arbitrary continuous function.

1) Find the exact value of the definite integral.

2) Compute the numerical approximations of the definite integral applying the classical midpoint and trapezoidal quadrature formulas.

3) Compute numerical approximations of the definite integral applying the composite midpoint and trapezoidal quadrature formulas.

4) Determine the relative errors in the approximate values.

5) Compare the results with the results obtained applying the predefined functions in both systems.

Maple:

```
with(Student[Calculus1]): f:=x->x*exp(-x); a:=0; b:=2.*Pi;
IExt:=int(f(x),x=a..b); M:=evalf((b-a)*f(0.5*(a+b)));
EM:=evalf(abs((M-IExt)/IExt)); evalf(ApproximateInt(f(x),
 x=a..b,method=midpoint,partition=1));
nMC:=6; hMC:=(b-a)/nMC; X:=[seq(a+i*hMC,i=0..nMC)];
XMC:=array(0..nMC,X):
MC:=evalf(hMC*add(f(0.5*(XMC[i-1]+XMC[i])),i=1..nMC));
EMC:=evalf(abs((MC-IExt)/IExt)); evalf(ApproximateInt(f(x),
 x=a..b,method=midpoint,partition=nMC));
T:=evalf(0.5*(b-a)*(f(a)+f(b)));
ET:=evalf(abs((T-IExt)/IExt)); evalf(ApproximateInt(f(x),
 x=a..b,method=trapezoid,partition=1));
nTC:=10; hTC:=(b-a)/nTC; X:=[seq(a+i*hTC,i=0..nTC)];
XTC:=array(0..nTC,X): TC:=evalf(0.5*hTC*(f(a)+f(b))
 +hTC*add(f(XTC[i]),i=1..nTC-1));
ETC:=evalf(abs((TC-IExt)/IExt)); evalf(ApproximateInt(f(x),
 x=a..b,method=trapezoid,partition=nTC));
```

Mathematica:

```
f[x_]:=x*Exp[-x]; {a=0, b=2*Pi,nD=10,
IExt=Integrate[f[x],{x,a,b}], N[IExt,nD]}
{M=N[(b-a)*f[(a+b)/2],nD],EM=N[Abs[(M-IExt)/IExt],nD]}
{nMC=6, hMC=(b-a)/nMC, XMC=Table[a+i*hMC,{i,0,nMC}],
 MC=N[hMC*Sum[f[(XMC[[i]]+XMC[[i+1]])/2],{i,1,nMC}],nD],
 EMC=N[Abs[(MC-IExt)/IExt],nD]}
{T=N[(b-a)/2*(f[a]+f[b]),nD],ET=N[Abs[(T-IExt)/IExt],nD]}
{nTC=10, hTC=(b-a)/nTC, XTC=Table[a+i*hTC,{i,0,nTC}],
 TC=N[hTC/2*(f[a]+f[b])+hTC*Sum[f[XTC[[i+1]]],
 {i,1,nTC-1}],nD], ETC=N[Abs[(TC-IExt)/IExt],nD]}
NIntegrate[f[x],{x,a,b},Method->"TrapezoidalRule",
WorkingPrecision->nD]
```

$\Box$

Problem 12.31 *The Simpson quadrature formulas.* Let $f : [a, b] \to \mathbb{R}$ be an arbitrary continuous function.

1) Find the exact value of the definite integral.

2) Compute the numerical approximations of the definite integral applying the classical Simpson's 1/3 and Simpson's 3/8 quadrature formulas.

3) Compute numerical approximations of the definite integral applying the composite Simpson's 1/3 and Simpson's 3/8 quadrature formulas.

4) Determine the relative errors in the approximate values.

5) Compare the results with the results obtained applying the predefined functions in both systems.

Maple:

```
with(Student[Calculus1]): f:=x->x*exp(-x); a:=0; b:=2.*Pi;
IExt:=int(f(x),x=a..b);
S:=evalf((b-a)/6.*(f(a)+4.*f(0.5*(a+b))+f(b)));
ES:=evalf(abs((S-IExt)/IExt));
evalf(ApproximateInt(f(x),x=a..b,method=simpson,partition=1));
nSC:=10; hSC:=(b-a)/nSC; X:=[seq(a+i*hSC,i=0..nSC)];
XSC:=array(0..nSC,X):
SC:=evalf(hSC/6.*(add(f(XSC[i-1])+4.*f(0.5*(XSC[i-1]+XSC[i]))
 +f(XSC[i]),i=1..nSC))); ESC:=evalf(abs((SC-IExt)/IExt));
evalf(ApproximateInt(f(x),x=a..b,method=simpson,partition=nSC));
hS38:=(b-a)/3; X138:=a+hS38; X238:=a+2*hS38; X338:=a+3*hS38;
S38:=evalf(3/8*hS38*(f(a)+3*f(X138)+3*f(X238)+f(b)));
```

```
ES38:=evalf(abs((S38-IExt)/IExt));
ApproximateInt(f(x),x=a..b,method=simpson[3/8],partition=1);
m:=11; hSC38:=(b-a)/(3*m); SC38:=0:
X38:=[seq(a+i*hSC38,i=0..3*m)]; XSC38:=array(0..3*m,X38):
SC38:=evalf(3./8.*hSC38*add(f(XSC38[3*i-3])+3*f(XSC38[3*i-2])
 +3*f(XSC38[3*i-1])+f(XSC38[3*i])),i=1..m));
ESC38:=evalf(abs((SC38-IExt)/IExt));
ApproximateInt(f(x),x=a..b,method=simpson[3/8],partition=m);
```

Mathematica:

```
f[x_]:=x*Exp[-x]; {a=0, b=2*Pi, nD=10,
 IExt=Integrate[f[x],{x,a,b}], N[IExt,nD]}
{S=N[(b-a)/6*(f[a]+4*f[(a+b)/2]+f[b]),nD],
 ES=N[Abs[(S-IExt)/IExt],nD]}
{nSC=10, hSC=(b-a)/nSC, XSC=Table[a+i*hSC,{i,0,nSC}],
 SC=N[hSC/6*(Sum[f[XSC[[i]]]+4*f[(XSC[[i]]+XSC[[i+1]])/2]
 +f[XSC[[i+1]]],{i,1,nSC}]),nD],ESC=N[Abs[(SC-IExt)/IExt],nD]}
{hS38=(b-a)/3, X138=a+hS38, X238=a+2*hS38, X338=a+3*hS38,
 S38=N[3/8*hS38*(f[a]+3*f[X138]+3*f[X238]+f[b]),nD],
 ES38=N[Abs[(S38-IExt)/IExt],nD]}
X38[k_]:=a+k*hSC38; {m=11, hSC38=(b-a)/(3*m), SC38=N[3/8*hSC38*
 Sum[f[X38[3*i-3]]+3*f[X38[3*i-2]]+3*f[X38[3*i-1]]+f[X38[3*i]],
 {i,1,m}],nD], ESC38=N[Abs[(SC38-IExt)/IExt],nD]}
```

$\square$

Problem 12.32 *Derivation of equidistant interpolatory quadrature formulas.* Let $f:[a,b]\to\mathbb{R}$ be an arbitrary function sufficiently many times continuously differentiable.

1) Derive symbolically equidistant interpolatory quadrature formulas for approximating a definite integral.

2) Derive symbolically the discretization errors in the approximate formulas using the corresponding series expansions.

Maple:

```
L:=proc(x,i,n) local j,P; P:=1:
 for j from 0 to n do if j<> i then
  P:=P*(XA-X[j])/(X[i]-X[j]); fi; od; RETURN(P) end:
Subs:=k->{seq(X[i]=a+i*h,i=0..k)};
n:=1: fT:=add(L(x,i,n)*f(X[i]),i=0..n);
IT:=int(fT,XA=X[0]..X[n]); T:=sort(factor(subs(Subs(1),IT)));
ET:=series(subs(b=a+h,int(f(x),x=a..b)-T),h,4);
```

```
n:=2: fS:=add(L(x,i,n)*f(X[i]),i=0..n);
IS:=int(fS,XA=X[0]..X[n]); S:=sort(factor(subs(Subs(2),IS)));
ES:=series(subs(b=a+2*h,int(f(x),x=a..b)-S),h);
n:=3: fS38:=add(L(x,i,n)*f(X[i]),i=0..n);
IS38:=int(fS38,XA=X[0]..X[n]);
S38:=sort(factor(subs(Subs(3),IS38)));
ES38:=series(subs(b=a+3*h,int(f(x),x=a..b)-S38),h);
```

Mathematica:

```
L[x_,i_,n_]:=Module[{j}, P=1;
  Do[If[j!=i, P=P*(XA-X[j])/(X[i]-X[j])],{j,0,n}];P];
Subs[k_]:=Table[X[i]->a+i*h,{i,0,k}];
{n=1, fT=Sum[L[x,i,n]*f[X[i]],{i,0,n}], IT=Integrate[fT,
  {XA,X[0],X[n]}], T=Sort[Factor[IT/.Subs[1]]],
  ET=Series[((Integrate[f[x],{x,a,b}]-T)/.{b->a+h}),{h,0,3}]}
{n=2, fS=Sum[L[x,i,n]*f[X[i]],{i,0,n}], IS=Integrate[fS,
  {XA,X[0],X[n]}], S=Sort[Factor[IS/.Subs[2]]],
  ES=Series[((Integrate[f[x],{x,a,b}]-S)/.{b->a+2*h}),{h,0,5}]}
{n=3, fS38=Sum[L[x,i,n]*f[X[i]],{i,0,n}],
  IS38=Integrate[fS38,{XA,X[0],X[n]}],
  S38=Sort[Factor[IS38/.Subs[3]]], ES38=Series[((
  Integrate[f[x],{x,a,b}]-S38)/.{b->a+3*h}),{h,0,5}]}
```

$\square$

Problem 12.33 *Derivation of Newton–Cotes quadrature formulas.* Let $f:[a,b] \to \mathbb{R}$ be an arbitrary function sufficiently many times continuously differentiable.

1) Derive symbolically Newton–Cotes quadrature formulas of order n for approximating a definite integral.

2) Derive symbolically the discretization errors in the approximate formulas using the corresponding series expansions.

Maple:

```
with(CurveFitting): NewtonCotesInt:=n->subs(a-b=-n*h,
  sort(simplify((b-a)/(n*h)*int(PolynomialInterpolation(
  [seq(i*h,i=0..n)],[seq(f[i],i=0..n)],z),z=0..n*h))));
for i from 1 to 11 do NewtonCotesInt(i); od;
ENC:=(n,Ord)->simplify(series(subs({seq(f[i]=f(a+i*h),
  i=0..n)},h=a+n*h,int(f(x),x=a..b)-NewtonCotesInt(n)),h,Ord));
ENC(1,4); ENC(2,6); ENC(3,6); ENC(4,8); ENC(5,8); ENC(6,10);
ENC(7,10); ENC(8,12); ENC(9,12); ENC(10,14); ENC(11,14);
```

Mathematica:

```
NewtonCotesInt[n_]:=Sort[
 Simplify[(b-a)/(n*h)*Integrate[InterpolatingPolynomial[
 Table[{i*h,f[i]},{i,0,n}],z],{z,0,n*h}]]]/.{a-b->-n*h};
Do[Print[NewtonCotesInt[i]],{i,1,11}];
ENC[n_,Ord_]:=Simplify[Series[(Integrate[f[x],{x,a,b}]
 -NewtonCotesInt[n])/.(Flatten[{Table[f[i]->f[a+i*h],{i,0,n}],
 {b->a+n*h}}]),{h,0,Ord}]];
{ENC[1,3],ENC[2,5],ENC[3,5],ENC[4,7],ENC[5,7],ENC[6,9],
 ENC[7,9],ENC[8,11],ENC[9,11],ENC[10,13],ENC[11,13]}
```
□

Problem 12.34 *Romberg integration.* Let $f : [a, b] \to \mathbb{R}$ be an arbitrary function sufficiently many times continuously differentiable. Let us consider the Romberg integration method suggested by Romberg in 1955.

Obtain numerical approximations of the definite integral applying the Romberg integration rule, i.e. applying the Richardson extrapolation method repeatedly on the trapezoidal rule.

Maple:

```
Digits:=30: f:=x->x*exp(-x); n:=6; a:=0; b:=2*Pi;
RombergInt:=proc(f::algebraic,a,b,N) local R,h,k,i,j;
 R:=array(0..N,0..N); h:=b-a; R[0,0]:=evalf(0.5*h*(f(a)+f(b)));
 for i from 1 to N do h:=0.5*h; R[i,0]:=evalf(0.5*R[i-1,0]
 +add(h*f(a+(2*k-1)*h),k=1..2^(i-1))); for j from 1 to i do
 R[i,j]:=((4^j)*R[i,j-1]-R[i-1,j-1])/(4^j-1); od; od;
 for i from 0 to N do for j from 0 to i do
 printf("%30.25e\n",R[i,j]); od; od; RETURN(R[N,N]); end:
RombergInt(f,a,b,n);
```

Mathematica:

```
f[x_]:=x*Exp[-x]; {n=6,a=0,b=2*Pi,nD=30}
RombergInt[a_,b_,nN_]:=Module[{h,k,i,j}, h=N[b-a,nD];
 R[0,0]=N[h/2*(f[a]+f[b]),nD]; Do[h=h/2; R[i,0]=
 N[R[i-1,0]/2+Sum[h*f[a+(2*k-1)*h],{k,1,2^(i-1)}],nD];
 Do[R[i,j]=N[((4^j)*R[i,j-1]-R[i-1,j-1])/(4^j-1),nD],{j,1,i}],
 {i,1,nN}]; Do[Print[PaddedForm[R[i,j],{30,26}]],{i,0,nN},
 {j,0,i}]; R[nN,nN]]; RombergInt[a,b,n]
NIntegrate[f[x],{x,a,b},Method->{"TrapezoidalRule","Points"->8,
 "RombergQuadrature"->True},AccuracyGoal->7,PrecisionGoal->7,
 WorkingPrecision->nD]
```
□

Problem 12.35 *Derivation of Gaussian quadrature formulas.*
Let $f : [a, b] \to \mathbb{R}$ be an arbitrary function sufficiently many times continuously differentiable and $w : [a, b] \to \mathbb{R}$ is a continuous and positive function so that the integral $\int_a^b w(x)\,dx$ exists.
The main idea of Gaussian quadrature is to find the nodes x_i and the weights w_i $(i = 1, \ldots, n)$ in $[a, b]$ such that the general quadrature formula

$$\int_a^b w(x)f(x)\,dx \approx \sum_{i=1}^n w_i f(x_i)$$

is exact for all polynomials of degree less than or equal to $2n - 1$.

The fundamental theorem of Gaussian quadrature states that the optimal nodes x_i of the n-point Gaussian quadrature formulas are the roots of the orthogonal polynomial for the same interval and the weight function.

1) Let $a = -1$, $b = 1$, and $w(x) = 1$. Obtain the numerical values of the nodes x_i and the weights w_i and Gaussian quadrature formula of order n for approximating a definite integral using the definition of Gaussian quadrature, i.e. solving the system of $2n$ nonlinear equations with respect to the unknowns x_i and w_i $(i = 1, \ldots, n)$.

2) Let $a = -1$, $b = 1$, and $w(x) = 1$. Obtain the numerical values x_i, w_i and the Gauss–Legendre quadrature formulas.

3) Let $a = -1$, $b = 1$, and $w(x) = (1 - x^2)^{-1/2}$. Obtain the numerical values x_i, w_i and the Gauss–Chebyshev quadrature formulas.

Maple:

```
with(CurveFitting):with(orthopoly): a:=-1;b:=1;n:=3;w1:=x->1;
Eqs:={seq(add(w[i]*X[i]^j,i=1..n)=int(x^j,x=a..b),
  j=0..2*n-1)}; Sol:=[solve(Eqs,indets(Eqs,name))];
SG:=[seq(map(convert,Sol[i],radical),i=1..nops(Sol))];
SGf:=evalf(SG); GaussianInt:=Int(w1(x)*f(x),x=a..b)=
  sort(add(op(2,SGf[1,i+n])*F[i],i=1..n));
a:=-1; b:=1; n:=4; w1:=x->1;
X1:=[fsolve(P(n,x)=0,x)]; Y1:=[seq(F[i],i=1..n)];
GaussLegendreInt:=Int(w1(x)*f(x),x=a..b)=
  sort(int(w1(z)*PolynomialInterpolation(X1,Y1,z),z=a..b));
a:=-1; b:=1; n:=5; w2:=x->1/sqrt(1-x^2);
X2:=[fsolve(T(n,x)=0,x)];Y2:=[seq(F[i],i=1..n)];
GaussChebyshevInt:=Int(w2(x)*f(x),x=a..b)=
sort(int(w2(z)*PolynomialInterpolation(X2,Y2,z),z=a..b));
```

Mathematica:

```
nD=10; w1[x_]:=1; {a=-1,b=1,n=3}
Eqs1=Table[Sum[w[i]*X[i]^j,{i,1,n}]
            -Integrate[x^j,{x,a,b}],{j,0,2*n-1}]
{Eqs=Thread[Eqs1==0,Equal], vars=Variables[Eqs],
 SG=Solve[Eqs,vars]}
GaussianInt=HoldForm[Integrate[w1[x]*f[x],{x,a,b}]]==
 Sort[Sum[N[SG[[1,i,2]],nD]*F[i],{i,1,n}]]
w2[x_]:=1;{nD=11,a=N[-1,nD],b=N[1,nD],n=4}
{X11=FindRoot[LegendreP[n,x]==0,{x,{a,a/2,b/2,b}},
 WorkingPrecision->nD], Y1=Table[F[i],{i,1,n}],
 X1=X11[[1,2]], XY1=Table[{X1[[i]],Y1[[i]]},{i,1,n}]}
GaussLegendreInt=HoldForm[Integrate[w2[x]*f[x],{x,a,b}]]==
 Sort[Integrate[Expand[w2[z]*InterpolatingPolynomial[XY1,z]],
 {z,-1.,1.}]]
w2[x_]:=1/Sqrt[1-x^2]; {a=-1,b=1,n=5}
{X21=FindRoot[ChebyshevT[n,x]==0,{x,{a,a/2,0,b/2,b}},
 WorkingPrecision->nD], X2=X21[[1,2]], Y2=Table[F[i],{i,1,n}],
 XY2=Table[{X2[[i]],Y2[[i]]},{i,1,n}]}
GaussChebyshevInt=HoldForm[Integrate[w2[x]*f[x],{x,a,b}]]==
 ComplexExpand[Re[Sort[Integrate[Expand[w2[z]*
 InterpolatingPolynomial[XY2,z]],{z,-1.,1.}]]]]
```

$\square$

Problem 12.36 *Derivation of the Gauss–Lobatto quadrature formulas.*
Let $a=-1$, $b=1$, $w(x)=1$. Let $f:[-1,1]\to\mathbb{R}$ be an arbitrary function
sufficiently many times continuously differentiable.

1) For $n=3,\dots,6$, construct the Gauss–Lobatto quadrature formulas

$$\int_{-1}^1 f(x)\,dx \approx w_1 f(1) + w_n f(-1) + \sum_{i=2}^{n-1} w_i f(x_i),$$

obtaining the numerical values of the free abscissas x_i $(i=2,\dots,n-1)$
and the weights w_i.

2) Construct the Gauss–Radau quadrature formulas

$$\int_{-1}^1 f(x)\,dx \approx w_1 f(-1) + \sum_{i=2}^n w_i f(x_i),$$

obtaining the numerical values of the free abscissas x_i $(i=2,\dots,n)$ and
the weights w_i.

Maple:

```
with(orthopoly): a:=-1; b:=1; w:=x->1; n:=3;
X3:=[fsolve(diff(P(n-1,x),x)=0,x)];
G:=f->int(w(x)*f(x),x=a..b)=w[1]*(f(-1)+f(1))+w[2]*f(0);
sys:={seq(G(unapply(z^(2*(i-1)),z)),i=1..n-1)};
vars:={seq(w[i],i=1..n-1)}; Sol3:=evalf(solve(sys,vars));
GaussLobatto3:=subs(Sol3,G(F));
n:=4; X4:=[fsolve(diff(P(n-1,x),x)=0,x)];
G:=f->int(w(x)*f(x),x=a..b)=w[1]*(f(-1)+f(1))
  +w[2]*(f(-X4[1])+f(X4[1]));
sys:={seq(G(unapply(z^(2*(i-1)),z)),i=1..n-2)};
vars:={seq(w[i],i=1..n-2)}; Sol4:=evalf(solve(sys,vars));
GaussLobatto4:=subs(Sol4,G(F));
n:=5; X5:=[fsolve(diff(P(n-1,x),x)=0,x)];
G:=f->int(f(x),x=a..b)=w[1]*(f(-1)+f(1))+w[2]*f(0)
  +w[3]*(f(-X5[1])+f(X5[1]));
sys:={seq(G(unapply(z^(2*(i-1)),z)),i=1..n-2)};
vars:={seq(w[i],i=1..n-2)}; Sol5:=evalf(solve(sys,vars));
GaussLobatto5:=subs(Sol5,G(F));
n:=6; X6:=[fsolve(diff(P(n-1,x),x)=0,x)];
G:=f->int(f(x),x=a..b)=w[1]*(f(-1)+f(1))
  +w[2]*(f(-X6[1])+f(X6[1]))+w[3]*(f(-X6[2])+f(X6[2]));
sys:={seq(G(unapply(z^(2*(i-1)),z)),i=1..n-3)};
vars:={seq(w[i],i=1..n-3)}; Sol6:=evalf(solve(sys,vars));
GaussLobatto6:=subs(Sol6,G(F));
n:=4; XR:=simplify(fnormal([evalf(solve((P(n-1,x)+P(n,x))/
  (1+x)=0,x))])); kR:=nops(XR); GR:=f->int(w(x)*f(x),x=a..b)=
  w[0]*f(-1)+add(w[i]*f(XR[i]),i=1..kR);
sys:={seq(GR(unapply(z^(2*(i-1)),z)),i=1..n)};
vars:={seq(w[i],i=0..n-1)}; SolR:=evalf(solve(sys,vars));
GaussRadau:=subs(SolR,GR(F));
```

Mathematica:

```
w[x_]:=1; nD=10; {a=-1, b=1, n=3,
 X3=FindRoot[D[LegendreP[n-1,x],x]==0,{x,a,b},
 WorkingPrecision->nD][[1,2]]}
G[f_,x_]:=Integrate[w[x]*f[x],{x,a,b}]==
 W[1]*(f[-1]+f[1])+W[2]*f[0]; f3[z_]:=t^(2*(i-1))/.{t->z};
{G[f,x], sys=Table[G[f3,z],{i,1,n-1}],
 vars=Table[W[i],{i,1,n-1}], Sol3=Solve[sys,vars][[1]]}
```

```
GaussLobatto3=Expand[G[F,x]/.Sol3]
{n=4, X4=FindRoot[D[LegendreP[n-1,x],x]==0,{x,{a,b}},
 WorkingPrecision->nD][[1,2]]}
G[f_,x_]:=Integrate[w[x]*f[x],{x,a,b}]==
 W[1]*(f[-1]+f[1])+W[2]*(f[X4[[1]]]+f[X4[[2]]]);
 f4[z_]:=t^(2*(i-1))/.{t->z};
{G[f,x], sys=Table[G[f4,z],{i,1,n-2}],
 vars=Table[W[i],{i,1,n-2}], Sol4=Solve[sys,vars][[1]]}
GaussLobatto4=Expand[G[F,x]/.Sol4]
{n=5, X5=FindRoot[D[LegendreP[n-1,x],x]==0,{x,{a,0,b}},
 WorkingPrecision->nD][[1,2]]}
G[f_,x_]:=Integrate[w[x]*f[x],{x,a,b}]==W[1]*(f[-1]+f[1])+
 W[2]*f[X5[[2]]]+W[3]*(f[X5[[1]]]+f[X5[[3]]]);
 f5[z_]:=t^(2*(i-1))/.{t->z};
{G[f,x], sys=Table[G[f5,z],{i,1,n-2}],
 vars=Table[W[i],{i,1,n-2}], Sol5=Solve[sys,vars][[1]]}
GaussLobatto5=Expand[G[F,x]/.Sol5]
{n=6, X6=FindRoot[D[LegendreP[n-1,x],x]==0,{x,{a,a/2,b/2,b}},
 WorkingPrecision->nD][[1,2]]}
G[f_,x_]:=Integrate[w[x]*f[x],{x,a,b}]==
 W[1]*(f[-1]+f[1])+W[2]*(f[X6[[1]]]+f[X6[[4]]])+
 W[3]*(f[X6[[2]]]+f[X6[[3]]]);
 f6[z_]:=t^(2*(i-1))/.{t->z};
{G[f,x], sys=Table[G[f6,z],{i,1,n-3}],
 vars=Table[W[i],{i,1,n-3}], Sol6=Solve[sys,vars][[1]]}
GaussLobatto6=Expand[G[F,x]/.Sol6]
{n=4, XR=FindRoot[Expand[
 (LegendreP[n-1,x]+LegendreP[n,x])/(1+x)]==0,{x,{0.8,-0.6,0.2}},
 WorkingPrecision->nD][[1,2]]}
G[f_,x_]:=Integrate[w[x]*f[x],{x,a,b}]==W[0]*f[-1]+
 W[1]*f[XR[[1]]]+ W[2]*f[XR[[2]]]+W[3]*f[XR[[3]]];
fR[z_]:=t^(2*(i-1))/.{t->z};
{G[f,x],sys=Table[G[fR,z],{i,1,n}],vars=Table[W[i],{i,0,n-1}],
 SolR=Solve[sys,vars]}
GaussRadau=Expand[G[F,x]/.SolR]
```

□

Problem 12.37 *Derivation of the Gauss–Kronrod quadrature formulas.*
Let $a = -1$, $b = 1$, $w(x) = 1$. Let $f : [-1,1] \to \mathbb{R}$ be an arbitrary
function sufficiently many times continuously differentiable. Using the
results obtained in the previous problem, construct the Gauss–Kronrod
quadrature formulas.

Maple:

```
with(orthopoly): with(CurveFitting): a:=-1; b:=1; w:=x->1; n:=4;
X4:=[fsolve(diff(P(n-1,x),x)=0,x)];
GL:=f->int(w(x)*f(x),x=a..b)=w[1]*(f(-1)+f(1))+
 w[2]*(f(-X4[1])+f(X4[1]));
sys:={seq(GL(unapply(z^(2*(i-1)),z)),i=1..n-2)};
vars:={seq(w[i],i=1..n-2)}; Sol4:=evalf(solve(sys,vars));
GaussLobatto4:=subs(Sol4,GL(F));
GK:=f->int(w(x)*f(x),x=a..b)=w[1]*(f(-1)+f(1))+
 w[2]*(f(-A)+f(A))+w[3]*(f(-X4[1])+f(X4[1]))+w[4]*f(0);
sys:={seq(GK(unapply(z^(2*(i-1)),z)),i=1..n+1)};
vars:={seq(w[i],i=1..n),A}; Sol:=evalf(solve(sys,vars));
GaussKronrod:=subs(Sol,GK(F));
```

Mathematica:

```
w[x_]:=1; nD=10; {a=-1,b=1}
{n=4, X4=FindRoot[D[LegendreP[n-1,x],x]==0,{x,{a,b}},
 WorkingPrecision->nD][[1,2]]]}
GL[f_,x_]:=Integrate[w[x]*f[x],{x,a,b}]==W[1]*(f[-1]+
 f[1])+W[2]*(f[X4[[1]]]+f[X4[[2]]]); fL[z_]:=t^(2*(i-1))/.
 {t->z}; {GL[f,x],sys=Table[GL[fL,z],{i,1,n-2}],
 vars=Table[W[i],{i,1,n-2}],Sol4=Solve[sys,vars][[1]]}
GaussLobatto4=Expand[GL[F,x]/.Sol4]
GK[f_,x_]:=Integrate[w[x]*f[x],{x,a,b}]==W[1]*(f[-1]+f[1])+
 W[2]*(f[-A]+f[A])+W[3]*(f[X4[[1]]]+f[X4[[1]]])+W[4]*f[0];
fK[z_]:=t^(2*(i-1))/.{t->z}; {G[f,x], sys=Table[GK[fK,z],
 {i,1,n+1}], vars=Flatten[{Table[W[i],{i,1,n}],A}],
 Sol=Solve[sys,vars][[1]]}
GaussKronrod=Expand[GK[F,x]/.Sol]
```

□

12.4   Linear Systems of Equations

The problem of solving systems of linear equations arises in various areas of mathematics, sciences, and engineering. The explicit formulas for solving linear systems of equations, the Cramer rule, requires $2(n+1)!$ operations, so the computational cost is very high for the large values of n. We have to consider alternative approaches, discussing the two basic classes of methods for solving linear systems:

1) *Direct methods or elimination methods or matrix decompositions methods*, giving theoretically exact solutions that can be determined through a finite number of arithmetic operations (in computational practice, there is a cost due to roundoff errors). Direct methods are best suited for full matrices.

2) *Iterative methods*, giving a sequence of approximations to the solution (that under some assumptions converges to the solution of the system) by repeating the application of the same computational procedure at each step of the iteration. Iterative methods are best suited for large and sparse matrices.

The best-known and most widely used direct method was proposed by Gauss in 1801. The classical iterative methods, the Jacobi and the Gauss–Seidel iterations, go back also to Gauss, in 1822 he published a version of Gauss–Seidel method for solving linear systems.

Direct methods or elimination methods or matrix decompositions methods for solving linear systems

```
with(LinearAlgebra):          LinearSolve(A,B); Determinant(A);
ReducedRowEchelonForm(A);                GaussianElimination(A);
LUDecomposition(A,ops);                     QRDecomposition(A);
ForwardSubstitute(A);                    BackwardSubstitute(A);
          CramerRule:=x[i]=Determinant(A)/Determinant(A[i]);
RowOperation(A,[i1,i2],s);      with(Student[LinearAlgebra]);
AddRow(A,i1,i2,s);        SwapRow(A,i1,i2); Pivot(A,i,j,ops);
AddColumn(A,j1,j2,s);          GaussianEliminationTutor(A);
```

LUDecomposition with the option **method**, performing LU-factorization or decomposition of a matrix using the given method (**RREF**, **GaussianElimination**, **FractionFree**, **Cholesky**,

LUDecomposition with the option **output**, performing LU-factorization of a matrix and determining the form of the results, e.g. **U**, upper triangular factor, **R**, reduced row echelon form,

ForwardSubstitute, BackwardSubstitute, computing all unknowns according to the forward and backward substitutions algorithms (where the matrix A is represented, respectively, in upper and lower row echelon form).

Mathematica:

```
LinearSolve[a,b]   RowReduce[a]      {lu,p,c}=LUDecomposition[m]
luBack1=LinearSolve[u,LinearSolve[l,b[[p]]]]                Det[a]
luBack2=LinearSolve[a]   x=luBack2[b] {q,r}=QRDecomposition[m]]
LinearSolve[r,q.b]                        CholeskyDecomposition[m]
  elimGauss[m_,b_]:=Last/@RowReduce[Flatten/@Transpose[{m,b}]];
```

Note. More detailed descriptions and information on linear algebra in both systems, we refer to Chapter 5.

Problem 12.38 *Direct classical methods.* Let A and B be an $n \times n$ nonsingular matrix and n-dimensional vector with real entries, respectively. Solve the linear system of equations $Ax = B$ using floating-point arithmetic and various direct methods:

1) Gaussian elimination performing a finite number of elementary row and column operations (see Sect. 5.2) and compare with the predefined functions.

2) Gauss–Jordan method performing a finite number of elementary row and column operations and compare with the predefined functions (reduced row-echelon form).

3) The factorization LU.

Maple:

```
with(LinearAlgebra): A:=<<-12,1,-2>|<1,-6,-1>|<-7,4,10>>;
B:=<-80,-2,92>; M:=Matrix(3,3,A,datatype=float[4]);
V:=Vector(3,B,datatype=float[4]); MV:=<M|V>; Determinant(M);
M1:=RowOperation(MV,[2,1],1./12.);
M2:=RowOperation(M1,[3,1],-1./6.);
M3:=RowOperation(M2,[3,2],-14./71.);
GE:=GaussianElimination(MV); M3=GE; X:=BackwardSubstitute(GE);
simplify(fnormal(M.X))=V; M4:=RowOperation(M3,[1,2],12./71.);
M5:=RowOperation(M4,[1,3],456./745.);
M6:=RowOperation(M5,[2,3],-41.*71./(745.*12.));
M7:=RowOperation(M6,1,1./M6[1,1]);
M8:=RowOperation(M7,2,1./M6[2,2]);
MA9:=RowOperation(M8,3,1./M6[3,3]);M9=ReducedRowEchelonForm(MV);
LUSol:=LUDecomposition(MV); LUSol[3]=GE;
LUDecomposition(MV,method='GaussianElimination');
```

Mathematica:

```
{A={{-12,1,-7},{1,-6,4},{-2,-1,10}},B={-80,-2,92},
 M=N[A], V=N[B], n=MatrixRank[M], id=IdentityMatrix[n],
 MV=Transpose[Join[Transpose[M],{V}]], Det[M]}
{M1=MV,M1[[2]]=M1[[2]]+M1[[1]]*(1./12.),M1//MatrixForm}
{M2=M1,M2[[3]]=M2[[3]]+M2[[1]]*(-1./6.),M2//MatrixForm}
{M3=M2,M3[[3]]=M3[[3]]+M3[[2]]*(-14./71.),M3//MatrixForm}
elimGauss[m_,b_]:=Last/@RowReduce[Flatten/@Transpose[{m,b}]];
{X=elimGauss[M, V], M.X==V, luBack1=LinearSolve[M],
 X1=luBack1[V], M.X1==V, {lu,p,c}=LUDecomposition[M]}
 l=lu*SparseArray[{i_,j_}/;j<i->1,{n,n}]+id;
 u=lu*SparseArray[{i_,j_}/;j>=i->1,{n,n}];
Map[MatrixForm,{l,u,l.u, l[[p]], l[[p]].u}]
luBack2=LinearSolve[u,LinearSolve[l,V[[p]]]]
Map[MatrixForm,{u,M3[[All,-4;;-2]]}]
{M4=M3,M4[[1]]=M4[[1]]+M4[[2]]*(12./71.)}
{M5=M4,M5[[1]]=M5[[1]]+M5[[3]]*(456./745.)}
{M6=M5,M6[[2]]=M6[[2]]+M6[[3]]*-41.*71./(745.*12.)}
{M7=M6,M7[[1]]=M7[[1]]*1./M6[[1,1]]}
{M8=M7,M8[[2]]=M8[[2]]*1./M6[[2,2]]}
{MA9=M8,MA9[[3]]=MA9[[3]]*1./M6[[3,3]]}
M9=RowReduce[MV]
Map[MatrixForm,Chop[{M4,M5,M6,M7,M8,MA9,M9}]]
{AB=Transpose[Join[Transpose[A],{B}]],AG=AB}
Do[Do[M=N[AG[[i,k]]/AG[[k,k]]];
Do[AG[[i,j]]=AG[[i,j]]-M*AG[[k,j]],{j,k,n+1}],{i,k+1,n}];
Print[MatrixForm[AG]],{k,1,n-1}]; AG//MatrixForm
```

Additionally, in *Maple*, there is the Student[LinearAlgebra] subpackage that can be useful for understanding the basic material in linear algebra, for example a group of interaction functions, GaussianEliminationTutor, GaussJordan EliminationTutor, etc.   □

Problem 12.39 *The pivoting strategy.* Let A and B be a given $n \times n$ nonsingular matrix and n-dimensional vector with integer entries. Gaussian elimination exists if and only if the principle $k \times k$ minors A_k of A $(k = 1, \ldots, n-1)$, formed by the intersection of the first k rows and columns in A, are nonsingular, otherwise the corresponding *pivot elements* are zero, $a_{jj}^{(i)} = 0$. Moreover, Gaussian elimination is unique if A is nonsingular.

1) Verify if the principle minors of A are nonsingular.

2) If it is true, solve the linear system of equations $Ax = B$ in exact arithmetic and applying the pivoting strategy, i.e. generating new *pivot elements* by exchanging two rows (or columns) of the system.

Maple:

```
with(LinearAlgebra): with(Student[LinearAlgebra]): n:=4;
A:=Matrix(n,n,[[1,-1,2,-1],[2,-2,3,-3],[1,1,1,0],[1,-1,4,3]]);
B:=Vector(n,[-8,-20,-2,4]); AB:=<A|B>;
for i from 1 to n-1 do
 SMPrinc:=SubMatrix(A,[1..i],[1..i]);Determinant(SMPrinc);od;
A1:=AddRow(AB,2,1,-2); A2:=AddRow(A1,3,1,-1);
A3:=AddRow(A2,4,1,-1);A4:=SwapRow(A3,3,2);A5:=AddRow(A4,4,3,2);
X:=BackwardSubstitute(A5); GE:=GaussianElimination(AB);
X1:=BackwardSubstitute(GE);(A.X=B)=(A.X1=B); P1:=Pivot(AB,1,1);
P2:=SwapRow(P1,3,2);P3:=Pivot(P2,3,3);BackwardSubstitute(P3)=X;
```

Mathematica:

```
{n=4, A={{1,-1,2,-1},{2,-2,3,-3},{1,1,1,0},{1,-1,4,3}},
 B={-8,-20,-2,4}, AB=Transpose[Join[Transpose[A],{B}]]}
Do[SMPrinc=Take[A,{1,i},{1,i}]; Print[SMPrinc//MatrixForm];
   Print[Det[SMPrinc]],{i,1,n-1}];
{A1=AB,A1[[2]]=A1[[2]]+A1[[1]]*(-2),A1//MatrixForm}
{A2=A1,A2[[3]]=A2[[3]]+A2[[1]]*(-1),A2//MatrixForm}
{A3=A2,A3[[4]]=A3[[4]]+A3[[1]]*(-1),A3//MatrixForm}
{A4=A3,A4[[{3,2}]]=A4[[{2,3}]],A4//MatrixForm}
{A5=A4,A5[[4]]=A5[[4]]+A5[[3]]*2,A5//MatrixForm}
{A55=A5[[All,-5;;-2]], B5=A5[[All,-1]], X5=Table[0,{i,1,n}]}
Do[X5[[i]]=(B5[[i]]-Sum[A55[[i,j]]*X5[[j]],
 {j,i+1,n}])/A55[[i,i]],{i,n,1,-1}]; X5//MatrixForm
elimGauss[m_,b_]:=Last/@RowReduce[Flatten/@Transpose[{m,b}]];
{X=elimGauss[A,B], GE=RowReduce[AB], X1=GE[[All,-1]],
 (A.X==B)==(A.X1==B)==(A.X5==B)}
{AG=AB, n=4}
Do[Do[If[AG[[k,k]]==0,AG[[{k,i}]]=AG[[{i,k}]]];
 M=N[AG[[i,k]]/AG[[k,k]]]; Do[AG[[i,j]]=AG[[i,j]]-
 M*AG[[k,j]],{j,k,n+1}],{i,k+1,n}]; Print[MatrixForm[AG]],
 {k,1,n-1}]; AG[[1;;n,1;;n]].X==AG[[All,-1]]
Map[MatrixForm,{AG[[1;;n,1;;n]]},X,AG[[All,-1]]}]
LinearSolve[AG[[1;;n,1;;n]],AG[[All,-1]]]//MatrixForm
```

□

Iterative methods for solving linear systems

Applying an iterative method for solving a linear system, we start from a given initial vector $x^{(0)}$ and construct a sequence of vectors $x^{(i)}$ which must converge to the exact solution as $i \to \infty$. The necessary and sufficient condition for convergence of an iterative method in the finite-dimensional case is that, it converges for any initial vector $x^{(0)}$ if and only if $\rho(A) < 1$ for the spectral radius of the iteration matrix A.

We consider the classical iterative methods, the Gauss–Seidel and the Jacobi iterative methods, where the updating of the components are performed, respectively, *sequentially* and *simultaneously*.

Problem 12.40   *The Gauss–Seidel iterative method.* Let A and B be a given $n \times n$ nonsingular matrix and n-dimensional vector with real entries.

1) Compute an approximate solution, with an error less than a given ε in maximum norm, to the linear system of equations $Ax = B$ in a given d-figure floating-point arithmetic applying the Gauss–Seidel method.

2) Compare the results with the exact solution obtained by a direct method.

Maple:

```
with(LinearAlgebra): Digits:=30: n:=3;
A:=Matrix(n,n,[[3,0.1,0.2],[0.1,7,0.3],[0.2,0.3,10]]);
B:=Vector(n,[7.85,-19.3,71.4]); X:=Vector(n);
X0:=Vector(n,fill=0.); epsilon:=10^(-30); Nmax:=100;
for k from 1 to Nmax do for i from 1 to n do
 X[i]:=(-add(A[i,j]*X[j],j=1..i-1)-add(A[i,j]*X0[j],j=i+1..n)
 +B[i])/A[i,i]; od: E:=Norm(X-X0,infinity)/Norm(X,infinity);
 if E<epsilon then break; fi; print(k,E);
 for m from 1 to n do X0[m]:=X[m]; od: print(X); od:
printf("The Gauss-Seidel method converges to the solution");
print(X); printf("after %d iterations",k);
printf("The solution obtained by the direct method:");
BackwardSubstitute(GaussianElimination(<A|B>));
```

Mathematica:

```
{nD=30, n=3, A={{3,1/10,1/5},{1/10,7,3/10},{1/5,3/10,10}},
 B={157/20,-193/10,357/5}, X0=Table[0,{i,1,n}],
 X=Table[0,{i,1,n}], epsilon=N[10^(-30),nD], Nmax=100}
```

```
Do[Do[X[[i]]=N[(-Sum[A[[i, j]]*X[[j]],{j,1,i-1}]
 -Sum[A[[i,j]]*X0[[j]],{j,i+1,n}]+B[[i]])/A[[i,i]],nD],{i,1,n}];
Er=N[Norm[X-X0,Infinity]/Norm[X,Infinity],nD];
If[Er<epsilon,{NIter=k,Break[]}];
Print[k," ",Er]; Do[X0[[m]]=X[[m]],{m,1,n}];
Print[MatrixForm[X]],{k,1,Nmax}];
Print["The Gauss-Seidel method converges to the solution"];
Print[MatrixForm[X]];
Print["after"," ",NIter," ","iterations"];
Print["The solution obtained by the direct method:"];
N[Last/@RowReduce[Flatten/@Transpose[{A,B}]],nD]
```

□

Problem 12.41 *The Jacobi iterative method.* Let A and B be a given $n \times n$ nonsingular matrix and n-dimensional vector with real entries.

1) Compute an approximate solution to the same system of linear equations as in the previous problem and with the same parameters.

2) Compare the results with the exact solution obtained by a direct method and with results obtained by applying the Gauss–Seidel method.

Maple:

```
with(LinearAlgebra): Digits:=30: n:=3;
A:=Matrix(n,n,[[3,0.1,0.2],[0.1,7,0.3],[0.2,0.3,10]]);
B:=Vector(n,[7.85,-19.3,71.4]); X:=Vector(n); Nmax:=100;
R:=1..n; X0:=Vector(n,fill=0.); epsilon:=10^(-30);
for k from 1 to Nmax do
 for i from 1 to n do S:=0.:
 for j from 1 to n do if j<>i then S:=evalf(S+A[i,j]*X0[j]);
 fi; od: X[i]:=evalf((-S+B[i])/A[i,i]); od:
E:=evalf(max('abs(X[i]-X0[i])'$'i'=R)/max('abs(X[i])'$'i'=R));
 if E<epsilon then break; fi; print(k,E);
 for m from 1 to n do X0[m]:=X[m]; od: print(X); od:
printf("The Jacobi method converges to the solution");
print(X); printf("after %d iterations.",k);
printf("The solution obtained by the direct method:");
BackwardSubstitute(GaussianElimination(<A|B>));
```

Mathematica:

```
{nD=30, n=3, A={{3,1/10,1/5},{1/10,7,3/10},{1/5,3/10,10}},
 B={157/20,-193/10,357/5}, X0=Table[0,{i,1,n}],
 X=Table[0,{i,1,n}], epsilon=N[10^(-30),nD], Nmax=100}
```

```
Do[Do[S=0; Do[If[j!=i, S=N[S+A[[i,j]]*X0[[j]],nD]],{j,1,n}];
 X[[i]]=N[(-S+B[[i]])/A[[i,i]],nD],{i,1,n}];
 Er=N[(Max[Table[Abs[X[[i]]-X0[[i]]],{i,1,n}]])/(Max[
  Table[Abs[X[[i]]],{i,1,n}]]),nD];
 If[Er<epsilon,{NIter=k,Break[]}]; Print[k," ",Er];
 Do[X0[[m]]=X[[m]],{m,1,n}]; Print[MatrixForm[X]],{k,1,Nmax}];
Print["The Jacobi method converges to the solution"];
Print[MatrixForm[X]];
Print["after"," ",NIter," ","iterations."];
Print["The solution obtained by the direct method:"];
N[Last/@RowReduce[Flatten/@Transpose[{A,B}]],nD]
```

We can observe that in this example the convergence with the Gauss–Seidel method is about twice as fast as that of the Jacobi method. Frequently, this is the case (but not always). □

Problem 12.42 *Convergence analysis of the Gauss–Seidel iterative method.* Let A and B be a given $n \times n$ nonsingular matrix and n-dimensional vector with real entries.

1) Construct the iterative matrix of the Gauss–Seidel method T_{GS}, i.e. reduce the original system $Ax = B$ to the form $x^{(i+1)} = T_{GS}x^{(i)} + C_{GS}$ ($i \geq 0$). This reduction is based on the following splitting of A:

$$P = D - L, \quad A = P - (P - A), \quad T_{GS} = P^{-1}(P - A), \quad C_{GS} = P^{-1}B,$$

where P is the preconditioner of A, $D = \text{diag}(a_{ii})$, and L is the lower triangular matrix formed from A.

2) Verify that the necessary and sufficient conditions for the Gauss–Seidel method to converge for an arbitrary initial approximation is that the spectral radius of iteration matrix is less than 1, i.e. $\rho(T_{GS}) < 1$.

3) Verify the sufficient condition for convergence, i.e. the system matrix A is strictly diagonally dominant by row, or the corresponding norm of the iteration matrix is less than 1, i.e. $\|T_{GS}\|_\infty < 1$, or the system matrix A is a symmetric positive-definite matrix (according to the Sylvester criterion).

Maple:

```
with(LinearAlgebra): n:=3; LA:=Matrix(n,n);
A:=Matrix(n,n,[[3,0.1,0.2],[0.1,7,0.3],[0.2,0.3,10]]);
B:=Vector(n,[7.85,-19.3,71.4]); X:=Vector(n,symbol=x);
X1:=convert(X,list); A.X=B; DA:=DiagonalMatrix(Diagonal(A));
```

```
for i from 2 to n do for j from 1 to i-1 do
 LA[i,j]:=-A[i,j]; od: od: LA; P:=DA-LA; A=P-(P-A);
 T_GS:=MatrixInverse(P).(P-A); C_GS:=MatrixInverse(P).B;
 rho('T_GS')=max(op(convert(abs(Eigenvalues(T_GS)),list)));
 rho('T_GS')<1; N1:=Norm(T_GS,infinity); 'N1'<1;
 for i from 1 to n do S1:=0: for j from 1 to n do
  if j<>i then S1:=S1+abs(A[i,j]); fi;   od:
  print(abs(A[i,i]),S1,abs(A[i,i])>S1); od:
 A.A^%T = A^%T.A; for i from 1 to n do
 SMPrinc:=SubMatrix(A,[1..i],[1..i]);
 Sylvester:=Determinant(SMPrinc);
 print(i,Sylvester,type(Sylvester,positive)); od;
```

Mathematica:

```
{n=3, A={{3,0.1,0.2},{0.1,7,0.3},{0.2,0.3,10}},
 B={7.85,-19.3,71.4}, X=Table[Subscript[x,i],{i,1,n}],
 LA=Table[0,{i,1,n},{j,1,n}]}
{A.X==B,DA1=Table[A[[i,i]],{i,1,n}],DA=DiagonalMatrix[DA1]}
Do[Do[LA[[i,j]]=-A[[i,j]],{j,1,i-1}],{i,2,n}]; LA
{P=DA-LA,A=P-(P-A),TGS=Inverse[P].(P-A),CGS=Inverse[P].B}
{HoldForm[rho[TGS]]==Max[Abs[Eigenvalues[TGS]]],
 HoldForm[rho[TGS]]<1, HoldForm[N1]==Norm[TGS,Infinity],
 HoldForm[N1]<1}
Do[S1=0; Do[If[j!=i, S1=S1+Abs[A[[i,j]]]],{j,1,n}];
 Print[Abs[A[[i,i]]],", ",S1,", ",Abs[A[[i,i]]]," ",">",S1],
 {i,1,n}]; A.Transpose[A]==Transpose[A].A
Do[SMPrinc=Take[A,{1,i},{1,i}]; Print[MatrixForm[SMPrinc]];
 Sylvester=Det[SMPrinc];
 Print[i,", ",Sylvester,", ",Sylvester," ",">0"],{i,1,n}];
```

□

Problem 12.43 *Convergence analysis of the Jacobi iterative method.*
Let A and B be a given $n \times n$ nonsingular matrix and n-dimensional vector with real entries.

1) Construct the iterative matrix of the Jacobi method T_J, i.e. reduce the original system $Ax = B$ to the form $x^{(i+1)} = T_J x^{(i)} + C_J$ $(i \geq 0)$. This reduction is based on the following splitting of A:

$$P = D, \quad A = P - (P - A), \quad T_J = P^{-1}(P - A), \quad C_J = P^{-1}B,$$

where P is the preconditioner of A and $D = \text{diag}(a_{ii})$.

2) Verify that the necessary and sufficient conditions for the Jacobi method to converge for an arbitrary initial approximation is that the spectral radius of iteration matrix is less than 1, i.e. $\rho(T_J) < 1$.

3) Verify that the sufficient condition for convergence, i.e. the system matrix A is strictly diagonally dominant by row or the corresponding norm of the iteration matrix is less than 1, i.e. $\|T_J\|_\infty < 1$.

Maple:

```
with(LinearAlgebra): n:=3; A:=Matrix(n,n,[[3,0.1,0.2],
 [0.1,7,0.3],[0.2,0.3,10]]); B:=Vector(n,[7.85,-19.3,71.4]);
X:=Vector(n,symbol=x); X1:=convert(X,list); A.X=B;
Eqs:=GenerateEquations(A,X1,B); Eqs1:=NULL:
for i from 1 to n do Eqs1:=Eqs1,solve(Eqs[i],X[i]); od;
(TJ,CJ):=GenerateMatrix([Eqs1],X1); 'TJ'=TJ; 'CJ'=-CJ;
DA:=DiagonalMatrix(Diagonal(A)); P:=DA; A=P-(P-A);
T_J:=MatrixInverse(P).(P-A); C_J:=MatrixInverse(P).B;
rho('T_J')=max(op(convert(abs(Eigenvalues(T_J)),list)));
rho('T_J')<1; N1:=Norm(T_J,infinity); 'N1'<1;
for i from 1 to n do S1:=0: for j from 1 to n do
 if j<>i then S1:=S1+abs(A[i,j]); fi;  od:
 print(abs(A[i,i]),S1,abs(A[i,i])>S1); od:
```

Mathematica:

```
{n=3, A={{3,0.1,0.2},{0.1,7,0.3},{0.2,0.3,10}},
 B={7.85,-19.3,71.4}, X=Table[Subscript[x,i],{i,1,n}]}
{A.X==B, Eqs=Thread[A.X==B], Eqs11={}}
Do[Eqs11=Append[Eqs11,Solve[Eqs[[i]],X[[i]]]],{i,1,n}];
Eqs1=X/.(Flatten[Expand[Eqs11]])
{CJ,TJ}=Normal[CoefficientArrays[Eqs1,X]]
{Map[MatrixForm,{TJ,CJ}],HoldForm[TJ]==TJ,HoldForm[CJ]==-CJ}
{DA1=Table[A[[i,i]],{i,1,n}], DA=DiagonalMatrix[DA1],
 P=DA, A=P-(P-A), TJ=Inverse[P].(P-A), CJ=Inverse[P].B}
{HoldForm[rho[TJ]]==Max[Abs[Eigenvalues[TJ]]],
 HoldForm[rho[TJ]]<1, N1=Norm[TJ,Infinity], HoldForm[N1]<1}
Do[S1=0; Do[If[j!=i,S1=S1+Abs[A[[i,j]]]],{j,1,n}]; Print[
 Abs[A[[i,i]]]," ", S1," ",Abs[A[[i,i]]],">",S1],{i,1,n}];
```

□

12.5   Differential Equations

The study of differential equations goes back to the 17 th century, the beginnings of calculus with Newton and Leibniz and now plays the central role within general development of mathematics, since differential equations can describe the evolution of many phenomena in various fields of science and engineering.

Although there exist various analytic solution methods for special classes of differential equations, in general there is no explicit solution of a differential equation. Moreover, the functions and data in differential equation problems are frequently defined in discrete points. Therefore we have to study numerical approximation methods for differential equations.

We will consider the classical numerical approximation methods for initial value problems and boundary value problems for ordinary differential equations and partial differential equations.

Numerical and graphical solutions of differential equations. Initial value problems. Boundary value problems. Initial boundary value problems

Maple:

```
dsolve(ODEs,numeric,vars,ops);      dsolve(numeric,proc_ops,ops);
dsolve(ODEs,numeric,method=m);      dsolve(ODEs,numeric,output=n);
dsolve[interactive](ODEs,ops);    NS:=dsolve(ODEs,numeric,vars);
with(plots):                              odeplot(NS,vars,tR,ops);
with(DETools);              DEplot(ODEs,vars,tR,ICs,xR,yR,ops);
                         DEplot3d(ODEs,vars,tR,xR,yR,ICs,ops);
dfieldplot(ODEs,vars,tR,xR,yR);phaseportrait(ODEs,vars,tR,ICs);
                   Sol:=pdsolve(PDEs,ICsBCs,numeric,vars,ops);
Num_vals:=Sol:-value();          Sol:-plot3d(var,t=t0..t1,ops);
Num_vals(num1,num2);    Sol:-animate(var,t=t0..t1,x=x0..x1,ops);
```

where ODEs is an ordinary differential equation (ODE) or a set or list of ordinary differential equations (ODEs) and initial or boundary conditions (ICs, BCs), and tR, xR, yR is, respectively, the range of the independent and dependent variables, t=t1..t2, x=x1..x2, y=y1..y2.

dsolve, numeric, finding numerical solutions to ODEs problems,

dsolve, method, finding numerical solutions to ODEs problems using one of the numerical methods:

rkf45, Runge–Kutta Fehlberg fourth-fifth order method (the predefined method),

classical[name], classical numerical methods: **foreuler**, the forward Euler method, **heunform**, the Heun method (or the improved Euler method), **rk2**, the second-order classical Runge–Kutta method, **rk3**, the third-order classical Runge–Kutta method, **rk4**, the fourth-order classical Runge–Kutta method, **adambash**, the Adams–Bashforth method (or a predictor method), **abmoulton**, the Adams–Bashforth–Moulton method (or a predictor-corrector method),

rosenbrock, implicit Runge–Kutta Rosenbrock third-fourth order method,

dverk78, seventh-eighth order continuous Runge–Kutta method,

lsode, Livermore stiff ODE method,

gear, Gear single-step extrapolation method,

taylorseries, Taylor series method,

dsolve[interactive], interactive numeric solving ODEs,

odeplot (of the **plots**) package, constructing graphs or animations of 2D and 3D solution curves obtained from the numerical solution,

various functions of the **DETools** package can be used for working with differential equations, e.g.,

DEplot, DEplot3d, constructing graphs or animations of 2D and 3D solutions to a system of differential equations using numerical methods,

phaseportrait, constructing phase portraits for a system of first order differential equations or a single higher order differential equation using numerical methods,

pdsolve[numeric], finding numerical solutions to a partial differential equation PDE or a system of PDEs and obtaining various visualizations (**plot, plot3d, animate, animate3d**) and numerical values (**value**), in more detail, see **?pdsolve[numeric]**.

Note. For a more comprehensive details on finding closed-form solutions of ordinary differential equations we refer to Sect. 11.2.

Mathematica:

```
s=NDSolve[{ODEs,ICs},y,{t,t1,t2},ops]        s1=Evaluate[y[t]/.s]
NDSolve[{ODEs,ICs},y[t],{t,t1,t2},ops]           Plot[s1,{t,t1,t2}]
                    NDSolve[{ODEs,ICs},y,{t,t1,t2},Method->m]
   NDSolve[{ODEs,ICs},y,{t,t1,t2},Method->{m,Method->subMeth}]
Options[NDSolve`ExplicitEuler]                  <<VectorFieldPlots`
            VectorFieldPlot[{ft,fy},{t,t1,t2},{y,y1,y2},ops]
                    NDSolve[PDE,u,{x,x1,x2},{t,t1,t2},...]
            NDSolve[PDE,{u1,...,un},{x,x1,x2},{t,t1,t2},...]
```

NDSolve, finding numerical solutions to ODEs and PDEs problems,

NDSolve, Method->m, finding numerical solutions to ODEs problems using one of the numerical methods:

Adams, predictor-corrector Adams method of orders 1–12,

BDF, implicit backward differentiation formulas of orders 1–5,

ExplicitRungeKutta, adaptive embedded pairs of Runge–Kutta methods of orders 2(1)—9(8),

ImplicitRungeKutta, families of implicit Runge–Kutta methods of arbitrary-order,

SymplecticPartitionedRungeKutta, interleaved Runge–Kutta methods for separable Hamiltonian systems,

NDSolve, Method->{c,Method->subM}, finding numerical solutions to ordinary differential equation problems using one of the numerical controller methods:

Composition (compose a list of submethods), DoubleStep (adapt step size by the double-step method), EventLocator (respond to specified events), Extrapolation (adapt order and step size using polynomial extrapolation), FixedStep (use a constant step size), OrthogonalProjection (project solutions to fulfill orthogonal constraints), Projection (project solutions to fulfill general constraints), Splitting (split equations and use different submethods), StiffnessSwitching (switch from explicit to implicit methods if stiffness is detected),

and one of the submethods: ExplicitEuler, forward Euler method, Explicit Midpoint, midpoint rule method, ExplicitModifiedMidpoint, midpoint rule method with Gragg smoothing, LocallyExact, numerical approximation to locally exact symbolic solution, LinearlyImplicitEuler, linearly implicit Euler method, LinearlyImplicitMidpoint, LinearlyImplicit ModifiedMidpoint, linearly implicit Bader-smoothed midpoint rule method,

VectorFieldPlot of the VectorFieldPlots package or the new VectorPlot function (for $ver \geq 7$), constructing vector fields (see Sect. 6.10 and 11.2).

Problem 12.44 *Initial value problems.* In general, an ordinary differential equation $y'(t) = f(t, y(t))$ can admit an infinite number of solutions $y(t)$ (where $y : [a, b] \to \mathbb{R}$ and $y'(t) \equiv y'_t(t)$). In order to find one of them we have to add a condition of the form $y(t_0) = y_0$ ($t_0 = a$), where y_0 is a given value called the *initial data*. Thus, we consider the *Cauchy problem*:

$$\begin{cases} y'(t) = f(t, y(t)), & \forall\, t \in [t_0, b] \in \mathbb{R} \\ y(t_0) = y_0, \end{cases}$$

According to the fundamental Picard–Lindelöf existence and uniqueness theorem for initial value problems (with the assumptions that $f(t, y(t))$,

$f : [t_0, b] \times \mathbb{R} \to \mathbb{R}$, is a given continuous function with respect to $t, y(t)$ and Lipschitz-continuous with respect to $y(t)$), the Cauchy problem has a unique solution.

1) Find infinitely many solutions that admits the given ordinary differential equation and plot some of them.

2) Plot the unique solution of the Cauchy problem with the vector field.

3) Find the exact and approximate numerical solutions of the Cauchy problem and plot them. Compare the results.

Maple:

```
with(DETools):with(plots):setoptions(axes=boxed,numpoints=200);
ODE1:=D(y)(t)=-y(t)*cos(t); ICs:=y(0)=1; IVP1:={ODE1,ICs};
ExSol1:=dsolve(ODE1,y(t)); Sols:={seq(subs(_C1=i,rhs(ExSol1)),
 i=-10..10)}; plot(Sols,t=0..4*Pi);
DEplot(ODE1,y(t),t=0..10,[[0,1]],y=0..3);
ExSol2:=dsolve(IVP1,y(t)); NSol:=dsolve(IVP1,numeric,y(t));
G:=array(1..2); G[1]:=odeplot(NSol,[t,y(t)],0..10,color=blue):
G[2]:=plot(rhs(ExSol2),t=0..10,color=red): display(G);
```

Mathematica:

```
SetOptions[Plot,PlotRange->{0,3},PlotStyle->Thickness[0.01],
ImageSize->300]; {ODE1=y'[t]==-y[t]*Cos[t],ICs=y[0]==1,
IVP1={ODE1,ICs}}
exSol1=DSolve[ODE1,y[t],t]
sols=Table[exSol1[[1,1,2]]/.{C[1]->i},{i,-10,10}]
Plot[sols,{t,0,4*Pi},PlotRange->Automatic]
eq2=DSolve[IVP1,y[t],t]; exSol2=eq2[[1,1,2]]
eq3=NDSolve[IVP1,y,{t,0,10}]; NSol=eq3[[1,1,2]]
Table[{t,NSol[t]},{t,0,10}]//TableForm
g1=Plot[exSol2,{t,0,10},PlotStyle->Hue[0.6]];
g2=Plot[NSol[t],{t,0,10},PlotStyle->Hue[0.8]];
GraphicsRow[{g1,g2}]
<<VectorFieldPlots`
g3=VectorFieldPlot[{1,-Y*Cos[T]},{T,0,10},{Y,0,3},
 Axes->Automatic,AspectRatio->1]; Show[g2,g3]
```

In *Mathematica 7*, the new VectorPlot function has been introduced, see in more detail Sect. 6.10.   □

Problem 12.45 *Single-step methods. The Euler method and convergence analysis.* Let us consider one of the classical methods, the *forward Euler method* or *explicit Euler method.* This method belongs to a family of *single-step methods* that compute the numerical solution Y_{i+1} at the node T_{i+1} knowing the information related only to the previous node T_i.

The strategy of these methods consists of dividing the integration interval $[a, b]$ into N subintervals of length $h = (b - a)/N$, h is called the *discretization step.* Then at the nodes T_i $(0 \le i \le N)$ we compute the unknown value Y_i which approximates the exact value $y(T_i)$, i.e. $Y_i \approx y(T_i)$. The set of values $\{Y_0 = y_0, Y_1, \ldots, Y_N\}$ is the *approximate numerical solution.* The formula for the explicit Euler method reads:

$$Y_{i+1} = Y_i + hF(T_i, Y_i), \quad Y_0 = y(T_0), \quad i = 0, \ldots, N - 1.$$

1) Find the exact solution and the approximate numerical solution of a given Cauchy problem on the interval $[a, b]$ using the explicit Euler method. Compare the results and plot the exact and numerical solutions.

2) Find the absolute computational error at each step and plot it in $[a, b]$.

3) The Euler method converges with order 1, i.e. $|Y_i - y(T_i)| \le C(h)$ $(i = 0, \ldots, N)$, where $C(h) = (M/2L)(e^{L(T_i - a)} - 1)h \to 0$ as $h \to 0$. Here $|y''(t)| \le M$ for any $t \in [a, b]$ and L is the Lipschitz constant. Analyze the convergence and compare the real error and its estimated value at each node T_i.

Maple:

```
with(plots): ODE1:=diff(y(t),t)=y(t)+t^2;
ICs:=y(0)=0.5; IVP1:={ODE1,ICs};
ExSol:=unapply(rhs(dsolve(IVP1,y(t))),t);
F:=(t,y)->y+t^2;
a:=0; b:=2; N:=40; h:=evalf((b-a)/N); T:=x->a+x*h;
Y:=proc(n) option remember; Y(n-1)+h*F(T(n-1),Y(n-1)) end;
Y(0):=0.5; F1:=[seq([T(i),Y(i)],i=0..N)]; Array(F1);
for i from 0 to N do
 print(i,T(i),Y(i),evalf(ExSol(T(i))),
  evalf(abs(Y(i)-ExSol(T(i))))); od;
G1:=plot(ExSol(t),t=a..b):
G2:=plot(F1,style=point,color=red):
display({G1,G2});
```

```
F2:=[seq([T(i),abs(Y(i)-evalf(ExSol(T(i))))],i=0..N)];
G3:=plot(F2): M1:=maximize(diff(ExSol(t),t$2),t=a..b);
L1:=solve(abs(F(t,y1)-F(t,y2))=L*abs(y1-y2),L);
F3:=[seq([T(i),evalf(h*M1/(2*L1)*(exp(L1*(T(i)-a))-1))],
    i=0..N)]; plot([F2,F3],color=[red,blue]);
```

Mathematica:

```
nD=10; {a=0, b=2, n=40,
 h=N[(b-a)/n,nD], Y=Table[0,{i,1,n+1}]}
{ODE1=D[y[t],t]==y[t]+t^2, ICs=y[0]==1/2, IVP1={ODE1,ICs}}
ExtSol[t1_]:=DSolve[IVP1,y[t],t][[1,1,2]]/.{t->t1}; ExtSol[t]
F[t_,y_]:=y+t^2; T=Table[a+i*h,{i,0,n}]; Y[[1]]=N[1/2,nD];
Do[Y[[i+1]]=N[Y[[i]]+h*F[T[[i]],Y[[i]]],nD],{i,1,n}];
F1=Table[{T[[i+1]],Y[[i+1]]},{i,0,n}]
Do[Print[PaddedForm[i,3]," ",PaddedForm[T[[i]],{15,10}]," ",
 PaddedForm[Y[[i]],{15,10}]," ",
 PaddedForm[N[ExtSol[T[[i]]],nD],{15, 10}]," ",
 PaddedForm[N[Abs[Y[[i]]-ExtSol[T[[i]]]],nD],{15,10}]],
 {i,1,n+1}];
g0=Plot[ExtSol[x],{x,a,b},PlotStyle->Blue];
g1=ListPlot[F1,PlotStyle->{PointSize[.02],Hue[0.9]}];
Show[{g0,g1}]
F2=Table[{T[[i+1]],Abs[Y[[i+1]]-N[ExtSol[T[[i+1]]],nD]]},
 {i,0,n}];
g2=ListPlot[F2,Joined->True,PlotStyle->Hue[0.99],
 PlotRange->All]; f1[x1_]:=D[ExtSol[x],{x,2}]/.{x->x1};
Plot[Evaluate[f1[x]],{x,a,b}]
{M1=Evaluate[f1[b]],
 L11=Solve[Abs[F[t,y1]-F[t,y2]]==L*Abs[y1-y2],L],
 L1=Flatten[L11][[1,2]]}
F3=Table[{T[[i+1]],N[h*M1/(2*L1)*(Exp[L1*(T[[i+1]]-a)]-1),nD]},
 {i,0,n}]; g3=ListPlot[F3,Joined->True,PlotStyle->Hue[0.8],
 PlotRange->All]; Show[{g2,g3}]
```

                                                                    □

Problem 12.46 *Single-step methods. The convergence analysis of the Euler method.* There is a general way to determine the order of convergence of a numerical method. If we know the errors E_i $(i = 1, \ldots, N)$ relative to the values h_i of the discretization parameter (in our case, h_i is the discretization step of the Euler method) and assume that $E_i = Ch_i^p$ and $E_{i-1} = Ch_{i-1}^p$, then $p = \dfrac{\log(E_i/E_{i-1})}{\log(h_i/h_{i-1})}$, $i = 2, \ldots, N$.

1) Solve the same Cauchy problem as in the previous problem by explicit Euler method for various values of the discretization step h.

2) Apply the above formula and verify that the order of convergence of the explicit Euler method is 1.

Maple:

```
ODE1:=diff(y(t),t)=y(t)+t^2; ICs:=y(0)=0.5; IVP1:={ODE1,ICs};
Euler:=proc(IVP::set,a,b,N) local h,x,T,Y,F,F1,ExSol,EN;
 h:=(b-a)/N; T:=x->a+x*h; F:=(t,y)->y+t^2;
 ExSol:=unapply(rhs(dsolve(IVP,y(t))),t);
 Y:=proc(x) option remember; Y(x-1)+h*F(T(x-1),Y(x-1)) end;
 Y(0):=0.5; EN:=[seq(abs(Y(i)-evalf(ExSol(T(i)))),i=0..N)];
RETURN(EN); end:
L1:=NULL: N1:=4: for k from 1 to 12 do
E||k:=Euler(IVP1,0,2,N1):
 print(E||k[N1+1]); L1:=L1,E||k[N1+1]; N1:=N1*2; od:
Ers:=[L1]; NErs:=nops(Ers);
p:=[seq(evalf(abs(log(Ers[i]/Ers[i-1])/log(2))),i=2..NErs)];
```

Mathematica:

```
$RecursionLimit=Infinity; $HistoryLength=0; nD=10;
F[t_,y_]:=y+t^2; {a=0,b=2}
{ODE1=D[y[t],t]==y[t]+t^2,ICs=y[0]==1/2,IVP1={ODE1,ICs}}
ExtSol[t1_]:=DSolve[IVP1,y[t],t][[1,1,2]]/.{t->t1};
ExtSol[x]
Euler[a_,b_,n_]:=Module[{h,x,Y,EN},h=N[(b-a)/n,nD];
 Y[0]=N[1/2,nD]; T[x_]:=a+x*h;
 Do[Y[i_]:=Y[i]=N[Y[i-1]+h*F[T[i-1],Y[i-1]],nD],{i,1,n}];
 EN=Table[N[Abs[Y[i]-N[ExtSol[T[i]],nD]],nD],{i,0,n}]; EN];
{L1={},n1=4}
Do[Er[k]=Euler[a,b,n1]; Print[Last[Er[k]]];
 L1=Append[L1,Last[Er[k]]]; n1=n1*2,{k,1,12}];
{Ers=L1,NErs=Length[Ers]}
p=Table[N[Abs[Log[Ers[[i]]/Ers[[i-1]]]/Log[2]],nD],
 {i,2,NErs}]
```

□

Problem 12.47 *High order methods. Derivation of explicit Runge–Kutta methods.* Runge–Kutta methods are single-step methods which involve several evaluations of the function $f(t,y)$ and none of its derivatives on every interval $[T_i, T_{i+1}]$. In general, explicit or implicit Runge–Kutta formulas can be constructed with arbitrary order according to the formulas. Let us consider an s-stage explicit Runge–Kutta method:

$$k_1 = f(t_n, y_n), \quad k_2 = f(t_n + c_2 h, y_n + a_{2,1} k_1 h), \quad \dots,$$

$$k_s = f\left(t_n + c_s h, y_n + \sum_{i=1}^{s-1} a_{s,j} k_j\right),$$

$$Y_{n+1} = Y_n + h \sum_{i=1}^{s} b_i k_i, \quad Y_0 = y_0, \quad n = 0, \dots N - 1.$$

Let us perform analytical derivation of the most well-known Runge–Kutta methods.

1) Let $s = 1$, obtain the *Euler method,* $b_1 = 1$.

2) Let $s = 2$, obtain the *2-stage modified Euler method,* $a_{2,1} = 1/2$, $b_1 = 0$, $b_2 = 1$, $c_2 = 1/2$.

3) Let $s = 2$, obtain the *2-stage improved Euler method,* $a_{2,1} = 1$, $b_1 = b_2 = 1/2$, $c_2 = 1$.

4) Let $s = 2$, obtain the *2-stage Heun method,* $a_{2,1} = 2/3$, $b_1 = 1/4$, $b_2 = 3/4$, $c_2 = 2/3$.

5) Let $s = 3$, obtain the *3-stage Heun method,* $a_{2,1} = 1/3$, $a_{3,1} = 0$, $a_{3,2} = 2/3$, $b_1 = 1/4$, $b_2 = 0$, $b_3 = 3/4$, $c_2 = 1/3$, $c_3 = 2/3$.

6) Let $s = 4$, obtain the *Runge–Kutta* of order four (introduced by Runge in 1895), $a_{2,1} = 1/2$, $a_{3,1} = 0$, $a_{3,2} = 1/2$, $a_{4,1} = 0$, $a_{4,2} = 0$, $a_{4,3} = 1$, $b_1 = 1/6$, $b_2 = 1/3$, $b_3 = 1/3$, $b_4 = 1/6$, $c_2 = 1/2$, $c_3 = 1/2$, $c_4 = 1$.

Maple:

```
h0:=h=0; alias(F=f(t,y(t)),Ft=D[1](f)(t,y(t)),
 Fy=D[2](f)(t,y(t)),Ftt=D[1,1](f)(t,y(t)),Fty=D[1,2](f)(t,y(t)),
 Fyy=D[2,2](f)(t,y(t)),Fyyy=D[2,2,2](f)(t,y(t)),
 Fttt=D[1,1,1](f)(t,y(t)),Ftyy=D[1,2,2](f)(t,y(t)),
 Ftty=D[1,1,2](f)(t,y(t))); D(y):=t->f(t,y(t)); s:=1;
P1:=convert(taylor(y(t+h),h0,s+1),polynom);
P2:=expand((P1-y(t))/h); k1:=taylor(f(t,y(t)),h0,s);
P3:=expand(convert(taylor(add(b[i]*k||i,i=1..s),h0,s),polynom));
Eq1:=P2-P3; Eq2:={coeffs(Eq1,[h,F])};Sol:=solve(Eq2,indets(Eq2));
s:=2;P1:=convert(taylor(y(t+h),h0,s+1),polynom);
P2:=expand((P1-y(t))/h); k1:=taylor(f(t,y(t)),h0,s);
k2:=taylor(f(t+c[2]*h,y(t)+h*add(a[2,i]*k||i,i=1..2-1)),h0,s);
P3:=expand(convert(taylor(add(b[i]*k||i,i=1..s),h0,s),polynom));
Eq1:=P2-P3; Eq2:={coeffs(Eq1,[h,F,Ft,Fy])}; Eq3:={}:
```

```
for i from 2 to s do
 Eq3:=Eq3 union{c[i]=add(a[i,j],j=1..i-1)}; od;
Sol1:=solve(Eq2 union Eq3 union {c[2]=1/2},indets(Eq2));
Sol2:=solve(Eq2 union Eq3 union {b[2]=1/2},indets(Eq2));
Sol3:=solve(Eq2 union Eq3 union {b[2]=3/4},indets(Eq2));
s:=3;P1:=taylor(y(t+h),h0,s+1); P2:=expand(convert(
 expand((P1-y(t))/h),polynom)); k1:=taylor(f(t,y(t)),h0,s):
k2:=taylor(f(t+c[2]*h,y(t)+h*(add(a[2,i]*k||i,i=1..2-1))),h0,s);
k3:=taylor(f(t+c[3]*h,y(t)+h*(add(a[3,i]*k||i,i=1..3-1))),h0,s);
P3:=expand(convert(taylor(add(b[i]*k||i,i=1..s),h0,s),polynom)):
Eq1:=P2-P3: Eq2:={coeffs(Eq1,[h,F,Ft,Fy,Ftt,Fty,Fyy])}; Eq3:={}:
for i from 2 to s do Eq3:=Eq3 union {c[i]=add(a[i,j],j=1..i-1)};
od; Sol:=solve(Eq2 union Eq3 union {b[1]=1/4,c[2]=1/3},
 indets(Eq2)); s:=4;P1:=taylor(y(t+h),h0,s+1);
P2:=expand(convert(expand((P1-y(t))/h),polynom));
k1:=taylor(f(t,y(t)),h0,s):
k2:=taylor(f(t+c[2]*h,y(t)+h*(add(a[2,i]*k||i,i=1..2-1))),h0,s);
k3:=taylor(f(t+c[3]*h,y(t)+h*(add(a[3,i]*k||i,i=1..3-1))),h0,s);
k4:=taylor(f(t+c[4]*h,y(t)+h*(add(a[4,i]*k||i,i=1..4-1))),h0,s);
P3:=expand(convert(taylor(add(b[i]*k||i,i=1..s),h0,s),polynom)):
Eq1:=P2-P3: Eq2:={coeffs(Eq1,[h,F,Ft,Fy,Ftt,Fty,Fyy,Fttt,Ftty,
 Ftyy,Fyyy])}; Eq3:={}: for i from 2 to s do Eq3:=Eq3 union
 {c[i]=add(a[i,j],j=1..i-1)}; od; Sol:=solve(Eq2 union Eq3 union
 {b[1]=1/6,c[2]=1/2,a[3,2]=1/2},indets(Eq2));
```

Mathematica:

```
SubsF={Dt[F]->Dt[f[t,y[t]]]}; Subs1={D[y[t],t]->F};
Subs2={D[y[t],{t,2}]->Dt[f[t,y[t]]]}; Subs3={D[y[t],{t,3}]->
 Dt[Dt[f[t,y[t]]]]}; Subs4={D[y[t],{t,4}]->
 Dt[Dt[Dt[f[t,y[t]]]]]}; SubsD={f[t,y[t]]->F,Dt[t]->1,
 (D[f[t,t1],t]/.{t1->y[t]})->Ft, (D[f[t,t1],t1]/.
 {t1->y[t]})->Fy, (D[f[t,t1],{t,2}]/.{t1->y[t]})->Ftt,
 (D[f[t,t1],t1,t]/.{t1->y[t]})->Fty, (D[f[t,t1],{t1,2}]/.
 {t1->y[t]})->Fyy, (D[f[t,t1],{t1,3}]/.{t1->y[t]})->Fyyy,
 (D[f[t,t1],{t,3}]/.{t1->y[t]})->Fttt, (D[f[t,t1],t,t1,t1]/.
 {t1->y[t]})->Ftyy, (D[f[t,t1],t,t,t1]/.{t1->y[t]})->Ftty};
{s=1, P1=Normal[Series[y[t+h],{h,0,s}]]/.Subs1, P2=Expand[
 (P1-y[t])/h], k[1]=Series[f[t,y[t]],{h,0,s}]/.SubsD, P3=Expand[
 Normal[Series[Sum[b[i]*k[i],{i,1,s}],{h,0,s}]]]/.SubsD, Eq1=
 P2-P3, Eq2=DeleteCases[Flatten[CoefficientList[Eq1,{h,F}]],_0]}
Sol=Solve[Eq2==0,Variables[Eq2]]
```

```
{s=2, P1=Normal[Series[y[t+h],{h,0,s}]]/.Subs2/.Subs1/.SubsD,
 P2=Expand[(P1-y[t])/h]}
{k[1]=Series[f[t,y[t]],{h,0,s}]/.SubsD, k[2]=Series[f[t+c[2]*h,
 y[t]+h*Sum[a[2,i]*k[i],{i,1,2-1}]],{h,0,s-1}]/.SubsD,
 P3=Expand[Normal[Series[Sum[b[i]*k[i],{i,1,s}],
 {h,0,s}]]]/.SubsD, Eq1=P2-P3, Eq21=DeleteCases[Flatten[
 CoefficientList[Eq1,{h,F,Ft,Fy}]],_0],
 Eq2=Map[Thread[#1==0,Equal]&,Eq21]}
Eq3={}; Do[Eq3=Append[Eq3,{c[i]==Sum[a[i,j],{j,1,i-1}]}],
 {i,2,s}]; {S1=Flatten[{Eq2,Eq3,{c[2]==1/2}}], S2=Flatten[
 {Eq2,Eq3,{b[2]==1/2}}],S3=Flatten[{Eq2,Eq3,{b[2]==3/4}}]}
{Sol1=Solve[S1,Variables[Eq2]], Sol2=Solve[S2,Variables[Eq2]],
 Sol3=Solve[S3,Variables[Eq2]]}
{s=3,
 P1=Normal[Series[y[t+h],{h,0,s}]]/.Subs3/.Subs2/.Subs1/.SubsD,
 P2=Expand[Normal[Expand[(P1-y[t])/h]]]}
{k[1]=Series[f[t,y[t]],{h,0,s}]/.SubsD, k[2]=Series[f[t+c[2]*h,
 y[t]+h*(Sum[a[2,i]*k[i],{i,1,2-1}])],{h,0,s-1}]/.SubsD,
 k[3]=Series[f[t+c[3]*h,y[t]+h*(Sum[a[3,i]*k[i],{i,1,3-1}])],
 {h,0,s-1}]/.SubsD, P3=Expand[Normal[Series[Sum[b[i]*k[i],
 {i,1,s}],{h,0,s}]]], Eq1=P2-P3, Eq21=DeleteCases[Flatten[
 CoefficientList[Eq1,{h,F,Ft,Fy,Ftt,Fty,Fyy}]],_0],
 Eq2=Map[Thread[#1==0,Equal]&,Eq21]}
Eq3={}; Do[Eq3=Append[Eq3,{c[i]==Sum[a[i,j],{j,1,i-1}]}],
 {i,2,s}]; S1=Flatten[{Eq2,Eq3,{b[1]==1/4},{c[2]==1/3}}]
Sol1=Solve[S1,Variables[Eq2]]
{s=4, P1=Normal[Series[y[t+h],{h,0,s}]]/.Subs4/.Subs3/.Subs2/.
 Subs1/.SubsD, P2=Expand[Normal[Expand[(P1-y[t])/h]]]}
{k[1]=Series[f[t,y[t]],{h,0,s}]/.SubsD,
 k[2]=Series[f[t+c[2]*h,y[t]+h*(Sum[a[2,i]*k[i],{i,1,2-1}])],
 {h,0,s-1}]/.SubsD, k[3]=Series[f[t+c[3]*h,y[t]+
 h*(Sum[a[3,i]*k[i],{i,1,3-1}])],{h,0,s-1}]/.SubsD,
 k[4]=Series[f[t+c[4]*h,y[t]+h*(Sum[a[4,i]*k[i],{i,1,4-1}])],
 {h,0,s-1}]/.SubsD, P3=Expand[Normal[Series[Sum[b[i]*k[i],
 {i,1,s}],{h,0,s}]]]/.SubsD, Eq1=P2-P3,
 Eq21=DeleteCases[Flatten[CoefficientList[Eq1,
 {h,F,Ft,Fy,Ftt,Fty,Fyy,Fttt,Ftty,Ftyy,Fyyy}]],_0],
 Eq2=Map[Thread[#1==0,Equal]&,Eq21]}
Eq3={}; Do[Eq3=Append[Eq3,{c[i]==Sum[a[i,j],{j,1,i-1}]}],
 {i,2,s}]; S1=Flatten[{Eq2,Eq3,{b[1]==1/6},{c[2]==1/2},
 {a[3,2]==1/2}}]; Sol1 = Solve[S1, Variables[Eq2]]
```

☐

Problem 12.48   *High order methods.   Multi-step methods.   The explicit Adams–Bashforth method.*   There are more sophisticated methods that achieve a high order of accuracy by considering several values $(Y_i, Y_{i-1}, \dots)$ to determine Y_{i+1}. One of the most notable methods is the explicit four-steps fourth-order Adams–Bashforth method (AB4):

$$Y_{i+1} = Y_i + \frac{h}{24}\left(55F(T_i, Y_i) - 59F(T_{i-1}, Y_{i-1}) + 37F(T_{i-2}, Y_{i-2}) - 9F(T_{i-3}, Y_{i-3})\right).$$

1) Find the exact and approximate numerical solutions of the same Cauchy problem as in the previous problem in $[a, b]$ by applying the explicit Adams–Bashforth method. Compare the results and plot the exact and numerical solutions.

2) Find the absolute computational error at each step and plot it in $[a, b]$.

Maple:

```
with(plots): ODE1:=diff(y(t),t)=y(t)+t^2; ICs:=y(0)=0.5;
IVP1:={ODE1,ICs}; ExSol:=unapply(rhs(dsolve(IVP1,y(t))),t);
F:=(t,y)->y+t^2; a:=0; b:=2; N:=40; h:=evalf((b-a)/N);
T:=x->a+x*h; Y_AB:= proc(n) option remember;
  Y_AB(n-1)+h/24*(55*F(T(n-1),Y_AB(n-1))-59*F(T(n-2),Y_AB(n-2))
  +37*F(T(n-3),Y_AB(n-3))-9*F(T(n-4),Y_AB(n-4))); end:
Y_AB(0):=0.5: Y_AB(1):=evalf(ExSol(T(1)));
Y_AB(2):=evalf(ExSol(T(2))); Y_AB(3):=evalf(ExSol(T(3)));
F1:=[seq([T(i),Y_AB(i)],i=0..N)];
for i from 0 to N do print(T(i),evalf(ExSol(T(i))),Y_AB(i),
  evalf(abs(Y_AB(i)-ExSol(T(i))))): od:
G1:=plot(ExSol(t),t=a..b): G2:=plot(F1,style=point,color=red):
F2:=[seq([T(i),abs(Y_AB(i)-evalf(ExSol(T(i))))],i=0..N)];
display({G1,G2}); plot(F2);
```

Mathematica:

```
nD=10; {a=0,b=2,n=40,h=N[(b-a)/n,nD],
  YAB=Table[0,{i,1,n+1}]}
{ODE1=D[y[t],t]==y[t]+t^2,ICs=y[0]==1/2,IVP1={ODE1,ICs}}
ExtSol[t1_]:=DSolve[IVP1,y[t],t][[1,1,2]]/.{t->t1}; ExtSol[t]
F[t_,y_]:=y+t^2; T=Table[a+i*h,{i,0,n}]; YAB[[1]]=N[1/2,nD];
YAB[[2]]=N[ExtSol[T[[2]]],nD]; YAB[[3]]=N[ExtSol[T[[3]]],nD];
```

```
YAB[[4]]=N[ExtSol[T[[4]]],nD];
Do[YAB[[i+1]]=N[YAB[[i]]+h/24*(55*F[T[[i]],YAB[[i]]]-
  59*F[T[[i-1]],YAB[[i-1]]]+37*F[T[[i-2]],YAB[[i-2]]]-
  9*F[T[[i-3]],YAB[[i-3]]]),nD],{i,4,n}];
F1=Table[{T[[i+1]],YAB[[i+1]]},{i,0,n}]
Do[Print[PaddedForm[i,3]," ",PaddedForm[T[[i]],{15,10}]," ",
  PaddedForm[YAB[[i]],{15,10}]," ",PaddedForm[N[ExtSol[T[[i]]],
  nD],{15, 10}]," ",PaddedForm[N[Abs[YAB[[i]]-ExtSol[T[[i]]]],
  nD],{15,10}]],{i,1,n+1}];
g0=Plot[ExtSol[x],{x,a,b},PlotStyle->Blue];
g1=ListPlot[F1,PlotStyle->{PointSize[.02],Hue[0.9]}];
Show[{g0,g1}]
F2=Table[{T[[i+1]],Abs[YAB[[i+1]]-N[ExtSol[T[[i+1]]],nD]]},
  {i,0,n}];
ListPlot[F2,Joined->True,PlotStyle->Hue[0.99],PlotRange->All]
```

$$\square$$

Problem 12.49 *High order methods. Multistep methods. The implicit Adams–Moulton method.* Another important example of multistep methods is the implicit three-steps fourth-order Adams–Moulton method (AM4):

$$Y_{i+1} = Y_i + \frac{h}{24}(9F(T_{i+1},Y_{i+1}) + 19F(T_i,Y_i) - 5F(T_{i-1},Y_{i-1}) + F(T_{i-2},Y_{i-2})).$$

1) Find the exact and approximate numerical solutions of the same Cauchy problem as in the previous problem in $[a,b]$ by applying the implicit Adams–Moulton method. Compare the results and plot the exact and numerical solutions.

2) Find the absolute computational error at each step and plot it in $[a,b]$.

Maple:

```
with(plots): with(codegen): ODE1:=diff(y(t),t)=y(t)+t^2;
ICs:=y(0)=0.5; IVP1:={ODE1,ICs};
ExSol:=unapply(rhs(dsolve(IVP1,y(t))),t); F:=(t,y)->y+t^2;
a:=0; b:=2; N:=20; h:=evalf((b-a)/N); T:=x->a+x*h;
Eq1:=Y_AM(i)-Y_AM(i-1)-h/24*(9*F(T(i),Y_AM(i))+19*F(T(i-1),
  Y_AM(i-1))-5*F(T(i-2),Y_AM(i-2))+F(T(i-3),Y_AM(i-3)));
Eq2:=solve(Eq1,Y_AM(i)); Y_AM:=makeproc(Eq2,i); Y_AM(0):=0.5;
Y_AM(1):=evalf(ExSol(T(1))); Y_AM(2):=evalf(ExSol(T(2)));
```

```
Y_AM(3):=evalf(ExSol(T(3))); F1:=[seq([T(i),Y_AM(i)],i=0..N)];
for i from 0 to N do print(T(i), evalf(ExSol(T(i))),
 Y_AM(i),evalf(abs(Y_AM(i)-ExSol(T(i))))): od:
G1:=plot(ExSol(t),t=a..b): G2:=plot(F1,style=point,color=red):
display({G1,G2}); F2:=[seq([T(i),
 abs(Y_AM(i)-evalf(ExSol(T(i))))],i=0..N)]; plot(F2);
```

Mathematica:

```
nD=10; {a=0,b=2,n=20,h=N[(b-a)/n,nD],
 YAM=Table[0,{i,1,n+1}]}
{ODE1=D[y[t],t]==y[t]+t^2, ICs=y[0]==1/2, IVP1={ODE1,ICs}}
ExtSol[t1_]:=DSolve[IVP1,y[t],t][[1,1,2]]/.{t->t1}; ExtSol[t]
F[t_,y_]:=y+t^2; T[x_]:=a+x*h; T1=Table[a+i*h,{i,0,n}];
{Eq1=Y[i]-Y[i-1]-h/24*(9*F[T[i],Y[i]]+19*F[T[i-1],Y[i-1]]-
 5*F[T[i-2],Y[i-2]]+F[T[i-3],Y[i-3]]),
 Eq21=Solve[Eq1==0,Y[i]][[1,1,2]], Eq2=Expand[Eq21]}
YAM[[1]]=N[1/2,nD]; YAM[[2]]=N[ExtSol[T1[[2]]],nD];
YAM[[3]]=N[ExtSol[T1[[3]]],nD];YAM[[4]]=N[ExtSol[T1[[4]]],nD];
Do[YAM[[i+1]]=N[Eq2/.{Y[i-1]->YAM[[i]],Y[i-2]->YAM[[i-1]],
 Y[i-3]->YAM[[i-2]]},nD],{i,4,n}];
F1=Table[{T1[[i+1]],YAM[[i+1]]},{i,0,n}]
Do[Print[PaddedForm[i,3]," ",PaddedForm[T1[[i]],{15,10}]," ",
 PaddedForm[YAM[[i]],{15,10}]," ",
 PaddedForm[N[ExtSol[T1[[i]]],nD],{15, 10}]," ",
 PaddedForm[N[Abs[YAM[[i]]-ExtSol[T1[[i]]]],nD],{15,10}]],
 {i,1,n+1}];
g0=Plot[ExtSol[x],{x,a,b},PlotStyle->Blue];
g1=ListPlot[F1,PlotStyle->{PointSize[.02],Hue[0.9]}];
Show[{g0,g1}]
F2=Table[{T1[[i+1]],Abs[YAM[[i+1]]-N[ExtSol[T1[[i+1]]],nD]]},
 {i,0,n}];
ListPlot[F2,Joined->True,PlotStyle->Hue[0.99],PlotRange->All]
```

□

Problem 12.50 *Systems of ordinary differential equations.* Let us consider the system of first-order ordinary differential equations $y_i'(t) = f_i\big(t, y_1(t), \ldots, y_n(t)\big)$ $(i = 1, \ldots, n)$ with the initial conditions $y_i(t_0) = y_{0i}$ $(i = 1, \ldots, n)$. Here the unknowns are $y_1(t), \ldots, y_n(t))$ and $t \in [t_0, b]$. To obtain numerical solutions we can apply one of the above methods (developed for a single equation) to each equation of the system.

1) Let $n = 2$, $a = t_0 = 0$, $b = 2$. Find the exact and approximate numerical solutions of the Cauchy problem, $y_1' = y_2$, $y_2' = t - y_1 - 2y_2$, $y_1(t_0) = 1$, $y_2(t_0) = 1$, in $[a, b]$ by applying the explicit Runge–Kutta of order four method (RK4).

2) Compare the results and plot the exact and numerical solutions.

Maple:

```
with(plots): F1:=(t,u,v)->v; F2:=(t,u,v)->t-u-2*v;
N:=10: a:=0: b:=2: h:=evalf((b-a)/N); T:=x->a+x*h;
ODEsys:=diff(u(t),t)=v(t),diff(v(t),t)=t-u(t)-2*v(t);
ICs:=u(0)=1,v(0)=1; IVP1:={ODEsys,ICs};
ExSol:=sort(dsolve(IVP1,{u(t),v(t)},method=laplace));
uEx:=unapply(rhs(ExSol[1]),t); vEx:=unapply(rhs(ExSol[2]),t);
RK41:=proc(i,F1,F2,K) local k1,k2,k3,k4,m1,m2,m3,m4;
 option remember;
 k1:=h*F1(T(i-1),RK41(i-1,F1,F2,RK41),RK41(i-1,F1,F2,RK42));
 m1:=h*F2(T(i-1),RK41(i-1,F1,F2,RK41),RK41(i-1,F1,F2,RK42));
 k2:=h*F1(T(i-1)+h/2,RK41(i-1,F1,F2,RK41)+k1/2,
 RK41(i-1,F1,F2,RK42)+ m1/2); m2:=h*F2(T(i-1)+h/2,
 RK41(i-1,F1,F2,RK41)+k1/2,RK41(i-1,F1,F2,RK42)+ m1/2);
 k3:=h*F1(T(i-1)+h/2,RK41(i-1,F1,F2,RK41)+k2/2,
 RK41(i-1,F1,F2,RK42)+ m2/2); m3:=h*F2(T(i-1)+h/2,
 RK41(i-1,F1,F2,RK41)+k2/2,RK41(i-1,F1,F2,RK42)+ m2/2);
 k4:=h*F1(T(i-1)+h,RK41(i-1,F1,F2,RK41)+k3,
 RK41(i-1,F1,F2,RK42)+m3); m4:=h*F2(T(i-1)+h,
 RK41(i-1,F1,F2,RK41)+k3,RK41(i-1,F1,F2,RK42)+m3);
 if K=RK41 then
  evalf(RK41(i-1,F1,F2,RK41)+1/6*(k1+2*k2+2*k3+k4)); else
  evalf(RK41(i-1,F1,F2,RK42)+1/6*(m1+2*m2+2*m3+m4));fi;end;
RK41(0,F1,F2,RK41):=1: RK41(0,F1,F2,RK42):=1:
array([seq([T(i),RK41(i,F1,F2,RK41),evalf(uEx(T(i))),
 RK41(i,F1,F2,RK42),evalf(vEx(T(i)))],i=0..N)]);
uF1:=[seq([T(i),RK41(i,F1,F2,RK41)],i=0..N)];
uG1:=plot(uEx(t),t=a..b,color=red):
vF1:=[seq([T(i),RK41(i,F1,F2,RK42)],i=0..N)];
vG1:=plot(vEx(t),t=a..b,color=blue):
uG2:=plot(uF1,style=point,color=red):
vG2:=plot(vF1,style=point,color=blue):
display({uG1,uG2,vG1,vG2});
```

Mathematica:

```
F1[t_,u_,v_]:=v; F2[t_,u_,v_]:=t-u-2*v; nD=10;
{n=10,a=0,b=2,h=N[(b-a)/n,nD],T=Table[a+i*h,{i,0,n}],
 RK41=Table[0,{i,0,n}], RK42=Table[0,{i,0,n}]}
{ODEsys={D[u[t],t]==v[t],D[v[t],t]==t-u[t]-2*v[t]},
 ICs={u[0]==1,v[0]==1}, IVP1=Flatten[{ODEsys,ICs}]}
ExtSol=Sort[DSolve[IVP1,{u[t],v[t]},t]]
uExt[t1_]:=ExtSol[[1,1,2]]/.{t->t1};
vExt[t1_]:=ExtSol[[1,2,2]]/.{t->t1};
{RK41[[1]]=1,RK42[[1]]=1}
Do[k1=h*F1[T[[i]],RK41[[i]],RK42[[i]]];
 m1=h*F2[T[[i]],RK41[[i]],RK42[[i]]]; k2=h*F1[T[[i]]+h/2,
 RK41[[i]]+k1/2,RK42[[i]]+m1/2]; m2=h*F2[T[[i]]+h/2,
 RK41[[i]]+k1/2,RK42[[i]]+m1/2]; k3=h*F1[T[[i]]+h/2,
 RK41[[i]]+k2/2,RK42[[i]]+m2/2]; m3=h*F2[T[[i]]+h/2,
 RK41[[i]]+k2/2,RK42[[i]]+m2/2]; k4=h*F1[T[[i]]+h,
 RK41[[i]]+k3,  RK42[[i]]+m3];   m4=h*F2[T[[i]]+h,
 RK41[[i]]+k3,  RK42[[i]]+m3];
 RK41[[i+1]]=N[RK41[[i]]+1/6*(k1+2*k2+2*k3+k4),nD];
 RK42[[i+1]]=N[RK42[[i]]+1/6*(m1+2*m2+2*m3+m4),nD],
{i,1,n}];
Do[Print[PaddedForm[i, 3]," ",PaddedForm[T[[i+1]],{12,10}]," ",
 PaddedForm[RK41[[i+1]],{12,10}]," ",
 PaddedForm[N[uExt[T[[i+1]]],nD],{12,10}]," ",
 PaddedForm[RK42[[i+1]],{12,10}]," ",
 PaddedForm[N[vExt[T[[i+1]]],nD],{12,10}]],
{i,0,n}];
uF1=Table[{T[[i+1]],RK41[[i+1]]},{i,0,n}]
uG1=Plot[uExt[t],{t,a,b},PlotStyle->Red,PlotRange->All];
vF1=Table[{T[[i+1]],RK42[[i+1]]},{i,0,n}]
vG1=Plot[vExt[t],{t,a,b},PlotStyle->Blue,PlotRange->All];
uG2=ListPlot[uF1,PlotStyle->{PointSize[.02],Hue[0.99]}];
vG2=ListPlot[vF1,PlotStyle->{PointSize[.02],Hue[0.7]}];
Show[{uG1,uG2,vG1,vG2}]
```

High order ordinary differential equations. If we consider an ordinary differential equation of order n $(n > 1)$, $y^{(n)}(t) = f(t, y, y', \ldots, y^{n-1})$ $(t \in [t_0, b])$ with n initial conditions $y(t_0) = y_0$, $y'(t_0) = y_1$, $\ldots$, $y^{(n-1)}(t_0) = y_n$, we can always approximate the solution of this high order differential equation by transforming it to the equivalent system of n first-order equations and by applying an appropriate numerical method to this system of differential equations. □

Problem 12.51 *Boundary value problems. Linear shooting methods.*
Let us consider the boundary value problem for the ordinary differential
equation of the second order:

$$y'' = f(x, y, y'), \quad x \in [a, b], \quad y(a) = \alpha, \quad y(b) = \beta.$$

Let $f : [a, b] \times \mathbb{R}^2 \to \mathbb{R}$ be a continuous function. Also we assume that the
functions $f_y(x, y, y')$, and $f_{y'}(x, y, y')$ are continuous in an open domain
$D = \{(x, y, y') |\ a \leq x \leq b,\ -\infty < y < \infty,\ -\infty < y' < \infty\}$.
If $f_y(x, y, y') > 0$ for all $(x, y, y') \in D$ and there exist the constants M
and K such that $|f_y(x, y, y')| \leq M$, $|f_{y'}(x, y, y')| \leq K$ for all $(x, y, y') \in D$,
then the boundary value problem has a unique solution.
For the special case, when the function $f(x, y, y')$ is linear, i.e.

$$f(x, y, y') = p(x)y' + q(x)y + r(x),$$

the boundary value problem has a unique solution if $p(x)$, $q(x)$, and $r(x)$
are continuous in $[a, b]$ and $q(x) > 0$ in $[a, b]$.
The linear shooting methods employ the numerical methods discussed above for solving initial value problems, e.g.

$$y_1'' = p(x)y_1' + q(x)y_1 + r(x), \quad x \in [a, b], \quad y_1(a) = \alpha, \quad y_1'(a) = 0,$$
$$y_2'' = p(x)y_2' + q(x)y_2, \quad x \in [a, b], \quad y_2(a) = 0, \quad y_2'(a) = 1,$$

and the approximate solution of the original boundary value problem is
$y(x) = y_1(x) + y_2(x)(\beta - y_1(b))/y_2(b)$.
1) Find the exact solution and the approximate numerical solution
of the boundary value problem $y'' = -(2/x)y' + (2/x^2)y + x^3$, $x \in [1, 2]$,
$y(1) = 1$, $y(2) = 2$ by applying the linear shooting method.
2) Compare the results and plot the exact and numerical solutions.

Maple:

```
Fu1:=(x,u1,u2)->u2:
Fu2:=(x,u1,u2)->-2/x*u2+2/x^2*u1+x^3:
Fv1:=(x,v1,v2)->v2:
Fv2:=(x,v1,v2)->-2/x*v2+2/x^2*v1:
N:=10: a:=1: b:=2: h:=evalf((b-a)/N); X:=i->a+h*i;
alpha:=1; beta:=2;
ODE1:=(D@@2)(y)(x)+2/x*D(y)(x)-2/x^2*y(x)-x^3;
BCs:=y(a)=alpha,y(b)=beta; BVP1:={ODE1,BCs};
ExSol:=unapply(rhs(dsolve(BVP1,y(x))),x);
```

```
RK41:=proc(i,F1,F2,K) local k1,k2,k3,k4,m1,m2,m3,m4;
 option remember;
 k1:=h*F1(X(i-1),RK41(i-1,F1,F2,RK41),RK41(i-1,F1,F2,RK42));
 m1:=h*F2(X(i-1),RK41(i-1,F1,F2,RK41),RK41(i-1,F1,F2,RK42));
 k2:=h*F1(X(i-1)+h/2,RK41(i-1,F1,F2,RK41)+k1/2,
  RK41(i-1,F1,F2,RK42)+m1/2);
 m2:=h*F2(X(i-1)+h/2,RK41(i-1,F1,F2,RK41)+k1/2,
  RK41(i-1,F1,F2,RK42)+m1/2);
 k3:=h*F1(X(i-1)+h/2,RK41(i-1,F1,F2,RK41)+k2/2,
  RK41(i-1,F1,F2,RK42)+m2/2);
 m3:=h*F2(X(i-1)+h/2,RK41(i-1,F1,F2,RK41)+k2/2,
  RK41(i-1,F1,F2,RK42)+m2/2);
 k4:=h*F1(X(i-1)+h,RK41(i-1,F1,F2,RK41)+k3,
 RK41(i-1,F1,F2,RK42)+m3);
 m4:=h*F2(X(i-1)+h,RK41(i-1,F1,F2,RK41)+k3,
 RK41(i-1,F1,F2,RK42)+m3);
 if K=RK41 then evalf(RK41(i-1,F1,F2,RK41)
  +1/6*(k1+2*k2+2*k3+k4)); else evalf(RK41(i-1,F1,F2,RK42)
  +1/6*(m1+2*m2+2*m3+m4)); fi; end;
RK41(0,Fu1,Fu2,RK41):=alpha; RK41(0,Fu1,Fu2,RK42):=0;
RK41(0,Fv1,Fv2,RK41):=0; RK41(0,Fv1,Fv2,RK42):=1;
C:=(beta-RK41(N,Fu1,Fu2,RK41))/RK41(N,Fv1,Fv2,RK41);
Y:=proc(i) option remember;
  evalf(RK41(i,Fu1,Fu2,RK41)+C*RK41(i,Fv1,Fv2,RK41));end:
array([seq([RK41(i,Fu1,Fu2,RK41),RK41(i,Fv1,Fv2,RK41),Y(i),
 evalf(ExSol(X(i))),abs(Y(i)-evalf(ExSol(X(i))))],i=0..N)]);
with(plots): F1:=[seq([X(i),Y(i)],i=0..N)];
G1:=plot(ExSol(t),t=a..b,color=red):
G2:=plot(F1,style=point,color=blue): display({G1,G2});
```

Mathematica:

```
nD=10; Fu1[x_,u1_,u2_]:=u2;
Fu2[x_,u1_,u2_]:=-2/x*u2+2/x^2*u1+x^3; Fv1[x_,v1_,v2_]:=v2;
Fv2[x_,v1_,v2_]:=-2/x*v2+2/x^2*v1; {n=10,a=1,b=2,
 h=N[(b-a)/n,nD],X=Table[a+h*i,{i,0,n}],alpha=1,beta=2,
 RK41u=Table[0,{i,0,n}],RK42u=Table[0,{i,0,n}],
 RK41v=Table[0,{i,0,n}],RK42v=Table[0,{i,0,n}],
 Y=Table[0,{i,0,n}]}
{ODE1=D[y[x],{x,2}]+2/x*D[y[x],x]-2/x^2*y[x]-x^3==0,
 BCs={y[a]==alpha,y[b]==beta}, BVP1={ODE1,BCs}}
ExtSol[x1_]:=Expand[DSolve[BVP1,y[x],x][[1,1,2]]]/.{x->x1};
```

```
{ExtSol[x], RK41u[[1]]=alpha, RK42u[[1]]=0, RK41v[[1]]=0,
RK42v[[1]]=1}
Do[k1=h*Fu1[X[[i]],RK41u[[i]],RK42u[[i]]];
 m1=h*Fu2[X[[i]],RK41u[[i]],RK42u[[i]]];
 k2=h*Fu1[X[[i]]+h/2,RK41u[[i]]+k1/2,RK42u[[i]]+m1/2];
 m2=h*Fu2[X[[i]]+h/2,RK41u[[i]]+k1/2,RK42u[[i]]+m1/2];
 k3=h*Fu1[X[[i]]+h/2,RK41u[[i]]+k2/2,RK42u[[i]]+m2/2];
 m3=h*Fu2[X[[i]]+h/2,RK41u[[i]]+k2/2,RK42u[[i]]+m2/2];
 k4=h*Fu1[X[[i]]+h,RK41u[[i]]+k3,RK42u[[i]]+m3];
 m4=h*Fu2[X[[i]]+h,RK41u[[i]]+k3,RK42u[[i]]+m3];
 RK41u[[i+1]]=N[RK41u[[i]]+1/6*(k1+2*k2+2*k3+k4),nD];
 RK42u[[i+1]]=N[RK42u[[i]]+1/6*(m1+2*m2+2*m3+m4),nD],{i,1,n}];
Do[k1=h*Fv1[X[[i]],RK41v[[i]],RK42v[[i]]];
 m1=h*Fv2[X[[i]],RK41v[[i]],RK42v[[i]]];
 k2=h*Fv1[X[[i]]+h/2,RK41v[[i]]+k1/2,RK42v[[i]]+m1/2];
 m2=h*Fv2[X[[i]]+h/2,RK41v[[i]]+k1/2,RK42v[[i]]+m1/2];
 k3=h*Fv1[X[[i]]+h/2,RK41v[[i]]+k2/2,RK42v[[i]]+m2/2];
 m3=h*Fv2[X[[i]]+h/2,RK41v[[i]]+k2/2,RK42v[[i]]+m2/2];
 k4=h*Fv1[X[[i]]+h,RK41v[[i]]+k3,RK42v[[i]]+m3];
 m4=h*Fv2[X[[i]]+h,RK41v[[i]]+k3,RK42v[[i]]+m3];
 RK41v[[i+1]]=N[RK41v[[i]]+1/6*(k1+2*k2+2*k3+k4),nD];
 RK42v[[i+1]]=N[RK42v[[i]]+1/6*(m1+2*m2+2*m3+m4),nD],{i,1,n}];
C1=(beta-RK41u[[n+1]])/RK41v[[n+1]]
Do[Y[[i]]=N[RK41u[[i]]+C1*RK41v[[i]],nD],{i,1,n+1}];
Do[Print[PaddedForm[i,3]," ",PaddedForm[RK41u[[i+1]],{12,10}],
 " ",PaddedForm[RK41v[[i+1]],{12,10}]," ",PaddedForm[Y[[i+1]],
 {12,10}]," ",PaddedForm[N[ExtSol[X[[i+1]]],nD],{12,10}]," ",
 PaddedForm[Abs[Y[[i+1]]-N[ExtSol[X[[i+1]]],nD]],{12,10}]],
 {i,0,n}]; F1=Table[{X[[i+1]],Y[[i+1]]},{i,0,n}]
g1=Plot[ExtSol[t],{t,a,b},PlotStyle->Hue[0.99]]; g2=ListPlot[
 F1,PlotStyle->{PointSize[.02],Hue[0.7]}]; Show[{g1,g2}]
```
                                                                      □

Problem 12.52 *Boundary value problems. Nonlinear shooting methods.*
In addition to the nonlinear boundary value problem,

$$y'' = f(x, y, y'), \quad x \in [a, b], \quad y(a) = \alpha, \quad y(b) = \beta,$$

let us consider the initial value problem, $y'' = f(x, y, y')$, $y(a) = \alpha$, $y'(a) = s$, where $x \in [a, b]$, $s \in \mathbb{R}$. The parameter s describes the initial slope of the solution curve. Let $f: [a, b] \times \mathbb{R}^2 \to \mathbb{R}$ be a continuous function and satisfies the Lipschitz condition with respect to y and y'. Then by the Picard–Lindelöf theorem, for each $s \in \mathbb{R}$ there exists a unique solution $y(x, s)$

of the above initial value problem. To find a solution of the boundary
value problem, we choose the parameter s such that $y(b, s) = \beta$, i.e. we
have to solve the nonlinear equation $F(s) = y(b, s) - \beta = 0$ that can be
solved by one of the numerical methods of Sect. 12.1.

1) Find the approximate numerical solution of the boundary value
problem $y'' = -y(x)^2$, $x \in [0, 2]$, $y(0) = 0$, $y(2) = 1$ by applying the shooting
method.

2) Plot the numerical results for various values of the parameter s.

Maple:

```
with(plots): a:=0.; b:=2.; alpha:=0.; beta:=1.;
ODE1:=diff(y(x),x$2)+y(x)^2=0; BCs:=y(a)=alpha,y(b)=beta;
shootNL:=proc(x,s) local yN,ICs;
 ICs:=y(0)=0,D(y)(0)=s; yN:=rhs(dsolve({ODE1,ICs},
 numeric,output=listprocedure)[2]);RETURN(evalf(yN(x)));end;
shootNL(b,0.1); shootNL(b,0.5);
plot(['shootNL(b,s)',beta],'s'=0.5..1);
R:=fsolve('shootNL(2,s)=1','s'=0.5..1); shootNL(b,R)=beta;
plot('shootNL(x,R)','x'=a..b,color=red,thickness=2);
IC:=[0.6,0.5,1,0.8,0.85,R]; k:=nops(IC);
for i from 1 to k do G||i:=plot('shootNL(x,IC[i])',
 'x'=a..b,axes=boxed,thickness=2,color=
 COLOR(RGB,rand()/10^12,rand()/10^12,rand()/10^12)): od:
display({seq(G||i,i=1..k)}); plot(rhs(dsolve({ODE1,BCs},y(x),
 numeric,output=listprocedure)[2]),a..b);
```

Mathematica:

```
{a=0,b=2,alpha=0,beta=1,epsilon=N[1/50000,10],h1=1/10000}
{ODE1={D[y[x],{x,2}]+y[x]^2==0}, BCs={y[a]==alpha,y[b]==beta}}
shootNL[s_]:=Module[{ICs},ICs={y[0]==0,(D[y[x],x]/.{x->0})==s};
 IVP1=Flatten[{ODE1,ICs}]; yN1=NDSolve[IVP1,y,{x,a,b}];
 yN=y[x]/.yN1];
{N[shootNL[1/10]/.{x->b}],N[shootNL[1/2]/.{x->b}]}
g1=Plot[beta,{x,1/2,1},PlotStyle->Green];
g2=Plot[N[shootNL[s]/.{x->b}],{s,1/2,1},PlotStyle->Red];
Show[{g1,g2},PlotRange->{0.,1.05}]
Do[R=N[shootNL[s]/.{x->b}]; If[Abs[N[R[[1]]-beta]]<epsilon,
 {sN=s,RN=R[[1]],Break[]}],{s,1/2,1,h1}];
Print["\n",PaddedForm[N[sN],{12,9}]," ",PaddedForm[RN,{12,9}]];
N[shootNL[sN]/.{x->b}]==N[beta]
```

```
Plot[Evaluate[N[shootNL[RN]]],{x,a,b},
 PlotStyle->{Red,Thickness[0.01]}]
{IC={6/10,5/10,1,8/10,85/100,RN},k=Length[IC]}
Do[g[i]=Plot[Evaluate[N[shootNL[IC[[i]]]]],{x,a,b},Frame->True,
 PlotStyle->{Hue[0.15*i],Thickness->0.01},
 PlotRange->{0.,1.2}],{i,1,k}]; Show[Table[g[i],{i,1,k}]]
{yNB1=NDSolve[Flatten[{ODE1,BCs}],y,{x,a,b}],
 yNB=Evaluate[y[x]/.yNB1]}
Plot[yNB,{x,a,b},PlotStyle->Red]
```

$\square$

Problem 12.53 *Approximations by finite differences. Linear boundary value problems.* Let us apply the finite difference method for approximating the solution of the linear boundary value problem,

$$y'' = p(x)y' + q(x)y + r(x), \quad y(a) = \alpha, \quad y(b) = \beta.$$

Let $y : [a, b] \to \mathbb{R}$ be a sufficiently smooth function in $[a, b]$, and $p(x)$, $q(x)$, and $r(x) \in C[a, b]$. The basic idea of finite difference methods consists in replacing the derivatives in the differential equations by appropriate finite differences. We choose on $[a, b]$ an equidistant grid $X_i = a + ih$ $(i = 0, \ldots, N+1)$, with the step size $h = (b - a)/(N + 1)$ $(N \in \mathbb{N})$, where $X_0 = a$, $X_{N+1} = b$. The differential equation must be satisfied at any internal node X_i $(i = 1, \ldots, n)$, and approximating this set of N equations and replacing the derivatives by the appropriate finite differences, we obtain the system of equations:

$$\frac{Y_{i+1} - 2Y_i + Y_{i-1}}{h^2} = p(X_i)\frac{Y_{i+1} - Y_{i-1}}{2h} + q(X_i)Y_i + r(X_i), \quad Y_0 = \alpha, \quad Y_{N+1} = \beta,$$

for approximate values Y_i to the exact solution $y(X_i)$. This linear system admits a unique solution, since the matrix of the system is the $N \times N$ tridiagonal matrix, symmetric and positive definite.

1) Find the exact solution and the approximate numerical solution of the boundary value problem, $y'' = -(2/x)y' + (2/x^2)y + x^3$, $x \in [1, 2]$, $y(1) = 1$, $y(2) = 2$ (considered in **Problem 12.51**) by applying the finite difference method.

2) Compare the results and plot the exact and numerical solutions.

Maple:

```
a:=1; b:=2; alpha:=1; beta:=2; N:=10; h:=(b-a)/(N+1);
ODE1:=(D@@2)(y)(x)=-2/x*diff(y(x),x)+2/x^2*y(x)+x^3;
BCs:=y(a)=alpha,y(b)=beta; BVP1:={ODE1,BCs};
ExSol:=unapply(rhs(dsolve(BVP1,y(x))),x); X:=i->a+i*h;
p:=x->-2/x; q:=x->2/x^2; r:=x->x^3; SEq:={}:
for i from 1 to N do
 SEq:=SEq union {-(1+h/2*p(X(i)))*Y(i-1)+
 (2+h^2*q(X(i)))*Y(i)-(1-h/2*p(X(i)))*Y(i+1)=-h^2*r(X(i))};
od:
SEq; Y_DF:=convert(solve(SEq,{'Y(i)'$'i'=1..N}),list);
Y_DF:=evalf(subs({Y(0)=alpha,Y(N+1)=beta},Y_DF));
array([seq([rhs(Y_DF[i]),evalf(ExSol(X(i)))],
 rhs(Y_DF[i])-evalf(ExSol(X(i)))],i=1..N)]); with(plots):
F1:=[seq([X(i),rhs(Y_DF[i])],i=1..N),[X(0),alpha],[X(N+1),beta]];
G1:=plot(ExSol(t),t=a..b,color=red):
G2:=plot(F1,style=point,color=blue): display({G1,G2});
```

Mathematica:

```
nD=10; {a=1,b=2,alpha=1,beta=2,
 n=10,h=(b-a)/(n+1), X=Table[a+i*h,{i,0,n}]}
{ODE1=D[y[x],{x,2}]==-2/x*D[y[x],x]+2/x^2*y[x]+x^3,
 BCs={y[a]==alpha,y[b]==beta}, BVP1={ODE1,BCs}}
ExtSol[x1_]:=Expand[(DSolve[BVP1,y[x],x]/.{x->x1})[[1,1,2]]];
p[x_]:=-2/x; q[x_]:=2/x^2; r[x_]:=x^3; ExtSol[x]
SEq={}; Do[SEq=Append[SEq,{-(1+h/2*p[X[[i+1]]])*Y[i-1]+
 (2+h^2*q[X[[i+1]]])*Y[i]-(1-h/2*p[X[[i+1]]])*Y[i+1]==
 -h^2*r[X[[i+1]]]}],{i,1,n}]; SEqs=Flatten[SEq]
{Yvars=Table[Y[i],{i,1,n}], YDF=Solve[SEqs,Yvars],
 YDF1=Yvars/.YDF, YDFN=N[YDF1/.{Y[0]->alpha,Y[n+1]->beta},nD]}
Do[Print[PaddedForm[YDFN[[1,i]],{12,10}]," ",
 PaddedForm[N[ExtSol[X[[i+1]]],nD],{12,10}]," ",
 PaddedForm[YDFN[[1,i]]-N[ExtSol[X[[i+1]]],nD],{12,10}]],
 {i,1,n}]; F1=Table[{X[[i+1]],YDFN[[1,i]]},{i,1,n}]
F11=Append[Append[F1,{a,alpha}],{b,beta}]
g1=Plot[Evaluate[ExtSol[t]],{t,a,b},PlotStyle->Hue[0.99]];
g2=ListPlot[F11,PlotStyle->{PointSize[.02],Hue[0.7]}];
Show[{g1,g2}]
```

□

Problem 12.54 *Approximations by finite differences. Nonlinear boundary value problems.* Let us apply the finite difference method for approximating the solution of the nonlinear boundary value problem,

$$y'' = f(x, y, y), \quad y(a) = \alpha, \quad y(b) = \beta.$$

Let $f : [a, b] \times \mathbb{R}^2 \to \mathbb{R}$ be a continuous function, and let $f_y(x, y, y')$, $f_{y'}(x, y, y')$ be continuous functions in an open domain $D = \{(x, y, y') \mid a \le x \le b, -\infty < y < \infty, -\infty < y' < \infty\}$.

If $f_y(x, y, y') > 0$ for all $(x, y, y') \in D$ and there exist the constants M and K such that $|f_y(x, y, y')| \le M$, $|f_{y'}(x, y, y')| \le K$ for all $(x, y, y') \in D$, then the boundary value problem has a unique solution.

As for the linear case, we choose on $[a, b]$ an equidistant grid $X_i = a + ih$ $(i = 0, \ldots, N + 1)$, with the step size $h = (b - a)/(N + 1)$ $(N \in \mathbb{N})$, where $X_0 = a$, $X_{N+1} = b$. Approximating the nonlinear boundary value problem, we arrive at the system of nonlinear equations:

$$\frac{Y_{i+1} - 2Y_i + Y_{i-1}}{h^2} = f\left(X_i, Y_i, \frac{Y_{i+1} - Y_{i-1}}{2h}\right), \quad Y_0 = \alpha, \quad Y_{N+1} = \beta,$$

for approximate values Y_i to the exact solution $y(X_i)$. For solving this system of nonlinear equation, we apply the Newton method considered in Sect. 12.1.

1) Find the approximate numerical solution by using the predefined functions in both systems.

2) Find the approximate numerical solution of the nonlinear boundary value problem, $y'' = -y(x)^2$, $x \in [0, 2]$, $y(0) = 0$, $y(2) = 1$ (considered in **Problem 12.52**) by applying the finite difference method.

2) Compare the results and plot the numerical solutions.

Maple:

```
with(plots): with(LinearAlgebra): with(codegen): Nmax:=100:
epsilon:=10^(-4); f:=(x,y,dy)->y(x)^2; a:=0: b:=2: N:=20:
h:=evalf((b-a)/(N+1));  alpha:=0: beta:=1:
ODE1:=diff(y(x),x$2)=-y(x)^2; BCs:=y(a)=alpha,y(b)=beta;
BVP1:={ODE1,BCs}; Sol:=rhs(dsolve(BVP1,y(x),numeric,
 output=listprocedure)[2]); Z||0:=alpha; Z||(N+1):=beta;
for i from 1 to N do X||i:=a+i*h; Eq||i:=(Z||(i+1)-2*Z||i
 +Z||(i-1))/(h^2)+f(X||i,Z||i,(Z||(i+1)-Z||(i-1))/(2*h)); od:
SeqEq:=seq(Eq||i,i=1..N): SeqVar:=seq(Z||i,i=1..N);
```

```
for i from 1 to N do F[i]:=unapply(Eq||i,[SeqVar]): od:
J:=JACOBIAN([seq(F[i],i=1..N)],result_type=array):
FNewton:=W->W-convert(J(seq(W[i],i=1..N)),
  Matrix)^(-1).<seq(F[k](seq(W[i],i=1..N)),k=1..N)>;
Y[0]:=<seq(0,i=1..N)>:
for i from 1 to Nmax do Y[i]:=FNewton(Y[i-1]);
 if max(seq(abs(F[m](seq(Y[i][k],k=1..N))),m=1..N))>=epsilon
  then print(i,seq(Y[i][k],k=1..N)):
 else Iend:=i: lprint(`the results is`); print(Iend);
 for k from 1 to N do X:=k->a+k*h;
  print(X(k),Y[i][k],Sol(X(k)),evalf(abs(Y[i][k]-Sol(X(k))))):
od: break: fi: od:
F1:=[seq([X(k),Y[Iend][k]],k=1..N),[X(0),alpha],[X(N+1),beta]]:
G1:=plot(Sol(t),t=a..b,color=red):
G2:=plot(F1,style=point,color=blue): display({G1,G2});
```

Mathematica:

```
nD=20; JacobianMatrix[f_List?VectorQ,x_List]:=
 Outer[D,f,x]/;Equal@@(Dimensions/@{f,x});
{a=0,b=2,n=8,h=N[(b-a)/(n+1),nD],alpha=0,beta=1,
 Nmax=100,epsilon=N[10^(-4),nD],
 X=N[Table[a+i*h,{i,0,n+1}],nD]}
{ODE1=D[y[x],{x,2}]==-y[x]^2, BCs={y[a]==alpha,y[b]==beta},
 BVP1={ODE1,BCs}}
{Sol=NDSolve[BVP1,y,{x,a,b}], SolN=Evaluate[y[x]/.Sol][[1]],
 N[SolN/.{x->2},nD]}
Subs0={y[X[[1]]]->alpha,y[X[[n+2]]]->beta}
SeqEq1=Expand[Table[(y[X[[i+1]]]-2*y[X[[i]]]+
 y[X[[i-1]]])/(h^2)+(y[X[[i]]])^2,{i,2,n+1}]]
{SeqEq=SeqEq1/.Subs0, SeqVar=Variables[SeqEq],
 nV=Length[SeqVar], SeqVar1=Table[Subscript[Z,i],{i,1,nV}]}
Subs=Table[SeqVar[[i]]->SeqVar1[[i]],{i,1,nV}]
F[SeqVar1_List]:=Table[SeqEq[[i]]]/.Table[
 SeqVar[[i]]->SeqVar1[[i]],{i,1,nV}],{i,1,nV}]; F[SeqVar]
J=JacobianMatrix[F[SeqVar],SeqVar]
JInv[SeqVar1_List?VectorQ]:=(Inverse[JacobianMatrix[F[SeqVar],
 SeqVar]])/.Table[SeqVar[[i]]->SeqVar1[[i]],{i,1,nV}];
FNewton[W_List?VectorQ]:=W-JInv[Table[W[[i]],
 {i,1,nV}]].Table[F[Table[W[[i]],{i,1,nV}]][[k]],{k,1,nV}];
Y=Table[Table[0,{k,1,nV}],{i,1,Nmax}];
Y[[1]]=Table[N[10^(-10),nD],{i,1,nV}];
```

```
Do[Y[[i]]=FNewton[Y[[i-1]]]; Print[i," ",Y[[i]]];
  If[Max[N[Abs[Table[F[Y[[i]]][[m]],{m,1,nV}]],nD]]<epsilon,
  {Iend=i,Break[]}],{i,2,7}]; {Print[Iend],
  Print["\n The results is"]}
Do[Print[X[[k+1]]," ",Y[[Iend]][[k]]," ",N[SolN/.{x->X[[k+1]]},
  nD]," ",N[Abs[Y[[Iend]][[k]]-N[SolN/.{x->X[[k+1]]},nD]],nD]],
  {k,1,nV}];
F1=Table[{X[[k+1]],Y[[Iend]][[k]]},{k,1,n}]
F11=Append[Append[F1,{a,alpha}],{b,beta}]
g1=Plot[Evaluate[SolN],{x,a,b},PlotStyle->Hue[0.99]];
g2=ListPlot[F11,PlotStyle->{PointSize[.02],Hue[0.7]}];
Show[{g1, g2}]
```

We note that the above approximate numerical solution, obtained with the aid of symbolic-numerical computations in *Mathematica*, can be produced for small values of the partition parameter n, e.g., $n = 8$. For $n > 10$, we have written the another version of the solution:

```
nD=20; {a=0,b=2,n=20,h=N[(b-a)/(n+1),nD],alpha=0,
  beta=1,Nmax=100,epsilon=N[10^(-4),nD],
  X=N[Table[a+i*h,{i,0,n+1}],nD]}
{ODE1=D[y[x],{x,2}]==-y[x]^2,BCs={y[a]==alpha,y[b]==beta},
  BVP1={ODE1,BCs}}
{Sol=NDSolve[BVP1,y,{x,a,b}],SolN=Evaluate[y[x]/.Sol][[1]],
  N[SolN/.{x->2},nD]}
Subs0={y[X[[1]]]->alpha,y[X[[n+2]]]->beta}
SeqEq1=Expand[Table[(y[X[[i+1]]]-2*y[X[[i]]]+
  y[X[[i-1]]])/(h^2)+(y[X[[i]]])^2,{i,2,n+1}]]
{SeqEq=SeqEq1/.Subs0, SeqVar=Variables[SeqEq],
  SubsInitial=Table[{SeqVar[[i]],0.},{i,1,n}]}
Eqs=Map[Thread[#1==0,Equal]&,SeqEq]
Sol1=FindRoot[Eqs,SubsInitial,WorkingPrecision->nD]
F1=Table[{Sol1[[k,1,1]],Sol1[[k,2]]},{k,1,n}]
F11=Append[Append[F1,{a,alpha}],{b,beta}]
g1=Plot[Evaluate[SolN],{x,a,b},PlotStyle->Hue[0.99]];
g2=ListPlot[F11,PlotStyle->{PointSize[.02],Hue[0.7]}];
Show[{g1,g2}]
```

□

Problem 12.55 *Boundary value problems. Variational methods. The Ritz method.* Let us consider the linear boundary value problem,

$$-(p(x)y')' + q(x)y = f(x), \quad y(a) = \alpha, \quad y'(b) = \beta,$$

where $x \in [a, b]$. We assume that $p(x) \in C^1[a, b]$, $q(x), f(x) \in C[a, b]$ and $p(x) \geq 0$, $q(x) \geq 0$ for all $x \in [a, b]$. Then there exists a unique solution $y(x) \in C^2[a, b]$ if and only if $y(x)$ is the unique function that minimizes the functional

$$J[y(x)] = \int_a^b \{p(x)[y'(x)]^2 + q(x)[y(x)]^2 - 2f(x)y(x)\} \, dx.$$

In the Ritz method (proposed by Ritz in 1909), by introducing the approximate solution $y(x)$ in the form of the linear combination of the basis functions $\phi_i(x)$, $y(x) = c_1\phi_1(x) + \ldots + c_N\phi_N(x)$, we obtain a quadratic form in the unknown coefficients c_i. Minimizing this quadratic form, we determine the coefficients c_i and the approximate solution $y(x)$.

    1) Find the exact solution and the approximate numerical solution of the boundary value problem $y'' = y - \sin x$, $x \in [0, \pi]$, $y(0) = 0$, $y'(\pi) = 0$ by applying the Ritz method.

    2) Compare the results and plot the exact and numerical solutions.

Maple:

```
with(plots): a:=0: b:=Pi: N:=8: h:=evalf((b-a)/N);
alpha:=0: beta:=0: f:=x->sin(x); p:=x->1; q:=x->1;
ODE1:=-p(x)*diff(y(x),x$2)+q(x)*y(x)-f(x);
BCs:=y(a)=alpha,D(y)(b)=beta; BVP1:={ODE1,BCs};
ExSol1:=rhs(dsolve(BVP1,y(x)));
ExSol:=unapply(simplify(convert(ExSol1,trig)),x);
for i from 1 to N do phi[i]:=cos(i*X)-1: od:
Y:=X->add(c[i]*phi[i],i=1..N);
J:=int(expand(p(x)*diff(Y(X),X)^2+q(x)*Y(X)^2-2*f(X)*Y(X)),
  X=a..b); V:=VectorCalculus[Gradient](J,[op(indets(J))]);
CL:=[op(solve({seq(V[i],i=1..N)},indets(J)))];
AppSol:=unapply(subs(CL,Y(X)),X); plot([ExSol(x),AppSol(x),
  AppSol(x)-ExSol(x)],x=a..b,color=[red,blue,green]);
```

Mathematica:

```
nD=10; {a=0,b=Pi,n=8,h=N[(b-a)/n,nD], alpha=0, beta=0}
f[x_]:=Sin[x]; p[x_]:=1; q[x_]:=1;
{ODE1={-p[x]*D[y[x],{x,2}]+q[x]*y[x]-f[x]==0},
  BCs={y[a]==alpha,(D[y[x],x]/.{x->b})==beta},BVP1={ODE1,BCs}}
ExtSol[x1_]:=FullSimplify[ExpToTrig[(DSolve[BVP1,y[x],x]/.
  {x->x1})[[1,1,2]]]]; ExtSol[x]
```

```
Do[phi[i]=Cos[i*X]-1,{i,1,n}];
Y[X1_]:=Sum[c[i]*phi[i],{i,1,n}]/.{X->X1}; Y[X]
{J=Integrate[Expand[p[x]*D[Y[X],X]^2+q[x]*Y[X]^2-2*f[X]*Y[X]],
 {X,a,b}], V=D[J,{Variables[J]}]}
CL=Flatten[Solve[Map[Thread[#1==0,Equal]&,V],Variables[J]]]
AppSol[X1_]:=(Y[X]/.CL)/.{X->X1}; AppSol[x]
g1=Plot[Evaluate[ExtSol[x]],{x,a,b},PlotStyle->Red];
g2=Plot[Evaluate[AppSol[x]],{x,a,b},PlotStyle->Blue];
g3=Plot[Evaluate[AppSol[x]-ExtSol[x]],{x,a,b},
 PlotStyle->Green]; Show[{g1,g2,g3},PlotRange->{-0.1,0.6}]
```

$\square$

Problem 12.56 *Boundary value problems. Projection methods. The Galerkin method.* Let us consider the same linear boundary value problem as in the previous problem,

$$L[y(x)] + f(x) = -(p(x)y')' + q(x)y + f(x) = 0, \quad y(a) = \alpha, \quad y'(b) = \beta,$$

where $x \in [a, b]$ and $L[y(x)]$ is the linear differential operator. We assume that all the functions are square-integrable. We note that not every differential equation admits a minimization of a functional, and we consider a more powerful and general method for solving differential equations, the Galerkin method (proposed by Galerkin in 1915) that belongs to the *projection methods*, i.e. the equation to be approximated is projected onto a finite-dimensional function subspace.

In the Galerkin method (as in the Ritz method) we also introduce the approximate solution $y(x)$ in the form of the linear combination of the basis functions $\phi_i(x)$, $y(x) = c_1\phi_1(x) + \ldots + c_N\phi_N(x)$, and we choose the unknown coefficients c_i such that the residual $r(x) = L[y(x)] + f(x)$ is orthogonal to the space spanned by the basis functions $\phi_i(x)$:

$$\int_a^b r(x)\phi_i(x)\, dx = 0, \quad i = 1, \ldots, N,$$

i.e. the Galerkin equations reduces to the solution of a system of linear equations.

1) Find the exact solution and the approximate numerical solution of the boundary value problem

$$y'' = y - \sin x, \quad x \in [0, \pi], \quad y(0) = 0, \quad y'(\pi) = 0$$

by applying the Ritz method.

2) Compare the results and plot the exact and numerical solutions.

Maple:

```
with(plots): a:=0: b:=Pi: N:=8: h:=evalf((b-a)/N);
alpha:=0: beta:=0: f:=x->-sin(x); p:=x->1; q:=x->1;
ODE1:=unapply(p(x)*diff(y(x),x$2)-q(x)*y(x)-f(x),x);
BCs:=y(a)=alpha,D(y)(b)=beta; BVP1:={ODE1(x),BCs};
ExSol:=unapply(rhs(dsolve(BVP1,y(x))),x);
for i from 1 to N do phi[i]:=cos(i*t)-1: od:
Y:=t->add(c[i]*phi[i],i=1..N); Y(t);
Eq1:=combine(subs(y(t)=Y(t),ODE1(t)));
Eqs:=evalf({seq(int(Eq1*phi[i],t=a..b),i=1..N)});
CL:=[op(solve(Eqs,indets(Eqs)))];
AppSol:=unapply(subs(CL,Y(t)),t);
plot([ExSol(t),AppSol(t),AppSol(t)-ExSol(t)],
   t=a..b,color=[red,blue,green]);
```

Mathematica:

```
nD=10; f[x_]:=-Sin[x]; p[x_]:=1; q[x_]:=1;
ODE1[x1_]:=p[x]*D[y[x],{x,2}]-q[x]*y[x]-f[x]==0/.{x->x1};
{a=0,b=Pi,n=8,h=N[(b-a)/n,nD],alpha=0,beta=0}
{BCs={y[a]==alpha,(D[y[x],x]/.{x->b})==beta},BVP1={ODE1[x],BCs}}
ExtSol[x1_]:=FullSimplify[ExpToTrig[(DSolve[BVP1,y[x],x]/.
  {x->x1})[[1,1,2]]]]; ExtSol[x]
Do[phi[i]=Cos[i*t]-1,{i,1,n}];
Y[t1_]:=Sum[c[i]*phi[i],{i,1,n}]/.{t->t1}; Y[t]
Eq1=Expand[ODE1[t]/.{y[t]->Y[t],D[y[t],{t,2}]->D[Y[t],{t,2}]}]
Eqs=Table[Integrate[Expand[Eq1[[1]]*phi[i]],{t,a,b}],{i,1,n}]
CL=Flatten[Solve[Map[Thread[#1==0,Equal]&,Eqs],Variables[Eqs]]]
AppSol[t1_]:=(Y[t]/.CL)/.{t->t1}; Expand[N[AppSol[t],nD]]
g1=Plot[Evaluate[ExtSol[t]],{t,a,b},PlotStyle->Red];
g2=Plot[Evaluate[AppSol[t]],{t,a,b},PlotStyle->Blue];
g3=Plot[Evaluate[AppSol[t]-ExtSol[t]],{t,a,b},
 PlotStyle->Green]; Show[{g1,g2,g3},PlotRange->{-0.1,0.6}]
```

We note that the Galerkin method is a *semidiscrete method* (or semianalytic method), for a fully discrete method it is necessary to add the approximations of integrals or in a general case, approximations of differential or integral operators. Also we note that the art of application of Galerkin and Ritz methods begins with the analysis and appropriate choice of the approximating function subspace and the basic functions ϕ_i.   □

Problem 12.57 *Initial boundary value problems for partial differential equations. Numerical and graphical solutions.* Let us consider the initial boundary value problem for the wave equation,

$$u_{tt} = c^2 u_{xx}, \quad u(0,t) = 0, \quad u(L,t) = 0,$$
$$u(x,0) = f(x), \quad u_t(x,0) = g(x),$$

in the domain $\mathcal{D} = \{0 \le x \le L, \ 0 < t < \infty\}$. Here $f(x) = 0$, $g(x) = \frac{1}{2}\sin(2\pi x)$, $L = 1$, $c = 1/4$.

1) Find the approximate numerical solution of the initial boundary value problem and compare with the exact solution (in terms of the Green function, e.g. see [34], p. 1279):

$$u(x,t) = \frac{\partial}{\partial t} \int_0^L f(\xi) G(x,\xi,t)\,d\xi + \int_0^L g(\xi) G(x,\xi,t)\,d\xi,$$

where

$$G(x,\xi,t) = \frac{2}{c\pi} \sum_{n=1}^{\infty} \frac{1}{n} \sin\left(\frac{n\pi x}{L}\right) \sin\left(\frac{n\pi\xi}{L}\right) \sin\left(\frac{n\pi ct}{L}\right).$$

2) Plot the exact and numerical solutions for $(x_k, t_k) \in \mathcal{D}$ and their difference. Compare the numerical results.

3) Construct various graphical illustrations of the exact and numerical solutions.

Maple:

```
with(plots):with(PDEtools):Ops1:=spacestep=1/100,timestep=1/100;
Ops2:=numpoints=100,thickness=3,color=blue,scaling=constrained:
c:=1/4; L:=1; N:=4; f:=x->0; g:=x->sin(2*Pi*x)/2;
Eq1:=diff(u(x,t),t$2)=c^2*diff(u(x,t),x$2);
ICs:=u(x,0)=f(x),D[2](u)(x,0)=g(x); BCs:=u(0,t)=0,u(L,t)=0;
IBCs:={ICs,BCs}; G:=(x,xi,t,N)->2/(c*Pi)*add(1/n*sin(n*Pi*x/L)
 *sin(n*Pi*xi/L)*sin(n*Pi*c*t/L),n=1..N);
ExSol:=unapply(int(expand(g(xi)*G(x,xi,t,N)),xi=0..L),x,t);
Sol1:=pdsolve(Eq1,IBCs,numeric,u(x,t),Ops1,time=t,range=0..L):
Sol1:-animate(u(x,t),t=0..2*Pi,x=0..L,frames=30,Ops2);
Sol1:-plot3d(u(x,t),t=0..2*Pi,x=0..L); G:=array(1..3);
G[1]:=Sol1:-plot(t=Pi,color=red):
G[2]:=plot(ExSol(x,Pi),x=0..L,color=blue):
G[3]:=Sol1:-plot(u-ExSol(x,Pi),t=Pi,color=green):
```

```
display(G,scaling=constrained);
uPi:=Sol1:-value(t=Pi,output=listprocedure);
uVal:=rhs(op(3,uPi)); plot(uVal(x),x=0..L);
for x from 0 to L by 0.1 do print(uVal(x),
  evalf(ExSol(x,Pi),20),abs(uVal(x)-evalf(ExSol(x,Pi)))); od;
```

Mathematica:

```
nD=10; {c=1/4,L=1,nN=4}
f[x_]:=0; g[x_]:=Sin[2*Pi*x]/2;
{Eq1=D[u[x,t],{t,2}]==c^2*D[u[x,t],{x,2}],
 ICs={u[x,0]==f[x],(D[u[x,t],t]/.{t->0})==g[x]},
 BCs={u[0,t]==0,u[L,t]==0}, IBCs={ICs,BCs}}
ExtSol[x1_,t1_]:=Integrate[Evaluate[Expand[g[xi]*2/(c*Pi)*
 Sum[1/n*Sin[n*Pi*x/L]*Sin[n*Pi*xi/L]*Sin[n*Pi*c*t/L],
 {n,1,nN}]]],{xi,0,L}]/.{x->x1,t->t1}; ExtSol[x,t]
Sol=NDSolve[{Eq1,IBCs},u,{x,0,L},{t,0,2*Pi}];fN=u/.Sol[[1]]
Plot3D[fN[x,t],{x,0,L},{t,0,2*Pi}]
Animate[Plot[fN[x,t],{t,0,2*Pi},PlotRange->{-1,1}],{x,0,L}]
g1=Plot[Evaluate[fN[x,Pi]],{x,0,L},PlotStyle->Red];
g2=Plot[Evaluate[ExtSol[x,Pi]],{x,0,L},PlotStyle->Blue];
g3=Plot[Evaluate[fN[x,Pi]-ExtSol[x,Pi]],{x,0,L},
 PlotStyle->Green]; GraphicsArray[{{g1},{g2},{g3}}]
Plot[Evaluate[fN[x,Pi]],{x,0,L},PlotStyle->Red]
PaddedForm[Table[{N[fN[x,Pi],nD], N[ExtSol[x,Pi],nD],
 Abs[N[fN[x,Pi],nD]-N[ExtSol[x,Pi],nD]]},
 {x,0,L,0.1}]//TableForm,{12,10}]
```

□

Problem 12.58 *Initial boundary value problems for partial differential equations. Fourier series expansions and the application to the solution of PDEs by the method of separation of variables.* Let us consider the initial boundary value problem for the wave equation,

$$u_{tt} = c^2 u_{xx}, \quad u(0,t) = 0, \quad u(L,t) = 0,$$
$$u(x,0) = f(x), \quad u_t(x,0) = g(x),$$

in the domain $\mathcal{D} = \{0 \le x \le L, \ 0 < t < \infty\}$. Here $f(x) = 0$, $g(x) = \sin(4\pi x)$, $L = 0.5$, $c = 1/(4\pi)$.

1) Find the approximate solution to this initial boundary value problem by applying the Fourier method.

2) Plot the results obtained.

Maple:

```
L:=0.5: c:=1/(4*Pi): N:=4; f:=x->0; g:=x->sin(4*Pi*x);
an:=proc(n) option remember;
 2/L*evalf(int(f(x)*sin(n*Pi*x/L),x=0..L)); end;
bn:=proc(n) option remember;
 evalf(2/L*evalf(int(g(x)*sin(n*Pi*x/L),x=0..L))); end;
seq([i,an(i),bn(i)],i=1..N); un:=proc(x,t,n)
 (an(n)*cos(n*Pi*c*t/L)+bn(n)/(c*n*Pi/L)*sin(n*Pi*c*t/L))
 *sin(n*Pi*x/L); end; un(x,t,1);
u:=proc(x,t) evalf(add(un(x,t,i),i=1..N)); end; u(x,t);
plots[animate](u(x,t),x=0..L,t=0..2*Pi,frames=100,
 scaling=constrained,color=blue,thickness=3);
```

Mathematica:

```
nD=10; f[x]:=0; g[x_]:=Sin[4*Pi*x];
{L=1/2, c=1/(4*Pi), nN=4}
an[n_]:=N[2/L*Integrate[f[x]*Sin[n*Pi*x/L],{x,0,L}],nD];
bn[n_]:=N[2/L*Integrate[g[x]*Sin[n*Pi*x/L],{x,0,L}],nD];
Table[{i,an[i],bn[i]},{i,1,nN}]
un[x_,t_,n_]:=(an[n]*Cos[n*Pi*c*t/L]+bn[n]/(c*n*Pi/L)*
 Sin[n*Pi*c*t/L])*Sin[n*Pi*x/L]; un[x,t,1]
u[x1_,t1_]:=Evaluate[N[Sum[un[x,t,i],{i,1,nN}],
 nD]/.{x->x1,t->t1}]; u[x,t]
Animate[Plot[u[x,t],{x,0,L},PlotRange->{-1,1},
 AspectRatio->2,PlotStyle->{Blue,Thickness[0.015]}],{t,0,2*Pi},
 AnimationRate->0.1]
```

□

Problem 12.59 *Boundary value problems for partial differential equations. Fourier series expansions and the application to the solution of PDEs by the method of separation of variables.* Let us consider the initial boundary value problem for the Laplace equation inside a circle of radius R,

$$\nabla^2 u = \frac{1}{r}\frac{\partial}{\partial r}\left(\frac{r\partial u}{\partial r}\right)\frac{1}{r^2}\frac{\partial^2 u}{\partial \phi^2} = 0, \quad u(R,\phi) = f(\phi),$$

1) Find the approximate solution to the boundary value problem by applying the Fourier method for various functions $f(\phi)$ and R.

2) Plot the results obtained.

Maple:

```
LapIntD:=proc(f,R,N) local a, b, Sol;
 a:=n->1/Pi*Int(f*cos(n*phi),phi=-Pi..Pi);
 b:=n->1/Pi*Int(f*sin(n*phi),phi=-Pi..Pi);
 Sol:=a(0)/2+add(r^n/R^n*(a(n)*cos(n*phi)+b(n)*sin(n*phi)),
  n=1..N); RETURN(map(simplify,value(Sol)));
end;
with(VectorCalculus):
f:=cos(phi)^2; R:=1; s1:=LapIntD(f,R,3);
combine(Laplacian(s1,'polar'[r,phi]));
simplify(subs(r=R,s1)-f);
f:=sin(phi)^3; R:=1; s2:=LapIntD(f,R,3);
combine(Laplacian(s2,'polar'[r,phi]));
simplify(subs(r=R,s2)-f);
f:=sin(phi)^6+cos(phi)^6; R:=1;  s3:=LapIntD(f,R,5);
Laplacian(s3,'polar'[r,phi]);
simplify(subs(r=R,s3)-f);
f:=randpoly([x,y],terms=10); R:=9;
f:=subs(x=r*cos(phi),y=r*sin(phi),r=R,f); s4:=LapIntD(f,R,6);
combine(Laplacian(s4,'polar'[r,phi]));
simplify(subs(r=R,s4)-f);
f:=phi^2+phi+1; R:=1;  s6:=LapIntD(f,R,6);
combine(Laplacian(s6,'polar'[r, phi]));
Err6:=combine(simplify(subs(r=R,s6)-f)); s20:=LapIntD(f,R,20);
combine(Laplacian(s20,'polar'[r,phi]));
Err20:=combine(simplify(subs(r=R,s20)-f));
plot([Err6,Err20],phi=-Pi..Pi,color=[green,blue]);
R:=rand(1..10)(); g:=randpoly([x,y],degree=rand(2..4)(),terms=3);
n:=degree(g); g1:=subs(x=r*cos(phi),y=r*sin(phi),r=R,g);
g2:=combine(g1,trig);
C:=subs([seq(sin(i*phi)=0,i=1..n),seq(cos(i*phi)=0,i=1..n)],g2);
for i from 1 to n do
 a||i:=coeff(g2,cos(i*phi)); b||i:=coeff(g2,sin(i*phi)); od;
SolP:=C+add((R/r)^i*(a||i*cos(i*phi)+b||i*sin(i*phi)),i=1..n);
Sol1:=expand(SolP); str:=[cos(phi)=x/r,sin(phi)=y/r];
SolC:=subs(r=sqrt(x^2+y^2),collect(subs(str,Sol1),r));
print(Laplacian(u(x,y),[x,y])=0,`if r>R`); u=g,`if r`=R;
SolP; SolC; simplify(Laplacian(SolP,'polar'[r,phi]));
simplify(Laplacian(SolC,[x,y]));
simplify(subs(str,expand(SolP)-g),{x^2+y^2=R^2,r=R});
simplify(SolC-g,{x^2+y^2=R^2});
```

Mathematica:

```
<<VectorAnalysis`
nD=10;
LapIntD[f_,R_,n_]:=Module[{a,b,Sol},
 a[i_]:=1/Pi*Integrate[f*Cos[i*phi],{phi,-Pi,Pi}];
 b[i_]:=1/Pi*Integrate[f*Sin[i*phi],{phi,-Pi,Pi}];
 Sol=a[0]/2+Sum[r^i/R^i*(a[i]*Cos[i*phi]+b[i]*Sin[i*phi]),
{i,1,n}]; FullSimplify[Sol]];
{f=Cos[phi]^2, R=1}
s1[r1_,phi1_]:=LapIntD[f,R,3]/.{r->r1,phi->phi1};
s1[r,phi]
LapPolar[F_,r_,phi_]:=D[F,{r,2}]+1/r*D[F,r]+1/r^2*D[F,{phi,2}];
{LapPolar[g[r,phi],r,phi], LapPolar[s1[r,phi],r,phi],
 Simplify[s1[R,phi]-f], Expand[Laplacian[g[Rr,Ttheta],
 Cylindrical]], Laplacian[s1[Rr,Ttheta],Cylindrical]}
{f=Sin[phi]^3, R=1}
s2[r1_,phi1_]:=LapIntD[f,R,3]/.{r->r1,phi->phi1}; s2[r,phi]
{FullSimplify[Laplacian[s2[Rr,Ttheta],Cylindrical]],
 Simplify[s2[R,phi]-f]}
{f=Sin[phi]^6+Cos[phi]^6, R=1}
s3[r1_,phi1_]:=LapIntD[f,R,5]/.{r->r1,phi->phi1}; s3[r,phi]
{FullSimplify[Laplacian[s3[Rr,Ttheta],Cylindrical]],
 Simplify[s3[R,phi]-f]}
{R=9, n=2, m=2, d=10,
 cR=Table[RandomInteger[{-d,d}],{i,0,n},{j,0,m}]//Flatten,
 u=Table[x^i*y^j,{i,0,n},{j,0,m}]//Flatten, f1=cR.u//Sort,
 f=f1/.{x->r*Cos[phi],y->r*Sin[phi]}/.{r->R}}
s4[r1_,phi1_]:=Expand[LapIntD[f,R,6]]/.{r->r1,phi->phi1};
{s4[r,phi], Together[Laplacian[s4[Rr,Ttheta], Cylindrical]],
 Together[TrigExpand[s4[R,phi]-f ]]}
{f=phi^2+phi+1, R=1}
s6[r1_,phi1_]:=Expand[LapIntD[f,R,6]]/.{r->r1,phi->phi1};
s6[r,phi]
Together[TrigExpand[Laplacian[s6[Rr,Ttheta],Cylindrical]]]
Err6=Simplify[Together[TrigExpand[s6[R,phi]-f]]]
s20[r1_,phi1_]:=Expand[LapIntD[f,R,20]]/.{r->r1,phi->phi1};
s20[r,phi]
Together[TrigExpand[Laplacian[s20[Rr,Ttheta],Cylindrical]]]
Err20=Simplify[Together[TrigExpand[s20[R,phi]-f]]]
Plot[{Err6,Err20},{phi,-Pi,Pi},PlotRange->Full,
 PlotStyle->{Green,Blue}]
```

```
{R=RandomInteger[{1,10}], d=10, n=RandomInteger[{2,4}],
 m=RandomInteger[{2,4}],
 cR=Table[RandomInteger[{-d,d}],{i,0,n},{j,0,m}]//Flatten,
 u=Table[x^i*y^j,{i,0,n},{j,0,m}]//Flatten,
 g=cR.u//Sort, n1=Exponent[g,{x,y},Max],
 nDeg=n1[[1]]*n1[[2]],
 gPol=TrigReduce[g/.{x->R*Cos[phi],y->R*Sin[phi]}]}
C1=gPol/.Flatten[Table[{Sin[i*phi]->0,Cos[i*phi]->0},{i,1,nDeg}]]
Do[a[i]=Coefficient[gPol,Cos[i*phi]];
 b[i]=Coefficient[gPol,Sin[i*phi]];
 Print[a[i]," ",b[i]],{i,1,nDeg}];
{SolPolar=TrigExpand[C1+
 Sum[(R/r)^i*(a[i]*Cos[i*phi]+b[i]*Sin[i*phi]),{i,1,nDeg}]],
 SolCartesian=SolPolar/.
  {Cos[phi]->x/r,Sin[phi]->y/r}/.{r->Sqrt[x^2+y^2]}}
Print[HoldForm[D[u[x,y],{x,2}]+D[u[x,y],{y,2}]=0]," if r>R",
 " u=",g," if r=", R];
FSolP[r1_,phi1_]:=SolPolar/.{r->r1,phi->phi1};
FSolC[x1_,y1_]:=SolCartesian/.{x->x1,y->y1};
{FSolP[r,phi], FSolC[x,y],
 Together[TrigExpand[Laplacian[FSolP[Rr,Ttheta],Cylindrical]]],
 Together[TrigExpand[Laplacian[FSolC[Xx,Yy]]]]}
{FullSimplify[Expand[SolCartesian-g],x^2+y^2==R^2],
 TrigExpand[SolPolar-gPol]/.{r->R}}
```

□

Problem 12.60 *Initial boundary value problems for partial differential equations. Approximation by finite differences. The Crank–Nicolson method.* Let us consider the initial boundary value problem for the heat equation,

$$u_t = c^2\, u_{xx}, \quad u(0,t) = 0, \quad u(L,t) = 0, \quad 0 < t < T,$$
$$u(x,0) = f(x), \quad 0 \le x \le L,$$

in the domain $\mathcal{D} = \{0 \le x \le L, \ 0 < t < T\}$. Here $f(x) = \sin(\pi x)$, $L = 1$, $c = 1$.

1) Find the approximate numerical solution of the initial boundary value problem by applying the Crank–Nicolson method.

2) Compare the approximate numerical solution with the exact solution $u(x,t) = e^{-\pi^2 t} \sin(\pi x)$ for $(x_k, t_k) \in \mathcal{D}$.

Maple:

```
f:=x->sin(Pi*x);
L:=1; T:=0.2: c:=1: m:=20: n:=20:
m1:=m-1: m2:=m-2: h:=L/m: k:=T/n:
lambda:=c^2*k/(h^2): W[m-1]:=0:
for i from 1 to m1 do
 W[i-1]:=evalf(f(i*h));
od:
l[0]:=1+lambda: u[0]:=-lambda/(2*l[0]):
for i from 2 to m2 do
 l[i-1]:=1+lambda+lambda*u[i-2]/2;
 u[i-1]:=-lambda/(2*l[i-1]);
od:
l[m1-1]:=1+lambda+0.5*lambda*u[m2-1]:
for j from 1 to n do t:=j*k;
 z[0]:=((1-lambda)*W[0]+lambda*W[1]/2)/l[0];
 for i from 2 to m1 do
  z[i-1]:=((1-lambda)*W[i-1]+0.5*lambda*
  (W[i]+W[i-2]+z[i-2]))/l[i-1];
 od:
 W[m1-1]:=z[m1-1]:
 for i1 from 1 to m2 do
  i:=m2-i1+1; W[i-1]:=z[i-1]-u[i-1]*W[i];
 od:
od:
ExSol:=(x,t)->exp(-Pi^2*t)*sin(Pi*x):
printf(`Crank-Nicolson Method\n`);
for i from 1 to m1 do
 X:=i*h;
 printf(`%3d %11.8f %13.8f %13.8f,%13.8f\n`, i,X,W[i-1],
 evalf(ExSol(X,0.2)),W[i-1]-evalf(ExSol(X,0.2)));
od:
```

Mathematica:

```
nD=10; f[x_]:=Sin[Pi*x];
{L=1,T=0.2,c=1,m=20,n=20,m1=m-1,m2=m-2,h=L/m,k=T/n,
 lambda=c^2*k/(h^2)}
{W=Table[0,{i,0,m}],l=Table[0,{i,0,m}],u=Table[0,{i,0,m}],
 z=Table[0,{i,0,m}]}
W[[m]]=0;
```

```
Do[W[[i]]=N[f[i*h],nD],{i,1,m1}];
{l[[1]]=1+lambda, u[[1]]=-lambda/(2*l[[1]])}
Do[l[[i]]=1+lambda+lambda*u[[i-1]]/2;
  u[[i]]=-lambda/(2*l[[i]]),{i,2,m2}];
l[[m1]]=1+lambda+0.5*lambda*u[[m2]]];
Do[t=j*k;
 z[[1]]=((1-lambda)*W[[1]]+lambda*W[[2]]/2)/l[[1]]);
 Do[z[[i]]=((1-lambda)*W[[i]]+0.5*lambda*
  (W[[i+1]]+W[[i-1]]+z[[i-1]]))/l[[i]],{i,2,m1}];
 W[[m1]]=z[[m1]];
 Do[i=m2-i1+1;
  W[[i]]=z[[i]]-u[[i]]*W[[i+1]],{i1,1,m2}],
{j,1,n}];
ExtSol[x1_,t1_]:=Exp[-Pi^2*t]*Sin[Pi*x]/.{x->x1,t->t1};
Print["Crank-Nicolson Method"]
Do[X=i*h;
 Print[i," ",PaddedForm[N[X,nD],{12,10}]," ",
 PaddedForm[W[[i]],{12,10}]," ",PaddedForm[N[ExtSol[X,0.2],nD],
 {12,10}]," ",PaddedForm[W[[i]]-N[ExtSol[X,0.2],nD],{12,10}]],
{i,1,m1}];
```

$\square$

Problem 12.61 *Initial boundary value problems for partial differential equations. Approximations by finite differences.* Let us consider the initial boundary value problem for the wave equation describing the motion of a fixed string,

$$u_{tt} = c^2\, u_{xx}, \quad u(0,t) = 0, \quad u(L,t) = 0,$$
$$u(x,0) = f(x), \quad u_t(x,0) = g(x),$$

in the domain $\mathcal{D} = \{0 \le x \le L,\ 0 < t < \infty\}$. Here $f(x) = 0$, $g(x) = \sin(4\pi x)$, $L = 0.5$, $c = 1/(4\pi)$.

1) Find the approximate numerical solution of the initial boundary value problem by applying the explicit finite difference method.

2) Plot the numerical solution for $(x_k, t_k) \in \mathcal{D}$. Compare the results with the approximate solution obtained in **Problem 12.58**.

Maple:

```
L:=0.5; c:=1/(4*Pi); T:=0.5; m:=40; n:=40;
F:=X->0; G:=X->sin(4*Pi*X); m1:=m+1;
m2:=m-1;  n1:=n+1; n2:=n-1;
h:=L/m; k:=T/n; lambda:=evalf(c*k/h);
```

```
for j from 2 to n1 do
 w[0,j-1]:=0; w[m1-1,j-1]:=0; od:
w[0,0]:=evalf(F(0)); w[m1-1,0]:=evalf(F(L));
for i from 2 to m do
 w[i-1,0]:=F(h*(i-1)); w[i-1,1]:=(1-lambda^2)*F(h*(i-1))
   +lambda^2*(F(i*h)+F(h*(i-2)))/2+k*G(h*(i-1)); od:
for j from 2 to n do for i from 2 to m do
 w[i-1,j]:=evalf(2*(1-lambda^2)*w[i-1,j-1]
   +lambda^2*(w[i,j-1]+w[i-2,j-1])-w[i-1,j-2]); od;
 od; printf(`  i       X(i)         w(X(i),n)\n`);
for i from 1 to m1 do X[i-1] := (i-1)*h:
 printf(`%3d %11.8f %13.8f\n`,i,X[i-1],w[i-1,n1-1]); od:
Points:=[seq([X[i-1], w[i-1, n1-1]], i=1..m1)];
plot(Points,style=point,color=blue,symbol=circle);
```

Mathematica:

```
{l=0.5, c=1/(4*Pi)//N, t=0.5, m=41, n=41,
 h=l/(m-1), k=t/(n-1), \[Lambda]=c*k/h//N}
f[x_]:=0; g[x_]:=Sin[4*Pi*x]; fi[i_]:=f[h*(i-1)];
gi[j_]:=g[k*(j-1)]; w=Table[0,{n},{m}];
For[i=1,i<=n,i++,w[[i,1]]=fi[i]]; For[i=2,i<=n-1,i++,
 w[[i,2]]=(1-\[Lambda]^2)*fi[i]+\[Lambda]^2*
 (fi[i+1]+fi[i-1])/2+k*gi[i]];
For[j=3,j<=m,j++, For[i=2,i<=n-1,i++,
 w[[i,j]]=2*(1-\[Lambda]^2)*w[[i,j-1]]+\[Lambda]^2*
 (w[[i+1,j-1]]+w[[i-1,j-1]])-w[[i,j-2]]//N];];
Print[" i       x(i)       w(x(i),n)", "\n"];
For[i=1,i<=n,i++,Print[PaddedForm[i,2],
 PaddedForm[h*(i-1),7], "    ",PaddedForm[w[[i,n]],10]]];
points=Table[{h*(i-1),w[[i,n]]},{i,1,n}]
ListPlot[points,PlotStyle->{Blue,PointSize[0.02]}]
Print[NumberForm[TableForm[Transpose[Chop[w]]],3]];
ListPlot3D[w,ViewPoint->{3,1,3},ColorFunction->Hue]
```

                                                                                □

References

[1] Abel, M. L., and Braselton, J. P.: Maple V by Example. AP Professional, Boston, MA 1994

[2] Abramowitz, M. and Stegun, I. A. (Editors): Handbook of Mathematical Functions with Formulas, Graphs and Mathematical Tables. National Bureau of Standards Applied Mathematics, Washington, D.C. 1964

[3] Akritas, A. G.: Elements of Computer Algebra with Applications. Wiley, New York 1989

[4] Andreev, V. K.: Stability of Unsteady Motions of a Fluid with a Free Boundary. Nauka, Novosibirsk 1992

[5] Andrew, A. D., and Morley, T. D.: Linear Algebra Projects Using Mathematica. McGraw-Hill, New York 1993

[6] Aoki, H.: Higher-order Calculation of Finite Periodic Standing Waves by Means of the Computer. *J. Phys. Soc. Jpn.* 49, 1598–1606 (1980)

[7] Bahder, T. B.: Mathematica for Scientists and Engineers. Addison-Wesley, Redwood City, CA 1995

[8] Bronshtein, I. N., Semendyayev, K. A., Musiol, G., and Muehlig, H.: Handbook of Mathematics, 5th Edition. Springer-Verlag, Berlin 2007

[9] Calmet, J., and van Hulzen, J. A.: Computer Algebra Systems. Computer Algebra: Symbolic and Algebraic Computations (Buchberger, B., Collins, G. E., and Loos, R., Eds.), 2d ed. Springer, New York 1983

I.K. Shingareva, C. Lizárraga-Celaya, *Maple and Mathematica*, 2nd ed.,
DOI 10.1007/978-3-211-99432-0_BM2, © Springer-Verlag Vienna 2009

[10] Carter, J. D.: Maple and Mathematica: A Problem Solving Approach. *SIAM Review*, 50, 149–152 (2008)

[11] Char, B. W., Geddes, K. O., Gonnet, G. H., Monagan, M. B., and Watt, S. M.: Maple Reference Manual, 5th ed. Waterloo Maple Publishing, Waterloo, Ontario, Canada 1990

[12] Concus, P.: Standing Capillary-gravity Waves of Finite Amplitude. *J. Fluid Mech.* 14, 568–576 (1962)

[13] Corless, R. M.: Essential Maple. Springer, Berlin 1995

[14] Davenport, J. H., Siret, Y., and Tournier, E.: Computer Algebra Systems and Algorithms for Algebraic Computation. Academic Press, London 1993

[15] Heck, A.: Introduction to Maple. Springer, New York 1996

[16] Higham, N. J.: Functions of Matrices. Theory and Computation. SIAM, Philadelphia 2008

[17] Geddes, K. O., Czapor, S. R., and Labahn, G.: Algorithms for Computer Algebra. Kluwer Academic Publishers, Boston 1992

[18] Getz, C., and Helmstedt, J.: Graphics with Mathematica: Fractals, Julia Sets, Patterns and Natural Forms. Elsevier Science & Technology Book, Amsterdam Boston 2004

[19] Gradshteyn, I. S. and Ryzhik, I. M.: Tables of Integrals, Series, and Products. Academic Press, New York 1980

[20] Gray, J. W.: Mastering Mathematica: Programming Methods and Applications. Academic Press, San Diego 1994

[21] Gray, T., and Glynn, J.: Exploring Mathematics with Mathematica: Dialogs Concerning Computers and Mathematics. Addison-Wesley, Reading, MA 1991

[22] Green, E., Evans, B., and Johnson, J.: Exploring Calculus with Mathematica. Wiley, New York 1994

[23] Grosheva, M. V., and Efimov, G. B.: On Systems of Symbolic Computations (in Russian). Applied Program Packages. Analytic Transformations (Samarskii, A. A., Ed.). Nauka, Moscow 1988, pp. 30–38

[24] Krasnov, M., Kiselev, A., and Makarenko, G.: Problems and Exercises in Integral Equations. Mir Publishers, Moscow 1971

[25] Klerer, M. and Grossman, F.: A New Table of Indefinite Integrals Computer Processed. Dover, New York 1971

[26] Korn, G. A., and Korn, T. M.: Mathematical Handbook for Scientists and Engineers: Definitions, Theorems, and Formulas for Reference and Review, 2nd ed. Dover Publications, New-York 2000

[27] Kreyszig, E., and Normington, E. J.: Maple Computer Manual for Advanced Engineering Mathematics. Wiley, New York 1994

[28] Maeder, R. E.: Programming in Mathematica, 3rd ed. Addison-Wesley, Reading, MA 1996

[29] Nayfeh, A.: Perturbation Methods. Wiley, New York 1973

[30] Okamura, M.: Resonant Standing Waves on Water of Uniform Depth. *J. Phys. Soc. Jpn.* 66, 3801–3808 (1997)

[31] Penney, W. G., and Price, A. T.: Finite Periodic Stationary Gravity Waves in a Perfect Liquid, Part 2. *Phil. Trans. R. Soc. Lond.* A 224, 254–284 (1952)

[32] Polyanin, A. D., and Manzhirov, A. V.: Handbook of Integral Equations. CRC Press, Boca Raton 1998

[33] Polyanin, A. D.: Handbook of Linear Partial Differential Equations for Engineers and Scientists. Chapman & Hall/CRC Press, Boca Raton 2002

[34] Polyanin, A. D., and Manzhirov, A. V.: Handbook of Mathematics for Engineers and Scientists. Chapman & Hall/CRC Press, Boca Raton-London 2006

[35] Ramírez, V., Barrera, D., Posadas, M. y González, P.: Cálculo numérico con Mathematica. Ariel Ciencia, Barcelona 2001

[36] Rayleigh, Lord: Deep Water Waves, Progressive or Stationary, to the Third Order of Approximation. *Phil. Trans. R. Soc. Lond.* A 91, 345–353 (1915)

[37] Richards, D.: Advanced Mathematical Methods with Maple. Cambridge University Press, Cambridge 2002

[38] Ross, C. C.: Differential Equations: An Introduction with Mathematica. Springer, New York 1995

[39] Sekerzh-Zenkovich, Y. I.: On the Theory of Standing Waves of Finite Amplitude. *Doklady Akad. Nauk. USSR* 58, 551–554 (1947)

[40] Shingareva, I.: Investigation of Standing Surface Waves in a Fluid of Finite Depth by Computer Algebra Methods. PhD thesis, Institute for Problems in Mechanics, Russian Academy of Sciences, Moscow 1995

[41] Shingareva, I., and Lizárraga-Celaya, C.: High-order Asymptotic Solutions to Free Standing Water Waves by Computer Algebra. Proc. Maple Summer Workshop (Lopez, R. J., Ed.). Waterloo, Ontario, Canada 2004, pp. 1–28

[42] Shingareva, I., Lizárraga Celaya, C., and Ochoa Ruiz, A. D.: Maple y Ondas Estacionarias. Problemas y Soluciones. Editorial Unison, Universidad de Sonora, Hermosillo, México 2006

[43] Shingareva, I., and Lizárraga-Celaya, C.: On Frequency-amplitude Dependences for Surface and Internal Standing Waves. *J. Comp. Appl. Math.* 200, 459–470 (2007)

[44] Schultz, W. W., Vanden-Broeck, J. M., Jiang, L., and Perlin, M.: Highly Nonlinear Standing Water Waves with Small Capillary Effect. *J. Fluid Mech.* 369, 253–272 (1998)

[45] Spiegel, M. R.: Matemáticas Avanzadas para Ingeniería y Ciencias, 1ra Edición. McGraw-Hill, México 2001

[46] Spiegel, M. R., Liu, J., and Abellanas, L.: Fórmulas y Tablas de Matemática Aplicada, 2da Edición. McGraw-Hill, Madrid 2005

[47] Tadjbakhsh, I., and Keller, J. B.: Standing Surface Waves of Finite Amplitude. *J. Fluid Mech.* 8, 442–451 (1960)

[48] Vvedensky, D. D.: Partial Differential Equations with Mathematica. Addison-Wesley, Wokingham 1993

[49] Weisstein, E. W.: CRC Concise Encyclopedia of Mathematics, 2nd Edition. CRC Press, Boca Raton 2003

[50] Whittaker, E. T.: A Treatise on Analytical Dynamics of Particles and Rigid Bodies. Dover Publications, New York 1944

[51] Wickam-Jones, T.: Mathematica Graphics: Techniques and Applications. Springer, New York 1994

[52] Wester, M. J.: Computer Algebra Systems: A Practical Guide. Wiley, Chichester, UK 1999

[53] Wolfram, S.: A New Kind of Science. Wolfram Media, Champaign, IL 2002

[54] Wolfram, S.: The Mathematica Book, 5th ed. Wolfram Media, Champaign, IL 2003

[55] Yosihara, H.: Gravity Waves on the Free Surface of an Incompressible Perfect Fluid of Finite Depth. *Kyoto Univ. Math.* 18, 49–96 (1982)

[56] Zimmerman, R. L., and Olness, F.: Mathematica for Physicists. Addison-Wesley, Reading, MA 1995

[57] Zwillinger, D.: CRC Standard Mathematical Tables and Formulae, 31st Edition. CRC Press, Boca Raton 2002

General Index

Maple Index

Mathematica Index

CD-ROM Contents

The present CD-ROM (packaged with this edition) contains all *Maple* and *Mathematica* solutions (i.e. segments of code) of the problems and examples as a set of the following PDF files:

ChapterN.pdf — all solutions and examples described in the individual ChapterN of the book, where $N = 1, 2, ..., 12$.

Chapter1-12.pdf — a single file with the whole set of solutions and examples described in the book.